IN CASE OF ACCIDENT

In case of any accident, notify the instructor immediately and call for help from those around you. Remaining calm and acting quickly are essential. Here are some specific actions to take:

Organic chemicals on the skin: Wash thoroughly with soap and water.

Caustic chemicals on the skin: Rinse with a large volume of cold water for 10 to 15 minutes.

Small cuts or minor heat burns: Rinse with cold water. Apply pressure and bandage to stop bleeding.

Chemicals in the eye: While holding eyelid open, flush with water from an eyewash fountain or faucet for at least 15 minutes. Consult a physician promptly, especially if the chemical is caustic.

Fire: Cut off the oxygen supply to the burning material if possible by covering it with a watch glass, asbestos wire gauze square, or large, inverted beaker. Remove solvents or other flammable material from area. Turn off the gas supply to any nearby Bunsen burners.

If necessary, use a fire extinguisher by dragging it on its base (*not* carrying it) to the scene, removing the pin with a sharp jerk, and directing the spray at the *base* of the flames. Be careful not to knock over anything that may feed the flames.

Burning clothing: Extinguish flames by dousing with water in a safety shower, wrapping in a fire blanket, or simply rolling on the floor. To treat shock, a burned person should be kept warm and quiet until an ambulance arrives.

Large cuts: Stop the bleeding by applying pressure with fingers until a pressure bandage can be applied. Do not use a tourniquet. Elevate the injured limb and treat for shock by keeping the injured person warm and quiet until an ambulance arrives.

For more detail, see references throughout text.

EXPERIMENTS IN ORGANIC CHEMISTRY

From Microscale to Macroscale

Jonathan S. Nimitz
University of New Mexico

PRENTICE HALL, Englewood Cliffs, New Jersey 07632

Library of Congress (Cataloging-in-Publication) Data

Nimitz, Jonathan (date)
   Experiments in organic chemistry: from microscale to macroscale/
Jonathan Nimitz.
     p.   cm.
   Includes index.
   ISBN 0-13-295718-3
   1. Chemistry, Organic—Experiments.   I. Title.
QD257.5.N56   1990
547'.078—dc20   89-8402
                              CIP

Editorial/production supervision: Edward Thomas
Interior design: Judy Matz-Coniglio
Cover design: Bruce Kenselaar/Judy Matz-Coniglio
Manufacturing buyer: Paula Massenaro

The experiments in this book have been tested in college laboratories (over the course of several years) by hundreds of organic chemistry students and their teaching assistants, and by the author and his colleagues. The author believes that, if performed as described in text, if good laboratory practices and the correct quantities of reagents are used, and if all safety precautions are taken, then the experiments will be safe and valuable as educational experiences. However, those who conduct these experiments do so strictly at their own risk; the author does not warrant or guarantee the safety of individuals carrying out the procedures described herein. The author hereby disclaims any liability for any loss, damage, or injury claimed to have resulted from or to have been related in any way to these experiments.

Illustration credits: pp. 6, 14, 21—John H. Nelson/Kenneth C. Kemp, *Laboratory Experiments for Chemistry: The Central Science, 4e,* © 1988, pp. xvii, xxi, xxii, 3, 4. Reprinted by permission of Prentice Hall, Inc., Englewood Cliffs, New Jersey; pp. 143, 157—John R. Dyer, *Applications of Absorption Spectroscopy of Organic Compounds,* © 1965, pp. 52, 84-85. Reprinted by permission of Prentice Hall, Inc., Englewood Cliffs, New Jersey; p. 110—Bausch and Lomb, Inc; p. 165—Hewlett-Packard Analytical Group, Palo Alto, CA.

Printed in the United States of America

10   9   8   7   6   5   4   3   2   1

ISBN 0-13-295718-3   01

Prentice-Hall International (UK) Limited, *London*
Prentice-Hall of Australia Pty. Limited, *Sydney*
Prentice-Hall Canada Inc., *Toronto*
Prentice-Hall Hispanoamericana, S. A., *Mexico*
Prentice-Hall of India Private Limited, *New Delhi*
Prentice-Hall of Japan, Inc., *Tokyo*
Simon & Schuster Asia Pte. ltd., *Singapore*
Editora Prentice-Hall do Brasil, Ltda., *Rio de Janeiro*

This work is dedicated to all
students and teachers of organic chemistry

CONTENTS

PREFACE

While teaching organic laboratory and lecture courses for the last several years, I became aware of the need for a lab text incorporating the following features:

- A full range of micro- to macroscales
- Relatively low-risk and low-cost reagents in as small quantities as practical and using standard equipment to minimize dangers, waste, and expense
- A wide variety of experiments to allow for individual preferences of instructors and to allow easy correlation of lab with lecture
- Reliable, tested procedures that contain no unnecessary steps and can be completed in the time allowed
- A mechanism and clear explanation for every reaction so that students can prepare independently (for the most part), and lengthy prelab lectures by the instructor can be avoided
- Real-life examples to make the experiments more relevant to hold the student's interest
- Features designed to help with lecture as well as lab, such as a glossary of organic chemistry, a summary of mechanisms, and numerous practice problems including many involving spectroscopy

For several years, hundreds of organic chemistry students and teaching assistants at the University of New Mexico have tested and refined many of the procedures in this book. All the advice and encouragement from them are greatly appreciated.

I am grateful to David Todd, Shrewsbury, Massachusetts, for contributing the experiment he developed on enzymatic reduction of vanillin. Dr. Todd also provided many helpful suggestions during his proofreading of the manuscript.

Much valuable advice and support have also come from members of the organic chemistry faculty, including previous laboratory supervisors, at UNM. In particular, E. Paul Papadopoulos, Ulrich Hollstein, Cary Morrow, and Richard Holder made many useful suggestions.

Dorothy Cunningham, editorial assistant at UNM, did an outstanding job of typing most of the manuscript, a task that had been begun by her predecessor, Wanda Kartchner.

Finally, this text could not have come into being without the dedicated efforts of Dan Joraanstad, senior chemistry editor, and the rest of the staff at Prentice Hall.

To all these people I am very grateful.

JONATHAN S. NIMITZ

Laboratory Essentials: The First Steps

The laboratory may well prove to be your favorite part of the organic chemistry course because of the thrill of learning new techniques and making useful chemicals. Hundreds of industries around the world rely on the techniques of extraction, recrystallization, and distillation that you will be learning in this course. These methods find application in such diverse areas as the manufacturing of paints, plastics, and drugs; some types of food processing; and mining and medical research.

This lab manual presents these techniques as well as a wide range of the most common types of chemical reactions in a style designed to be as easy to understand and as safe and enjoyable as possible. The author recognizes the significant costs and potential dangers in operating an organic chemistry laboratory, and the 98 experiments described in this manual have been chosen to use inexpensive, low-risk reagents whenever feasible, while illustrating the full range of laboratory techniques and reactions. Several materials can be recycled for reuse, specifically in the recrystallization, distillation, and extraction experiments. This further lowers costs and cuts down on wastes to be disposed of.

This work is designed from an educational point of view, to enable you to learn and remember the maximum amount of organic chemistry in the time available. The write-ups all provide thorough explanations in clear, conversational language. Every write-up explains the mechanism of reaction, minimizing the need for lengthy preparation and talks by the instructor. The overviews often possess the additional advantage that reading them exposes you to the theory behind several other experiments in addition to the one you will be performing. Extensive illustrations clarify theory and proper use of equipment. ''Cookbook'' situations are avoided by requiring prelab calculations and planning for every experiment. Common mistakes and pitfalls encountered by beginning organic chemistry students are described.

The author believes that you, as a beginning organic chemistry student, should become familiar with the full range of scales commonly used by research chemists and should experience the satisfaction both of obtaining tangible products early in your career and of successfully carrying out reactions on a microscale.

In accord with the trend toward smaller-scale reactions and the many advantages they possess, most of the reactions in this manual are conducted on a small scale (approximately 1 g for solids and 5–10 g for liquids, to allow convenient distillation). In addition to four microscale structure determination experiments and one microscale chromatography, there are 10 microscale syntheses that are performed using normal glassware.

These microscale syntheses are experiments

 9.2: Synthesis of diphenylacetylene from stilbene dibromide
18.4: Friedel-Crafts alkylation of *p*-dimethoxybenzene
20.4: Preparation of ethyl trityl ether
22.2: Addition of dibromocarbene to 1,1–diphenylethylene
26.1: Synthesis of benzophenone oxime
37.4: Acetylation of benzoin with acetic anhydride
38.4: Sulfonamide formation from *p*-toluenesulfonyl chloride and methylamine
39.1: Aldol condensation of acetone and benzaldehyde
39.3: Condensation of a nitrile and an aromatic aldehyde
40.2: Gabriel synthesis of phthalimidoacetophenone

These experiments develop skill in handling small quantities of material without requiring the purchase of expensive, specialized microscale equipment. Microscale experiments have the advantages of being less expensive, less polluting, and using less equipment than larger-scale reactions. They are also faster to carry out than large-scale reactions since additions, heating, and cooling take less time. So if anything goes wrong you will normally have plenty of time to repeat the experiment. When researchers have a very valuable starting material a microscale reaction may be desirable in order to conserve materials.

Working on a microscale has the disadvantages that you must measure reagents and products more carefully and work a little more carefully since losses in transferring materials make a greater difference in the yield. Be aware, though, that even though a few flakes of product may look like a lot, they may weigh only 1 or 2 mg, and you can still afford that kind of loss.

There are several multistep sequences included to expose you to realistic synthetic schemes, including a four-step Grignard sequence (Experiment 33.1), a three-step preparation of camphor (Experiment 23.3), and a three-step sequence involving a Diels–Alder reaction, hydrolysis of an acid anhydride, and catalytic hydrogenation (Experiments 15.1 and 16.1). In addition, the product of one experiment can sometimes be used as the starting material for another; these cases are clearly marked in the text. Every experiment can be finished comfortably in less than 3 hours with reasonable preparation. Some can be finished in as little as 1 hour; approximate working times are given for each experiment. At the instructor's option, two short experiments may be combined in one lab period. These shorter experiments (or lengthy reflux periods) are the natural times to include extra discussions and/or lab quizzes.

The author believes that the laboratory should fit closely with the lecture so that they can reinforce one another. To make the correlation of lab and lecture easier, this work contains both a comprehensive choice of topics and some features normally found in workbooks and reference books. For example, there are several tables of summarized information that may be helpful in learning lecture material as well as lab. Questions are also designed to aid learning of lecture material. A glossary is provided to help familiarize you with the language of organic chemistry. As another bonus, there is an extensive section on structure determination practice, including hydrogen and carbon-13 NMR and mass spectroscopy. In addition to four structure determination experiments, a section of spectroscopy practice is included as a *learning lab*, to allow extra practice and strengthening of knowledge.

In my experience, students often struggle with spectroscopy and structure determination more than with any other area of organic chemistry. Most of this struggle is unnecessary and due primarily to a lack of adequate practice. It is hoped that hearing another explanation of the principles and working 20–30 additional problems in this book may help to provide a quicker grasp of this important area.

Since heterocycles (compounds containing rings with one or more non-carbon atoms, such as nitrogen, oxygen, or sulfur) make up a substantial fraction of compounds of interest to modern organic chemists, there is a section devoted to their synthesis. Section 43 includes six experiments which illustrate formation of heterocycles. Molecules containing pyridine and furan rings occur in several other experiments as well.

While biochemists have long been fascinated with enzymes, and with good reason, synthetic organic chemists have relatively recently started to take advantage of the remarkable properties of enzymes in both laboratory and industrial processes. An experiment on enzymic reduction of vanillin (Experiment 46.1) is included to familiarize you with the synthetic use of enzymes. This experiment also illustrates the process of monitoring the progress of a reaction by thin-layer chromatography.

Because of the rapidly increasing use of computers by organic chemists, two experiments using computers are included. In Experiment 50.1, you can learn to retrieve chemical information online using the tremendous database of *Chemical Abstracts*. In Experiment 50.2, educational software can help you with practice questions and experimental simulations.

Adequate preparation for lab is essential. Most of the labs involve some prelab calculations or questions, and, in most cases, the lab cannot be done without these, since the calculations tell you how much material to use. You are encouraged to talk over the prelabs with your classmates and to check your calculations carefully by yourself, with your classmates, and with the instructor before starting the experiment. *Attempting to carry out reactions using incorrect quantities of reagents could be dangerous; be sure your calculations are correct before beginning each experiment*. It is also essential to outline the procedure and, for the synthesis labs, to complete a reagent table *before* coming to lab. This planning is described in the section "Guidelines for Notebooks." Thorough descriptions of the common techniques used in introductory organic chemistry labs are included in the first five sections, eliminating the need for a separate section on methods. It is expected that you will remember all the techniques in labs done before the one you are currently doing; if necessary, you can always refer back to previous experiments for details.

Another feature in this manual is the occasional use of *concept maps*. Concept mapping is a method developed by learning theorists to show connections between important ideas graphically. Sometimes students are slow to make some of the fundamental connections that are obvious to an experienced chemist, and the concept maps may speed up this process. These illustrate the connections between important ideas, giving you the underlying intellectual framework on which to build more detailed knowledge later. On occasion, you will have an opportunity to make some of your own concept maps. A description of the principles and applications of concept mapping in chemical education is given by Joseph D. Novakin, "Applications of Advances in Learning Theory and Philosophy of Science to the Improvement of Chemistry Teaching," *Journal of Chemical Education* (1984) *61*, 607. In my experience, students' reactions to concept maps vary widely but are usually positive. If you find them helpful, great; but if not, they can simply be ignored.

If you follow all safety precautions scrupulously and prepare well ahead of time, I feel sure that you will enjoy the course and gain skill and confidence in the synthesis and analysis of organic compounds. Note that all temperatures in this manual are expressed in degrees centigrade (°C).

SAFETY PRACTICES IN THE ORGANIC LABORATORY

- Because of the danger of following unsafe practices and having nobody available in an emergency, no work is to be done without supervision.
- All accidents of any kind are to be reported to the instructor immediately so that appropriate action can be taken.
- Carry out only the assigned experiments following the designated procedures. Change procedures only when directed to do so by your instructor. Because of the dangers involved in unauthorized experiments, only authorized experiments are to be performed.
- Familiarize yourself with the locations and proper use of the emergency showers, eyewashes, fire blankets, fire extinguishers, and exits, so that you will know how to react in an emergency and can give help to others.
- Goggles or explosion-proof glasses must be worn whenever *anyone* is doing any work in the lab. Contact lenses should not be worn because chemicals splashed in the eye can get under them and do irreparable damage before the chemicals can be washed out.
- To avoid burns, do not touch glassware or equipment that may be hot. Proceed cautiously and use tongs or a cloth wrap if necessary.
- Whenever possible, avoid flames in the lab. If one is required, check all around first to be sure that any flammable solvents are at least 15 feet away, and do not leave a flame unattended. Be ready to shut off the flame quickly if necessary. Do not reach over flames.
- If there is a fire, notify the instructor immediately, get everyone away, and remove any nearby flammable materials. If the fire is confined to a flask or beaker, extinguish it by cutting off the oxygen by covering it with an asbestos wire gauze square, a watch glass, or a large beaker. If the fire is too large to put out in this way, get a fire extinguisher, pull the pin, and direct the spray at the base of the flames. Be aware that the force of the extinguisher will knock over any nearby glassware and be careful that their contents do not feed the flames. Water is not to be used on laboratory fires in most cases because many organic solvents would float on top and spread the flames.
- Unless you know differently, treat every chemical as flammable, toxic, and carcinogenic. Avoid contact with eyes, skin, and clothing, avoid prolonged breathing of vapors, and above all, avoid ingestion. To avoid chemical poisoning and fire dangers, eating, drinking, and smoking in the laboratory are prohibited.
- Whenever using a compound that gives off an irritating or harmful vapor, keep it in the fume hood to avoid exposure.
- Check carefully the labels of any chemicals used. Many compounds have similar names.
- To avoid contamination, never return unused material to reagent bottles. Use clean spatulas and avoid taking too much in the first place. Dispose of any excess reagents properly as directed by the instructor.

- When performing an experiment, be sure that you are familiar with the properties and special hazards of the chemicals involved, along with the appropriate safety precautions and first aid. This information is available in the "Special Hazards" section of each experiment as well as in the following sources:

ALDRICH CHEMICAL CO., *Aldrich Catalog/Handbook of Fine Chemicals*, Aldrich Chemical Co., Milwaukee, Wis., published annually.

BRETHERICK, L., ed., *Hazards in the Chemical Laboratory*, Royal Society of Chemistry, London, 1986.

BRETHERICK, L., ed., *Handbook of Reactive Chemical Hazards*, 2nd ed., Butterworth, Woburn, Mass., 1979.

KENKYUJO, K., *Toxic and Hazardous Industrial Chemicals Safety Manual*, International Technical Information Institute, Tokyo, 1978.

POJASKE, R., ed., *Toxic and Hazardous Waste Disposal*, Ann Arbor Science, Ann Arbor, Mich., 1979.

SAX, N. I., and R. LEWIS, *Hazardous Chemicals Desk Reference*, Van Nostrand Reinhold, New York, 1987.

SITTIG, M., *Handbook of Toxic and Hazardous Chemicals and Carcinogens*, 2nd ed., Noyes Data Corporation, Park Ridge, N.J., 1981.

WEISS, G., ed., *Hazardous Chemicals Data Book*, 2nd ed., Noyes Data Corporation, Park Ridge, N.J., 1986.

WINDHOLZ, M., ed., *The Merck Index: An Encyclopedia of Chemicals, Drugs, and Biologicals*, 10th ed., Merck and Co., Rahway, N.J., 1983.

- To avoid contact with chemicals or flames, loose hair and clothing must be confined while working in the laboratory.
- Shoes and long pants should always be worn in the lab to protect against spilled chemicals and broken glass. A lab coat is highly recommended for additional protection and to prevent holes and stains on clothes. One shirt or pair of pants saved may repay the cost of the lab coat.
- Horseplay endangers everyone's safety and is cause for immediate expulsion from the lab.
- When strong acids and water are mixed, heat is evolved. To diffuse this heat more rapidly through the mixture, the more dense liquid (the acid) should be added to the less dense liquid (the water). It is also much safer for water to absorb this heat than for acid to absorb it, so *always add acid to water*, not vice versa. If an acid solution absorbs too much heat, it can vaporize and splatter; many serious accidents have occurred with hot acids. If mixing large quantities, use some ice in place of water and/or an ice bath for cooling.
- Chromic acid baths (sometimes used for cleaning glassware) are highly corrosive and contain cancer suspect agents. Avoid skin contact and wash immediately with a large volume of water if contact occurs.
- If a closed system is heated sufficiently, it will explode due to high internal pressure, so always be sure that distillations and other heated setups have openings to relieve pressure. Never heat a stoppered container.
- Never apply pressure to a glass stirring rod, because it can break easily and cut deeply. Scraping material out of a round-bottomed flask should always be done with a narrow metal spatula, which can be bent to reach the sides of the flask easily. A glass stirring rod should not be used for this purpose because it may break or punch a hole in the bottom of the flask.

- When inserting glass tubing into a rubber or cork stopper, use extreme care. To proceed properly, employ the following steps:
1. Make sure that the end of the glass tube is fire-polished and cool. Check the stopper for flexibility and to be sure that it has the correct size hole for the tubing.
2. Lubricate the glass tubing with glycerin.
3. Wrap both tubing and stopper in cloth (to protect your hands, as shown in Figure 1) and *gently* twist the stopper onto the tubing.

Figure 1   Inserting glass tubing into a rubber stopper

4. If resistance is encountered, *STOP; do not use force*. Remove the tubing, check for proper lubrication and fit, and try again.
- Never use chipped, cracked, or otherwise defective equipment. Damaged equipment may break while in use and shower you with hot glass and chemicals. Some defects to look out for include:
  (a) Chipped or broken rims on beakers, flasks, funnels, graduated cylinders, and test tubes
  (b) Cracks in beakers, flasks, graduated cylinders, test tubes, and crucibles
  (c) Star-shaped cracks in the bottoms of test tubes, beakers, and flasks
  (d) Sharp edges on glass tubing and glass rods
  (e) Inflexibility in rubber stoppers
  (f) Inoperable parts on screw clamps, burette clamps, and rings
  (g) Loose or exposed wiring on electrical equipment
- Any water spills or leaks on the floor are to be cleaned up immediately. They pose a danger of causing a fall and can leak through the floor, damaging equipment below.
- To avoid exposure to potentially harmful chemicals and fire hazards, keep the lab clean at all times.
- To note the odor of a chemical, *do not sniff the open bottle*. This could expose you to a large dose of a highly irritating compound. Instead, set the bottle down (preferably in the fume hood), remove the cap, and with your other hand gently waft the odor from the lid toward your nose.
- Mercury spills (for example, from broken thermometers) are to be sprinkled with sulfur to decrease the vapor pressure of the toxic metal, then swept up as thoroughly as possible and placed in a special container for mercury wastes.

- Dispose of organic chemicals as instructed in the waste containers provided. Each type of waste substance should be collected in its own container and clearly labeled to facilitate later disposal. In general, small amounts of water-soluble materials that are not particularly toxic in highly dilute solution (such as aqueous acids and bases) can be rinsed down the drain with lots of water. If in doubt about how to dispose of waste chemicals, ask your instructor.
- If you have a medical condition that might affect your ability to carry out work in the lab, inform your instructor right away so that he or she can respond appropriately if needed.
- Keep emergency phone numbers on hand in case of need.

Campus Student Health Service _____

Ambulance/Paramedics _____

FIRST AID

Chemical Burns

In general, any chemical spilled on the skin or in eyes should be rinsed *thoroughly* with cold water for 10–15 minutes. Base is worse than acid to get in the eyes or on skin because it keeps dissolving its way in, while acid "cauterizes" a wound. The slippery feeling of strong base on the skin is due to the skin dissolving. A base should *not* be used to "neutralize" an acid burn (and vice versa) because it causes more damage. However, very dilute acetic acid (approximately 2% in water) can be used on base spills on the skin. This solution may be kept in the lab in a bottle labeled "For Base Burns Only." If this solution is used, it should be followed by a rinse with copious amounts of water. *Only water* should be used for rinsing eyes.

There are also some specific reagents to neutralize particularly dangerous chemicals. For example, an aqueous solution of sodium thiosulfate applied to the skin quickly can prevent severe burns from bromine (Br_2). If you are ever working with cyanide (which is not used in any of the experiments in this book), an ampoule of amyl nitrite should be kept nearby to stimulate the heart in case of cyanide poisoning.

Cuts

These are the most common injuries in the lab. Remove any foreign objects, wash with water, and apply pressure using fingers (and perhaps a clean cloth) to stop the bleeding. Elevate the injured part. If the cut is severe, wrap the injured person in a blanket to avoid shock and get medical attention.

Burning Clothing

Do not run, or, if another person's clothing is on fire, prevent that person from running, which fans the flames. The best course of action is to drop immediately to the floor and roll to extinguish the flames. Once the person is rolling on the floor, a fire blanket may be used to roll into. A safety shower may also be used to extinguish flames and rinse away any chemicals. To avoid shock, the victim should be wrapped in a blanket and taken for medical attention.

GUIDELINES FOR NOTEBOOKS

Since the notebook is the main record of your work in the lab, and may determine a significant part of your grade, it is important to keep a well-organized, legible notebook. The purpose of a notebook is *reproducibility*, so that you or someone else could pick up that lab book in 10 years and reproduce exactly what you had done before.

Abraham Maslow, the humanistic psychologist, defined science as ''a way non-creative people can be creative.'' He did not mean to put down scientists, but rather to point out that through careful record keeping and reproducibility, small contributions add up over time. The function of the notebook is to keep a record of those small or large contributions. Bearing this purpose in mind, some general guidelines for notebooks are:

1. For durability they must be hard-bound, not spiral-bound.
2. All entries must be in waterproof ink.
3. Two or three blank pages left at the beginning will provide room for a table of contents.
4. Mistakes should be crossed out with one line through them, so they can still be read if necessary. No pages should ever be removed. If necessary, a page may be crossed out with a large X through it.
5. All data (such as weights) are to be entered in the notebook *when taken*, not copied from scraps of paper or looseleaf pages. This procedure minimizes errors and can prevent the frustration caused by losing data.
6. *The experiment name, date, purpose, relevant equations, prelab, table of reagents, and outline of procedure must be written in the notebook before the lab.* If these are not ready, you may be asked to come back to a later section when you are prepared because it is not safe to begin lab work without adequate preparation. A few minutes of planning can easily save you an hour in the lab. It is suggested that you put the procedure on one side of the page and note any observations or changes in procedure on the other side. Another recommended preparation that can save a lot of time and prevent mistakes is to take 5 minutes the evening or morning before the lab, find a quiet place, close your eyes, and visualize the next day's lab. This should include picturing yourself walking into the lab, taking out and setting up the equipment, measuring the chemicals, running the experiment, purifying the product, and cleaning up. This type of visualization has helped many great athletes to improve their performance, and in my experience it works well for lab preparation, too.
7. Lab notebooks are not to be recopied. This defeats the purpose of keeping thorough, neat records as you go along. Your instructor expects to see water blotches, acid holes, and stains in the notebook of the appropriate color for the experiment. A typical notebook page might look as shown in Figure 2.

After the lab is finished you will, where appropriate, calculate the percent yield rounded to the nearest 1%. A reported yield must be accompanied by some criterion of purity, such as melting range, boiling range, or index of refraction. A yield of 97% means nothing if the sample contains 15% water.

After the calculation of yield comes the discussion. The discussion should focus on such questions as: What were possible reasons for high or low yields? What are the major sources of error? How could the procedure be improved? How widely applicable is this reaction or procedure? Is it similar to any other reactions you know of? After the discussion the questions following the lab should be answered.

READ PROCEDURE 1ST

Today's Date

1. *Title*: Acetylation of Glycine with Acetic Anhydride

Reference: J. Nimitz, *Experiments in Organic Chemistry from Microscale to Macroscale*, expt. 38.2

Purpose: To synthesize acetylglycine, to acetylate an amino acid, to form an amide from an amine and an acid anhydride

2. *Reaction*: $HO\overset{O}{\overset{\|}{C}}-CH_2-NH_2 + CH_3\overset{O}{\overset{\|}{C}}O\overset{O}{\overset{\|}{C}}CH_3 \longrightarrow HO\overset{O}{C}CH_2NH\overset{O}{\overset{\|}{C}}CH_3$

EQUATION (BALANCED)

Glycine   Acetic anhydride   Acetylglycine
$+ CH_3COOH$
Acetic acid

PHYSICAL PROPERTIES — BOIL PT, MELT PT
$A+B \rightarrow X + 3D$
M.Wt.
wt used
wt, moles

Reagent Table:

REAGENT	MW	MP	BP	MMOL	EQUIV.	MASS	DENSITY	VOLUME	SOURCE OF CHEMICAL
Glycine	75.1	245°d	—	13.3	1.00	1.00g	—	—	Aldrich
Acetic anhydride	102	—	140°	24.5	1.84	2.50g	1.08 g/mL	2.31mL	Fisher
PRODUCT									
Acetyl glycine	117	207-209°	—						
Acetic acid	60.1	16°	116-118°						

WITH SOLID YOU MUST HAVE MELTING PT.

3.

Procedure:	*Observations or variations*
1. Place 1.00 g glycine in 25 mL Erlenmeyer	✓
2. Add 4.0 mL of water	✓
3. Swirl & warm slightly to dissolve	Dissolved after about 30 sec on hot plate
4. Add acetic anhydride (2-3 mL)	✓ by pipette
5. Swirl to mix, keep swirling for 10 min	Swirled 5-6 times over 12 min
6. Cool in ice bath 15 min	Heavy ppt of clear needle-shaped crystals
7. Collect ppt by suction filtration	✓
8. Wash with a few mL water	Used 5 mL water cooled in ice bath
9. Air dry	✓ 10 min on clean filter paper, stirred w/spatula occasionally
10. Record mass and melting point	1.14g, mp 205-207° Perhaps slightly wet still

Figure 2   A sample notebook page

4. OBSERVATION + DATA

$SO_3 + H_2O \rightarrow H_2SO_4$
MWT→80   18   98
6.5   1.8g   7.35   THEORETICAL YIELD (.075 × 98)
LIMITING REACT .075 mol .1 mol   .075   ACTUAL YIELD – WHAT YOU REALLY GET
PERCENT YIELD – ACTUAL YIELD ÷ THEORETICAL × 100 = %

For further guidelines on how to keep a lab notebook and on scientific writing in general, you may wish to consult *Writing the Laboratory Notebook* by Howard Kanare or *The ACS Style Guide* by Janet Dodd, both published by the American Chemical Society.

5. CONCLUSION

NOTE ON CALCULATING THEORETICAL YIELDS

The *theoretical yield* of a reaction is the number of grams of product that would be produced if the reaction went 100% to completion and formed only the desired product. In other words, it is the quantity of product expected under ideal conditions, where there is no leftover starting material, and no undesired side reactions have occurred.

This is the same kind of stoichiometric calculation you have often done in general chemistry. Here it has the practical use of telling you how much product to expect as a maximum. By comparing the *actual* yield obtained (in grams) to the *theoretical* yield (in grams) you can calculate the percent yield obtained:

$$\frac{\text{actual yield (in grams)}}{\text{theoretical yield (in grams)}} \times 100\% = \text{percent yield}$$

A high percentage yield means that the reaction followed mainly the desired path and that reaction conditions such as time and temperature were appropriate. A low percentage yield means that the conditions were not optimal and probably could be improved. Perhaps there are competing reactions occurring or some of the product is being lost in the purification steps.

There are four steps in calculating the theoretical yield:

1. Find the number of moles of each starting material used.
2. Determine which reagent is limiting.
3. Calculate the moles of product expected if the yield were 100% based on the limiting reagent.
4. Calculate the grams of product corresponding to this number of moles.

To find the limiting reagent, look at the ratios of reactants (stoichiometry) in the balanced reaction. For example, consider the reaction described on the sample notebook page, repeated below for convenience.

$$\underset{\substack{\text{Glycine}\\\text{MW 75.1}}}{\text{HO}\overset{\overset{\displaystyle O}{\|}}{\text{C}}\text{CH}_2\text{NH}_2} + \underset{\substack{\text{Acetic anhydride}\\\text{MW 102, dens. 1.08 g/mL}}}{\text{CH}_3\overset{\overset{\displaystyle O}{\|}}{\text{C}}-\text{O}-\overset{\overset{\displaystyle O}{\|}}{\text{C}}\text{CH}_3} \longrightarrow \underset{\substack{\text{Acetylglycine}}}{\text{HO}\overset{\overset{\displaystyle O}{\|}}{\text{C}}\text{CH}_2\text{NH}\overset{\overset{\displaystyle O}{\|}}{\text{C}}\text{CH}_3} + \underset{\substack{\text{Acetic acid}}}{\text{CH}_3\overset{\overset{\displaystyle O}{\|}}{\text{C}}-\text{OH}}$$

This equation is balanced as it stands, so the stoichiometry of the reactants is 1:1. Now from the masses or volumes listed in the reagent table on the sample notebook page we can calculate the moles of each that are in fact mixed. For the glycine:

$$1.00 \text{ g} \times \frac{1 \text{ mol}}{75.1 \text{ g}} = 0.0133 \text{ mol} = 13.3 \text{ mmol}$$

For the acetic anhydride:

$$2.5 \text{ g} \times \frac{1 \text{ mol}}{102 \text{ g}} = 0.0245 \text{ mol} = 24.5 \text{ mmol}$$

So, rather than being in the ratio 1:1, there are really in the ratio 24.5:13.3 or 1.84:1. This is why we say that relative to each other there are 1.84 equivalents of acetic anhydride to 1.0 equivalent of glycine. In other words, there is an excess of 0.84 equivalents of the anhydride. Glycine is, of course, the limiting reagent since the acetic anhydride is in excess; the glycine would run out long before the acetic anhydride would in this case. Usually, the cheaper reagent will be used in excess and the more valuable one will be the limiting reagent.

Now we have completed the first two steps toward arriving at a theoretical yield. We have calculated the number of moles of reagents and decided which is the limiting reagent. The next two steps of calculating moles and grams of product can be combined into one calculation. Since the mole ratio is 1 mole of product per mole of glycine, we get

$$0.0133 \text{ mol glycine} \times \frac{1 \text{ mol acetylglycine}}{1 \text{ mol glycine}} \times \frac{117 \text{ g acetylglycine}}{\text{mol}}$$

$$= 1.56 \text{ g acetylglycine} \qquad \text{(theoretical yield)}$$

Since on the sample notebook page the actual yield reported is 1.38 g, this would represent a yield of

$$\frac{1.38 \text{ g (actual yield)}}{1.56 \text{ g (theoretical yield)}} \times 100\% = 88\%$$

In this course always remember to round yields to the nearest whole percentage point, since greater precision than that will not be justified.

Practice Questions

1. What is the theoretical yield of ethene obtainable from 100 g of ethanol in the following acid-catalyzed dehydration?

$$CH_3CH_2OH \xrightarrow{H^+} CH_2{=}CH_2 + H_2O$$

Ethanol   Ethene

2. How many grams of 1,1,2,2–tetrabromopropane could theoretically be produced by mixing 10 g of propyne with 60 g of Br_2? If 58.3 g of product were obtained, what is the percent yield?

$$H{-}C{\equiv}C{-}CH_3 \quad + \quad 2Br_2 \quad \longrightarrow \quad H{-}\overset{\displaystyle Br}{\underset{\displaystyle Br}{\overset{|}{\underset{|}{C}}}}{-}\overset{\displaystyle Br}{\underset{\displaystyle Br}{\overset{|}{\underset{|}{C}}}}{-}CH_3$$

Propyne   Bromine

1,1,2,2-Tetrabromopropane

3. An experiment involves the reaction

$$X_2(l) + 2YZ(s) \longrightarrow 2XY(s) + Z_2(g)$$

(a) Complete the following table.

Reagent	MW	Equiv.	mmol	Mass	Density	Volume
X_2	102				1.53 g/mL	0.75 mL
YZ	58			2.0 g		

(b) If the theoretical yield of XY is 1.38 g, what is its MW?

(c) If the actual yield of XY is 1.23 g, what is the percent yield?

TYPICAL ORGANIC CHEMISTRY DESK EQUIPMENT

Note: Equipment will vary between institutions.

1 adapter, vacuum
6 beakers,
   50, 100, 150, 250, 400, and 600 mL
5 bottles, 10 dram (for samples)
1 bowl, plastic (for ice baths)
1 clamp, screw (for tubing)
1 clamp, test tube
2 condensers
1 cork ring (to support round-bottomed flasks)
2 cylinders, graduated, 10 and 100 mL
1 distilling head, Claisen
2 droppers, medium
1 drying tube
6 flasks, Erlenmeyer,
   25, two 50, 125, 250, 500 mL
1 flask, filtering, 250 mL
3 flasks, round-bottomed boiling
   50, 100, and 250 mL
1 funnel, Büchner
1 funnel, Hirsch
1 funnel, separatory, 125 mL
1 funnel, short stem
1 funnel, stemless
1 funnel, powder
5 Pasteur pipettes, 9 inch
1 tongs, flask
2 rubber pipette bulbs,
   2 and 25 mL
1 scoopula
1 spatula, micro
1 stir bar, magnetic, Teflon-coated
1 stopper, pennyhead
6 test tubes, 15 x 125 mm
6 test tubes, 10 x 75 mm
1 test tube rack
1 thermometer, −10 to 260°C
1 thermometer adapter with rubber holder
1 tube, connecting, three-way
2 watch glasses, 50 mm
1 watch glass, 155 mm

GENERAL HINTS ON USING GLASSWARE AND EQUIPMENT

Every experiment you will be carrying out involves the use of glassware. If you can understand the advantages and limitations of each item ahead of time, you will be able to work much more safely and efficiently. Also, some of the glassware you will be using is quite expensive and you will be responsible for it. For all of these reasons, it is wise to understand some general guidelines before beginning work. Illustrations of some common glassware and equipment appear in Figure 3; it would be a good idea to look over the illustrations carefully so that you can identify each item easily.

This section provides information on how to clean laboratory glassware and how to use the most common items. In addition, the standard sources of heat used in the laboratory are discussed. Specific techniques such as filtration, recrystallization, distillation, extraction, and chromatography will be discussed in the appropriate techniques lab in the first five experiments.

Cleaning Glassware

As is true with your dishes at home, it is much easier to clean laboratory glassware immediately after use and it helps to make this a habit. Tarry residues left on glassware will tend to stick more tightly the more time passes. An additional advantage of making sure that everything is clean at the end of each lab period is that you can start the next experiment with clean glassware that has had time to dry. There are several experiments that would be ruined if wet glassware were used.

The steps in proper cleaning of glassware are as follows:

1. Empty out as much foreign material as possible. Organic materials should go into specially marked waste containers, while relatively nontoxic water-soluble materials such as common acids or bases can go down the sink with lots of water.
2. Rinse the item out with tap water.
3. Scrub it with a bottle brush and some laboratory detergent, using tap water.
4. Rinse it under tap water.
5. Rinse it quickly under distilled or deionized water to remove any minerals from the tap water. This should suffice for normal washing.
6. If organic materials remain, rinse them out with acetone from a squirt bottle.
7. If a brown or black tarry residue remains after the foregoing steps, the following special measures can be tried. If one method does not work, try the next *after disposing of any chemicals properly and rinsing the glassware thoroughly. Extreme caution* must be used with all of these methods because they involve strong acids or bases.
    (a) Rinse the item with concentrated HNO_3.
      or
    (b) Soak the item in a chromic acid bath made by adding 1 L of concentrated H_2SO_4 *very slowly* to 35 mL of saturated aqueous $Na_2Cr_2O_7$ or $K_2Cr_2O_7$. Use great care with chromic acid baths since they are extremely corrosive and many chromium compounds are carcinogenic.
      or
    (c) Add a few pellets of solid NaOH and a little water. The heat evolved plus the corrosive alkali may loosen the residue.

Use of Glassware and Equipment

Pyrex glass possesses the valuable properties of being heat resistant and chemically inert toward most reagents, with a few exceptions. The main exceptions you might encounter in the lab are hydrofluoric acid (HF, not used in these experiments), which etches glass quickly, and sodium hydroxide (NaOH), which dissolves glass surfaces very slowly (over months or years). The main disadvantage of glass, of course, is its fragility, leading to the dangers of breakage and cuts. Glassware obviously should never be

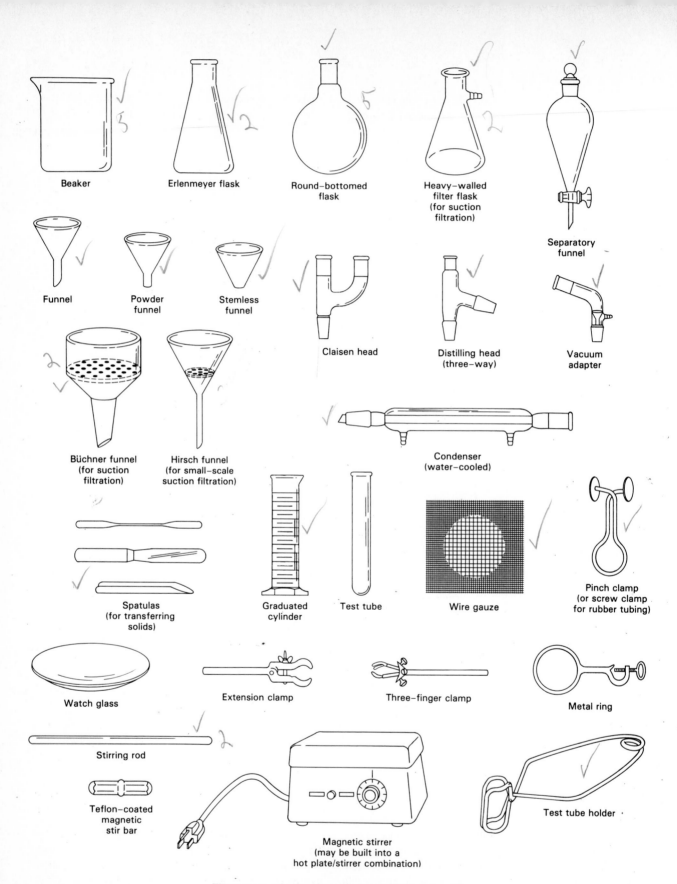

Figure 3 Common equipment in the organic chemistry lab

Beaker

Erlenmeyer flask

Round-bottomed flask

Heavy-walled filter flask (for suction filtration)

Separatory funnel

Funnel

Powder funnel

Stemless funnel

Claisen head

Distilling head (three-way)

Vacuum adapter

Büchner funnel (for suction filtration)

Hirsch funnel (for small-scale suction filtration)

Condenser (water-cooled)

Spatulas (for transferring solids)

Graduated cylinder

Test tube

Wire gauze

Pinch clamp (or screw clamp for rubber tubing)

Watch glass

Extension clamp

Three-finger clamp

Metal ring

Stirring rod

Teflon-coated magnetic stir bar

Magnetic stirrer (may be built into a hot plate/stirrer combination)

Test tube holder

dropped. More subtly, it should never be exposed to force or torque. For example, glass stirring rods should never be pressed with force, and when fitting glassware together you should be careful that pieces are aligned properly before clamping to avoid causing twisting forces. Another example is that while inserting glass tubing in rubber stoppers the tubing should always be lubricated with glycerin, the hands protected with a cloth, and excessive force avoided.

Some items of your equipment apply mostly for use with solids. Spatulas of various shapes are used for adding and weighing solids. Watch glasses provide places to hold small amounts of solids and serve as covers for beakers or Erlenmeyer flasks. Since most reactions occur in solution, most of your glassware is designed for handling liquids. Beakers hold liquids or solutions and are easy to pour from. Erlenmeyer flasks fulfill a similar role; they are more stable and less likely to spill. The narrower mouth of an Erlenmeyer gives slower evaporation; if one wants to evaporate a liquid, a beaker is faster.

Graduated cylinders are for measuring liquids. You will have several types of funnels available for pouring liquids or for separating solids from solutions using a filter paper. In a simple gravity filtration a filter paper is folded in a fluted pattern (Figure 4) and placed in a glass funnel atop a beaker or Erlenmeyer flask (Figure 5). A solution containing a solid material is poured through. The liquid passes through as the solid remains on the filter paper.

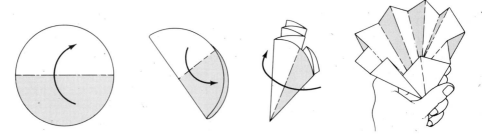

Figure 4   Fluting a filter paper

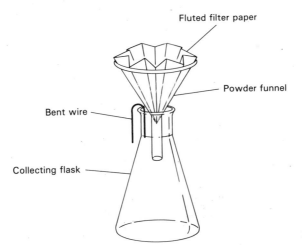

Figure 5   Gravity filtration apparatus

A variation of this technique is suction or vacuum filtration, in which a filter flask (basically a thick Erlenmeyer flask with a sidearm) is attached to an aspirator to provide suction (Figure 6). An aspirator provides a vacuum of about 20–60 torr by the Venturi

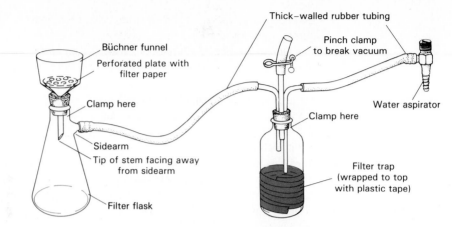

Figure 6   Apparatus for suction filtration

effect, passing a fast stream of water past a small opening (Figure 7). Placing a Büchner funnel with a round filter paper inside and a rubber adapter on top of the filter flask completes the setup. This apparatus, like any other, should be clamped firmly to prevent it from tipping over. When assembling an apparatus, use clamps generously in strategic positions. Clamp them *loosely* until the glassware is assembled in the proper arrangement and fit, then adjust them to minimize the strain on the glassware, and tighten finger-tight so that clamps are secure but do not cause strain.

Traps are used to prevent the undesired back-flow of liquids or gases into experimental setups. In this case a trap prevents water from flowing back from the aspirator into the filter flask. To break the suction after filtration, simply open the pinch clamp on the trap and, while leaving it open, shut off the aspirator.

When a solution containing a suspended solid is poured in the top of the Büchner funnel, the suction pulls the liquid through, leaving a cake of solid material. This solid can then be rinsed by pouring a solvent over it and letting this washing solvent be

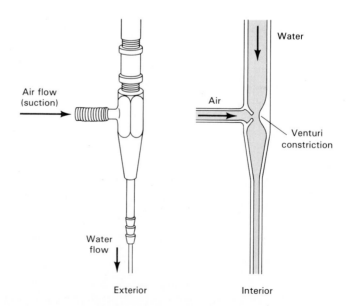

Figure 7   Detailed view of aspirator

sucked into the filter flask. Filtration techniques will be discussed more thoroughly in the first experiment.

Round-bottomed flasks are for holding solutions. They have standard-taper ground-glass joints on top so that an adapter or condenser can fit snugly. Ground-glass joints (Figure 8) have the advantages of being easy to assemble and virtually airtight. They have the disadvantages that any item containing ground glass is fairly expensive and that sometimes these joints tend to stick, especially if not lubricated before assembly or if exposed to basic solutions and not cleaned. To prevent sticking, ground-glass joints should always be coated with a light film of stopcock grease before assembly. A properly greased and fitted joint appears transparent, with no bubbles or gaps. When assembling a joint, one of the pieces should be rotated to spread the grease evenly and to make sure that the pieces fit properly.

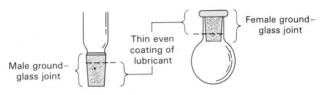

Figure 8 Ground-glass joints

Other than round-bottomed flasks, the main items containing ground glass you will use will be adapters and condensers. The function of adapters is to connect other pieces in the proper arrangement. A condenser has a jacket through which cool water flows, so that vapor entering the condenser becomes liquid. Condensers are used for distillations, in which case they recondense vapor for collection (Figure 9). The appa-

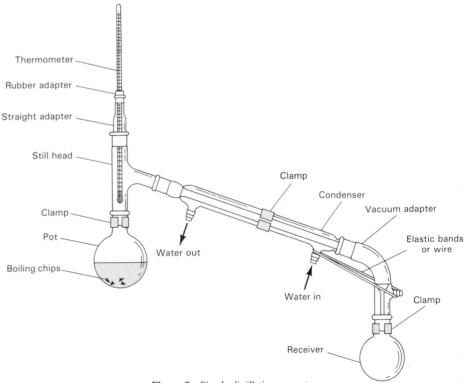

Figure 9 Simple distillation apparatus

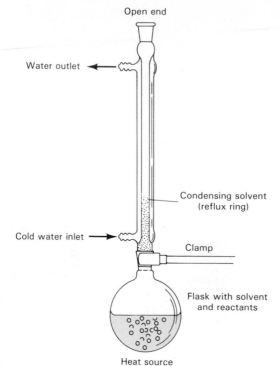

Open end

Water outlet ←

Condensing solvent
(reflux ring)

Cold water inlet →

Clamp

Flask with solvent
and reactants

Heat source

Figure 10   Reflux apparatus

ratus and procedure for distillation will be discussed in Experiment 2.1. A condenser can also be placed directly on top of a round-bottomed flask to form a *reflux apparatus* (Figure 10). The purpose of a reflux apparatus is to be able to boil a solution indefinitely without losing the solvent through evaporation. As the solvent vaporizes it rises into the condenser, where it condenses and drips back down into the reaction flask.

Heat Sources

Sources of heat found in the organic laboratory include Bunsen burners, hot plates, steam baths, and heating mantles. The Bunsen burner (Figure 11) was developed in the nineteenth century and is the oldest of these heat sources. In modern versions it has a valve to control gas flow and one for air, so that the size and temperature of the flame can be controlled somewhat.

A lot of gas and little air lead to a large, cool flame. More air will make a smaller, hotter flame. Within a flame, the tip of the inner blue cone is the hottest spot. Hot flames are needed to melt soft glass, such as stirring rods, while cool flames are better for heating and drying glassware or for freeing frozen glass stopcocks. The art of loosening stuck glass stopcocks and stoppers is discussed in the *note on storing separatory funnels* in Section 3.

The safety precautions to use with a Bunsen burner are as follows:
1. To light it, cut down on the air and strike a match over it before turning on the gas. Turning on the gas before lighting a match can lead to a dangerous buildup

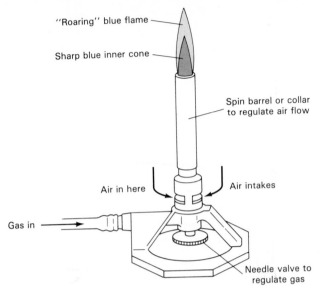

"Roaring" blue flame

Sharp blue inner cone

Spin barrel or collar
to regulate air flow

Air in here   Air intakes

Gas in

Needle valve to
regulate gas

Figure 11   Bunsen burner

of gas which will be ignited. If there is too much air coming in, it will be difficult
to light the burner.

2. *Always* keep the burner well away (at least 15 feet) from any flammable organic
materials, particularly solvents such as ether or toluene. Solvent fires can be very
dangerous, and preventing one is much easier than fighting one.

See the section "Safety Practices in the Organic Laboratory" on safety precautions relating to fires, in case one should start. Because of the fire danger we will avoid
using Bunsen burners as much as possible. We will use them only for conducting a
Beilstein test (which involves heating a copper wire), for drying glassware for a Grignard experiment, and for taking micro boiling points.

The standard heat sources in the modern introductory organic laboratory are the
hot plate, steam bath, and heating mantle (Figure 12). These are all less likely to lead
to fires than Bunsen burners, although hot plates and heating mantles certainly can cause

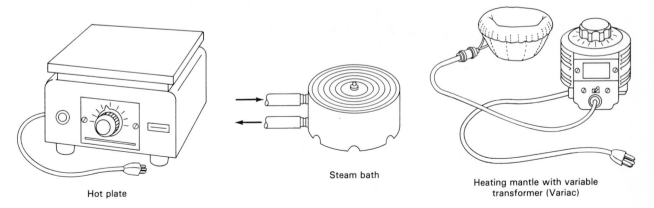

Hot plate   Steam bath   Heating mantle with variable
                                                           transformer (Variac)

Figure 12   Heat sources

fires if they contact flammable materials such as paper or solvents. As the heat cycles on and off, the electrical contacts in a hot plate produce sparks that can ignite solvent vapors.

Hot plates are used primarily for heating liquids or solutions in beakers or Erlenmeyer flasks. Although they take a few minutes to warm up, hot plates are usually quite strong and can boil water on fairly low settings. They have chemically inert ceramic surfaces that are somewhat brittle and can crack if dropped. If a flask of flammable solvent boils over onto a hot plate, it is likely to ignite; anything left on a hot plate should be watched closely. If a solution is left to evaporate on a hot plate, it must be watched carefully because any solid residue left is likely to overheat and decompose.

One way to avoid the overheating problem is by using a steam bath. A steam bath uses steam flowing in at the top and water out at the bottom to provide heat which, by definition, cannot rise above a temperature of 100°C. It is the method of choice for evaporating low-boiling solvents or heating small samples quickly. Of course, it may get a little water in your product because of all the steam around, but in general the amount is negligible. Steam baths come with rings on top which can be removed to give maximum exposure of the bottom of the flask to be heated, while supporting it securely.

For round-bottomed flasks the most convenient heat source usually proves to be a *heating mantle*. A heating mantle consists of some high-resistance heating wire enclosed in a glass fiber material, of a shape designed to fit a particular size of round-bottomed flask. To control it, a heating mantle must be plugged into a variable transformer (sometimes called a Variac). By controlling the voltage to the heating mantle the rate of heating can be adjusted with good control. Of course, it takes time to heat up and cool down a heating mantle, so if the setting is found to be too high, it may need to be shut off and lowered away from the round-bottomed flask to let it cool for a few minutes.

To work properly, a heating mantle must always fit the flask snugly enough so that heat can be transferred from the mantle to the glass effectively. This means that the correct-size mantle for the flask must always be used—otherwise, the heating mantle may overheat and burn out or melt while the flask does not get enough heat. Never turn on a heating mantle with nothing in it to conduct the heat away; it can overheat easily. A heating mantle also must never be operated in contact with paper or other flammable materials, which may scorch or ignite. This means that if an iron ring is used to support a heating mantle, it must not have paper or tape on it.

Balances

Several types of balances are commonly found in chemistry laboratories; some of these are shown in Figure 13.

The triple-beam balance, accurate to 0.01 g, requires manual adjustment of weights on three balance beams. A top-loading digital automatic balance is faster to operate because the tare (empty mass of the container) can be obtained at the touch of a button and displays the reading in a few seconds. Depending on the sensitivity, these may measure to 0.01 g (two-digit) or 0.001 g (three-digit) precision. An analytical balance (which may be a beam or top-loading instrument) measures mass to closer than 1 mg. In chemistry the term ''weight'' is often used to refer to mass. Of course, mass (measured in g, kg, etc.) is a measure of the quantity of matter an object contains, while weight (a force, measured in dynes, newtons, etc.) measures the gravitational attraction of the earth for an object. Although it is technically incorrect to refer to mass as weight, in our case, always operating at normal gravity, it should not lead to confusion. Balances are calibrated in grams because the scale used factors out gravity.

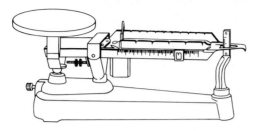

Laboratory platform balance (one knife edge) for crude weighing (0.1 g)

Digital electronic balance; the balance gives the mass directly when an object to be weighed is placed on the pan (0.001 g)

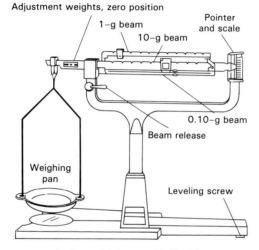

Adjustment weights, zero position

1–g beam

10–g beam

Pointer and scale

0.10–g beam

Beam release

Weighing pan

Leveling screw

Agate or stainless steel knife–edge (two knife edges) triple–beam balance (0.1g)

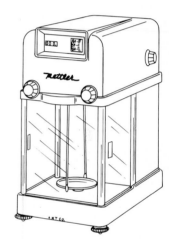

One–pan type analytical balance, for rapid weighing by substitution on a single beam arm (0.0001 g)

Figure 13 Some common types of balances and their sensitivities

To use a beam-type balance:
1. Check that it is clean and in proper operating condition.
2. Set the weights to zero. If it does not zero properly, use the zero adjustment screw provided.
3. If it has a stop, place it in the stopped position. If not, simply proceed.
4. Place the empty container on the pan.
5. Adjust the weights to approximate the expected weight, release the stop (if necessary), and check.
6. Adjust weights to obtain equilibrium, then take the reading.

To use a top-loader:
1. Check for cleanliness and operability.
2. Place the empty container on it.
3. Punch the tare button. Zeros should appear, indicating that the mass of the container has been subtracted out.
4. Carefully add the compound you want to weigh, a little at a time, until the desired mass is obtained.

Analytical balances require much greater care and extra steps to avoid damaging them. If you use these, follow the instructor's directions closely.

Recrystallization and Melting Points

Overview

Recrystallization is the most common method for purifying an organic solid. This technique relies on the fact that the solubility of an organic compound often increases greatly as the solvent is heated to its boiling point. When an impure organic solid is heated in an appropriate solvent to dissolve it, then cooled down to decrease its solubility, it will usually recrystallize from solution in much purer form. The graph of solubility versus temperature of solvent in Figure 1–1 illustrates three possible scenarios.

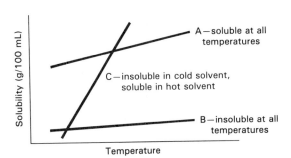

Figure 1–1 Graph of Solubility vs. Temperature for Recrystallization

If the compound is quite soluble at all temperatures, curve A is obtained, and the compound cannot be made to recrystallize satisfactorily from the solution on cooling; it will remain dissolved. If the compound is insoluble at all temperatures, curve B is obtained, and the compound cannot be dissolved in the first place. Curve C is the desired situation for recrystallization, where a great difference in solubility exists between hot and cold solvent. For our purposes, the dividing line between soluble and insoluble occurs at roughly 3%. In other words, if more than 3 g of a solute dissolves in 100 g of solvent at a given temperature, it is considered soluble; if less than 3 g, it is considered insoluble.

During recrystallization, insoluble impurities do not dissolve on heating and are removed by gravity filtering or decanting the hot solution. Soluble impurities remain dissolved in the cold solution from which the desired compound has recrystallized (called the *mother liquor*).

If the hot solution is gravity filtered (Figure 1–2), it is important to do it quickly so that the solvent does not cool off, which would cause the product to precipitate in the funnel. To keep the solution hot during the filtration, it helps to use a stemless funnel and to prewarm the funnel in the oven or on a steam bath.

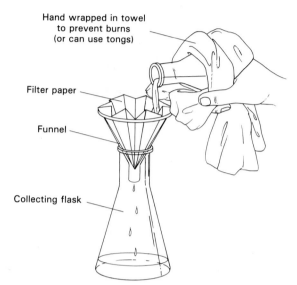

Hand wrapped in towel to prevent burns (or can use tongs)

Filter paper

Funnel

Collecting flask

Figure 1–2 Apparatus for Gravity Filtration

Recrystallization, then, is carried out by dissolving the solid in a minimum amount of boiling solvent, filtering while hot (if necessary), cooling the solution to room temperature or below, then filtering out the crystals formed using suction filtration, washing them with a little cold solvent, and allowing them to dry in air. The apparatus for suction filtration, also called vacuum filtration, is shown in Figure 1–3.

The faster a solution cools, the smaller the crystals will be, and if it is cooled too fast, solvent molecules or impurities may be trapped in the crystals. Impurities lead to

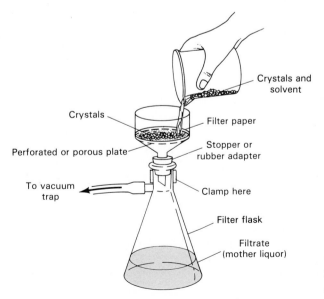

Crystals and solvent

Crystals

Filter paper

Perforated or porous plate

Stopper or rubber adapter

To vacuum trap

Clamp here

Filter flask

Filtrate (mother liquor)

Figure 1–3 Suction or Vacuum Filtration Apparatus

a low melting point, since regular intermolecular forces in the crystal are disrupted. One way to prevent too-rapid cooling is to insulate the beaker or Erlenmeyer flask containing the hot recrystallization solution by placing it inside another beaker with a paper towel or cotton wool as insulation and placing a watch glass on top.

If a crude sample of a compound known to be colorless is colored, the colored impurities can often be removed by adding some activated charcoal to the hot recrystallization solution. The solution should be below the boiling point to avoid boiling over when the charcoal is added. Large, colored molecules are adsorbed in the surface of the charcoal, which is then filtered out of the hot solution. When removing charcoal, a powdered filter aid (such as *Celite* or *Filter Cel*, consisting of diatomaceous earth, which is hydrated SiO_2) must be placed on the filter paper to prevent the tiny particles of charcoal from going through.

If no crystals form when the hot solution is cooled to room temperature, there may be too much solvent present or it may be in a supercooled state where the crystals would precipitate but do not have a "focus" to start forming. Standard tricks to induce crystallization in a supercooled solution include scratching the glass surface with a stirring rod, cooling in an ice bath, or adding a few "seed crystals" of the desired compound. If these methods do not work, too much solvent was used and some of it must be boiled off.

In some instances it may be desirable to collect a second or third crop by boiling down the mother liquor, the solution from which the first crop crystallized. The second and third crops are generally less pure than the first one, and their melting points should be checked to see how clean they are.

Another occasional problem encountered in recrystallization is the phenomenon of "oiling out." This means that rather than precipitating as a crystalline material, the product forms a syrupy, amorphous blob on the bottom of the flask. Low-melting compounds are more likely to oil out than are high-melting ones. Oiling out may occur as the solute becomes "contaminated" with solvent. You will recall that mixing an impurity with a solid lowers its melting point. If the melting point of the solute is lowered below the temperature of the solution, a supersaturated solution, instead of forming a precipitate, will form a second liquid phase, consisting of molten solute containing a small amount of solvent.

If you encounter an oil, it should be redissolved in hot solvent; then as it cools the glass surface should be scratched and the solution seeded if you have crystals of the compound available. Cooling the solution more slowly will also help prevent oiling out.

To choose the correct solvent one must be found in which the compound is fairly soluble when hot and fairly insoluble when cold. In Table 1–1 you can see that the best solvent listed for recrystallizing benzoic acid is water.

TABLE 1–1   SOLUBILITY DATA FOR BENZOIC ACID IN VARIOUS SOLVENTS

Solvent	Solubility (g/100 mL)	
	Hot	Cold
Water	6.8 at 95°	0.21 at 10°
Ethanol	67 at 78°	43 at 0°
Chloroform	Very soluble	Very soluble
ether, acetone, benzene		

If a suitable single solvent cannot be found, a solvent mixture can be used where the compound is fairly soluble in one of the solvents and fairly insoluble in the other. In this case the compound is dissolved in the solvent in which it is soluble at the boiling point, then the other solvent is added to the boiling solution until the "cloud point" is reached. The cloud point is the point at which the solution becomes cloudy because a precipitate just starts to form. As the mixture cools, the desired compound will often crystallize out. For example, naphthalene is reported to be soluble in methanol, ethanol, ether, acetone, and benzene, and insoluble in water. This suggests the procedure you will follow during the second period of this lab, which is to dissolve the naphthalene in hot methanol and drive it out of solution with water.

Mixed solvents must of course be miscible so that separate layers do not form. A very polar, hydrogen-bonding solvent such as water or methanol cannot mix with a very nonpolar solvent such as hexane or benzene. Lists of common solvents with their properties and of common solvent mixtures appear in Tables 1–2 and 1–3.

TABLE 1–2   PROPERTIES OF SOME COMMON CRYSTALLIZATION SOLVENTS

	Solvent	Boiling point (°C)	Density (g/mL)	Miscibility with water	Flammability
Most polar	Water	100	1.00	+	None
	Methanol	65	0.79	+	Low
	95% Ethanol	78	0.81	+	Moderate
	Acetic acid	118	1.05	+	Low
	Acetone	56	0.79	+	High
	Ethyl acetate	77	0.90	–	Moderate
	Ether	35	0.71	–	High
	Dichloromethane (methylene chloride)	41	1.34	–	None
	Chloroform	61	1.48	–	None
	Toluene	111	0.87	–	Moderate
	Benzene	80	0.88	–	High
	Carbon tetrachloride	77	1.59	–	None
	Petroleum ether	30–60	0.63	–	High
	Ligroin (low bp)	60–80	0.68	–	High
	Hexane	69	0.66	–	High
Least polar	Ligroin (high bp)	100–115	0.70	–	Moderate

TABLE 1–3   SOME COMPATIBLE SOLVENT PAIRS

Most polar	Ethanol–water
	Methanol–water
	Acetic acid–water
	Acetone–water
	Ethyl ether–methanol
	Ethanol–acetone
	Ethanol–petroleum ether
	Ethyl acetate–cyclohexane
	Chloroform–petroleum ether
	Toluene–ligroin
Least polar	Toluene–cyclohexane

Once a solid has been isolated, its melting range should be determined to establish its identity and purity. To find the melting range, take a closed-end capillary tube and press it open end down on the product to pack a little into the tube, as shown in Figure 1–4. If it is enough to see, it is enough to take a melting point. Tap the closed end of the tube on the bench top to slide the sample down to the closed end. If it does not slide down easily, drop the tube (closed end down) down the middle of a long piece of hollow glass tube and bounce it on a hard surface. This procedure is called using a ''bouncing tube.''

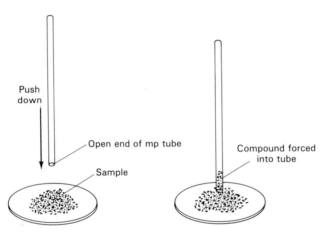

Figure 1–4 Filling a Capillary Tube for Melting Point Determination

Take the sample in its capillary tube over to the melting point apparatus, some common types of which are shown in Figure 1–5. Make sure that the temperature is below the melting point of your compound and insert your sample in the capillary port. Turn the apparatus *on*, observe your sample through the magnifying lens, and turn the knob to heat the sample. You can heat it up quickly to within 15–20° below the expected melting point, then slow the heating rate to about 2° per minute to get an accurate melting point. Record the temperature at which the first sign of melting appears and the temperature at which all the sample is melted. This is called the *melting range* and indicates the purity of a compound.

A pure compound will melt over a range of 1–2° near the melting point reported in the literature. An impure compound will melt over a range of anywhere from 3 to 80° or more and will melt below the literature value. Remember the freezing-point depression formula $\Delta T = k_f m$, where k_f is the freezing-point depression constant for the solvent (in this case, our compound) and m is the molality of an impurity.

This property can also be used in a test on the identity of an unknown called a *mixed melting point*. To take a mixed melting point a sample of the unknown is mixed with an equal quantity of the known compound that it is suspected to be, and after the two compounds are finely ground together, the melting range is determined. If it is the same as it was for each individual compound, this confirms the identity. If the mixed melting point is much lower than those of the individual compounds, it indicates that the compounds are not identical. They have ''contaminated'' each other and lowered the melting point of the mixture.

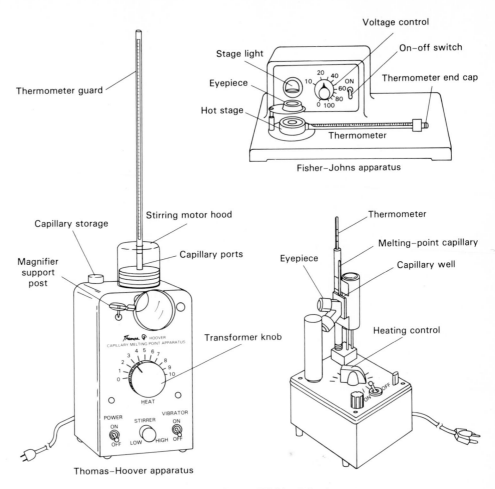

Figure 1–5 Common Types of Melting Point Apparatuses

Questions

1. What is the purpose of recrystallization?

2. What types of compounds are soluble in water? Insoluble?

3. How would you choose a solvent for recrystallizing a particular organic compound? Which properties are necessary and desirable for this solvent?

4. Why is suction filtration preferable to ordinary gravity filtration for separating purified crystals from the mother liquor after recrystallization?

5. In general, the addition of an impurity to a pure substance will cause its melting point to decrease in proportion to the amount of impurity. Explain this phenomenon.

6. Whenever a compound crystallizes from water, it has some water on the surface. Suggest three ways to get rid of this water and dry the product.

7. If, during a recrystallization, a solution is boiled for a long time and some of the liquid evaporates, what is likely to happen to some of the solute? How could you prevent or counteract this occurrence?

Concept Map

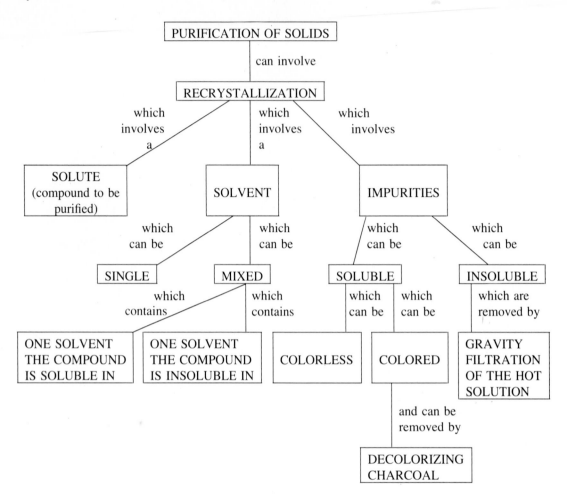

handwritten: 1st EXP.

EXPERIMENT 1.1 RECRYSTALLIZATION OF BENZOIC ACID FROM WATER

handwritten: COOH - ALWAYS ACID

handwritten: C₆H₅ COOH

Estimated Time:
1.5 hours

handwritten:
SOLUBILITY   4°C 0.18g/100ml
18°C 0.27g/100ml
75° 2.2g/100ml
ass 2.5g AT 100°C

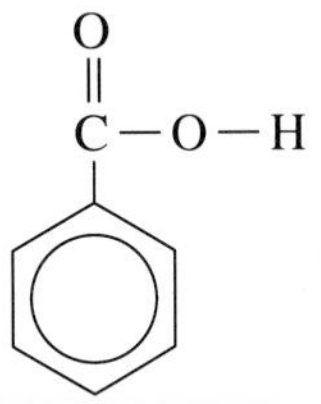

$$\underset{\substack{\text{O} \\ \| \\ \text{C}-\text{O}-\text{H}}}{}$$

Benzoic acid
MW 122
mp 122 – 123°

Prelab

1. Make a list of the equipment needed for this experiment. After the name of each item, describe its function.
2. What explanation can you give for the fact that benzoic acid is soluble in hot water and not in cold water?
3. Calculate the volume of boiling water expected to barely dissolve 1 g of benzoic acid.

Special Hazards

Benzoic acid is a mild irritant to skin, eyes, and mucous membranes, although its salt, sodium benzoate, is safe enough to be used as a common preservative in soft drinks.

In working with hot plates and boiling solutions, there is always the danger of burns and fires.

Procedure

In a small Erlenmeyer flask (25 or 50 mL), place benzoic acid (1.0 g) and the calculated volume of water that should dissolve it at the boiling point. Heat to boiling on a hot plate. Then, while keeping the solution boiling, add more water 1 mL at a time until the benzoic acid just dissolves. This may require up to 5 mL more water because some is evaporating during the solution process.

Remove the flask from the heat and insulate it by wrapping it with a paper towel and placing it inside a beaker topped with a watch glass. Allow the solution to cool slowly. In this case gravity filtration of the hot solution is unnecessary because there should be almost no insoluble impurities visible. When it has reached room temperature place it in an ice bath for 5 minutes to complete crystallization, then vacuum filter the crystals, wash them with a little cold water, and let them air-dry by placing them on a clean filter paper on a watch glass on the bench.

Examine your product carefully, noting the crystal shape and size. After weighing your dry product, place it in the collection container for reuse by the next group of students. After all, it is probably cleaner now than when you started.

Report your yield of recrystallized material, based on starting material, and its melting range. Comment on the purity of the crystals obtained.

QUESTION

1. Based on the volume of water you used to recrystallize the benzoic acid, how much would you expect to remain dissolved in the mother liquor after cooling? Report the percent recovery based on the amount you would have expected to crystallize from the volume of water used. (*Hint:* use Table 1–1.)

EXPERIMENT 1.2 RECRYSTALLIZATION OF NAPHTHALENE FROM METHANOL USING DECOLORIZING CHARCOAL

Estimated time:
2.0 hours

Naphthalene
MW 128
mp 81 – 83°

Prelab

1. What is the purpose of using mixed solvents for recrystallization?
2. What are the purpose and techniques for using decolorizing charcoal?
3. Suggest another method for recrystallizing naphthalene besides using methanol and water.

Special Hazards

Naphthalene is an irritant and is flammable. Methanol is toxic on ingestion and flammable as well. Avoid breathing fumes or spilling material on hot plates.

Procedure

In a 50–mL Erlenmeyer flask, place impure naphthalene (2.0 g) and methanol (20–30 mL). The naphthalene has been contaminated by adding about 0.1% of a dye such as Martius Yellow or Congo Red. Heat to boiling and note that all the naphthalene dissolves. Remove the flask from the heat, let the solution cool for a few seconds, and add some decolorizing charcoal. Be sure to let the solution cool for about 30 seconds and add the charcoal *cautiously*. Sudden addition of charcoal to a near-boiling solution may cause it to boil over, as the many small particles provide focal points for bubbles to form. Add enough charcoal until the edge of the solution, when held up to a light, appears colorless between the black particles. It does not hurt to add too much charcoal; at the worst, it may decrease the yield very slightly. Try to avoid long delays as you add the charcoal, since should the solution cool too much, the naphthalene may begin to precipitate. If it does, you can reheat the solution and/or add a little more hot methanol.

Since the charcoal particles are very small, some of them would pass through a filter paper. To prevent this, add a heaping spatula-full of Celite to your hot mixture and swirl it. The Celite will pack onto the filter paper in the next step as you filter the solution, and will trap the charcoal particles.

Preheat a stemless funnel by either placing it on a steam bath, pouring hot water through it, or warming it in a drying oven. Quickly gravity filter the hot solution using this preheated funnel containing a fluted filter paper. If any of your product precipitates in the funnel, rinse it through with a little hot methanol. Alternatively, you can set the collection flask on a hot plate as you are filtering and the hot vapors rising will help prevent the naphthalene from precipitating.

After filtering, check that the charcoal and most of the color are gone. If either remains, repeat the appropriate steps. Bring the solution to a boil again and add water (about 4–6 mL) dropwise until the mixture remains slightly cloudy. Insulate the flask and allow it to cool slowly to room temperature, then cool it in an ice bath. Vacuum filter the crystals and wash them with some water. Let them air dry and record the melting range. When you report your yield, comment on the purity of the naphthalene obtained. A gray color in the product means that not all the charcoal was removed, while a yellow or red tint means some of the dyestuff impurity remains. Place your product in the collection container provided.

EXPERIMENT 1.3 CHOOSING A RECRYSTALLIZATION SOLVENT

Estimated time:
2.0 hours

Special Hazards

Methanol, petroleum ether, toluene, and acetone are all highly flammable. Methanol is toxic on ingestion or prolonged breathing of vapor, while petroleum ether, toluene, and acetone are all irritants. Avoid exposure and handle boiling solvents only in the fume hood.

Procedure

Obtain approximately 1 g of a solid unknown to be recrystallized and record the unknown number. Record the melting range of the unpurified unknown, then into each of five 10 x 75 mm test tubes place approximately 100 mg (0.10 g, about enough to fill the tip of a micro spatula) of the unknown. Label these tubes W, M, P, T, and A (representing water, methanol, petroleum ether, toluene, and acetone, respectively). Into each tube place approximately 2 mL (one Pasteur pipette full) of the appropriate solvent. Swirl each tube for a minute or so and record in Table 1–4 whether the unknown is soluble in the solvent at room temperature (R.T.) To record results on the chart, use the abbreviations S = soluble and I = insoluble.

If the unknown is soluble in a particular solvent at room temperature, discard the contents of that tube into the appropriate waste container in the fume hood. For the other tubes, add a boiling chip and heat each briefly to boiling (or, for water, nearly boiling) in a steam bath or boiling-water bath in the fume hood. Be careful not to point the tubes at anyone while heating, since the contents may spurt out unexpectedly.

TABLE 1–4 RESULTS OF SOLUBILITY TESTS

Solvent	Water	Methanol	Petroleum ether	Toluene	Acetone
Solubility at R.T.					
Solubility near bp of solvent					
Recrystallizes on cooling?					

For each tube record whether the unknown is soluble near the boiling point. Discard (into an appropriate waste bottle in the fume hood) the contents of any tubes in which the unknown is still insoluble. The only remaining tube is that in which the unknown is insoluble at room temperature and soluble at the boiling point. Allow this tube to cool to room temperature and observe whether recrystallization occurs. If no crystals have appeared after 15 minutes, cool the tube in an ice bath and scratch the inside with a stirring rod to induce crystallization. If crystals still do not appear, try seeding the solution with a few small particles of the unknown.

When crystals are obtained, isolate them by suction filtration and wash with 1–2 mL of whichever solvent they were recrystallized from after cooling the wash solvent in an ice bath. Allow the crystals to dry in air and record the new melting range.

Questions

1. Your unknown is one of the compounds in Table 1–5 (listed in order of increasing melting range). In the *CRC Handbook of Chemistry and Physics*, look up all the compounds melting within 10° above or below your unknown. Record which solvent is listed as suitable for

TABLE 1–5 POSSIBLE UNKNOWNS

Melting range (°C)	Name(s)
1. 59–60	1-Nitronaphthalene
2. 69–72	Biphenyl
3. 78–80	Methyl 3-nitrobenzoate
4. 81–83	Naphthalene
5. 108–110	Anthranilamide
6. 113–115	Acetanilide
7. 113–115	Benzoquinone
8. 122–123	Benzoic acid
9. 122–124	*trans*-Stilbene
10. 140–142	2-Nitrobenzoic acid
11. 148–150	4-Nitroaniline
12. 152–154	Adipic acid
13. 160–163	Triphenylmethanol
14. 180–182	2-Methylbenzoic acid (*o*-toluic acid)
15. 182–185	4-Methoxybenzoic acid (anisic acid)
16. 188–189	4-Aminobenzoic acid (*p*-aminobenzoic acid, PABA)

recrystallizing each compound. How do these compare to your results? If you can identify your unknown, report its name and structure. If you cannot decide among several, explain your reasoning and describe a method that could be used for determining which possibility is the correct one.

2. What are some general guidelines for predicting the solubility of a compound?

3. Briefly describe how to choose a recrystallization solvent for an unknown organic compound.

Distillation

Overview

Distillation is the easiest and most common way of purifying organic liquids. Distillation, of course, means heating a liquid enough to vaporize it, then cooling and condensing the vapor back to a liquid. Any lower-boiling impurities vaporize before the desired material and are collected and discarded in the *forerun*, while any higher-boiling impurities remain as a *residue* in the distilling pot after the desired material has been distilled.

The process of distillation, then, involves placing an impure liquid in a *distilling pot* and heating it. As the mixture reaches its boiling point, the vapor rises through a distilling head, past a thermometer, and enters a *condenser*, where cool water flowing through an outer jacket causes the vapor to condense to liquid. It runs down the condenser, through an adapter, and into a collection flask. The apparatus for a simple distillation appears in Figure 2–1. An enlarged view of the still head is shown in Figure 2–2.

Any liquid collected at a thermometer reading *below* the expected boiling point of the desired material is the forerun and, in general, will be discarded. If there is only one desired component in the distilling pot, it is collected after the forerun at the desired component's boiling point and then, after it has come over, the residue in the pot is discarded.

If, as in this experiment, there are two desired components, the procedure changes slightly. After the forerun comes over and the collecting flask is changed, the first component is collected at its boiling point. Then, after switching collecting flasks again, an intermediate fraction is collected which is a *mixture* of both desired components. This distills over at temperatures *between* the boiling points of the two desired components. Once the boiling point of the second desired component is reached, the collecting flask is changed again and this component is collected in pure form. Anything remaining in the pot after this second component distills over is the residue.

Figure 2–3 shows the thermometer readings expected during the course of the distillation of a two-component mixture. As the lower-boiling component A vaporizes and strikes the thermometer bulb, the reading will increase from room temperature to the boiling point of A.

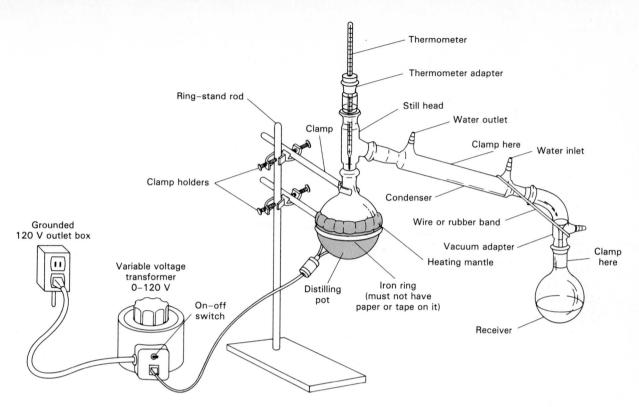

Figure 2–1 Apparatus for Simple Distillation

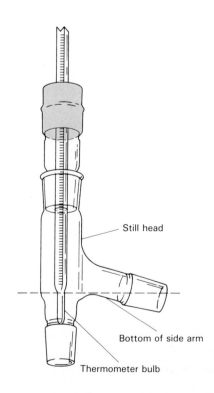

Figure 2–2 Enlarged View of Still Head
(Note: Thermometer bulb must be entirely below bottom of sidearm)

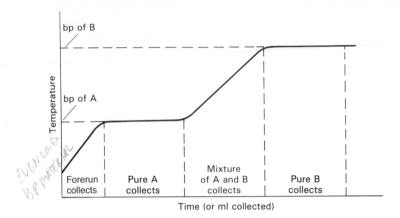

Figure 2–3   Thermometer Readings During Distillation of a Two-Component Mixture

Relatively pure A distills in a narrow temperature range near its boiling point. When most of A has come over, the thermometer reading rises as a mixture of A and B distills. When the thermometer temperature reaches the boiling point of B, it stabilizes and relatively pure B comes over. After component B is exhausted, the thermometer reading may fall (if little vapor reaches it) or may continue to rise if there is a significant volume of a somewhat-higher-boiling component in the residue.

Boiling occurs when the vapor pressure of a liquid equals the atmospheric pressure above it. Vapor pressure measures the tendency of molecules to leave the surface, while atmospheric pressure and intermolecular forces are holding the molecules down. Figure 2–4 illustrates this situation. As a consequence of this balance of forces, if the pressure above the liquid decreases (for example, if vacuum is applied), it is easier for the molecules to leave the surface, and the boiling point decreases. A handy rule of thumb states that every time the pressure is halved, the boiling point decreases by about 10°. For example, a liquid that boils at 125° at 760 torr (1 atm) would boil at approximately 115° at 380 torr, 105° at 190 torr, and so on.

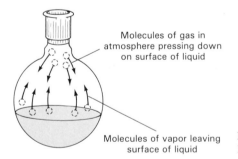

Molecules of gas in atmosphere pressing down on surface of liquid

Molecules of vapor leaving surface of liquid

Figure 2–4   Molecules in Atmosphere Pressing Down as Liquid Vaporizes

Another rule of thumb predicts that at higher altitudes of 5000 feet or so, where the atmospheric pressure is around 650 torr, boiling points will be about 4% below their normal values. For example, water boils at about 96° under these conditions.

High-boiling liquids are often distilled under reduced pressure to lessen the danger of decomposition. For example, cinnamaldehyde, one of the main components of cinnamon flavor, boils at 246° with some decomposition at 760 torr, but it is much easier and cleaner to distill under pump vacuum at 1 torr, where it boils at 76°.

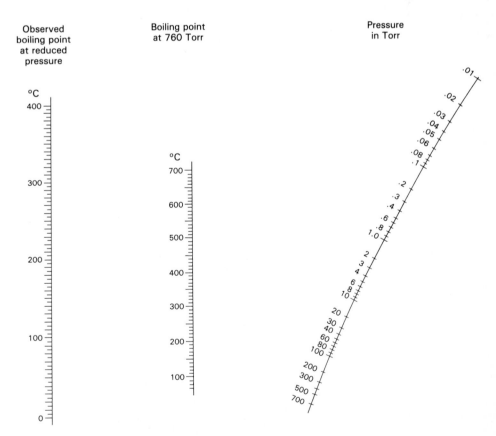

Cinnamaldehyde
(*trans*-3-phenylpropenal)
bp 246° at 760 torr or 246^{760}, 76° at 1.0 torr or $76^{1.0}$

To predict the boiling point of a liquid at any pressure you can use the pressure–temperature nomograph in Figure 2–5. Using a ruler, line up the atmospheric boiling point in the center column with the pressure at the right. The new boiling point at reduced pressure is read on the left-hand column.

In this experiment we will not use reduced pressure because it is unnecessary, but will separate and purify two components of a mixture by a technique called *fractional distillation*. Simple distillation can separate compounds cleanly that differ in boiling points by about 70° or more, but since the two components we will use (acetone, bp 56° and toluene, bp 110°) differ in boiling point by only 54°, we need to use fractional distillation to get a clean separation. Fractional distillation gives the same purification in one step that several simple distillations would give. The way to do this is by using

Figure 2–5 Pressure-Temperature Nomograph

a column packed with stainless steel sponge called a *fractionating column*. As the vapor rises through the column it condenses and vaporizes several times on the surface of the packing material, giving the equivalent of several simple distillations. The apparatus for fractional distillation differs from that for a simple distillation only in the column, as you can see in Figure 2–6.

Fractional distillation has two additional advantages. First, it is easier to purify large volumes of liquids by one fractional distillation than by repeated simple distillations. Second, the losses of material are smaller. These losses occur in the forerun, the mixture obtained between desired fractions, and the residue left in the pot.

There is a simple way to describe the efficiency of a fractional distillation. The number of simple distillations a fractional distillation is equivalent to is called the number of *theoretical plates* in the fractionating column. One plate equals one simple distillation. A typical laboratory fractionating column may possess two to four theoretical plates, while a large industrial column may contain 20 or more theoretical plates. Once we know the number of equivalent plates in a column, and we know its length, we can talk about the height equivalent to a theoretical plate (HETP), which is the length of a column divided by the number of theoretical plates in the column. If, for example, a laboratory column had three theoretical plates and a length of 24 cm, its HETP would be 8 cm. The lower the HETP, the more efficient the column.

To see what kind of purification we expect from a simple distillation of a mixture of acetone and toluene, we look at the temperature–composition diagram. Except for

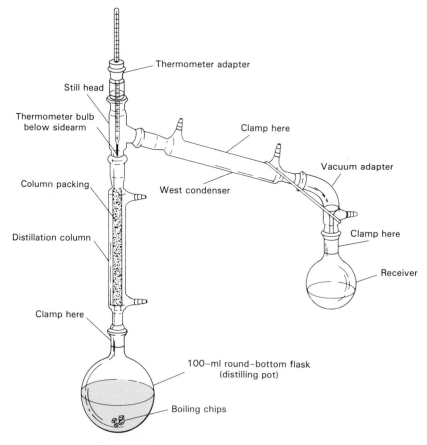

Figure 2–6 Fractional Distillation Setup

the special case of an azeotropic mixture, the vapor above a mixture always has a different composition from the liquid, since it is enriched in the more volatile component (acetone in this case). The curves in Figure 2–7 show, for any given composition of liquid acetone and toluene, the boiling point of the liquid and the composition of the vapor above it. For example, a liquid that is 10% (by moles) acetone and 90% toluene has a boiling point of about 107° (point A). The vapor at this temperature has the composition at point B, about 50:50.

Let's follow through the processes of simple and fractional distillation on this chart. For example, starting with a *liquid* of 10% acetone and 90% toluene (by moles) at point A, if we heat it to boiling and vaporize some of it, the *vapor* will have the composition at point B, and as this is cooled in the condenser we go to point C, so we now have a liquid that is approximately 50% acetone and 50% toluene, by moles. Since standard reagent grade needs to be at least 95% pure (preferably 99%) and even technical grade needs to be 85% pure, we need more theoretical plates. A second theoretical plate (or a second simple distillation) would put us at point E. You can see that for reasonable purity we need at least two theoretical plates, preferably three or four. Your fractionating column, packed with stainless steel sponge, has two or three theoretical plates. Thus you should be able to get a reasonably clean separation.

One additional factor to consider in boiling a solvent or solution is that for bubbles to form easily, it helps to have a site or nucleus for them to form. You may have noticed that bubbles usually form at scratches or imperfections in a surface. If a surface is very smooth, it may be difficult for the vapor to nucleate and "bumping" may occur. Bumping is uneven, violent boiling that may splash material out of the container. The presence of solid materials with rough surfaces encourages the formation of many small bubbles instead of fewer large ones. For this reason chemists often add inert granules of carborundum, ceramic, or Teflon (called boiling chips or boiling stones) or small pieces of wood (called boiling sticks). Wood will react with strong acids and some ceramic materials may dissolve in hot concentrated acids or bases, but usually these materials are unreactive. In most cases, when boiling a solution, the addition of a few (one to five) boiling stones is advisable to prevent bumping, and you will be using them in this experiment.

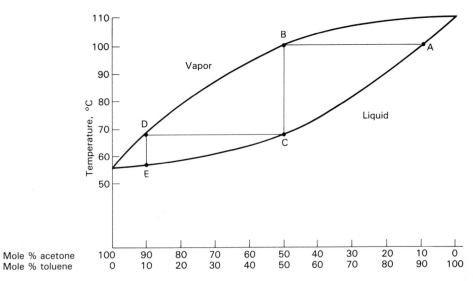

Figure 2–7 Composition of Vapor and Liquid During Distillation of an Acetone-Toluene Mixture

APPLICATION: FRACTIONAL DISTILLATION OF PETROLEUM

Pascagoula Refinery (Courtesy Chevron Corp.)

cracking (*or reforming*) process involves addition of a catalyst, heat, and pressure to break larger hydrocarbon molecules into smaller ones. Hydrogen gas may also be added, in which case only saturated hydrocarbons are obtained.

FRACTIONAL DISTILLATION OF CRUDE OIL

Petroleum fraction	Composition	Commercial use
Natural gas	C_1–C_4	Fuel for heating
Gasoline	C_5–C_{10}	Motor fuel
Kerosene	C_{11} and C_{12}	Jet fuel and heating
Light gas oil	C_{13}–C_{17}	Furnaces, diesel engines
Heavy gas oil	C_{18}–C_{25}	Motor oil, paraffin wax, petroleum jelly
Residuum	C_{26}–C_{60}	Asphalt, residual oils, waxes

Fractional distillation of crude oil yields gasoline and kerosene, as well as heavier oils as shown in table (opposite). Since initial distillation of crude oil yields only 19% gasoline (called *straight-run* gasoline) and a large quantity of less useful, larger hydrocarbons, the remaining oil is often subjected to catalytic cracking prior to another distillation. The

Large-scale industrial fractional distillations are often carried out using a *bubble-cap* column. The column is arranged so that lighter hydrocarbons vaporize and move up the column, where they recondense and are drawn off at appropriate levels. A schematic of a refinery is shown below.

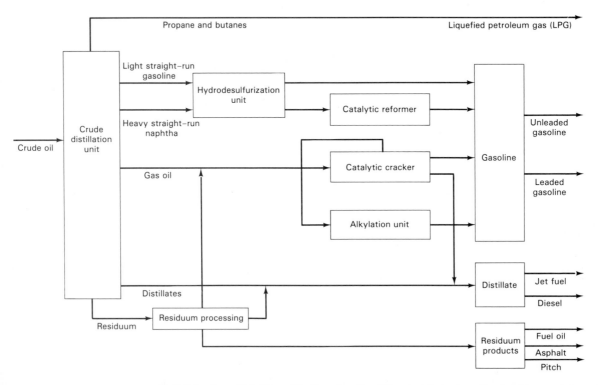

Oil Refining (from Kirk Othmer/The Encyclopedia of Chemical Technology. Reprinted by permission of John Wiley & Sons, Inc.)

Concept Map

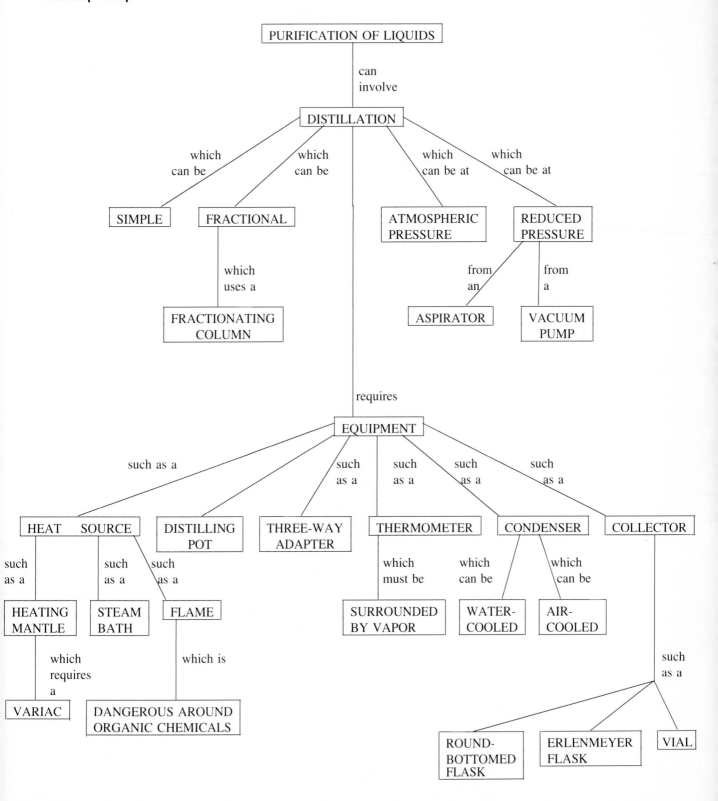

QUESTIONS

1. What is the purpose of distillation?
2. Explain in your own words the purpose of using a boiling chip. When would you not use one?
3. Clove oil (eugenol) is reported to have a boiling point of 255° at 1 atm. If you are distilling it under aspirator vacuum (about 40 torr), what would you expect the bp to be? Under pump vacuum (0.1 torr)? Under atmospheric pressure at an elevation of 5000 feet?
4. If you started with a mixture of 20 mol % acetone and 80 mol % toluene and your column had two theoretical plates, approximately how pure would the first acetone distilling over be? (*Hint:* Find the correct composition on the chart and draw a line up to the *liquid* curve. The temperature at this intersection point is the boiling point of this mixture. From the intersection point, draw a line left to intersect the *vapor* curve. The composition at this point is the composition of the vapor above the original mixture as it boils. Of course, it is enriched in the more volatile component. Now drop a line down from this point to the *liquid* curve. This represents cooling the vapor and condensing it. You have now completed one simple distillation, also known as one theoretical plate. Repeat this process for the second theoretical plate.)
5. Under what conditions could you separate cleanly a mixture of three or more liquids by fractional distillation?
6. If an industrial fractionating column has 40 theoretical plates and is 10 meters high, what is its HETP? Do you think it could separate a mixture of cyclohexane (bp 69°) and toluene cleanly?

EXPERIMENT 2.1 FRACTIONAL DISTILLATION OF AN ACETONE–TOLUENE MIXTURE

Estimated time:
2.0 hours

Prelab

1. Make a list of the equipment needed for this experiment. After the name of each piece, describe its function.
2. Describe the process of assembling a fractional distillation apparatus.
3. Describe the process of carrying out a fractional distillation once the apparatus is assembled.
4. What will happen if you leave gaps at the joints of your apparatus?
5. What is likely to happen if the clamps are too loose? Too tight?
6. At what thermometer readings do you expect to collect the acetone? The toluene? A mixture of the two?
7. If a mixture of acetone (MW 58.1, dens. 0.791 g/mL) and toluene (MW 92.1, dens. 0.867 g/mL) is 1:1 by volume, what percent is each by moles?

Special Hazards

Never *force* thermometers or other glassware; they are likely to break and cause cuts. If thermometers are lubricated with glycerin, joints are greased lightly and fitted properly, and clamps are secure but not *too* tight, danger of breakage will be minimized.

To avoid overheating, a heating mantle must be plugged into a variable transformer, never directly into a wall socket. A heating mantle should never be placed in contact with flammable materials such as paper or cork rings. Be sure that the iron ring you use for support does not have any paper or tape on it.

No closed system should ever be heated since it builds up pressure and may explode. Be sure that your system always has an opening to relieve pressure.

To avoid releasing organic fumes into the lab, be sure that all joints in the distillation apparatus remain tight.

Both acetone and toluene are flammable, and neither should be inhaled in large quantities.

Procedure

In a dry 100–mL round-bottomed flask, which will be your distilling pot, place three or four boiling chips, then add acetone (15 mL) and toluene (15 mL). Clamp the flask securely so that its bottom is about 6 inches (15 cm) above the lab bench. Now you will proceed to assemble the apparatus for fractional distillation shown previously by following the instructions below.

Pack your fractionating column by pulling a little stainless steel sponge through the column using a hooked copper wire. Never push it, because of the danger of breakage or of cutting your fingers on the sharp steel. The amount of stainless steel is not critical as long as it is not packed so tightly that the vapor cannot get through. If one of your condensers is larger than the other, use the larger one for the fractionating column. If there are glass protrusions inside the column, be careful not to break them. After greasing the joints lightly, mount the packed column on the flask. Be sure that the joint is tight, then clamp the column securely, but not *too* tightly.

Put a drop of glycerin on the thermometer and insert it into its holder on top of the Claisen distilling head. After greasing its joints, mount the still head on top of the fractionating column. Adjust the thermometer so that its bulb is *completely below the sidearm* in the distilling head. This ensures that it will be totally surrounded by vapor and give an accurate reading. If the bulb is too high, your temperature readings will be low.

Attach rubber tubing to the water inlet and outlet nipples on the condenser and, after greasing its joints lightly, slip it securely onto the distilling head and clamp it. Finally, add the adapter and a collecting flask, which is also clamped. The adapter should be secured to the condenser with a rubber band or wire to prevent it from falling when the collecting flask is changed. Double-check that all joints are tight and straight and that the apparatus is clamped securely. Adjust the clamps if needed for the best fit. Avoid overtightening clamps, which can cause torque on the glassware and break it. Fit the heating mantle tightly onto the bottom of the flask and support it with an iron ring. Plug the heating mantle into a variable transformer but do not turn it on yet.

Now you are ready to begin the distillation. Make sure that water is flowing through the condenser at a slow, constant rate. To avoid bubbles, water should go *in* at the bottom and *out* at the top. Then turn on the variable transformer controlling the voltage to the heating mantle. You can heat the mixture relatively strongly at first, but keep an eye on it. As soon as it nears the boiling point, adjust the variable transformer to get a good constant boil without overheating. You will know it is overheating if bubbles are reaching into the neck of the flask or liquid is splashing clear up into the distilling column. If this happens, immediately loosen the iron ring, lower the heating mantle away from the flask and shut it off. After the flask cools (either by itself or in an ice bath) you can begin heating it again, less vigorously. Actually, most students tend to err on the side of caution while heating the flask, and the distillation takes longer than necessary. You can maintain a fairly vigorous boil and aim for a distillation rate (once you start getting material distilling over) of one drop every second or two. After bringing the pot to a boil, but before any material comes over, you should see a ring of condensing vapor (called the reflux ring) rising in the flask and through the column. Adjust the heat if necessary so that the reflux ring rises at a constant rate to the top of the column.

As the reflux ring hits the thermometer the temperature reading will shoot up and vapor will start to condense in the condenser. Collect the first few drops of distillate (no more than 1 mL) and change receivers to another round-bottomed flask. By this time the thermometer reading should be up near the boiling point of acetone. In general, in any distillation the pure fraction(s) should be collected at a fairly constant boiling point, say a range of 3–4° or so. In this particular case the range may be up to 7°.

On general principles the first few drops (the forerun) in any distillation should be discarded, since this will contain any low-boiling impurities present. So note the volume of the forerun you collected, then discard it in the container marked "waste."

Collect the acetone as long as the thermometer reading remains relatively constant, within ± 3° of the expected boiling point. The volume will be about 10–15 mL. Adjust the heat so that your rate of distillation is approximately 1 drop collected every 1–2 seconds. When the thermometer reading has risen by about 3° above the boiling point of pure acetone, change the receiver because this means that a mixture of acetone and toluene is coming over. Collect this small middle fraction. The temperature of the thermometer may drop after almost all the acetone has distilled over because there is not enough vapor to surround the thermometer. It may be necessary to increase the voltage to the heating mantle in order to drive over the higher-boiling toluene. Now observe the toluene as the reflux ring rises through the column. After the first few drops of the toluene have come over, driving over any residual acetone with them, and the thermometer reading reaches the boiling point of toluene, change receivers again. Label the middle fraction collected and record its volume. Collect the pure toluene, keeping an eye on the distilling pot. When the volume remaining in the pot gets down to 2–3 mL shut off the variable transformer and lower the heating mantle away from the flask. A distillation should never be carried to dryness because once the pot is dry the glass becomes very hot, and if peroxides, diazo compounds, or other explosive impurities are present, they are likely to detonate.

Record the volumes of the acetone, toluene, and intermediate fraction collected. Feel free to admire the results of your work for a while, and then when you are ready to clean up, pour your clean acetone and toluene fractions into the appropriate collection containers. Discard the forerun, mixture fraction, and residue into the properly marked waste container.

Acid–Base Extraction

Overview

Along with distillation and recrystallization, extraction is one of the most useful separation techniques in organic chemistry. Because extractions are so important and you will be doing so many of them in this course, we take some time now to go through the theory thoroughly. The idea behind extraction is to use two liquids (or solutions), one an organic phase and the other an aqueous phase, so that the desired product dissolves in one while undesired by-products or starting materials dissolve in the other. Since the liquids are immiscible, they form two layers and are easily separated by drawing off the lower layer in a separatory funnel.

The process of extraction, then, consists of adding two liquids or solutions (one organic and one aqueous) to a separatory funnel, shaking with occasional venting to relieve pressure, then drawing off the lower layer through the stopcock. These steps are shown in Figures 3–1 through 3–3. When filling a separatory funnel, first make sure that the stopcock is *securely fastened and closed*, and place a beaker or Erlenmeyer flask underneath to catch the liquid in case the stopcock leaks. Note the use of an iron ring to support the separatory funnel. A clamp alone should not be used for support because it is inadequate for the task.

When using a solvent with a low boiling point such as ether (bp 35°) or dichloromethane (bp 41°), pressure builds up in a closed separatory funnel as some of the solvent vaporizes at room temperature. For this reason it is essential to vent the separatory funnel frequently to relieve the pressure. To vent the separatory funnel, hold it securely in both hands as shown in Figure 3–2 and open the stopcock briefly. Since you will be shaking the separatory funnel vigorously and often, it is important to know the safe way to do it.

For right-handed people, the separatory funnel is held securely with the stopper between the fingers of the right hand and the stopcock between the fingers of the left hand. Left-handed people will reverse this arrangement. This way of holding the separatory funnel makes it easy to vent it by opening the stopcock between shakings, and if the stopper does blow out due to internal pressure, you will catch it before it falls out completely and will not lose all of your product. Remember to vent frequently and *never* to point the stem of the separatory funnel at anyone, including yourself. After shaking

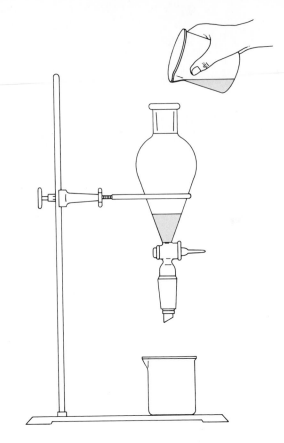

Figure 3–1 Filling a Separatory Funnel

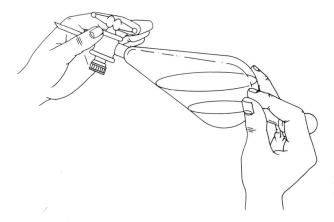

Figure 3–2 Venting Position Showing Correct Way of Holding Funnel for Shaking and Venting

and venting about three to five times, the lower layer can be drawn off as illustrated in Figure 3–3.

There are several points to keep in mind during an extraction. One is *always to be aware of which layer is the organic phase and which is the aqueous phase.* The more dense liquid or solution will of course always be on the bottom. For example, in using ether (density 0.71 g/mL) and water (density 1.00 g/mL), the ether layer is the top one. Using water and chloroform ($CHCl_3$, density 1.49 g/mL) or dichloromethane

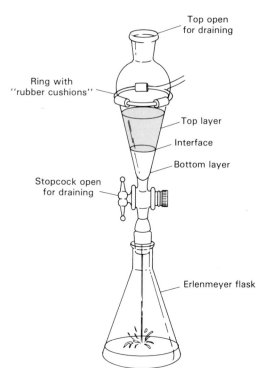

Top open
for draining

Ring with
"rubber cushions"

Top layer

Interface

Bottom layer

Stopcock open
for draining

Erlenmeyer flask

Figure 3–3 Drawing Off the Lower
Layer from a Separatory Funnel

(also known as methylene chloride, CH_2Cl_2, density 1.33 g/mL) the water will be the upper layer and the organic phase the lower. In general, halogenated hydrocarbons (CH_2Cl_2, $CHCl_3$, CCl_4, *n*-butyl bromide) have high densities and will be the lower layer, while nonhalogenated organics (diethyl ether, hexane, toluene) have low densities and will be the upper layer with water.

If you are ever in doubt about which layer is which, remove a drop of the layer in question and add it to a test tube containing about 1 mL of water. If it dissolves, that layer is aqueous; if not, it is organic. Also be aware that whenever you are draining the separatory funnel, the top stopper must be removed; otherwise, a vacuum will develop on top which prevents the liquid from flowing.

One important rule for novice organic chemists is to *save all discarded layers combined in a large beaker until the final product is in hand*. This precaution is taken so that if by chance you save the wrong layer, your product can be recovered by extracting the waste container. Many students in organic labs have poured their products straight down the drains while saving the wrong layers.

Let's take a moment and talk about which kinds of solvents will dissolve which compounds. To begin with, let's consider what makes a compound soluble in the most common solvent, water. Water molecules, which are strongly hydrogen bonded to each other, have their hydrogen bonding disrupted when something else dissolves. Therefore, to dissolve in water a molecule must be relatively small, to minimize disruption of the water–water hydrogen bonds. A rule of thumb states that in order to be water soluble, a compound must have less than six carbons if it contains only one polar functional group. Otherwise, the molecule must have an ionic or several polar functional groups to *replace* the bonding lost between the water molecules with some type of favorable bonding between water and the compound.

For example, benzoic acid has a very low solubility in water because it has seven carbons and only one polar functional group, whereas sodium benzoate (used as a preservative in soft drinks) is highly water soluble because it is ionic.

Benzoic acid
(low solubility in water)

Sodium benzoate
(high solubility in water)

Another example is that glucose is water soluble, but l-hexanol is not.

Glucose
(high solubility in water)

$HOCH_2CH_2CH_2CH_2CH_2CH_3$

1-Hexanol
(low solubility in water)

Glucose has *five* polar OH groups for six carbons, whereas l-hexanol has only one. Therefore, the glucose will hydrogen bond much better with water.

A third example is that aniline (aminobenzene) has low solubility in water, whereas its protonated (ionic) form, the anilinium ion, is highly soluble.

Aniline
(low solubility
in water)

Anilinium ion
(high solubility
in water)

The key to separating organic compounds by extraction consists of choosing the form of a compound with which you want to deal. You noticed above, for example, that by removing a proton from benzoic acid, which is ether soluble, we can "pull" it into the aqueous phase. This would be done by extracting with 5% aqueous NaOH. The base deprotonates the acid, making a carboxylate ion that is soluble in water and insoluble in ether. This would leave behind in the organic phase all organic-soluble compounds that are not acidic, in other words neutral or basic compounds.

If later we want to drive the benzoic acid back out of the aqueous phase, we can acidify (carefully, since heat will be evolved). This will reprotonate the benzoate ion and cause benzoic acid to precipitate or dissolve in a fresh organic phase if present. On the other side of the coin, an amine such as aniline can be pulled out of the organic phase (such as ether) by extracting with 5% HCl, which protonates it and makes it water soluble. If we want to recover the aniline, we can then basify the aqueous layer. These properties lead to the rule "*extract acids out of the organic phase with base, extract bases out of the organic phase with acid.*"

We can illustrate these concepts with the following diagrams, showing which form of a compound (ionic or neutral) will dissolve in which phase (aqueous or organic).

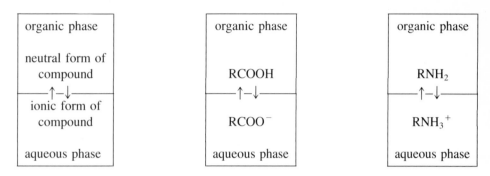

In this example the organic phase is shown on top of the aqueous phase, which would be the case with ether and water. The concept would, of course, apply equally well if the organic phase were on the bottom: for example, with dichloromethane and water.

Let's take a moment to define some terms. *Extraction* means pulling the desired compound out of one phase into another by shaking with a new solvent. *Washing* a solution means removing impurities by shaking the solution with a different solvent that will dissolve the impurities but not the desired compound. *Drying* to an organic chemist means removing water, either by washing with a hygroscopic solution such as saturated aqueous NaCl or by adding a drying agent such as Na_2SO_4, $MgSO_4$, or $CaCl_2$. Drying *does not* mean removing the solvent unless it is water; *evaporation* is the term used for removing an organic solvent.

Sometimes during shaking in the course of an extraction an emulsion will form that consists of finely divided droplets of both solvents and is slow to separate. If this happens, you can reach in with a stirring rod to try to break up the droplets, add some more of one or both of the solvents, and shake less vigorously next time. If an emulsion forms, you can also help to break it up by enhancing the difference in density between the phases. For example, if the organic phase is on top, adding low-density ether may help. If the aqueous phase is on the bottom, adding salt (NaCl) to increase its density may help. Adding salt also decreases the solubility of organics in water because it increases the polarity of the aqueous phase, which makes the nonpolar organic materials even less soluble. This process is known as "salting out" an organic compound from an aqueous solution.

The flowchart on the next page illustrates the separation of a mixture containing acidic, basic, and neutral components. In order to separate a mixture of acidic, basic, and neutral components, you would first dissolve the mixture in ether. Now to get the *basic* component out you would extract with 5% HCl. The basic component (amine) will be protonated and extracted into the aqueous phase, while the neutral and acidic components will remain in the organic phase. The layers are separated and the aqueous phase is carefully basified with 6 N NaOH. This will drive the amine out of solution, often causing it either to precipitate (if a solid) or to form its own layer (if a liquid). If it is easy to separate at this stage by filtration or separating layers, that is done. Otherwise, if it did not separate well, perhaps remaining as just a cloudy solution, a fresh portion of ether is added and the ether layer containing the amine is separated off, washed with saturated aqueous NaCl, dried over anhydrous Na_2SO_4, filtered, and evaporated to yield the amine. If no product is obtained from this procedure, it means that there was no basic component present.

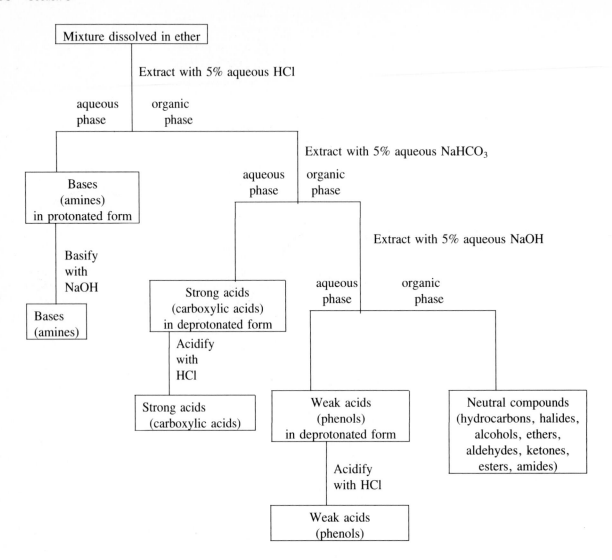

Now let's consider how to get the *acidic* component out of the original ether solution. The original ether layer, which may still contain acidic and neutral compounds, is extracted with 5% aqueous $NaHCO_3$. This weak base solution will deprotonate and dissolve strong acids (carboxylic acids) but not phenols, which are too weakly acidic to be deprotonated by $NaHCO_3$. On acidifying the aqueous layer (which contains $NaHCO_3$) with concentrated HCl, lots of CO_2 gas will be evolved according to the equation

$$HCO_3{}^- + H^+ \longrightarrow H_2CO_3 \longrightarrow H_2O + CO_2 \uparrow$$

and the carboxylic acid will be driven out of the acidic solution, to be collected in any of the ways described above for amines (filtration, separating layers, or extraction with a fresh portion of ether). Again, if no product is obtained, it means that there was no carboxylic acid component. In order to extract strong acids such as carboxylic acids, a weak base ($NaHCO_3$) suffices. To extract weaker acids, such as phenols, we need to use a stronger base, 5% aqueous NaOH. This is the next step. Extracting the ether layer with 5% aqueous NaOH deprotonates and dissolves weak acids (phenols). Upon acidifying the extract, any phenolic component can be collected.

Anything that has not been removed from the original ether layer by extracting with HCl, NaHCO$_3$, or NaOH must be neutral. Washing this ether layer with saturated aqueous NaCl to remove most of the water that has dissolved in the ether, drying over Na$_2$SO$_4$ to remove any final traces of water, decanting or filtering the ether solution away from the drying agent, and evaporating off the ether should give the neutral component.

Experiment 3.2 describes the extraction of the basic compound caffeine from tea leaves. Caffeine is perhaps the most widely used drug in the United States. It is present in tea, coffee, cola drinks, chocolate, and many common nonprescription drugs, including alertness, weight-loss, and pain-relieving products. Caffeine stimulates the respiratory, cardiac, and central nervous systems, and can become psychologically addictive. We all know (and maybe are) people who need a morning "fix" of caffeine. Table 3–1 lists the amount of caffeine found in some common drinks.

TABLE 3–1 APPROXIMATE CAFFEINE CONTENTS OF SOME COMMON BEVERAGES

Beverage	Caffeine content (mg/8-oz-cup)
Coffee	60–260
Tea	30–160
Cola drinks	20–40
Hot chocolate	3–30

Caffeine is classified as an *alkaloid*: a basic nitrogen-containing plant material with a bitter taste, pharmacological activity, and a name ending in *ine*. Some other examples of alkaloids are nicotine, morphine, and cocaine, shown below.

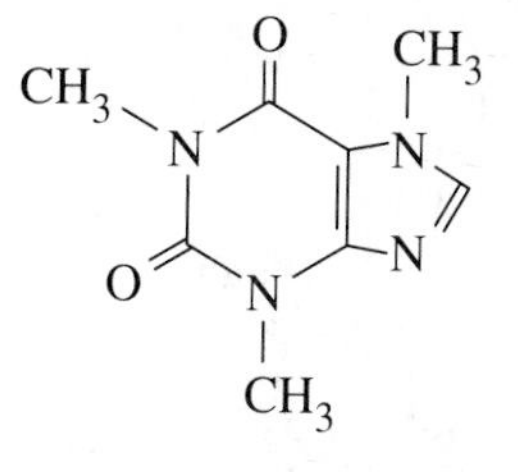

Caffeine
MW 194
mp 234 – 236°

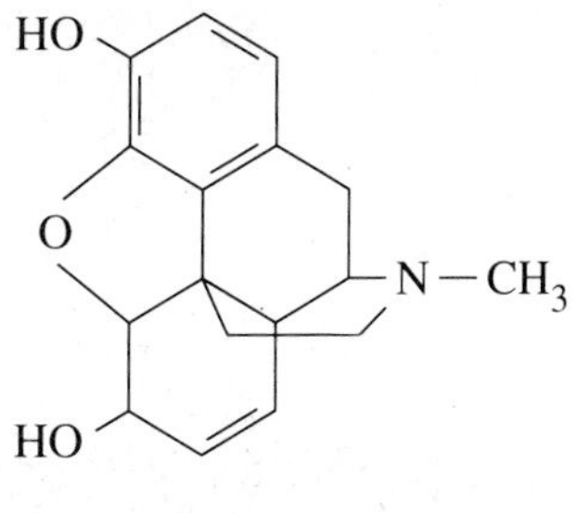

Morphine
MW 285
mp 254°d

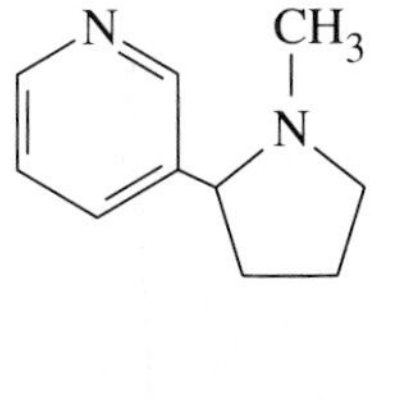

Nicotine
MW 162
bp 247°d

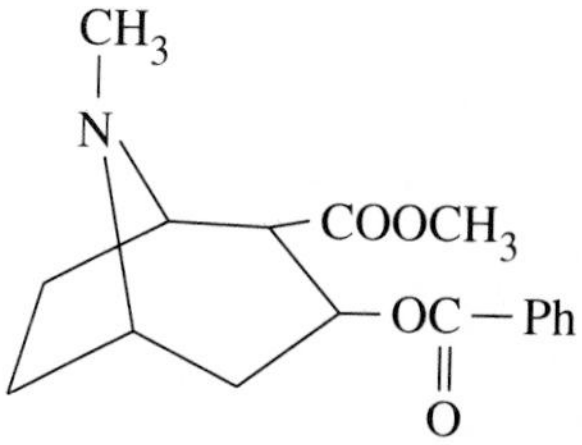

Cocaine
MW 303
mp 98°

Coca-Cola originally contained extracts from both coca leaves and cola nuts. Since 1903, Coke has not included cocaine from the coca leaves, although some probably wish that it did. It still contains significant quantities of caffeine from the cola nuts. In Experiment 3.2 caffeine is extracted from tea leaves, which contain about 2–4% caffeine by weight.

Since caffeine is a large basic molecule, it is insoluble in aqueous base. Adding sodium carbonate to an aqueous tea infusion makes it basic, driving caffeine molecules into the added organic phase (dichloromethane). Any acidic compounds present (such as tannic acid) will be deprotonated and dissolved in the base. Any neutral components will, of course, be left in the organic phase with the caffeine. Since caffeine is the major basic component of tea and there are no significant amounts of neutral compounds, this is a good way to separate it out of the mixture.

Caffeine can be purified by *sublimation*, going directly from the solid to a gas and then back to a solid. The idea is that the crude material is heated while under vacuum, so that it vaporizes, then the pure solid recondenses on a cool surface. Volatile impurities do not recondense and are sucked out of the system, while nonvolatile impurities do not sublime and are left behind at the bottom.

A typical sublimation apparatus is shown in Figure 3–4a. You can see that as the sample is heated under vacuum it will vaporize and then collect on the water-cooled condenser, which can be lifted out to obtain the crystals. Since this apparatus is quite

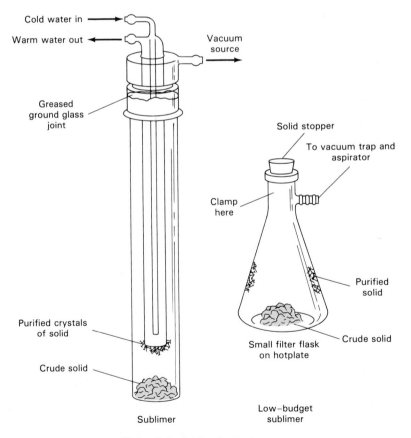

Figure 3–4 Sublimation Equipment

APPLICATION: HISTORY, CULTIVATION, AND COMPOSITION OF TEA

A Tea Plantation in India (Courtesy R. Twinings & Co., LTD.)

Legend holds that the use of the tea leaf was discovered in 2737 B.C. when the Chinese Emperor Shen-Nung was boiling his drinking water to purify it and leaves from a branch of a nearby tea plant were blown into the pot, imparting a delicate flavor the Emperor found pleasing.

English Afternoon Tea (Courtesy R. Twinings & Co., LTD.)

Sometime used as a medicine and stimulant, tea gained widespread acceptance as a soothing beverage. The tea ceremony developed in Japan to allow relief from daily stresses by emphasizing simplicity and harmony with nature. The British also developed tea drinking (usually flavored with milk and sugar) to a fine art, and morning and afternoon tea-times provided needed relaxation and opportunities for social gatherings. In Russia, tea was traditionally served in a glass with lemon slices and sugar or jam. A traditional English afternoon tea is shown at left (below).

In the 1830s the British introduced tea cultivation to India, and in 1869 to Ceylon (now Sri Lanka), when the coffee crop was destroyed by a fungus.

If allowed to grow freely, the Indian variety *Thea viridis* will grow to be a loosely branched tree over 30 feet high, while the predominant Chinese variety *Thea bohea* grows more as a bush or hedge. Many hybrid varieties are known. Most tea production today takes place in China, India, Sri Lanka, and Indonesia.

In addition to caffeine, tea contains significant quantities of tannins, proteins, and carbohydrates. Tannins consist of two types: (1) esters of a sugar, usually glucose, with one or more trihydroxyben-

Corilagin (a hydrolyzable tannin)

zenecarboxylic acids, the so-called hydrolyzable tannins; and (2) derivatives of flavonols, so-called condensed tannins.

Tannins find use in dyeing of fabrics, in manufacture of inks, paper, and silk, in tanning leather, in clarifying beer and wine, and in photography.

Tea also contains small amounts of theophylline, a xanthine that is a smooth muscle relaxant, myocardial stimulant, and diuretic, sometimes used in the treatment of asthma.

Theophylline

expensive and fragile, we will use a much simplified version consisting of a filter flask under vacuum on a hot plate, shown in Figure 3–4b. The caffeine will sublime onto the sides of the flask, and when the sublimation is finished it will be scraped off. Of course, all this discussion may be unnecessary, since as many psychologists have pointed out, college students are used to sublimating anyway.

There is a slight mathematical side to extractions as well. Whenever a compound has a choice of dissolving in two liquids (or solutions) there is a distribution coefficient (K) that describes the relative concentrations in the two phases:

$$K = \frac{\text{concentration of A in solvent 1}}{\text{concentration of A in solvent 2}}$$

The larger the value of K, the cleaner the separation will be.

For example, if the distribution coefficient K for compound A in ether versus water is 7, then A is seven times as soluble in ether as it is in water. That means that if a separatory funnel contains equal volumes of ether and water and and a total of 1 mole of A, the ether layer will contain 7/8 of a mole and the water will contain 1/8 of a mole. In other words, every extraction of the water with an equal volume of ether would remove 7/8 of the A present. How much would remain in the water layer after two extractions? Out of the 1/8 mole remaining after the first extraction, 7/8 was removed and 1/8 remains, so there is 1/8 x 1/8 = 1/64 mole left in the water. What if, instead of using two portions of equal volume, we had extracted with one portion of ether twice as large? In this case the ratio of the concentrations must still be 7, so

$$\frac{[A] \text{ in ether}}{[A] \text{ in water}} = 7$$

or

$$\frac{\dfrac{\text{moles in ether}}{\text{volume of ether}}}{\dfrac{\text{moles in water}}{\text{volume of water}}} = 7$$

Since the volume of ether is twice the volume of water, letting the volume of water equal X and the volume of ether equal $2X$, we obtain

$$\frac{\dfrac{\text{moles in ether}}{2X}}{\dfrac{\text{moles in water}}{X}} = 7$$

Canceling the X and moving the 2 to the other side yields

$$\frac{\text{moles in ether}}{\text{moles in water}} = 14$$

or $\qquad\qquad$ moles in ether $= 14$ times moles in water

or $\qquad\qquad$ moles in ether $= \dfrac{14}{15}$ of total moles

Therefore, 1/15 remains in the water, compared to 1/64 for extraction with two equal portions. You can see from this example that it is much more effective to extract with several small portions than with one large one.

In fact, there is a handy formula for computing the amount of material remaining in a layer after n extractions:

$$[A_{aq}]_n = \left(\frac{V_{aq}}{V_{org}K + V_{aq}}\right)^n [A_{aq}]_0$$

where $\quad [A_{aq}]_n =$ concentration of A remaining in the aqueous phase after n extractions

$\qquad\quad [A_{aq}]_0 =$ initial concentration of A in the aqueous phase

$\qquad\qquad V_{aq} =$ volume of the aqueous phase

$\qquad\qquad V_{org} =$ volume of the organic phase used in each extraction

$\qquad\qquad\quad K =$ distribution coefficient between organic and aqueous phases

$\qquad\qquad\quad n =$ number of extractions carried out

Questions

1. What is the purpose of an extraction?
2. What properties do you look for in a good solvent for extraction?
3. Which layer (upper or lower) would each of the following solvents usually form if used to extract an aqueous solution: diethyl ether, chloroform, acetone, and hexane?
4. If, when extracting an aqueous solution with an organic solvent, you are uncertain of which layer is the organic layer in the separatory funnel, how could you quickly settle the issue?
5. If the cap is left off a bottle of a drying agent such as anhydrous Na_2SO_4 for a long time, what will happen?
6. What makes a compound acidic? What types of functional groups are acidic?
7. What makes a compound basic? What types of functional groups are basic?
8. If you test an unknown organic compound and find it to be soluble in 5% aqueous $NaHCO_3$, what functional group does it probably contain?
9. If you accidentally poured one of your products, an amine, into your *waste* beaker, how could you recover it in reasonably clean form?

10. Describe in detail how you would separate the following mixtures.

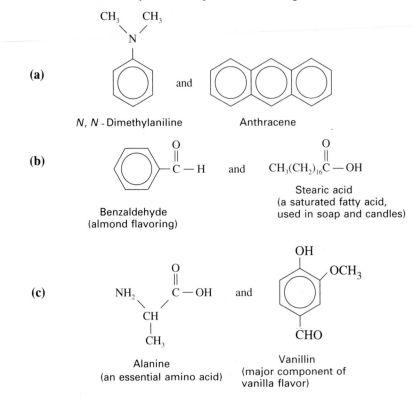

(a)

N, N - Dimethylaniline   Anthracene

(b)

Benzaldehyde
(almond flavoring)

and

Stearic acid
(a saturated fatty acid,
used in soap and candles)

(c)

Alanine
(an essential amino acid)

and

Vanillin
(major component of
vanilla flavor)

11. If you wanted to recrystallize the benzoic acid, *m*-nitroaniline, or naphthalene, how would you decide on a solvent?

12. Assume that the distribution coefficient for compound X between chloroform and water is 2.8. If 50 mL of water that is initially 0.1 M in X is extracted with three 25–mL portions of chloroform:

(a) What is the final concentration of X remaining in the aqueous phase?

(b) How many millimoles of X remain in the aqueous phase?

(c) If X has a MW of 156, how many grams remain in the aqueous phase?

(d) If instead of using three 25–mL portions of chloroform, one 75–mL portion had been used, how much of the solute would remain in the aqueous phase?

(e) What can you conclude about the desirability of using one large portion or several small portions of solvent for extractions?

Concept Map

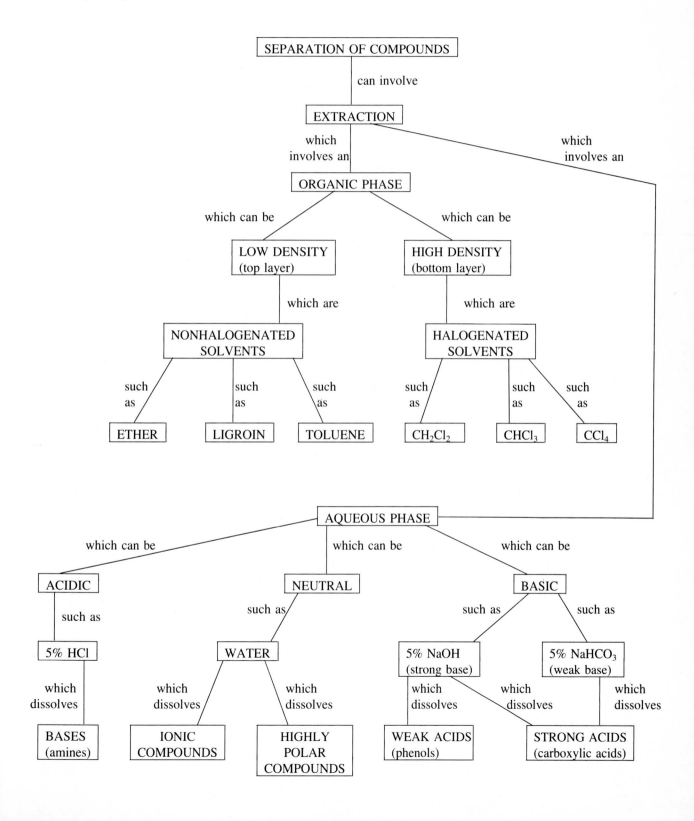

Concept Map

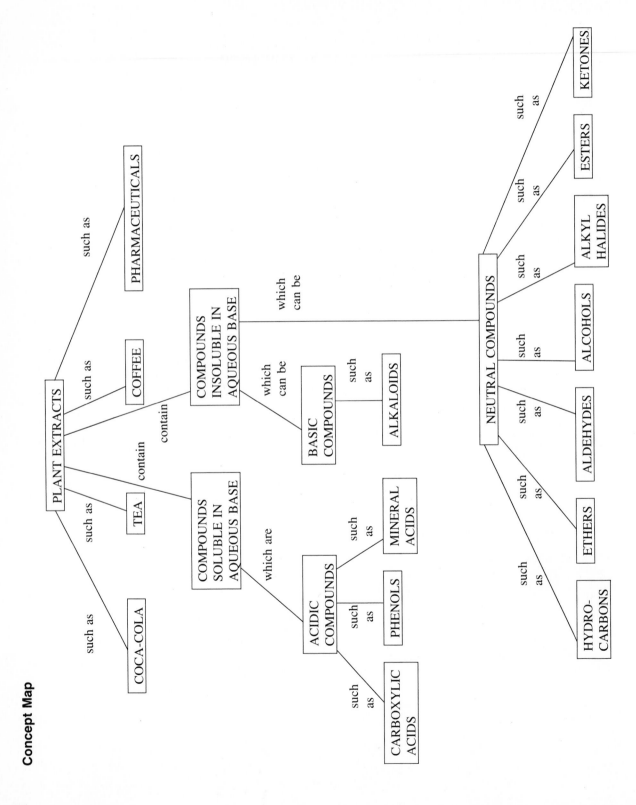

EXPERIMENT 3.1 SEPARATION OF BENZOIC ACID, *m*-NITROANILINE, AND NAPHTHALENE

Estimated Time:
2.5 hours

Prelab

1. Using a flow sheet similar to that in the overview, outline the separation of benzoic acid, *m*-nitroaniline, and naphthalene.
2. How many millimoles of benzoic acid, *m*-nitroaniline, and naphthalene are contained in 3 g of a mixture that is equal parts by weight?
3. How many millimoles of HCl are contained in 50 mL of 5% (by weight) aqueous HCl (density 1.02 g/mL)? What excess is this of the minimum needed to extract all the *m*-nitroaniline, assuming complete protonation? Similarly, in what excess is the 5% (by weight) aqueous NaOH used to extract the benzoic acid?
4. What volume of 5 *N* NaOH will be needed to neutralize 50 mL of 5% HCl? About what volume of 6 *N* HCl will be needed to neutralize 50 mL of 5% NaOH?

Special Hazards

Low-boiling solvents build up pressure easily, so be sure to vent the stopcock frequently. Ether is flammable, mildly irritating to skin, and an anesthetic if inhaled in large quantities. Use adequate ventilation. Like many aromatic amines, *m*-nitroaniline is toxic on ingestion, inhalation, or prolonged skin contact. Avoid contact with skin and clothing; wash immediately and thoroughly if contact occurs.

Procedure

You will separate a mixture of benzoic acid (an acidic compound), *m*-nitroaniline (a basic compound), and naphthalene (a neutral compound) by successive extractions from ether.

Benzoic acid	*m*-Nitroaniline	Naphthalene
MW 122	MW 138	MW 128
mp 122°	mp 112 – 114°	mp 80°

To a separatory funnel add ether (50 mL), 5% HCl (50 mL), and 3 g of a mixture of equal weights of benzoic acid, *m*-nitroaniline, and naphthalene. Stopper the funnel, hold it as shown in the diagram, being sure not to point it toward anyone, vent it, shake vigorously for 5–10 seconds, vent again, and repeat the shaking and venting two to three more times. Since ether has a boiling point of only 35° and heat is released as the layers mix, you must vent frequently to avoid building up pressure and blowing the stopper out. The aqueous layer now contains the protonated amine, while the ether layer contains the benzoic acid and naphthalene.

Place the separatory funnel in a ring supported on a ring stand, remove the stopper, and drain the lower (aqueous) layer into an Erlenmeyer flask. Label the flask "HCl extract." *Note:* The stopper must be removed to drain a separatory funnel. Otherwise, a vacuum develops on top and the flow is very slow. Also be sure that your stopcock has a plastic or rubber ring on it to prevent it from slipping out unexpectedly.

Carefully make basic the acid extract by cooling in an ice bath and adding 3 *N* NaOH, checking with pH paper to be sure that it becomes basic. To test the pH, dip a glass stirring rod in the solution and touch it to a strip of pH paper. In this way each piece of indicator paper can be used several times.

The free *m*-nitroaniline should precipitate out, at which point it is collected by vacuum filtration, rinsed with a little water, and allowed to air dry as you proceed with the rest of the experiment.

To the original ether layer still in the separatory funnel add 5% NaOH solution (50 mL). Stopper it, vent, then shake and vent several times. Remove the stopper, draw off the aqueous layer, and acidify with 6 *N* HCl, once again checking with pH paper. Collect the precipitate (benzoic acid) by vacuum filtration, wash with water, and allow it to air dry.

The neutral component should still be in the ether layer. To remove most of the water that has dissolved in this layer during the previous extractions, add saturated NaCl solution (50 mL), vent and shake a few times, then draw off the lower layer into a large beaker marked "waste." *Remember:* Do not throw away any layers until all your products are in hand, to avoid the danger of throwing away the wrong layer. Murphy's law, third corollary, states that if you throw away a layer before isolating your product, the product is in the layer you throw away.

Now place the ether layer in an Erlenmeyer flask and add enough anhydrous Na_2SO_4 until it stops clumping and some remains as a loose powder in the solution. The drying agent should be added slowly, a small amount at a time, in order to dry the solution efficiently and prevent the formation of large chunks of drying agent that are wet on the outside and still powdery on the inside. As long as all the drying agent clumps together, there is water left and more sodium sulfate should be added. Let it stand for a few minutes. Decant the solution away from the drying agent, add a boiling chip, and evaporate the ether on a steam bath in the hood to give the neutral component.

Take the melting point of each component isolated, compare with the literature values, and comment on the purity of each. Remember that the purpose of extraction is to separate the components in reasonably pure form, but you have not recrystallized or done any further purification. Weigh the three compounds isolated and report a crude percentage recovery for each. Place your three compounds in appropriate containers provided for recycling them.

Special note on storing separatory funnels. When a separatory funnel is used it should be washed thoroughly at the end of the lab period, including removing the stopcock. If the stopcock is glass, it should be stored *out* of the separatory funnel and greased thoroughly before being replaced the next time. Glass stopcocks left in separatory funnels tend to get stuck, especially if exposed to basic solutions and not cleaned thoroughly, or if the grease has dissolved. Teflon stopcocks do not have this problem, but the nut should still be loosened for storage so that pressure against the glass does not deform the Teflon. If your glass stopcock should ever freeze up, the following techniques are recommended for loosening it. If one method does not work, try the next:

1. Tap it gently with a wooden spatula handle.
2. Place it in an ultrasonic cleaner, if available. The vibration often shakes the stop-cock loose.
3. Trade the item for another with the stockroom manager, who will soak it in a bath of Coca-Cola. This procedure, recommended by several glass manufacturers, has proven remarkably effective, although it may take up to 6 months. Perhaps the effectiveness is due to the action of the weak phosphoric acid solution on the ground glass.
4. Soak the item in a soap solution for about a week.
5. Quickly heat up the stopcock area in a steam bath or a cool flame, causing the outer glass ring to expand more quickly than the stopcock itself, then tap it gently with a wooden handle.

Taken together these methods are about 90% effective in saving expensive separatory funnels. Even so, you can see that it would be easier to avoid the problem in the first place. These methods may also be used to loosen ground-glass stoppers stuck in round-bottomed flasks. However, a flask neck should not be heated in a flame if the flask contains ether or other volatile or flammable materials because of the danger of fire or explosion from pressure buildup.

Questions

1. What color was your pure naphthalene from Experiment 3.1? If the naphthalene from this extraction has a yellow tint, what does that mean? How could you further purify the naphthalene obtained here if you wanted to?
2. If you wanted to recrystallize the benzoic acid, *m*-nitroaniline, or napthalene, how would you decide on a solvent?
3. Organic layers are frequently washed with saturated aqueous NaCl and then placed over Na_2SO_4. What are the purposes of these two procedures?

EXPERIMENT 3.2 EXTRACTION OF CAFFEINE FROM TEA LEAVES

**Estimated time:
2.5 hours**

Prelab

1. If you have an aqueous phase and a dichloromethane phase in a separatory funnel, what are two ways to determine which is which?

Special Hazards

Methylene chloride is an anesthetic and narcotic if inhaled in large quantities. It is a carcinogen under conditions of daily exposure for several years. Use adequate ventilation and evaporate methylene chloride only in the fume hood.

Procedure

In a 400–mL beaker, place 100 mL of water, 15 g of sodium carbonate, and approximately 25 g of tea bags. Bring the mixture to a boil on a hot plate and boil for 15 minutes. Cool the beaker in an ice bath, then decant the liquid into a separatory funnel. Add 20 mL of dichloromethane.

Shake gently to get good extraction but avoid emulsions (an emulsion consists of droplets of one phase suspended in another phase). If an emulsion does form, it will separate in approximately 15 minutes. Decide which layer is which, separate the layers, then extract the aqueous layer with another portion of CH_2Cl_2 (20 mL). Combine the dichloromethane extracts, and to remove any water, add anhydrous sodium sulfate until it no longer all clumps together.

Save the aqueous layer in a beaker marked "waste." After allowing 5 min for the CH_2Cl_2 extracts to dry, decant or filter the solution (to remove the drying agent) into a tared beaker. Add a couple of boiling chips and evaporate the dichloromethane to dryness on a steam bath in the hood. Remove the boiling stones and record the weight and melting range of the crude caffeine.

Place the crude caffeine in the bottom of a small filter flask, stopper it, and pull a vacuum on it with the aspirator. Be sure that the aspirator has a trap on it to prevent water from backing up into the flask. Once a vacuum is established, lower the flask onto the hot plate, clamp it securely, and turn on the heat to a fairly high setting. Observe the caffeine as it starts to collect on the sides. The sublimation is done when there is only a small amount of residue left on the bottom. Turn off the heat, break the vacuum carefully, and let the flask cool. Reach in with a bent spatula and scrape the product off the sides, being careful that it does not fall back to the bottom. Weigh the purified product and record its melting point.

Questions

1. Suggest two other ways to isolate the caffeine from tea leaves, one by a different extraction method and one by another method entirely.
2. What property of caffeine makes it possible to sublime it? (*Hint:* Draw a phase diagram, putting temperature on the *x*-axis and pressure on the *y*-axis. Show regions of stability for solid, liquid, and gas phases and a line showing the path followed by the caffeine during sublimation.)
3. When brewing tea, if lemon juice is added before the tea leaves are removed, a bitter taste results. Explain.

REFERENCE

MITCHELL, R. H., W. A. SCOTT, and P. R. WEST, "The Extraction of Caffeine from Tea," *Journal of Chemical Education* (1974) *51*, 69.

Thin-Layer Chromatography

Overview

Chromatography is a separation method in which compounds are distributed between a stationary phase and a mobile phase. Separations occur because compounds have differing affinities for the stationary and mobile phases and therefore move at different speeds. For example, in thin-layer chromatography (TLC) molecules have a certain attraction to the silica on the slide surface (the stationary phase) and a different attraction to the solvent moving up the slide by capillary action (the mobile phase). Very polar compounds, such as alcohols, amines, or carboxylic acids, will stick tightly to the surface of the silica, forming hydrogen bonds with the silicic acid residues, one of which is

$$-\underset{|}{\overset{|}{Si}}-O{\diagdown}_{H}$$

Therefore, it requires a very polar solvent to pull these compounds off the surface and make them move. Nonpolar compounds will not be very attracted to the silica and will move up the slide quickly even with a nonpolar solvent.

Thin-layer chromatography is very similar to paper chromatography, which you may have done before. The main difference is that paper chromatography uses cellulose as the stationary phase and generally works best with colored compounds. In TLC we usually use colorless compounds and visualize them by a variety of techniques.

The overall process of TLC consists of applying a small droplet of dilute solution near one end of the slide, developing the chromatogram with solvent, then observing the results by a visualization technique. The ratio of the distance that a substance has moved to the distance traversed by the solvent is called the R_f value for the compound. R_f values can aid in the identification of substances when measurements are made under the same conditions; this usually means carrying out comparative chromatographs at the same time. The procedure of spotting a small droplet of dilute solution on the TLC slide, developing with solvent, then visualizing, is shown in Figure 4–1.

For good separation of components in the spot, the solvent should be polar enough so that the components move from the origin, but not so polar that they move with the

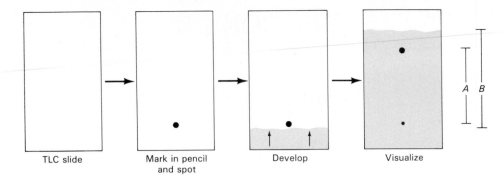

| TLC slide | Mark in pencil and spot | Develop | Visualize |

Figure 4–1 Developing and Visualizing a TLC Slide

solvent front. A list of solvents in approximate order of increasing polarity is shown in Table 4–1.

If a single solvent does not give good separation of components, a solvent mixture may work better. Solvent mixtures also give you a wider range of polarity. Common solvent mixtures may have two or three components, such as 6:1 chloroform–methanol or 90:5:5 benzene–ethyl acetate–acetic acid. Tables of the best solvents for separating certain types of compounds and of R_f values for particular compounds in various solvents can be found in the *CRC Handbook of Chromatography*.

TABLE 4–1 SOLVENTS LISTED IN APPROXIMATE ORDER OF POLARITY

Least polar	Cyclohexane
	Petroleum ether
	Pentane
	Carbon tetrachloride
	Benzene
	Toluene
	Choroform
	Ethyl ether
	Ethyl acetate
	Ethanol
	Acetone
	Acetic acid
	Methanol
Most polar	Water

To visualize the spots at the end we will use three techniques; the first two are nondestructive. First, looking at the slide under an ultraviolet light will show any compounds that absorb ultraviolet light, for example, compounds containing benzene rings or conjugated systems. Next, placing the slide in an iodine chamber allows iodine to collect on the spots by a weak electronic attraction. This will cause the spots to appear dark brown. If the slide is removed from the chamber, the iodine will sublime off over the course of a few minutes. The final visualization technique is to dip the slides in a dilute solution of phosphomolybdic acid and warm it on a hot plate. The formula for phosphomolybdic acid is (brace yourself) $20MoO_3 \cdot 2H_3PO_4 \cdot 48H_2O$.* The molybdenum in a $+6$ oxidation state oxidizes most organic compounds, leaving permanent colored spots that range in color from dark green to orange to brown.

*Now aren't you glad you're dealing mostly with organic chemicals?

APPLICATION: FORENSIC CHEMISTRY

Bomb in FBI Lab (Courtesy FBI)

Forensic chemists apply their analytical skills to physical evidence relating to crimes. They analyze chips of paint, fibers of cloth, strands of hair, drug samples, blood and urine samples or stains, and residues of bombs and burned substances. They attempt to answer such questions as: Was a fire started accidentally or by arson (and if arson, what was the accelerant)? Does any physical evidence link a particular suspect to the scene of a crime? Is an unknown white powder an illegal drug?

Forensic chemists use many analytical techniques, including TLC, gas chromatography, atomic absorption spectroscopy, and neutron activation analysis to identify traces of materials. TLC is useful for initial screening of urine samples for the presence of many drugs. If the results appear to be positive, further testing by gas chromatography or other means may be required to confirm the results.

Questions

1. What is likely to happen if you spot too little material on a TLC slide? Too much?
2. If you developed a slide with chloroform and the material did not move from the origin, what solvent(s) might you try next?
3. How could TLC be used to monitor the progress of a reaction that is being run?
4. Preparative TLC is a method using large, thick plates to separate up to 100 mg of compounds. Describe how you might run one of these, including spotting, developing, visualizing, and extracting the desired compound(s) off the plate. [*Hint:* In preparative TLC the material is spotted in a *line* near the bottom of the plate rather than at one point. It is developed in a large glass tank containing a thin layer of solvent on the bottom (Figure 4–2).]

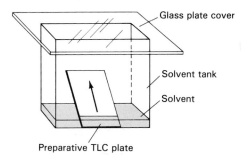

Glass plate cover

Solvent tank

Solvent

Preparative TLC plate

Figure 4–2 Developing a Preparative TLC Plate

5. Column chromatography is a method for separating anywhere from a few milligrams up to several grams of material using a column packed with silica gel (Figure 4–3). The impure material is placed at the top and solvent is run through the column to carry the components down the column at different speeds. Fractions are then collected at the bottom and tested by TLC to see if any of the components are present in that fraction. If you got the TLC results shown in Figure 4–4 from your fractions off a column, which fractions should be combined and evaporated to give the pure components?

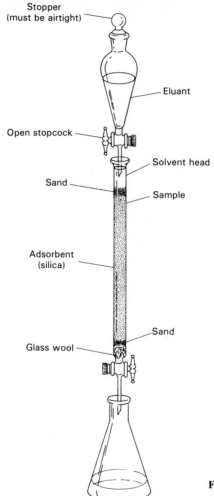

Stopper
(must be airtight)

Eluant

Open stopcock

Solvent head

Sand

Sample

Adsorbent
(silica)

Sand

Glass wool

Figure 4–3 Packed Column with Continuous-Feed Reservoir

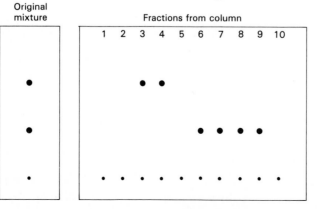

Original
mixture

Fractions from column

1 2 3 4 5 6 7 8 9 10

Figure 4–4 TLC of Original Mixture and of Fractions from Column

Concept Map

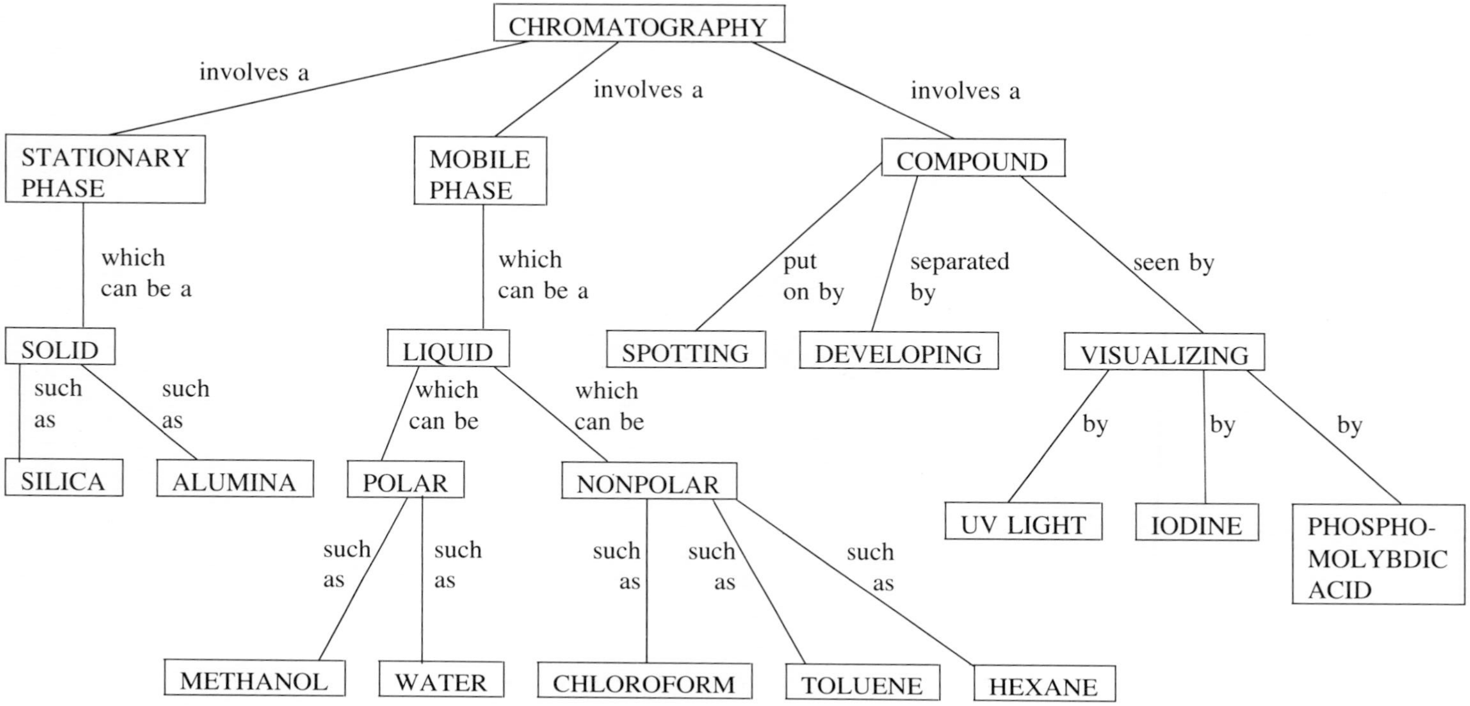
CHROMATOGRAPHY
involves a
involves a
involves a
STATIONARY PHASE
MOBILE PHASE
COMPOUND
which can be a
which can be a
put on by
separated by
seen by
SOLID
LIQUID
SPOTTING
DEVELOPING
VISUALIZING
such as
such as
which can be
which can be
by
by
by
SILICA
ALUMINA
POLAR
NONPOLAR
UV LIGHT
IODINE
PHOSPHO-MOLYBDIC ACID
such as
such as
such as
such as
such as
METHANOL
WATER
CHLOROFORM
TOLUENE
HEXANE

EXPERIMENT 4.1 DETERMINING R_f VALUES AND UNKNOWNS BY TLC

Estimated time:
2.5 hours

Prelab

1. If your solvent became contaminated with water or acetone, would your measured R_f values be too high or too low, and why?
2. Arrange the following solvent systems in order of increasing polarity: chloroform, 9:1 $CHCl_3$–MeOH, and 1:1 cyclohexane–toluene. In which would you expect the highest R_f value for a given compound?
3. Arrange the following compounds in order of increasing polarity: *m*-nitroaniline, *N,N*-dimethylaniline, and anthracene. Which would probably have the highest R_f value in a given solvent? The lowest?

Special Hazards

Looking directly at an ultraviolet (UV) light tends to cause cataracts in the eyes, and shining it on the skin promotes skin cancer. For these reasons the UV lamp should be kept in the hood and only held over the TLC slides for a few seconds to visualize them. The lamp should be pointed downward at all times and turned off immediately when done with.

Cyclohexane and toluene are flammable and are narcotic in high concentrations. Chloroform is an anesthetic and a possible carcinogen under conditions of daily exposure for several years. All organic solvents and solvent developing chambers should be kept in the fume hood.

Procedure

Choose one of the solvents in Table 4–2. Determine R_f values for the six compounds listed, using your solvent. In this manner the class will fill in the R_f values in a copy of this chart on the chalkboard. Be sure to measure your R_f values very carefully since your results will be used by the whole class in determining the unknowns. If your values are correct, you will be praised and esteemed; if they are in error, the wrath of your classmates may descend upon you.

Bear in mind also the importance of using clean, pure solvents to get accurate, reproducible R_f values. For this reason any glassware or syringe needles that need to be cleaned should be rinsed in acetone (not water) and allowed to dry thoroughly by standing in air for a few minutes before use.

The R_f values are determined in the following way: On a TLC slide mark a spot in *pencil* approximately 1 cm from the end of the slide, far enough so that it will stay above the solvent in the developing jar. You can make up to three of these spots per slide. Pens should never be used on TLC slides because the ink will develop. You will also want to be careful not to disturb the surface of the silica while marking. On top of the slide label the spots in pencil according to what you will put there (i.e., A = anthracene, C = cholesterol, etc.). Remember never to touch a TLC slide on the face—

TABLE 4–2 EXPERIMENTAL R_f VALUES

Compound	Solvent		
	Cyclohexane–toluene 1:1	Chloroform	Chloroform–methanol 9:1
Anthracene			
Cholesterol			
N,N-Dimethylaniline			
m-Nitroaniline			
1,4-Naphthoquinone			
Oleic acid			

only on the sides and back. Otherwise, you will get extra spots. The slide now looks like this:

Bottom of slide	·
	 ·
	 · A
	                 C
	                 T
		Top of slide

(sideways view of slide)

Now using a clean cutoff-flat syringe needle, dip it in the solution to be spotted (one of the 1% solutions of the known compounds in toluene). The liquid rises in the needle by capillary action.

Very briefly touch the needle to the TLC slide on the pencil mark to spot the material. The spot should be as small in diameter as possible to keep the spots sharp and must not run into another spot. Let the first spot dry, then touch the needle down on top of it two or three more times to ensure adequate sample. Between spotting new samples, the needle should be cleaned with acetone and shaken dry.

Now let the spot dry, then carefully place the TLC slide in a developing jar containing about 1/2 cm (1/4 inch) of the appropriate solvent on the bottom. Be sure to *label the jar* so that it can be reused by other students. If the solvent covers the spot, of course, it will wash it away and ruin the slide. Cover the jar to ensure saturation of the air in the chamber with solvent. Two or three slides may be developed at the same time, if desired.

When the solvent has reached about three-fourths of the way up the slide, take it out of the chamber and quickly mark the solvent front in pencil, before it evaporates. Let the slide dry, then visualize the spots in the following way. First, hold the slide under an ultraviolet light and mark the spots that fluoresce. (*REMEMBER: Never look directly at a UV light or shine it at anyone else.*) Next, place the slide in an iodine chamber and mark any new spots that become visible. Third, after removing the slide from the iodine chamber, dip it quickly into a 2% solution of phosphomolybdic acid in 95% ethanol, wipe off the back, and set it on a warm hot plate.

The hot plate should be on a setting low enough not to melt the plastic slide but high enough to cause the spots to appear. If the slide curls, it may be necessary to hold it down with tongs or a wire gauze. When the dark green, orange, or brown spots have appeared, remove the slide from the hot plate.

Decide which spot is the compound (this should be the major spot) and which, if any, are impurities. Measure the distance from the original position of spotting to the spot, and to the solvent front. Calculate the R_f value and record it in your notebook and on the board. When the table compiled by you and your classmates is complete, examine it to see which solvent looks as if it gives the best separation.

Now choose an unknown, which is a mixture of any two of the previous six compounds, also as a 1% solution in toluene. Prepare a TLC slide of just the unknown, develop it *in the solvent from which you expect the best separation*, and visualize it. If the two components are not well separated, you may wish to try another solvent. Measure and record the R_f values for your two unknowns in this solvent. Considering the R_f values of the two components should give you some idea of what they are.

Now prepare another slide with three spots: the unknown in the middle and the two suspected components on either side. This is shown below, using the example of oleic acid (abbreviated O) and *N,N*-dimethylaniline (abbreviated D) as the suspected components.

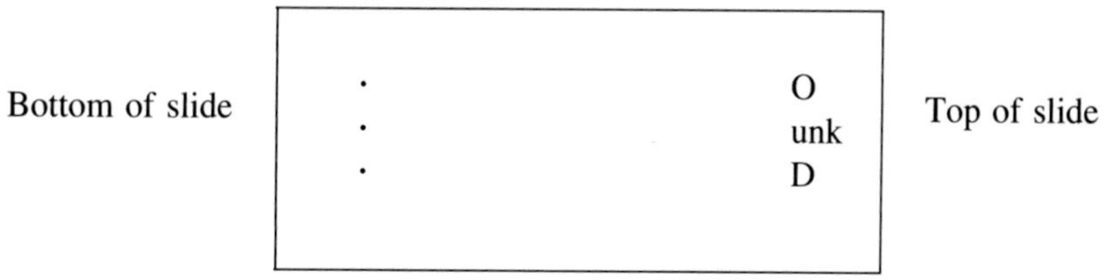

Bottom of slide

O

unk

D

Top of slide

(sideways view of slide)

Develop the slide in the solvent you expect to give the best separation and visualize it. Do the spots come at the same heights? If so, continue on to cospotting; if not, try again with other known compounds.

The final determination of the unknown is done with cospotting. Spot *directly on top of one another* the unknown and the two suspected components. If, after developing and visualizing, there are only two spots, this means that you have identified your unknown correctly. Tape the TLC slides into your notebook or make accurate drawings of them for future reference.

Column Chromatography

Overview

Frequently when isolating a natural product or after carrying out a reaction the investigator wishes to isolate one or more components of a mixture in pure form. If the desired product is easily purified by distillation or recrystallization, these are usually the methods of choice. However, many compounds decompose on heating or are present in such small quantities that distillation or recrystallization are impractical. In these cases column chromatography can often be used to carry out the desired separation.

Column chromatography operates on the same principles as gas or thin-layer chromatography: There is a stationary phase (the packing) and a mobile phase (the solvent or eluant), and a compound distributes itself between the two phases according to its affinity for each. The higher the affinity for the mobile phase, the larger the fraction of time or compound spends there and the faster it moves down the column.

A chromatography column consists of a tube packed with a solid adsorbent such as alumina (Al_2O_3), silica (SiO_2), or Florisil (magnesium silicate, Mg_2SiO_4). Usually, an adsorbent–compound weight ratio of 25:1 is used for easy separations and 100:1 or so for more difficult ones. A small plug of cotton or glass wool is placed at the bottom of the column to prevent the solids from trickling out, then a layer of sand to provide a flat base for the packing material. If the sand were not present, the adsorbent would be uneven at the bottom and would not provide as sharp a separation. Another layer of sand at the top of the column keeps the top of the adsorbent flat and protects the column from disruption as new solvent is poured in at the top. A column may be packed wet, with the adsorbent poured in as a slurry in the desired solvent. A column can also be packed by filling it with solvent, then slowly pouring a stream of solid adsorbent at the top. The latter method (adding dry powder to a solvent-filled column) is generally less messy for an inexperienced chromatographer. An illustration of a properly packed column appears in Figure 5–1, accompanied by drawings of some types of solvent reservoirs in Figure 5–2. It is important to keep the column wet at all times and not to let the solvent level drop below the top of the adsorbent. If it does, channels and bubbles will form in the packing and the efficiency of separation will be decreased.

After the column is packed the mixture to be separated is placed dropwise with a Pasteur pipette in a thin, even layer on the top. This layer is run down through the sand onto the column by letting a few drops out of the stopcock at the bottom. Solvent is

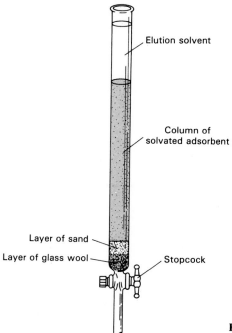

Elution solvent

Column of
solvated adsorbent

Layer of sand

Layer of glass wool

Stopcock

Figure 5–1 A Packed Chromatography Column

added at the top and the stopcock is opened to allow the solvent to move down the
column. The components of the mixture at the top of the columnm separate into bands
and move down the column at different speeds. Sometimes (as in this experiment) the
compounds are visible and it is easy to tell where they are on the column. In other cases
it may be necessary to use an ultraviolet absorption detector (which follows absorbance
of the eluant and indicates when organic compounds are coming off the column) or to
check the eluant by TLC.

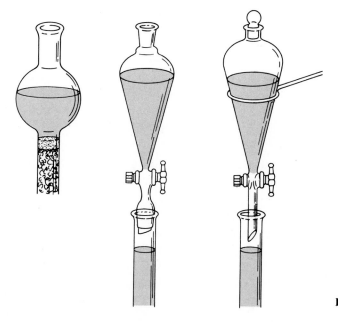

Figure 5–2 Types of Solvent Reservoirs

If the desired components are not visible on the column, the standard procedure is to collect numbered fractions of solvent coming off the column, then to examine them by TLC or other methods to determine which fraction(s) contain the desired compound. The desired fractions are then combined and the solvent is removed (often under reduced pressure) to yield the pure compound.

The correct choice of solvent is of course critical in achieving good separation in a reasonable time. The more polar the solvent, the faster it will move a compound down the column. The approximate order of solvents in terms of eluting power from least polar to most polar is listed in Table 5–1; this is the same list that was discussed with respect to TLC in Section 4.

TABLE 5–1   SOLVENTS LISTED IN APPROXIMATE ORDER OF ELUTING POWER

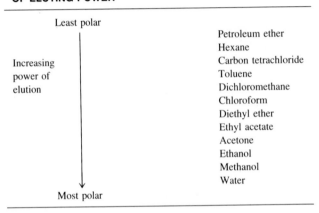

Least polar	Petroleum ether
	Hexane
Increasing power of elution	Carbon tetrachloride
	Toluene
	Dichloromethane
	Chloroform
	Diethyl ether
	Ethyl acetate
	Acetone
	Ethanol
	Methanol
Most polar	Water

Often a mixture of solvents is used, such as 5% chloroform in petroleum ether, to obtain the desired polarity. If the solvent is too polar, the compounds move down the column too quickly (at the solvent front) and are not separated. They come off the column still mixed together. If the solvent chosen is too nonpolar, the compounds will move down the column very slowly if at all. The separation may be very tedious and require enormous volumes of solvents.

If the separation has not been carried out before an educated guess must be made as to the choice of an appropriate solvent. One solution to this problem is to use gradient elution; starting with a nonpolar solvent and gradually changing the solvent to increasing

TABLE 5–2   STRENGTH OF ADSORPTION OF VARIOUS FUNCTIONAL GROUPS ON ALUMINA

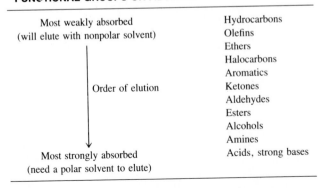

Most weakly absorbed (will elute with nonpolar solvent)	Hydrocarbons
	Olefins
	Ethers
	Halocarbons
	Aromatics
Order of elution	Ketones
	Aldehydes
	Esters
	Alcohols
	Amines
Most strongly absorbed (need a polar solvent to elute)	Acids, strong bases

polarity. This procedure allows nonpolar compounds to be eluted first, then gradually strips off the more polar ones from the column.

How strongly a compound adsorbs onto a column depends on its polarity, and therefore on its functional groups. Adsorption also depends on the nature of the packing material (silica, alumina, etc.). The approximate order of strength of adsorption for various functional groups on alumina is shown in Table 5–2. This order means that, for example, in a mixture containing a carboxylic acid, an ether, and a halide, the halide would elute first on an alumina column, followed by the ether, followed by the carboxylic acid. Very polar compounds such as carboxylic acids adsorb strongly to columns and may be difficult to elute.

Experiment 5.1 is the separation of a mixture of syn and anti isomers of azobenzene.

syn -Azobenzene   anti-Azobenzene

Both of these compounds are orange-colored and easy to see on a column. The *syn* isomer has the polar nitrogen-containing portion of the molecule more exposed than the *anti isomer,* so the *syn*-azobenzene will be more strongly adsorbed on the alumina and will move down the column much more slowly than the *anti* isomer.

Experiment 5.2 is the isolation of chlorophyll and β-carotene from spinach. Chlorophyll is a green plant pigment that contains a porphyrin ring system, a magnesium ion, and a long hydrophobic side chain, and catalyzes the photosynthesis of glucose from carbon dioxide. Without chlorophyll, probably none of us would be here!

Chlorophyll *a*

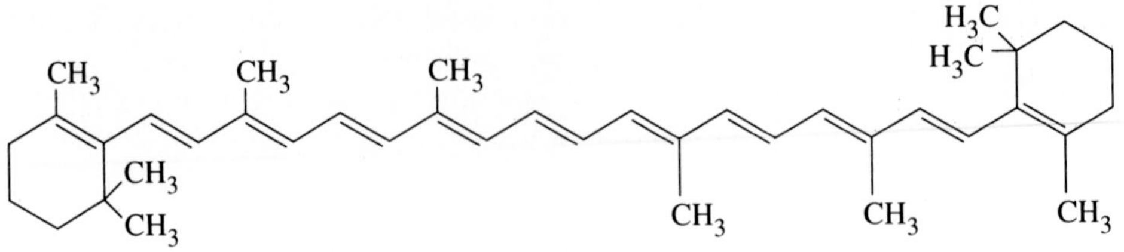

β-Carotene

β-Carotene is the orange pigment found in carrots that functions as a precursor to vitamin A. In the liver β-carotene is cleaved to form vitamin A, which is further converted to 11–*cis*-retinal within the eye.

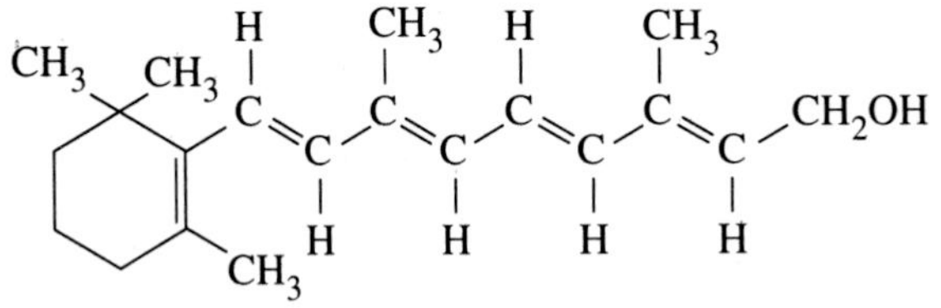

Vitamin A

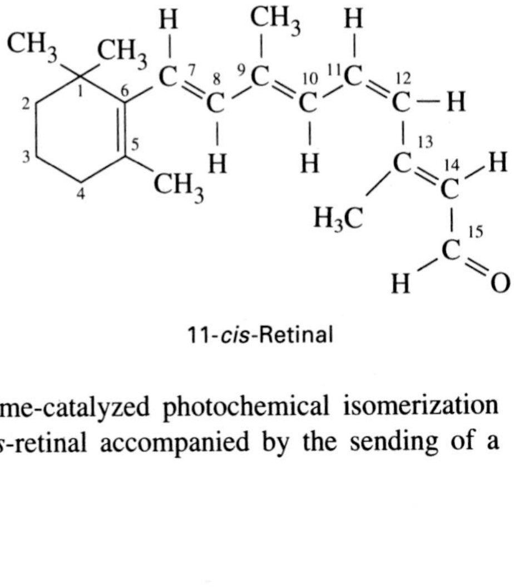

11-*cis*-Retinal

The mechanism of vision involves the enzyme-catalyzed photochemical isomerization of 11–*cis*-retinal to the more stable all-*trans*-retinal accompanied by the sending of a visual signal as a nerve impulse.

APPLICATION: HIGH-PERFORMANCE LIQUID CHROMATOGRAPHY (HPLC)

Column chromatography often proves time consuming, since low flow rates of solvents are required for equilibration and efficient separation. The need for faster and higher-resolution column chromatography has been addressed by a method called high-pressure (or high-performance) liquid chromatography (HPLC). In HPLC, the liquid phase enters a column packed with very fine particles (10 to 50 μm in diameter) of a solid adsorbent. Several materials, including silica, alumina, or small glass spheres coated with a thin layer of porous material (called pellicular beads) with or without a very thin layer of a liquid, may serve as the stationary phase. These small par-

ticles provide a larger surface area and the particles pack together more tightly, allowing shorter column lengths than in standard column chromatography. This tight packing restricts the flow of the solvent, though, and pressure must be applied to push the mobile phase through the column. A pump supplies a constant pressure of up to 10,000 psi (for a small, 2– to 3–mm diameter column) to force the solvent through. For chemical inertness and strength, the tubing and column used are constructed of thick polypropylene, Teflon, or stainless steel. A schematic diagram and photograph of an HPLC system are shown here.

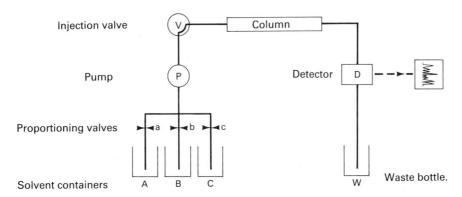

Schematic of a Liquid Chromatograph

Waters Delta-Prep 3000 HPLC System (Courtesy of Waters Chromatography Division of Millipore Corporation)

The liquid emerges from the column at atmospheric pressure. In analytical HPLC this eluant is then passed through a detector to measure compounds coming off the column. Common detectors measure ultraviolet light absorption, fluorescence, or refractive index. The signal from the detector is passed to a chart recorder, which provides a chromatogram similar to those obtained in gas chromatography (see figure).

In preparative HPLC, separate fractions are collected and evaporated to yield the purified components of the original mixture. By using large columns with diameters of 3–4 cm, it is sometimes possible to separate up to 15–20 g of material in one run.

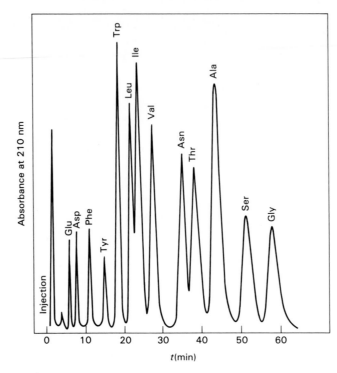

Separation of Free Amino Acids by HPLC

HPLC has the advantage over normal column chromatography of much faster separations and greater resolving power. In addition, it can be used on high-molecular-weight (nonvolatile) compounds not amenable to gas chromatography. Since the compounds do not undergo heating in HPLC and are on the column for only a short time, danger of decomposition is diminished. HPLC has been praised for greatly simplifying the work involved in analyzing and synthesizing complex natural products such as vitamin B_{12} and periplanone B, the sex attractant of the American cockroach.

Structure of coenzyme B_{12}
(vitamin B_{12})

Periplanone B (sex attractant
of the American cockroach)

Questions

1. How would you suggest eluting the *syn*-azobenzene?
2. If grease were present initially on the stopcock, where would you find it after performing the chromatography?
3. What is the purpose of column chromatography?
4. What are the advantages of this method over other purification methods? The disadvantages?
5. Arrange the following solvents in order of increasing polarity: dichloromethane, methanol, hexane.
6. Arrange the following compounds in expected order of elution: benzoic acid, 1–heptanol, 1–bromobutane.
7. What will happen if the sand layers are uneven or there are bubbles in the packing?

EXPERIMENT 5.1 MICROSCALE SEPARATION OF AZOBENZENES

Estimated Time:
2.5 hours

Special Hazards

Petroleum ether is highly flammable; keep it away from all flames or heat sources. Azobenzene is a cancer suspect agent; avoid skin contact or ingestion. Avoid breathing finely powdered inert materials such as alumina or silica; they can cause lung damage.

Procedure:

Packing the column.   Securely clamp a 50–mL buret vertically and close the stopcock. Push a small plug of glass wool to the bottom using a long glass rod. Obtain about 100 mL of petroleum ether (bp 60–80). Now place a funnel on top of the column and add about 30 mL of the petroleum ether. Pour a little sand through the top to settle in a layer about 1 cm thick on top of the glass wool. The exact thickness is not critical as long as the top is level. Tap the burette lightly with a pencil or plastic pen to ensure even packing. Weigh 30 g of alumina and gradually pour it into the top through the funnel while continuing intermittent tapping. If necessary, excess solvent may be run out into a beaker at the bottom to avoid spilling solvent out the top of the column. If wet alumina accumulates at the top and blocks the flow, push it down with a stirring rod. When finished packing the alumina, pour another 1–cm layer of sand on top. This packing procedure leaves about 15 cm of the column empty at the top for the solvent head. Place a beaker or flask underneath the burette and open the stopcock to drain the solvent to a point just below the top of the sand. The column is now ready for the sample.

Separation of *syn* and *anti* isomers of azobenzene.   Record the melting range of a mixture of practical grade *syn*- and *anti*-azobenzene. Weigh 100 mg of this mixture. Dissolve this sample in 1 mL of petroleum ether with slight warming. Carefully and evenly add this solution to the top of the column using a dropper. Open the stopcock to run the sample barely below the sand surface. Add about 1 mL of fresh petroleum ether to the top and similarly run this down to the sand level. Now carefully add petroleum ether almost to the top of the burette.

Place a beaker labeled *forerun* below the burette tip and open the stopcock to begin solvent flow down the column. Monitor the solvent level on the top carefully and keep adding more petroleum ether before the level drops to the sand. If by accident the solvent level falls below the level of the sand, simply refill the solvent on top and proceed. There may be bubbles in the packing, resulting in slower solvent flow and poorer separation, but the *anti*-azobenzene can still be collected.

Observe the orange band of *anti*-azobenzene moving down the column. The syn isomer remains in a band near the top of the column. When the *anti*-azobenzene begins to reach the bottom of the burette change to another collection beaker (labeled anti) and collect the band. When it has been collected shut off the column, add a boiling chip to the beaker, and evaporate the solvent on a steam bath in the hood. Record the mass and melting point of the material obtained (lit. mp for *anti*-azobenzene, 68°). Dispose of the forerun in a properly marked waste bottle in the fume hood.

EXPERIMENT 5.2 ISOLATION OF CHLOROPHYLL AND β-CAROTENE FROM SPINACH BY GRADIENT ELUTION

Estimated Time:
2.5 hours

Special Hazards

See Experiment 5.1. Ethyl acetate is also flammable.

Procedure

Read the procedure for Experiment 5.1 and prepare a column as described. Before the lab, the instructor will take a 10–oz (280–g) package of frozen spinach (which has been thawed), place it in a blender with 400 mL of absolute ethanol, and blend it thoroughly. This process extracts most of the water into the ethanol, while leaving the chlorophyll and β-carotene in the spinach. As an alternative, strained spinach baby food (8 g per person) may be used and stirred thoroughly with 10 mL of absolute ethanol.

Take approximately 20 mL of this spinach–ethanol paste (or the baby food–ethanol mixture) and place it in a funnel containing a small plug of glass wool instead of a filter paper. Place a piece of filter paper on the top and squeeze to remove the ethanol. Drain excess ethanol from the top as well.

Place the remaining residue on a clean paper towel, remove the glass wool and filter paper, and press to dry futher. Place the resulting pellet in a 100–mL beaker and add 20 mL of CH_2Cl_2. Stir thoroughly to extract the pigments into the dichloromethane. Decant the solution away from the bulk of the spinach residue, then filter into a clean beaker and concentrate it on a steam bath in the fume hood to a volume of 1–2 mL. Be careful not to heat to dryness because the large, complex pigment molecules may become oxidized and discolored if overheated.

Place the sample on the column as described in Experiment 5.1. Elute with petroleum ether and observe that the yellow band of β-carotene moves down the column slightly faster than the green chlorophyll. The quantity of β-carotene is not great so the band may be faint. Observe and record the shapes of the bands. Collect the yellow band of β-carotene as it emerges from the bottom of the column, then begin eluting with a 1:1 (by volume) mixture of petroleum ether and ethyl acetate. Collect the green band of chlorophyll in a tared beaker. On a steam bath in the fume hood carefully evaporate the fractions containing the two products and record their masses on an analytical balance. Scrape up and examine the green film of chlorophyll formed. At the instructor's option you may take ultraviolet–visible spectra of the two compounds. Compare the spectra obtained with the literature spectra shown in Figure 5–3.

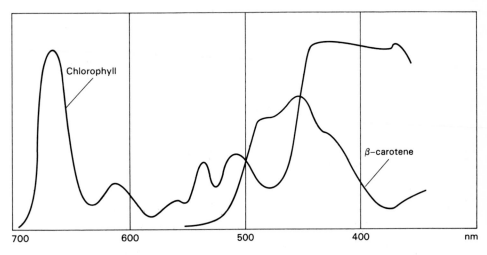

Figure 5–3   Visible absorption spectra of chlorophyll and β-carotene in petroleum ether.

Reference:

McKone, A. T., "The Rapid Isolation of Carotenoids from Foods," *Journal of Chemical Education* (1979) *56*, 676.

SECTION 6

Acid-Catalyzed Dehydration of Alcohols

Overview

One of the earliest organic reactions discovered was the elimination of water from an alcohol to form an alkene, and it is still a widely used reaction. Polyethylene bags for your supermarket, for example, are made by polymerizing ethene (ethylene), which can be made either by dehydration of ethanol (as shown below) or by thermal cracking of petroleum distillate fractions.

$$
\underset{\text{Ethanol}}{\overset{\displaystyle \begin{array}{cc} H & OH \\ | & | \\ H-C-C-H \\ | & | \\ H & H \end{array}}{}} \quad \xrightarrow{H^{+}} \quad \underset{\text{Ethene}}{\overset{\displaystyle \begin{array}{cc} H & H \\ \diagdown & \diagup \\ C=C \\ \diagup & \diagdown \\ H & H \end{array}}{}} \quad + \quad H_2O
$$

In recent years it has been less expensive to make ethylene from petroleum, a nonrenewable resource. However, when oil supplies run low someday and we are looking for more renewable resources, we will be able to make large quantities of ethene from ethanol. Ethanol, of course, is available cheaply from fermentation of sugarcane or cornstalks.

The reactions you will carry out in Experiments 6.1 and 6.2 are very similar to the dehydration of ethanol; they are the acid-catalyzed dehydration of cyclohexanol to cyclohexene and of 2-pentanol to a mixture of pentenes.

Overall reaction for Experiment 6.1:

Cyclohexanol $\xrightarrow{H_3PO_4}$ Cyclohexene

Cyclohexanol
MW 100, bp 161°
dens. 0.963 g/mL

Cyclohexene
MW 82, bp 83°
dens. 0.811 g/mL

The dehydration of cyclohexanol is a typical elimination reaction with the following mechanism:

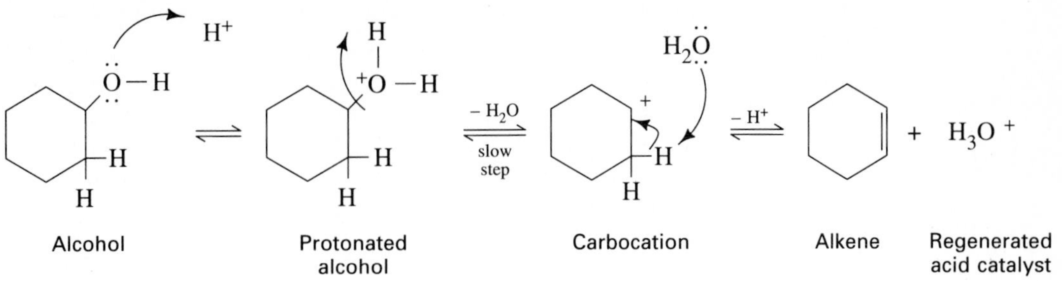

| Alcohol | Protonated alcohol | Carbocation | Alkene | Regenerated acid catalyst |

In Experiment 6.2, 2-pentanol is dehydrated to form a mixture of three products:

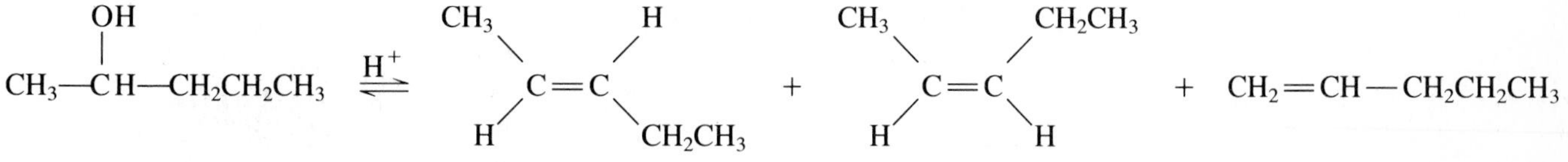

2-Pentanol	*trans*-2-Pentene	*cis*-2-Pentene	1-Pentene
MW 88.2	MW 70.1	MW 70.1	MW 70.1
bp 120°	bp 36°	bp 37–38°	bp 30°
dens. 0.812 g/mL			

This mixture forms because the intermediate carbocation can lose hydrogen atoms from two different types of adjacent carbon atoms to form alkenes. Furthermore, if the alkene is internal to the chain, it can be formed in either the *cis* or *trans* configuration.

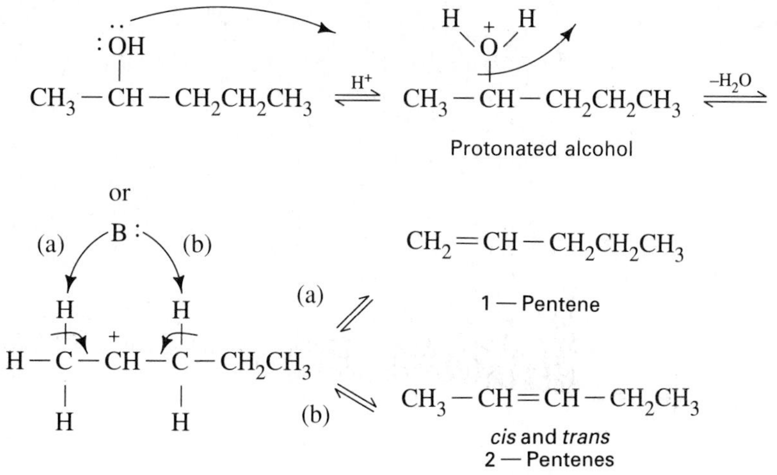

Protonated alcohol

or

(a)   B: (b)

$CH_2=CH-CH_2CH_2CH_3$

(a)   1 — Pentene

$H-C-CH-C-CH_2CH_3$

(b) $CH_3-CH=CH-CH_2CH_3$

cis and *trans*
2 — Pentenes

In general, the more highly substituted an alkene is (tetrasubstituted > trisubstituted > disubstituted > monosubstituted), the more thermodynamically stable and the lower the activation energy of formation. Because the activation energy is lower and the rate of reaction is higher, formation of 2-pentene is favored over 1-pentene. For steric reasons the *trans* isomer of 2-pentene is favored over the *cis*.

Dehydration of cyclohexanol and of 2-pentanol are both examples of what is called an E1 reaction, short for elimination, unimolecular. It is called an elimination because the elements of water are eliminated (H and OH from adjacent carbons) from the alcohol to make the alkene and it is unimolecular because the rate-determining step of the reaction involves only *one* species (the protonated alcohol). All the reactions are reversible, but in accordance with Le Châtelier's principle, the equilibrium can be driven to the right by removing the final product (the alkene) by distillation.

You will know that you have produced the right product(s) in each case because of the boiling ranges observed during the distillations. For cyclohexene, the literature boiling point is 83° at 1 atm, which is significantly different from the boiling point of cyclohexanol (161° at 1 atm). Similarly, the boiling points of *cis*- and *trans*-2-pentenes (37–38°) differ significantly from that of 2-pentanol (118–119°). Of course, these syntheses are not "normal" distillations in which you are simply purifying a liquid. These procedures will take longer because the alkenes are being formed in the flask as you heat the alcohols with acid, and the products can only distill as quickly as they form.

You will also know you have produced an alkene because you will do the bromine in carbon tetrachloride (Br_2/CCl_4) test. This test for alkenes relies on the fact that bromine (which is reddish brown) adds across double bonds to give a colorless product, as shown here:

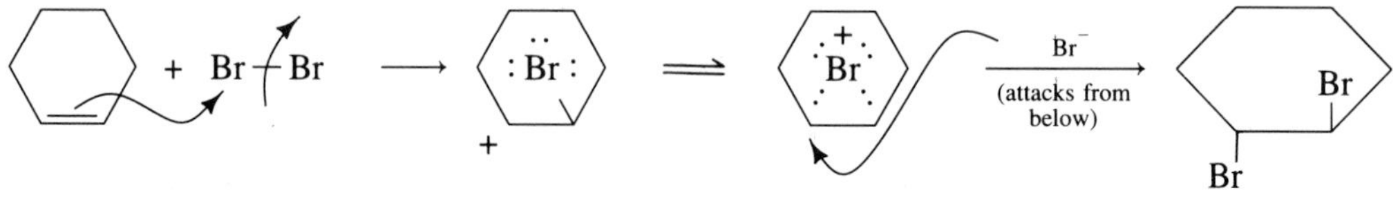

| Molecular bromine (red-brown) | Nonclassical bromonium ion | Vicinal dibromide (colorless) |

You can tell if your product contains an alkene by putting a few drops in a test tube, then adding a few drops of 0.2 M Br_2 in CCl_4. If the red-brown color disappears, it is positive for alkenes.

Another test for alkenes relies on the reaction they undergo with permanganate. An alkene is oxidized to a vicinal diol as the purple MnO_4^- ion is reduced to a precipitate of muddy-brown MnO_2.

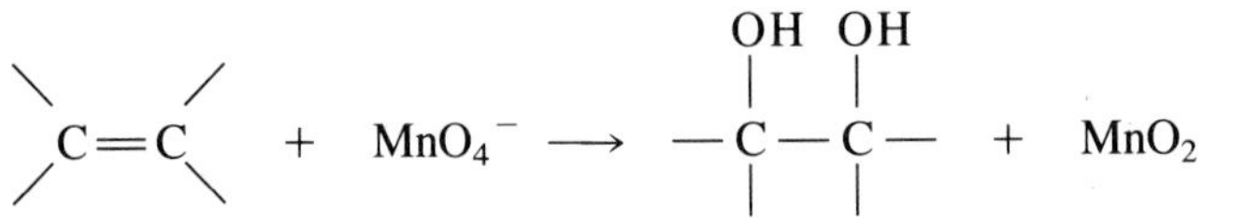

| Alkene (colorless) | Permanganate (purple) | Vicinal diol (colorless) | Manganese dioxide (brown) |

Questions

1. What would be the expected products from the following reactions?

(a)

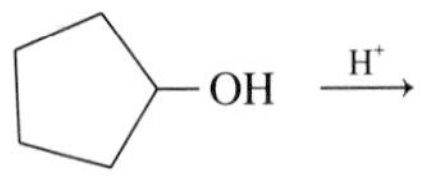

(b)

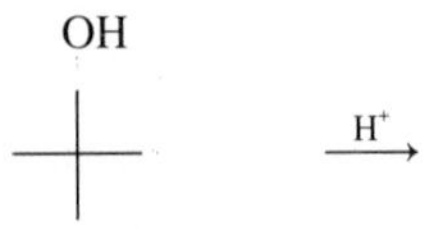

(c)

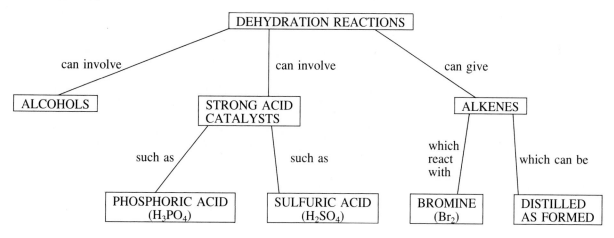

(d)   two products

2. Draw the mechanism for the following reaction.

1-Phenylethanol   Styrene

Concept Map

DEHYDRATION REACTIONS

can involve   can involve   can give

ALCOHOLS   STRONG ACID CATALYSTS   ALKENES

such as   such as   which react with   which can be

PHOSPHORIC ACID (H_3PO_4)   SULFURIC ACID (H_2SO_4)   BROMINE (Br_2)   DISTILLED AS FORMED

EXPERIMENT 6.1 CYCLOHEXENE FROM CYCLOHEXANOL

Estimated Time:
2.0 hours

Prelab

1. Calculate the volume and number of millimoles of cyclohexanol corresponding to 15 g. Fill in these figures on your table of reagents. Is it easier to measure cyclohexanol (a liquid) by weight or by volume?
2. Calculate the theoretical yield of cyclohexene.
3. During the distillation, at what thermometer reading do you expect to collect the cyclohexene?

Special Hazards

Concentrated phosphoric acid and bromine are highly corrosive; exercise caution. Carbon tetrachloride is toxic and a cancer suspect agent; use only in the fume hood and avoid exposure.

Procedure

In a 100-mL round-bottomed flask, place cyclohexanol (15.0 g, 15.6 mL, ___ mmol), 85% phosphoric acid (4 mL), and a boiling chip. To mix the layers swirl the flask and set up the flask for fractional distillation with an ice-cooled collector. The ice-cooled collector is necessary because cyclohexene has such a low boiling point that it can evaporate quickly. Check that the heating mantle fits properly and is not in contact with any flammable materials. Turn on the variable transformer to a moderately high setting and distill the cyclohexene that is formed in the flask.

Be aware that this is not a "normal" distillation in which we are simply purifying a liquid. In this case we are distilling the cyclohexene *as it is being formed*, so the process will be slower than a normal distillation. In addition, the cyclohexene may be in a superheated state when formed, so the vapor may give thermometer readings above the normal boiling point of cyclohexene.

Record the boiling range. Add anhydrous Na_2SO_4 to the product to remove any water that may have codistilled, gravity filter, and weigh your product. To verify that you have an alkene, add a few drops of your product to a test tube, then add a few drops of 0.2 M Br_2 in CCl_4 and observe whether it is decolorized. For comparison, test some of the starting cyclohexanol to see if it decolorizes bromine.

Similarly, test both the starting material and product with $KMnO_4$ by placing a few drops of each in separate test tubes, adding 2 mL of 95% ethanol to each (as a solvent to help mix organic and aqueous phases) followed by a few drops of 1% aqueous $KMnO_4$. Observe and record the results.

Reference

COLEMAN, G. H. and H. F. JOHNSTONE, *Organic Syntheses Collective Volume 1* (1941) 183.

EXPERIMENT 6.2: PENTENES FROM 2-PENTANOL

Estimated Time:
2.5 hours

Prelab

1. How many millimoles of 2-pentanol are contained in 10 g?
2. What concentration of H_2SO_4 (in M) is obtained by diluting 10 mL of concentrated H_2SO_4 to 20 mL with water?
3. What theoretical yield of pentenes is expected?

Special Hazards

Pentenes are highly volatile and flammable. Use care to keep the products of the distillation cold and away from heat sources. 2-Pentanol is also flammable and an irritant. Concentrated sulfuric acid is extremely corrosive; use due care. Bromine is highly corrosive and carbon tetrachloride is a suspected carcinogen; use only in the fume hood and avoid exposure. Keep a bottle of 5% aqueous sodium thiosulfate available for treating bromine burns.

Procedure

In a 50-mL round-bottomed flask cooled in an ice bath, place water (10 mL) and slowly add concentrated H_2SO_4 (10 mL; *use caution*). When the solution is cool, slowly add 2-pentanol (10 mL, __ g, __ mmol). Add a couple of boiling chips and set up for simple distillation. Because of the low boiling points of the products, the receiving flask must be well cooled in an ice-water bath. To direct the vapor into the flask, a vacuum adapter should be used, leading to a round-bottomed flask acting as the receiver. Make sure that water is flowing through the condenser, then gradually heat the flask until some material begins to distill. Continue the heating and distillation for about an hour or until no more product distills, whichever comes first. The residue in the distilling pot is strongly acidic. To dispose of it safely, cool the round-bottomed flask in an ice bath and cautiously pour the contents down the sink with a lot of water. Do not add water to the flask before emptying its contents since that would be adding water to strong acid, which may become very hot and splatter.

Weigh the collected product, then place it in a separatory funnel and wash with 10 mL of 5% aqueous NaOH to remove any SO_2 present. Watch for pressure buildup from the volatile product mixture. Dry the hydrocarbon layer over Na_2SO_4, decant and redistill it, after rinsing the distillation apparatus with acetone. Use a clean ice-cooled round-bottomed flask as a receiver. Collect the material boiling in the range 35–41°. Record the mass. The product must be stored in an airtight vial and kept cold or it will evaporate.

To verify that you have produced alkenes, place a few drops of the product in a test tube, then in the fume hood add a few drops of 0.2 *M* Br_2 in CCl_4. Observe whether the red-brown color of molecular bromine disappears. For comparison test a few drops of the starting material, 2-pentanol, with the Br_2/CCl_4 solution. Similarly test both starting material and product with permanganate by dissolving a few drops of each in separate test tubes in 2 mL of acetone, then adding a few drops of 1% aqueous $KMnO_4$. Record the results.

At the instructor's option, you may now analyze the composition of the product mixture by gas chromatography, which is described in Section 12. On a nonpolar column at room temperature these products should elute in order of their boiling points. Authentic *cis*- and *trans*-2-pentenes and 1-pentene can be injected for comparison. Dispose of unused product in the waste container provided in the fume hood.

Reference

NORRIS, F. J. and F. C. WHITMORE, *Organic Syntheses Collective Volume 1*, (1941) 430.

HYDRATION OF ALKENES AND ALKYNES

Overview

Addition of water across a multiple bond is one of the tried-and-true methods of organic synthesis. Carbon-to-carbon pi bonds are relatively weak (about 250 kJ/mol) and break easily, particularly when strong carbon-to-oxygen (340 kJ/mol) and carbon-to-hydrogen (420 kJ/mol) bonds can be formed. Since strong bonds are formed and weaker bonds are broken, this process is favored thermodynamically. The overall scheme appears here:

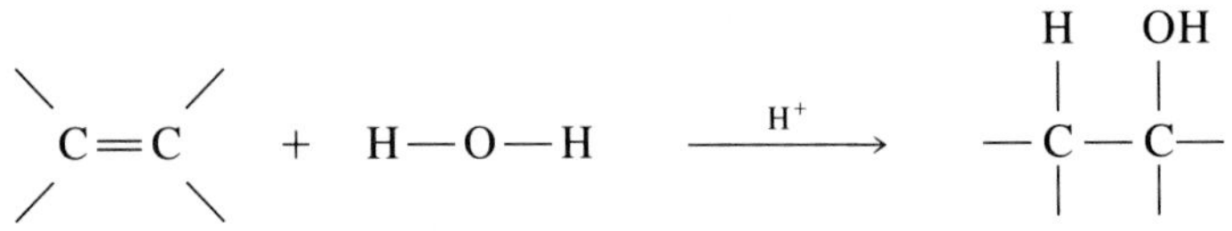

$$
\underset{\text{An Alkene}}{\diagdown C = C \diagup} \quad + \quad \underset{\text{Water}}{H - O - H} \quad \xrightarrow[\text{Acid catalyst}]{H^+} \quad \underset{\text{An alcohol}}{\overset{H \quad OH}{\underset{|}{-}C - \overset{|}{C} -}}
$$

You have doubtless already seen several ways to accomplish this addition. The oldest known method for achieving this transformation consists of simply heating an alkene in the presence of water and a strong acid catalyst. The acid protonates the alkene, leaving a carbocation that attracts and bonds to the oxygen of a water molecule. When finally the former water molecule loses a proton, the addition is complete and the acid catalyst is regenerated.

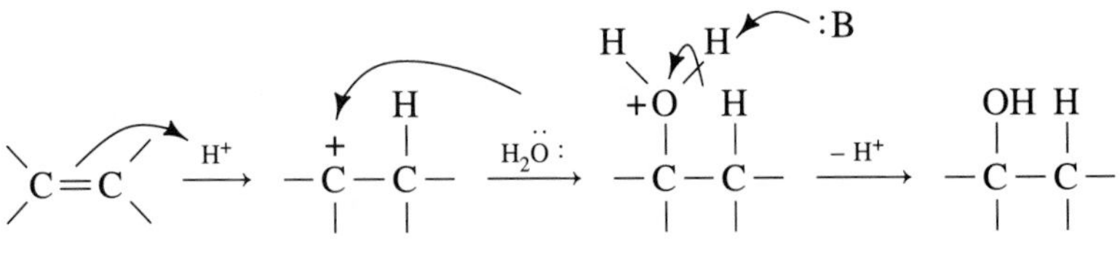

$$
\underset{\text{Alkene}}{\diagdown C = C \diagup} \xrightarrow{H^+} \underset{\text{Carbocation}}{-\overset{+}{\underset{|}{C}} - \overset{|}{\underset{|}{C}} -} \xrightarrow{H_2\ddot{O}:} \underset{\substack{\text{Protonated} \\ \text{alcohol}}}{-\overset{+O \; H}{\underset{|}{C}} - \overset{|}{\underset{|}{C}} -} \xrightarrow{-H^+} \underset{\text{Alcohol}}{-\overset{OH \; H}{\underset{|}{C}} - \overset{|}{\underset{|}{C}} -}
$$

The drawback to this method is that harsh conditions of high temperature and strong acid are required. These conditions may result in low yields due to unwanted side reactions, particularly if the molecule possesses other acid-sensitive functional groups.

Other, milder methods have been developed to avoid these problems. For example, use of mercuric acetate and water, followed by sodium borohydride, also results in an overall addition of H_2O to an alkene. The mechanism is similar to the acid-catalyzed addition of water in that the electrophilic mercuric ion attacks the electron-rich pi bond, forming a carbocation. The oxygen atom of a water molecule bonds to this carbocation, then loses H^+ as in the previous case.

$$Hg(OAc)_2 \rightleftharpoons HgOAc^+ + OAc^-$$

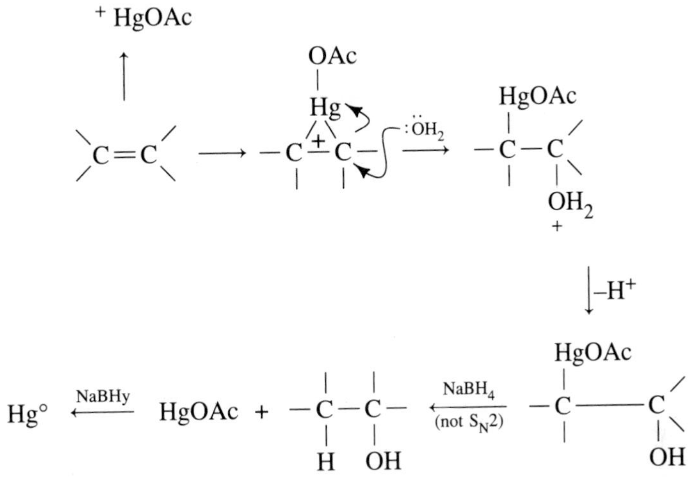

If an alcohol (ROH) is used in place of the water, an ether is formed. Once the mercury-containing intermediate has been formed, the mercury can be displaced using a hydride reagent, $NaBH_4$. Hydride (H^-) can displace the mercury, simultaneously causing the reduction of the mercury from the $+2$ to the $+1$ oxidation state. The mechanism is not completely understood; it is not an S_N2 reaction because it is not stereospecific. There is some evidence for radical intermediates.

After displacement, the mercurous ion undergoes further reduction by hydride to elemental mercury (zero oxidation state), so a small ball of mercury appears at the bottom of the reaction flask. The net result is the same addition of water across the double bond as in the acid-catalyzed addition, but the conditions are much milder and fewer by-products result.

One more important point to consider is the *direction of addition* (regiochemistry) of the addition to an unsymmetrical alkene. In the case of 1-pentene, for example, there are two conceivable products from hydration:

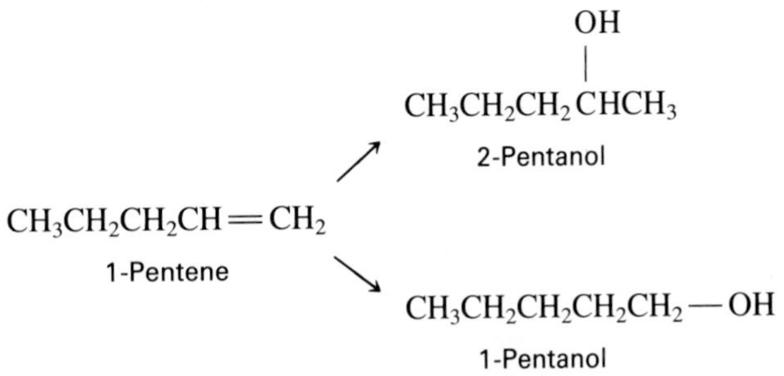

In the first case, the hydrogen added to the carbon having more hydrogens (the less substituted carbon), while the —OH bonded to the other carbon of the double bond.

This is called the Markovnikov product because it follows Markovnikov's rule. This rule can be stated in several forms:

1. In an electrophilic addition of HX across a double bond (where X represents a halogen or —OH group), the hydrogen atom bonds to the carbon with fewer alkyl substituents (the less substituted carbon, the one with more hydrogens on it). Conversely, the X group bonds to the more substituted carbon.

or, more succinctly,

2. Unto those who have, shall more be given.
   (Referring to the fact that the hydrogen bonds to the carbon having more hydrogens.)

or, more briefly still,

3. Them as has, gits.

In a "normal" electrophilic addition, Markovnikov's rule will be followed. This results from the greater stability of the more-substituted carbocation formed as hydrogen bonds to the less-substituted carbon. You recall that carbocations are more stable the more highly substituted they are (tertiary > secondary > primary). The acid-catalyzed addition of water forms the Markovnikov product, as does the mercuric acetate-promoted reaction.

For example, treatment of 1-hexene with mercuric acetate and water, followed by sodium borhydride forms 2-hexanol:

$$CH_2{=}CH{-}(CH_2)_3{-}CH_3 \xrightarrow[\text{(2) NaBH}_4]{\text{(1) Hg(OAc)}_2,\text{H}_2\text{O}} \underset{\displaystyle CH_3{-}\overset{\displaystyle OH}{\overset{|}{CH}}{-}(CH_2)_3{-}CH_3}{}$$

$$\text{1-Hexene} \qquad\qquad\qquad\qquad\qquad \text{2-Hexanol}$$

To obtain an anti-Markovnikov product, a different procedure involving a different mechanism must be used. For example, in hydroboration, hydrogen and boron add across a carbon-to-carbon double bond in a concerted, four-center syn addition:

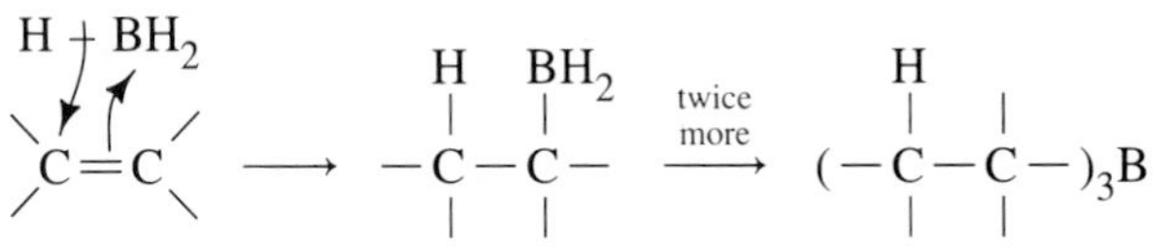

A trialkylborane

The hydride attacks the more substituted carbon, both for steric reasons and because any fleeting partial negative charge formed on carbon is more stable on the less-substituted carbon.

Once this intermediate has formed, it is only necessary to replace the boron with an —OH group, and this is accomplished by the use of hydrogen peroxide (H_2O_2) in base (OH^-). This combination oxidizes the boron to the level of boric acid (in its anionic form, $H_2BO_3^-$) and forms an alcohol.

The mechanism of oxidation of an alkylborane by alkaline peroxide (shown below) is believed to begin with attack of the anion of hydrogen peroxide on boron. Remember that a boron atom with three bonds has only six valence electrons and can accept another bonding pair. An alkyl group then migrates from boron to oxygen. Attack of hydroxide on boron followed by loss of an alkoxide group and proton transfer completes the reaction.

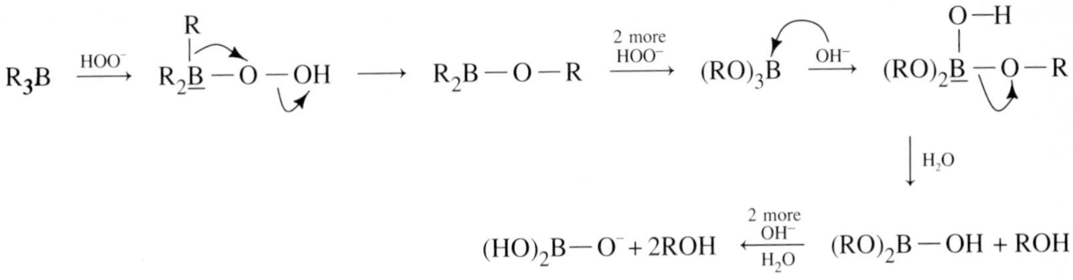

$$R_3B \xrightarrow{HOO^-} R_2\underline{B}-O-OH \longrightarrow R_2B-O-R \xrightarrow[HOO^-]{2\ more} (RO)_3B \xrightarrow{OH^-} (RO)_2\underline{B}-O-R$$

$$(HO)_2B-O^- + 2ROH \xleftarrow[H_2O]{\substack{2\ more \\ OH^-}} (RO)_2B-OH + ROH$$

As you may know, alkynes undergo many of the same types of reactions as alkenes. For example, they may be hydrogenated (once or twice), HBr or Br_2 can be added twice, and water can be added in either a Markovnikov or anti-Markovnikov fashion.

The addition of water across a triple bond occurs under slightly different conditions from addition to an alkene. For Markovnikov addition of H_2O to both alkenes and alkynes, Hg^{2+} can be used. While an alkene generally yields the best results using $Hg(OAc)_2$, for an alkyne the preferred mercuric salt is $HgSO_4$. Alkynes also require the presence of catalytic amounts of sulfuric acid.

In Experiment 7.1, the conversion you will carry out is the following overall reaction:

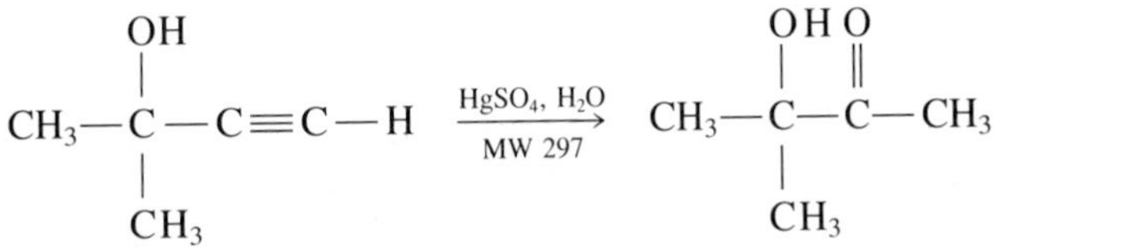

2-Methyl-3-butyn-2-ol
MW 84.1
bp 104°
dens. 0.87 g/mL

3-Hydroxy-3-methyl-2-butanone
MW 102
bp 140°
dens. 0.97 g/mL

Note that the presence of the alcohol group does not interfere with the reaction of the alkyne, and that a methyl ketone is formed instead of an alcohol.

The mechanism of the reaction is

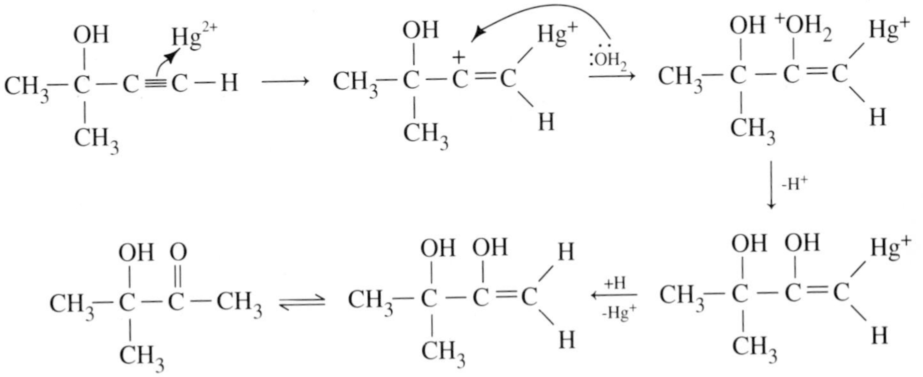

Keto form   Enol form

Addition of water to an alkyne forms an *enol*, $-\overset{H}{\underset{}{C}}=\overset{OH}{\underset{}{C}}-$, an unstable arrangement, which tautomerizes to the favored keto form. The driving force for the reaction is the formation of a stronger carbon-to-oxygen pi bond while breaking the weaker carbon-to-carbon pi bond. A terminal alkyne will therefore yield a methyl ketone as the ultimate product of hydration. If the alkyne is internal to the chain, there will be little or no preference for direction of addition of and, if the alkyne is unsymmetrical, a mixture of products can result:

$$CH_3-C\equiv C-CH_2CH_3 \xrightarrow[H_2SO_4]{HgSO_4} CH_3-\overset{\overset{O}{\|}}{C}-CH_2CH_2CH_3 + CH_3CH_2\overset{\overset{O}{\|}}{C}CH_2CH_3$$

2-Pentyne 2-Pentanone 3-Pentanone

The starting material, 2-methyl-3-butyn-2-ol, is made inexpensively from acetylene and acetone.

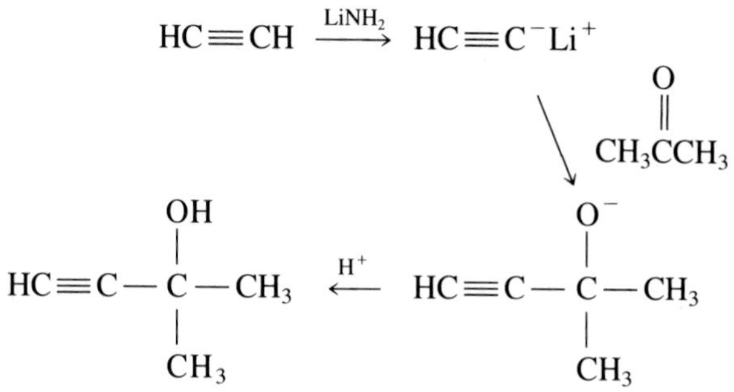

$$HC\equiv CH \xrightarrow{LiNH_2} HC\equiv C^-Li^+$$

$$\overset{\overset{O}{\|}}{CH_3CCH_3}$$

$$\underset{\underset{CH_3}{|}}{\overset{\overset{OH}{|}}{HC\equiv C-C-CH_3}} \xleftarrow{H^+} \underset{\underset{CH_3}{|}}{\overset{\overset{O^-}{|}}{HC\equiv C-C-CH_3}}$$

The anion of acetylene, formed by adding the strong base $LiNH_2$, attacks the carbonyl group of acetone, opening the $C=O$ double bond to form an alkoxide. After protonation, an alcohol results.

QUESTIONS

1. Show how to prepare the following molecules from alkyne precursors.

(a) $CH_3CH_2CH_2\overset{\overset{O}{\|}}{C}CH_3$

(b) $CH_3CH_2-\overset{\overset{O}{\|}}{C}-CH_2CH_2CH_3$

(c) 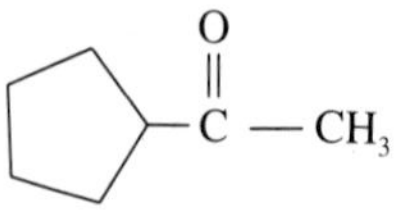 $\overset{\overset{O}{\|}}{C}-CH_3$

2. Draw the expected products of the following reactions.

(a) 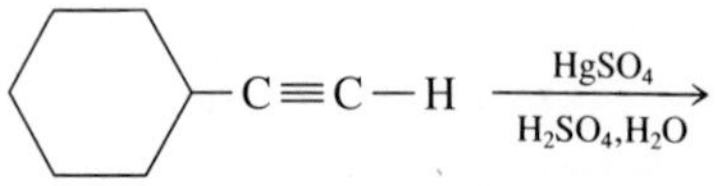 $C\equiv C-H \xrightarrow[H_2SO_4,H_2O]{HgSO_4}$

(b) $CH_2 = CH - CH_2CH_2CH_3$ $\xrightarrow[\text{(2) NaBH}_4]{\text{(1) Hg(OAc)}_2, \text{ H}_2\text{O}}$

$\xrightarrow[\text{(2) H}_2\text{O}_2, \text{ OH}^-]{\text{(1) BH}_3}$

3. What precautions will you use to avoid exposure to mercury and to potentially harmful organic materials while carrying out these procedures?

EXPERIMENT 7.1 HYDRATION OF 2-METHYL-3-BUTYN-2-OL

Estimated Time:
2.5 hours

Prelab

1. Calculate the mass and volume of 140 mmol of 2-methyl-3-butyn-2-ol and the mass of 140 mmol of mercuric sulfate.
2. What is the theoretical yield of 3-hydroxy-3-methyl-2-butanone?
3. How will you dispose of the mercury containing wastes?

Special Hazards

Mercuric sulfate, like most mercury compounds, is highly toxic if ingested or inhaled; avoid skin contact or breathing of the powder and wash hands after use. The starting material (2-methyl-3-butyn-2-ol), product, and ethyl ether used for extraction are all flammable and harmful to breathe or touch. Exercise appropriate caution.

Procedure

This procedure consists of three parts: an addition during reflux, an extraction, and a distillation. While waiting for each process to finish, assemble the apparatus for the next. Into a 100-mL round-bottomed flask, place 3 *M* sulfuric acid (50 mL) and mercuric sulfate (140 mmol, ___ g). Add a Claisen head, addition funnel, reflux condenser, and heating mantle with variable transformer. This setup appears in Figure 7–1.

Make sure that water is flowing through the condenser at a slow, steady rate and heat the mixture gently to reflux. While the flask is heating, measure 2-methyl-3-butyn-2-ol (140 mmol, ___ g, ___ mL) and place it in the addition funnel. Be sure to leave the top of the condenser open so that you are not heating a closed system. Once the flask is refluxing, add the alkynol from the addition funnel dropwise over the course of about 5 minutes. After the addition is complete, reflux for another 30 minutes.

At the end of the reflux period, cool the flask in an ice bath to below room temperature and pour the contents into a separatory funnel. Use caution since the mixture contains potentially toxic mercury residues. Add ether (30 mL), stopper the separatory funnel, shake gently, and vent. Remove and set aside the lower (aqueous) layer in a beaker labeled "Caution: Hg waste" and leave the upper (organic) layer in the separatory funnel. Wash this ether extract with water (20 mL) to remove traces of acid, then with saturated NaCl solution to remove residual water. Place the water and brine used for these washings in the beaker of mercury-containing waste, and pour it into a

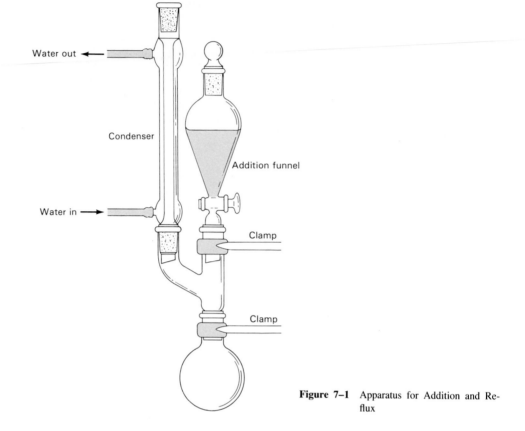

Water out

Condenser

Water in

Addition funnel

Clamp

Clamp

Figure 7–1 Apparatus for Addition and Reflux

properly marked waste jug in the fume hood for safe disposal. Dry the ether extract
over anhydrous $HgSO_4$, then decant or filter the solution into another beaker. Add a
boiling chip and evaporate the ether on a steam bath in the hood. Once it has stopped
bubbling vigorously, place the residue (your crude product) in a 50-mL round-bottomed
flask set up for simple distillation. *Caution:* Do not use a flame as a heat source, since
some ether may remain in the crude product at this stage. Distill and discard any remaining ether, then collect the product (3-hydroxy-3-methyl-2-butanone) near its expected boiling point (140–141°). If desired, a refractive index measurement may be
taken to compare with the literature value.

STEREOCHEMISTRY: ADDING BROMINE TO ALKENES

Overview

Many reactions involve additions across carbon-to-carbon double bonds. In the preceding section we discussed addition of the elements of water (H and OH) across an alkene. Other reagents that add to alkenes are hydrogen (H_2), hydrohalic acids (HX = HCl or HBr), halogens (Cl_2 or Br_2) hypohalous acids (HOCl or HOBr) and oxidants such as OsO_4 or peracids. The results of each of these reagents are as follows:

So far we have not specified the stereochemistry of the addition, in other words whether the incoming atoms arrive on the same side or opposite sides of the alkene molecule. While in some cases the product of addition would be the same whichever mechanism was operating, in many cases it makes a great difference in the structure of the product. For example, consider the addition of bromine to cyclohexene.

Two hypothetical products are possible, *cis*-1,2-dibromocyclohexane from syn (same side) addition and *trans*-1,2-dibromocyclohexane from anti (opposite side) addition.

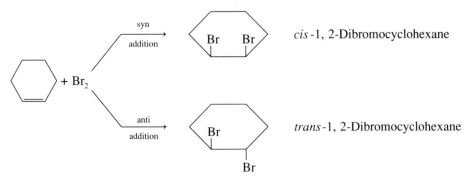

The cis product would be meso because it possesses an internal plane of symmetry. Properties of meso compounds are summarized here.

1. They possess an internal plane of symmetry.
2. They are identical to their mirror images.
3. If one asymmetric center has the R configuration, the corresponding opposite one is S.
4. They are optically inactive; they do not rotate plane-polarized light.

Note that properties 2, 3, and 4 follow automatically from 1.

In the addition of bromine to cyclohexene above, anti addition would give *trans*-1,2-dibromocyclohexane as a racemic mixture, since there would be equal quantities of the R,R and S,S isomers.

There are a couple of mnemonic devices helpful for predicting the stereochemical outcome of an addition to a symmetrically disubstituted alkene, based on whether the starting alkene is cis or trans and whether the addition is syn or anti. One such device utilizes the following abbreviations:

Abbreviation	Description
CAR	A cis alkene undergoing anti addition yields a racemic product
TAM	Trans-anti-meso
CSM (CHASM)	Cis-syn-meso
TSR (TASER)	Trans-syn-racemic

For example, addition of OsO_4 to a symmetrical cis alkene (*cis*-2-butene) in a syn fashion would yield a meso product (cis-syn-meso, or CHASM). Similarly, addition of Cl_2 to a symmetrical trans alkene (*trans*-2-butene) in an anti fashion would yield a meso product (trans-anti-meso, or TAM).

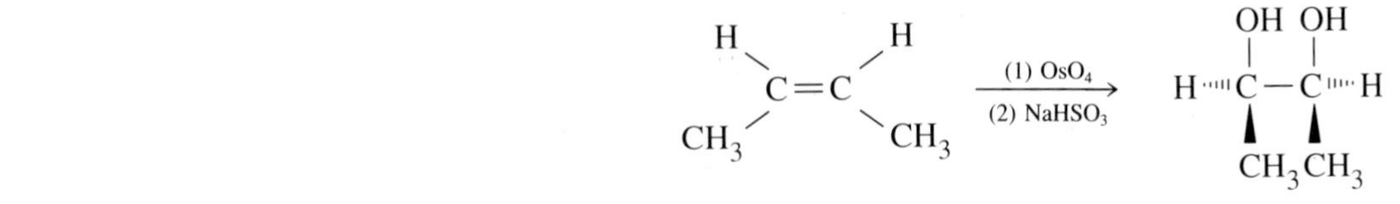

$$H_2C=CH_2 \xrightarrow[\text{(2) NaHSO}_3]{\text{(1) OsO}_4}$$

cis-2-Butene   meso 2, 3-Butanediol

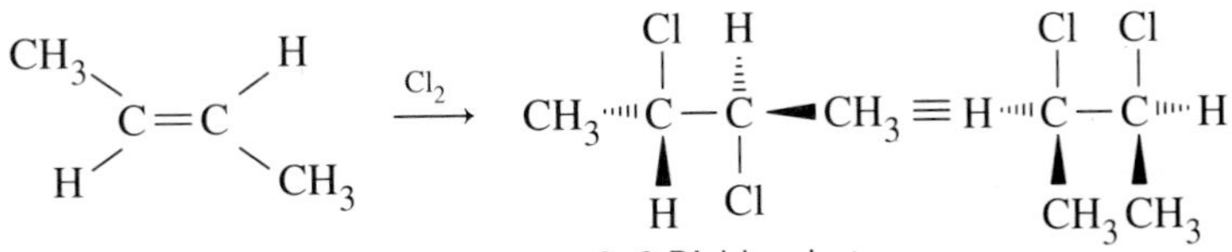

trans-2-Butene   meso 2, 3-Dichlorobutane

Another way to remember the sterochemical outcomes of additions to alkenes is to classify compounds as symmetric (cis or meso) or antisymmetric (trans or racemic). A syn reaction does not change the symmetry, while an anti reaction does.

Schematically, this can be represented

symmetric	$\xrightarrow{\text{syn}}$	symmetric	(cis	$\xrightarrow{\text{syn}}$	meso)
antisymmetric	$\xrightarrow{\text{syn}}$	antisymmetric	(trans	$\xrightarrow{\text{syn}}$	racemic)
symmetric	$\xrightarrow{\text{anti}}$	antisymmetric	(cis	$\xrightarrow{\text{anti}}$	racemic)
antisymmetric	$\xrightarrow{\text{anti}}$	symmetric	(trans	$\xrightarrow{\text{anti}}$	racemic)

Of course, these devices are not applicable to all alkenes: for example, terminal or trisubstituted alkenes (which are not cis or trans). It should also be borne in mind that a particular reaction may go by a mechanism that results in only syn addition or only anti or a mixture of the two. Examination of the structures of the starting material and product(s) will indicate which mode of addition actually took place.

In Experiment 8.1 you will be adding bromine to *trans*-cinnamic acid. The two conceivable products are as follows:

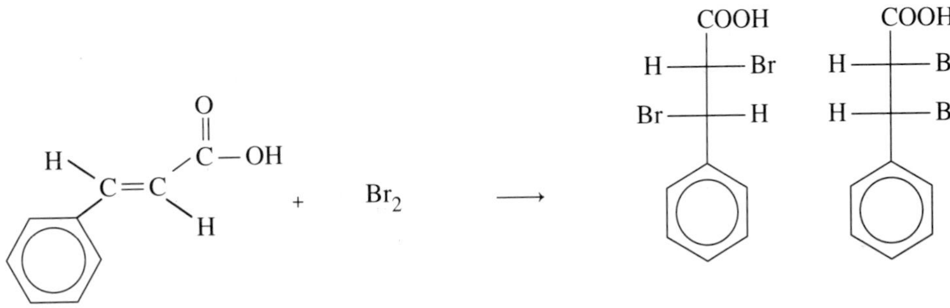

Cinnamic acid   Bromine   (±)Threo and/or (±)Erythro
(3-phenylpropenoic acid)   MW 160,   mp 93–95° mp 202–204°
MW 148, mp 133–134°   bp 60°   2, 3-Dibromo-3-phenylpropanoic acid
             dens. 3.10 g/mL   MW 308

Since these two products have significantly different melting points, you will be able to tell which one was formed and therefore which mode of addition occurred. Since it is sometimes difficult to visualize three-dimensional structures from drawings on paper, it is highly recommended that you build a model of *trans*-cinnamic acid and run through the formation of both potential products.

In cases where the original alkene is not symmetrically substituted, the products may be classified according to a different system, as being either threo (similar in configuration to the sugar threose, corresponding to racemic) or erythro (meso-like, similar to erythrose).

CHO

HO ————— H

H ————— OH

CH$_2$OH

Threose

CHO

H ————— OH

H ————— OH

CH$_2$OH

Erythrose

EXPERIMENT 8.1 ADDITION OF BROMINE TO CINNAMIC ACID

Estimated Time:
2.0 hours

Prelab

1. Calculate the mass of 4.0 mmol of *trans*-cinnamic acid.
2. What antidote will you use if you ever spill a solution containing molecular bromine on your skin?
3. Draw Fischer projections and label the stereochemistry of all conceivable products of addition of Br$_2$ to *cis*-2-butene.

Special Hazards

Molecular bromine is extremely toxic and corrosive; its vapors are damaging to skin, eyes, and the respiratory tract. Handle only in the hood, wear gloves, and avoid contact.

Sodium thiosulfate is a specific reagent for neutralization of bromine (it reduces Br$_2$ to Br$^-$). When working with Br$_2$ always keep a bottle of 5% aqueous sodium thiosulfate handy for rinsing the skin in case of contact. Exercise normal precautions with methylene chloride, cinnamic acid, and the product.

In very rare cases a person may be hypersensitive to the product, dibromocinnamic acid. If during the course of this experiment you experience a skin reaction or lung irritation, inform your instructor and remove yourself from exposure immediately.

Procedure

Assemble a 50-mL round-bottomed flask with a Claisen head, reflux condenser, and addition funnel as shown in Figure 8–1. Coming off the top of the condenser set up a gas trap containing 5% aqueous sodium thiosulfate. The gas trap is not necessary if you can work in a fume hood for the whole experiment.

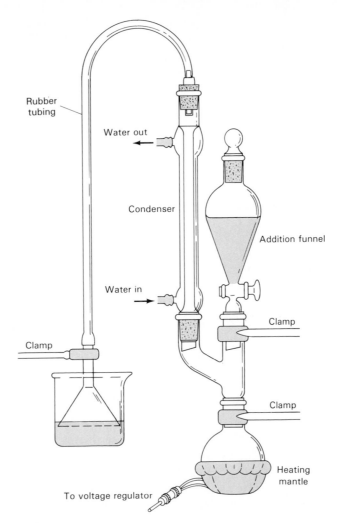

Figure 8-1 Apparatus for Addition and Reflux with Gas Trap

The trap consists of an inverted funnel clamped on the stem with its lip *barely* beneath the surface of some 5% aqueous $Na_2S_2O_3$ (sodium thiosulfate) in a beaker. This setup will allow the Br_2 gas to remain in contact with the sodium thiosulfate solution long enough to react and neutralize it. In case the pressure should decrease in the system, make sure there is not enough water above the funnel lip to be sucked clear back into your setup.

To prevent the escape of Br_2, make sure that all connections leading to the trap are airtight. If you are using a one-hole rubber stopper with a piece of glass tubing on top of the condenser, wire the rubber tubing securely onto the glass tubing using two turns of wire tightened with a pliers to ensure an airtight seal. As an alternative you can place a vacuum adapter on top of the condenser, plug the open end of the adapter with a solid stopper or cork, and attach the rubber tubing on the nipple of the adapter. The room should also be well ventilated in case any Br_2 escapes from traps improperly set up.

An alternative to a trap is to place a vacuum adapter packed with $CaCl_2$ pellets atop the condenser and a hose from it to the aspirator, as shown in Figure 8–2. When the aspirator is turned on gently, fumes are drawn down and mixed with the water going down the drain. For this purpose the aspirator can function well with a low flow rate of water.

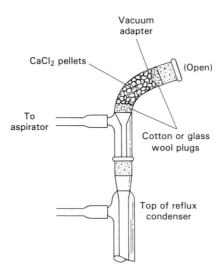

Figure 8-2 Gas Trap for Removal of Gases from a Reaction Mixture

Into the round-bottomed flask place *trans*-cinnamic acid (4.0 mmol, ___ g) and CH_2Cl_2(dichloromethane, methylene chloride, 10 mL). Add two or three boiling chips. Under the fume hood place 4.0 mL of a 1.0 *M* solution of Br_2 in CH_2CL_2 in the addition funnel; stopper the funnel securely before removing it from the hood and replacing it on the apparatus. Add a heating mantle and variable transformer, then heat the mixture to a gentle reflux. While it is refluxing, begin adding the solution of Br_2 dropwise at the rate of about two drops per second. Because of the high density of the CH_2CL_2 it is not necessary to loosen the stopper on the addition funnel to maintain the addition. The red-orange color of the Br_2 should disappear as it strikes the solution and reacts. After all of the Br_2 solution has been added, continue the reflux for an additional 10 minutes. If there is a red color remaining, indicating the presence of Br_2, add cyclohexene (or another low-molecular-weight alkene) dropwise through the top of the condenser until the reddish color disappears.

Remove the reaction flask from the apparatus, stopper it, and cool it in an ice bath for 10 minutes to allow the product (2,3-dibromo-3-phenylpropanoic acid, or cinnamic acid dibromide) to crystallize. While it is crystallizing, take the remaining apparatus apart in the fume hood and rinse it with the sodium thiosulfate from the trap to destroy any remaining bromine. Be careful not to breathe Br_2 fumes.

When the product has precipitated, collect it by suction filtration, rinse it with 10 mL of ice cold methylene chloride (dichloromethane, CH_2CL_2), and allow it to air dry. Record the mass and melting range of the product. If you will be carrying out Experiment 9.1, save your product in a vial clearly labeled with your name, the date, and the name and structure of the product.

Carefully place the waste methylene chloride in a special bottle provided by your instructor in the fume hood.

QUESTIONS

1. Based on the melting range, does your product have the erythro or threo structure? What does this tell you about the mechanism of bromine addition to this alkene?

2. Based on this result, predict the stereochemical results of addition of Br_2 to the following compounds.

    (a) *trans*-2-butenoic acid.

    (b) *cis*-Cyclooctene.

    (c) *E*-2-chloro-2-pentene.

REFERENCE

REIMER, M., ''Preparation of Phenylpropiolic Acid,'' *Journal of the American Chemical Society* (1942) *64*, 2510.

SECTION 9

ALKYNE FORMATION FROM VICINAL DIHALIDES

Overview

Alkynes often play important roles in our lives. For example, ignition of acetylene combined with oxygen provides a very hot flame for welding (approximately 2200°). Alkynes in general have a high energy content and heat of combustion because of the possibility of breaking two or more weak carbon-to-carbon pi bonds per molecule and forming strong carbon-to-oxygen double bonds (in CO_2).

An alkyne can be formed by loss of 2 moles of HX (where X = Cl or Br) from a dihalide, as shown below. Two moles of base are required to remove the 2 moles of hydrohalic acid.

$$\begin{array}{cc} X & X \\ | & | \\ -C - C- \\ | & | \\ H & H \end{array} \quad \xrightarrow[-2HX]{2B^-} \quad -C \equiv C- \;+\; 2BH \;+\; 2X^-$$

More specifically, a vicinal dibromide can undergo double dehydrobromination in the presence of a base such as KOH. The reaction occurs in two steps, both E2 eliminations. In each step hydroxide ion removes a (positive) hydrogen ion, accompanied by concurrent formation of a carbon-to-carbon pi bond and elimination of bromide ion.

$$\begin{array}{cc} Br & Br \\ | & | \\ -C - C- \\ | & | \\ H & H \end{array} \xrightarrow{OH^-} \quad \overset{Br}{\underset{H}{C=C}} \;+\; H_2O + Br^- \;\longrightarrow\; -C \equiv C- + H_2O + 2Br^-$$

$$HO^-$$

Depending on the structure of the dihalide, a stronger base such as $NaNH_2$ may be required to remove the second equivalent of HX. Hydroxide ion also tends to cause isomerization of internal alkynes to terminal alkynes, and in some case this might be a problem.

This conversion of a dihalide to an alkyne suggests a route to alkynes from alkenes:

$$-CH=CH- \xrightarrow{Br_2} \underset{\overset{|}{H} \quad \overset{|}{Br}}{\overset{\overset{Br \quad H}{|} \quad \overset{|}{}}{-C-C-}} \xrightarrow[\underset{(2)\ NaNH_2}{\overset{or}{(1)\ KOH}}]{2\ NaNH_2} -C\equiv C-$$

Alkene   Vicinal dibromide   Alkyne

Note that the stereochemistry is not specified here, since (except in special cases) it is not critical to alkyne formation.

In Experiment 9.1 you will take the product you made in Experiment 8.1 by bromination of cinnamic acid and dehydrobrominate it twice to form phenylpropynoic acid:

2,3-Dibromo-3-phenylpropanoic acid
(cinnamic acid dibromide)
MW 308

Phenylpropynoic acid
(phenylpropiolic acid)
MW 146, mp 136–138°

The purpose of the hydrochloric acid is to reprotonate the carboxylic acid salt.

Experiment 9.2 is a similar double dehydrobromination of stilbene dibromide to diphenylacetylene on a micro scale.

Stilbene dibromide
(1, 2-dibromo-1,2-diphenylethane)
MW 340, mp 241°d

Potassium
hydroxide
MW 56.1

Diphenylacetylene
MW 178, mp 61°

In this experiment the solvent triethylene glycol is used because it is moderately polar (allowing it to dissolve KOH) and has a boiling point of 285°, which allows it to be heated strongly without boiling.

$$HO_2CH_2CH_2OCH_2CH_2OCH_2CH_2OH$$

Triethylene glycol

APPLICATION: ALKYNES AND BIRTH CONTROL

Since ancient times women have dreamed of taking a medicine to enable them to prevent pregnancy until desired. In 1937, researchers discovered that rabbits given daily doses of progesterone (one of the two female sex hormones) experienced temporary and reversible cessation of ovulation. In 1949 an American chemist named Russell Marker at the University of Pennsylvania discovered a steroid in wild Mexican yams that can be converted to progesterone, making this substance available inexpensively in large quanitities.

Progesterone could not be taken orally, how-

Progesterone

Norethynodrel

Mestranol

ever, because the digestive juices in the stomach broke it down before absorption. By chemical modification, many derivatives of progesterone were prepared and tested, including some alkynes. Several were found that both suppressed ovulation effectively and could be taken orally.

The earliest birth control pill (Enovid) consisted of a mixture of two steroidal alkynes, norethynodrel and mestranol.

Since that time further studies have found even more effective steroid derivatives, which can be taken in smaller doses. Modern birth control pills contain 30–50 μg of an estrogen and 1 mg or less of a synthetic progestin (norgestrel or norethindrone).

EXPERIMENT 9.1 PHENYLPROPYNOIC ACID BY DOUBLE DEHYDROBROMINATION OF CINNAMIC ACID DIBROMIDE

Estimated Time:
1.5 hours

Prelab

1. Calculate the millimoles in 1.00 g of 2,3-dibromo-3-phenylpropanoic acid.
2. Calculate the millimoles in 2.5 g of KOH (technical grade, consisting of 85% KOH and 15% H_2O by weight).

Special Hazards

Potassium hydroxide is highly caustic; avoid contact and wash immediately and thoroughly if contact occurs. Methanol is flammable and toxic if ingested.

Procedure

In a 100-mL beaker place 2,3-dibromo-3-phenylpropanoic acid (cinnamic acid dibromide, the product from Experiment 8.1, ___ mmol, 1.00 g), methanol (15 mL), potassium hydroxide (___ mmol, 2.5 g), and a couple of boiling chips. Heat on a steam bath or hot plate in the fume hood, with stirring, until almost all the methanol has evaporated.

Collect the pale yellow granular product (the potassium salt of phenylpropynoic acid) by suction filtration, wash it with a little cold methanol, and dissolve it in 20 mL of water. Cool the solution in an ice bath and acidify cautiously with concentrated HCl to below pH 4 to ensure complete precipitation. Check the pH by dipping a stirring rod into the solution and touching it to a strip of indicator paper. Scratch and cool the resulting oily precipitate until it solidifies, then collect it by suction filtration and wash it with cold water. Air dry and record the mass and melting point (lit. mp 136–138°).

To convince yourself that you have formed carbon-to-carbon pi bonds in the molecule, take a few milligrams of the product in a test tube and add dropwise 0.2 M Br_2/ CCl_4 solution. Observe the results. For comparison carry out the same test on the starting material.

QUESTIONS

1. What by-products are formed during the reaction and workup?
2. What product is formed from reaction of phenylpropynoic acid with Br_2?

REFERENCE

REIMER, M., "Preparation of Phenylpropiolic Acid," *Journal of the American Chemical Society* (1942) *64*, 2510.

EXPERIMENT 9.2 MICROSCALE PREPARATION OF DIPHENYLACETYLENE FROM STILBENE DIBROMIDE

Estimated Time:
1.0 hour

Prelab

1. Calculate the millimoles corresponding to 90 mg of stilbene dibromide and 80 mg of KOH (85% purity).
2. What is the theoretical yield of diphenylacetylene?

Special Hazards

Potassium hydroxide is very corrosive, especially in hot solution. Be cautious of hot glassware and splattering hot oil. Stilbene dibromide is a lachrymator and irritant.

Procedure

In a 15 x 125 mm test tube, place *d,l*-stilbene dibromide (90 mg, ___ mmol) and powdered 85% KOH (80 mg, ___ mmol). By pipette add 0.50 mL of triethylene glycol. Clamp the tube so that its bottom is immersed in an oil bath maintained at a temperature of approximately 190°C. After heating for 5 minutes, carefully remove the tube using a test tube clamp, and place it for 5 minutes into a bath of water at room temperature in order to cool it.

After it has cooled to below 50°C, add 2.0 mL of water and place the tube in an ice bath. Collect the precipitated diphenylacetylene from the dark solution by suction filtration and wash with a little water.

Place this crude product in a clean test tube and recrystallize from a minimum amount of 95% ethanol. Collect the product, air dry, and record the mass and melting range.

QUESTIONS

1. Show the mechanism for this reaction.
2. Design a synthesis of phenylacetylene ($C_6H_5C \equiv CH$) from styrene ($C_6H_5CH = CH_2$).

REFERENCES

SMITH, L. I., and M. F. FALKOF, *Organic Syntheses Collective Volume 3* (1955) 350.
JACOBS, T. L., *Organic Reactions* (1949) *5*, 1.

SECTION **10**

S$_N$2 REACTIONS

Overview

In a nucleophilic substitution reaction, the nucleophile attacks a carbon and displaces a leaving group:

$$\text{Nu:} + \text{R} - \text{L} \longrightarrow \text{Nu} - \text{R} + \text{L}^-$$

The unshared electrons of the nucleophile form a new bond with carbon, while the leaving group takes the R — L bonding pair of electrons with it. Nucleophile means ''nucleus-loving,'' in other words attracted to a nucleus or positive charge. So a nucleophile must be electron-rich, often having a negative charge. Some typical nucleophiles are SH$^-$, I$^-$, CN$^-$, OH$^-$, CH$_3$O$^-$, H$_2\ddot{\text{O}}$, $\ddot{\text{N}}$H$_3$, and CH$_3\ddot{\text{O}}$H. During the reaction the charge on the nucleophile becomes *less* negative as the new bond forms, while the charge on the leaving group becomes *more* negative as it takes the previously bonding electrons with it.

In Experiment 10.1, the nucleophile is bromide ion (Br$^-$) and the leaving group is water. Mixing 1-butanol with a strong acid protonates the alcohol and creates a good leaving group.

Mechanism:

$$\underset{\substack{\text{1- Butanol}\\ \text{MW 74.1, bp 118}^\circ \\ \text{dens. 0.810 g/mL}}}{\text{CH}_3\text{CH}_2\text{CH}_2\text{CH}_2\ddot{\text{O}}\text{H}} \xrightarrow{\text{H}^+} \underset{\text{Protonated alcohol}}{\text{CH}_3\text{CH}_2\text{CH}_2\text{CH}_2 - \overset{+}{\text{O}}\text{H}_2} \xrightarrow{\text{Br}^-} \underset{\substack{\text{1-Bromobutane}\\(n\text{-butyl bromide})\\ \text{MW 137, bp 102}^\circ \\ \text{dens. 1.27 g/mL}}}{\text{CH}_3\text{CH}_2\text{CH}_2\text{CH}_2\text{Br} + \text{H}_2\text{O}}$$

The reaction would not proceed without a strong acid present, since OH$^-$ is a poor leaving group and cannot be displaced by Br$^-$.

This is a typical S$_N$2 reaction, where S$_N$2 stands for substitution, nucleophilic, bimolecular. It is bimolecular because the *slow step* is the attack of the bromide ion on the protonated alcohol and therefore involves two species. So in the kinetics the rate expression would be

$$\text{rate of reaction} = k[\text{ROH}_2^+][\text{Br}^-]$$

Of course, any S_N2 reaction involves back-side attack by the nucleophile and inversion at the carbon that was attacked. In our case, though, the first carbon of 1-butanol is not chiral, so we cannot tell if it has been inverted.

A possible by-product would be the ether formed if another molecule of 1-butanol attacks the protonated alcohol:

$$\text{CH}_3\text{CH}_2\text{CH}_2\text{CH}_2\overset{+}{-}\text{OH}_2 \longrightarrow \text{CH}_3\text{CH}_2\text{CH}_2\text{CH}_2 - \overset{+}{\underset{H}{\text{O}}} - \text{CH}_2\text{CH}_2\text{CH}_2\text{CH}_3 \xrightarrow{-H^+}$$

$$+ \quad \text{HOCH}_2\text{CH}_2\,\text{CH}_2\text{CH}_3$$

$$\text{CH}_3\text{CH}_3\text{CH}_2\text{CH}_2 - \text{O} - \text{CH}_2\text{CH}_2\text{CH}_2\text{CH}_3$$

di-*n*-Butyl ether
bp 141°

Another by-product, l-butene, could be formed by E2 elimination of the protonated alcohol:

$$\text{CH}_3\text{CH}_2\overset{\text{H}\quad\text{H}}{\underset{\underset{+}{\text{H}}\;\;\overset{}{\underset{\text{OH}_2}{}}}{\text{C}-\text{C}-\text{H}}} \longrightarrow \text{CH}_3\text{CH}_2\text{CH}=\text{CH}_2 + \text{BH}^+ + \text{H}_2\text{O}$$

B :

1-Butene
bp − 6°

Possible bases present are water or 1-butanol. Since the leaving group is on a primary (1°) carbon we expect substitution to predominate over elimination, and since bromide ion is a better nucleophile than 1-butanol, we expect more *n*-butyl bromide than *n*-butyl ether. So by-products should not be a serious problem.

The identity of the product (1-bromobutane) can be confirmed by carrying out three tests on it which are described in the procedure. One of these tests involves displacement of bromide by iodide. Since iodide is a stronger nucleophile than chloride or bromide, it can displace these ions from carbon. This reaction, known as the Finkelstein reaction, proceeds by the familiar S_N2 mechanism:

$$\text{CH}_3\text{CH}_2\text{CH}_2\text{CH}_2 - \text{Br} \xrightarrow[\text{acetone}]{\text{Na}^+\,\text{I}^-} \text{CH}_3\text{CH}_2\text{CH}_2\text{CH}_2\text{I} + \text{NaBr} \downarrow$$

This reaction is carried out by refluxing an alkyl halide with sodium iodide using acetone as the solvent. Acetone dissolves NaI, but not the products NaCl or NaBr, so these salts precipitate and help drive the reaction to completion in accord with Le Châtelier's principle. The appearance of this precipitate and the time it takes to form are the basis for a qualitative test for chlorides and bromides. If they are to be isolated, alkyl iodides must be protected from light and excessive heat, since the carbon-to-iodine bond is relatively weak (about 200 kJ/mol) and can be ruptured either thermally or photochemically. The reaction flask will be covered with foil to prevent photochemical decomposition of the product.

Another way to convert an alcohol to a halide is to convert the —OH to a good leaving group using a phosphorus reagent such as PBr$_3$ or PI$_3$. Phosphorus forms a strong bond to oxygen and helps pull it off as Br$^-$ or I$^-$ attacks from the rear.

You can see that this reaction actually involves two S$_N$2 displacements, one by oxygen on phosphorus and one by bromide on carbon. A similar reaction occurs with PI$_3$. Phosphorus trihalides can be formed *in situ* by mixing elemental phosphorus with the elemental halide. For example,

$$2P + 3I_2 \longrightarrow 2PI_3$$

Then

$$PI_3 + 3ROH \longrightarrow 3R\text{-}I + P(OH)_3$$

In Experiment 10.4, *n*-pentyl alcohol is converted to *n*-pentyl iodide by this method.

$$CH_3(CH_2)_4\!-\!OH \qquad + \qquad P \qquad + \qquad I_2 \qquad \longrightarrow \qquad CH_3(CH_2)_4\!-\!I$$

1-Pentanol	Red	Iodine	1-Iodopentane
(*n*-pentyl alcohol, *n*-amyl alcohol)	phosphorus	MW 254	(*n*-pentyl iodide)
MW 88.2, bp 136–138°	AW 31		MW 198, bp 154–155°
dens. 0.81 g/mL			dens. 1.52 g/mL

For your convenience and future reference, four of the common ionic mechanisms (S$_N$2, S$_N$1, E2, and E1) are summarized following Experiment 10.4.

Since the products of the experiments in this section are liquids, it would be helpful to have a quick method for testing the identity and purity of a liquid. One such method is the index of refraction, which is a characteristic physical property of a liquid useful for determining its identity and purity. Refractive index (*n*) is defined as the ratio of the velocity of light in air to the velocity of light in the substance in question (our liquid product in this case). This ratio is also equal to the ratio of the sine of the incident angle to the sine of the angle of refraction, as shown in Figure 10–1.

To measure the refractive index you will use an instrument called a refractometer, the most common type of which is shown in Figure 10–2. If you have this type of refractometer available, take a refractive index measurement following these steps:

1. Raise the hinged prism, wipe off both glass surfaces with a soft Kimwipe, and place 2–4 drops of your product on the glass surface of the lower prism. Lower the top prism and observe your sample spread out in a thin uniform layer between the glass surfaces.

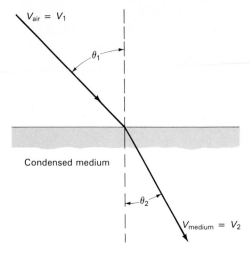

$V_{air} = V_1$

θ_1

Condensed medium

θ_2

$V_{medium} = V_2$

Figure 10-1 Refractive Index

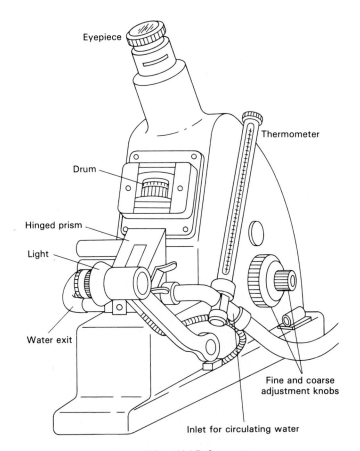

Eyepiece

Thermometer

Drum

Hinged prism

Light

Water exit

Fine and coarse
adjustment knobs

Inlet for circulating water

Figure 10-2 Abbé Refractometer

2. Switch on the lamp and position it near the prisms.
3. Look through the eyepiece, push the display button on the left of the instrument, and rotate the adjustment wheel to set the reading near the expected value (in this case, 1.4401).
4. Release the display button and rotate the adjustment wheel until you can see a light and a dark field approximately centered in the eyepiece (Figure 10–3). If necessary, adjust the lamp position for maximum brightness.

Figure 10-3 Visual Field for Properly Adjusted Refractometer

5. Rotate the compensating drum until the borderline of the light and dark fields is sharp and achromatic (has no colored fringes).
6. Readjust the adjustment wheel until the line is centered on the cross hairs.
7. Press the display button to show numbers through the eyepiece and read off the refractive index to four decimal places. Record the temperature of the instrument.
8. Clean and dry the prisms using a Kimwipe.

QUESTIONS

1. What is the purpose of a reflux apparatus? In other words, what difference does it make to place a reflux condenser on top of a flask of a boiling solution?
2. Predict the products of the following reactions.

   (a) CH_3CH_2OH $\xrightarrow{HCl}$

   (b) CH_3CH_2Cl $\xrightarrow{KCN}$

   (c)  $-CH_2Br$ $\xrightarrow{NaSH}$

   (d) R-2-bromobutane $\xrightarrow{NaI}$

3. Give a general discussion on how to predict which layer will be the upper layer and which will be the lower during an extraction.
4. If during an extraction you are unsure which layer is which, give three ways to determine their identities.
5. How would you make the following compounds from the corresponding alcohols?
   (a) 1-Chlorobutane
   (b) 1-Bromohexane
   (c) 1-Iodohexane
   (d) Benzyl cyanide   (benzyl $= C_6H_5-CH_2-$)
   (e) n-Butyl mercaptan ($CH_3CH_2CH_2CH_2SH$, skunk essence)

Concept Map

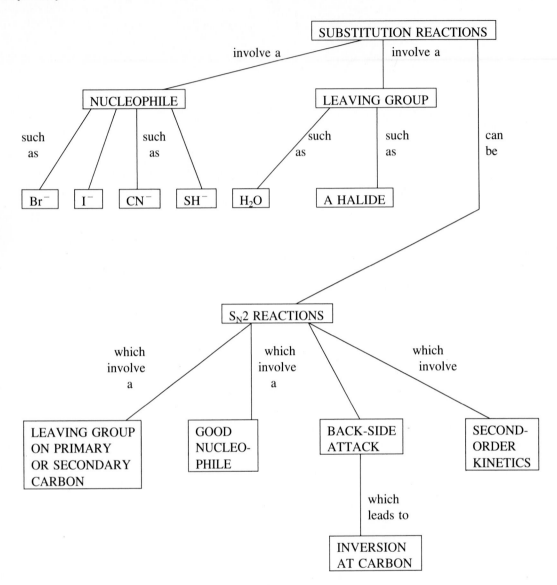

EXPERIMENT 10.1 CONVERSION OF 1-BUTANOL TO 1-BROMOBUTANE

Estimated Time:
2.5 hours

Prelab

1. Calculate the millimoles corresponding to 20.0 g of NaBr, 15 mL of *n*-butyl alcohol, and 15 mL of concentrated H_2SO_4.
2. If 10% NaOH solution has a density of 1.11 g/mL and is in a separatory funnel with 1-bromobutane, which layer will be on top?

3. At what thermometer reading do you expect to collect the *n*-butyl bromide?

4. Calculate the theoretical yield of 1-bromobutane.

5. If Br$_2$ is reduced, what does it become?

Special Hazards

Concentrated H$_2$SO$_4$ is extremely corrosive and it is advised that you use special care with it. Use of gloves and lab coats is recommended.

Procedure 10.1a: Preparation of Bromobutane from 1-Butanol

Estimated time:
2.0 hours

In a 250-mL round-bottomed flask place sodium bromide (20.0 g, __ mmol), water (15 mL), and *n*-butyl alcohol (15 mL, __ mmol). Cool the mixture in an ice-water bath. Measure concentrated sulfuric acid (15 mL, __ mmol) in a graduated cylinder, which must be clean to avoid forming colored by-products. Slowly add the sulfuric acid to the mixture with swirling and cooling, letting the acid run down the side of the flask. Place a condenser on top of the flask as shown in Figure 10–4 and add a heating mantle and variable transformer. Check that water is flowing through the condenser, then heat the mixture to reflux. The reflux ring should remain somewhere in the lower half of the condenser—if it rises above that you will need to turn down the heat.

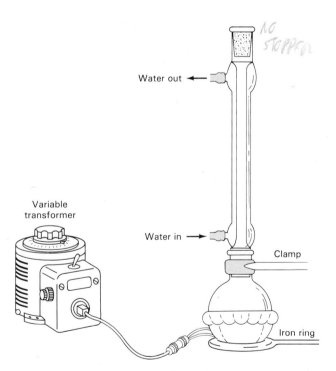

Figure 10-4 Reflux Apparatus

Allow the mixture to reflux for 45 minutes, during which time the reaction mixture may separate into two layers, the alkyl bromide above the more dense aqueous solution of inorganic salts. After the reflux period, remove the heat and allow the condenser to drain, then mount a still head on the flask and and set up the condenser and an adapter for simple distillation, collecting the product in a 50-mL round-bottomed flask or other suitable container. Refer to Experiment 2.1 if needed. Distill the mixture, watching the temperature carefully, and collect the distillate until no more water-insoluble droplets are coming over. To check this, collect a drop or two in a test tube and add some water to it. By this time the thermometer should read about 115°. During the distillation the boiling point gradually increases because n-butyl bromide forms an azeotrope with water which contains increasing amounts of sulfuric acid. After collecting the distillate, the residue in the distilling flask, which is strongly acidic, can be disposed of by cooling it to room temperature, pouring it onto ice, diluting with water, and rinsing down the drain with lots of water.

Shake the distillate in a separatory funnel with water (20 mL). If at any time a pink color forms due to the presence of Br_2, it can be removed by adding a pinch of sodium bisulfite ($NaHSO_3$, a mild reducing agent). Note that n-butyl bromide forms the lower layer since it is more dense than water. Drain off the n-butyl bromide. Record the volume and mass of this crude product. To the crude n-BuBr add enough anhydrous calcium chloride, sodium sulfate, or magnesium sulfate so that the drying agent does not all lump together and some remains as a loose powder. Seal the container in an airtight manner and store it in your drawer until the next class. A stoppered vial or flask will be needed since Parafilm alone will not prevent evaporation.

Procedure 10.1b Simple Distillation of 1-Bromobutane

**Estimated time:
1.5 hours**

Remove the drying agent from your n-butyl bromide by decanting or gravity filtration and place it in a 50-mL round-bottomed flask. Set up the apparatus for simple distillation. Discard the first few drops since they contain the low-boiling impurities, then collect the product in a tared 25-mL round-bottomed flask at the expected boiling point. Be sure to leave a few drops in the distilling pot.

Weigh the product and record its volume. To confirm the identity of the product, perform the following tests in any order.

Beilstein test.   The Beilstein test is an old-fashioned test for the presence of a halogen (Cl, Br, or I) in a molecule. This test relies on the fact that an organic molecule containing a halogen when heated with copper forms a copper halide which gives a green flame with a Bunsen burner.

To perform the test, take a piece of thick copper wire and clean it by holding it in the flame of a Bunsen burner in a corner of the room well away from any flammable materials. Remove the wire from the flame and let it cool for 30 seconds. Now dip it in your product and replace it in the flame. After the initial orange flame due to the burning of the product, a green flame indicates the presence of a halogen. For comparison, repeat this test using the starting material, 1-butanol.

Sodium iodide in acetone test.   This test relies on the fact that iodide ion is a good nucleophile and can displace other halides (Cl^- and Br^-) in an S_N2 reaction. The reaction is as follows:

$$R-X + Na^+I^- \longrightarrow RI + Na^+X^- \qquad X = Cl \text{ or } Br$$

Of course, RX must be a primary or secondary halide for this S_N2 reaction to work. Primary bromides react the fastest, within 3 minutes at 25°. The less reactive primary and secondary chlorides must be heated to 50° for the reaction to occur, as must secondary and tertiary bromides. Tertiary bromides, of course, must react by an S_N1 mechanism instead.

To perform the test, place 1 mL of the sodium iodide–acetone reagent in a small test tube. Add 2 drops of the suspected halogen-containing compound. Shake the mixture and let it stand for 3 minutes. Observe whether a red-brown color or precipitate forms. If there is no reaction after 3 minutes, place the tube in a beaker of water at 50°. After 6 minutes in the 50° bath, remove the tube and cool it to room temperature. Note whether any reaction has occurred.

Since sodium iodide is soluble in acetone, while sodium bromide and sodium chloride are not, a positive test consists of a cloudy solution or white precipitate of NaCl or NaBr. Compare your results to the guidelines stated above to see to which classification your product belongs. If available, you may want to run this test on some known compounds for comparison. Measure the refractive index of your product. How does your reading compare with the literature value of $n_D^{20} = 1.4401$? Bear in mind that the refractive index varies slightly with temperature and your readings may not be made exactly at 20°.

When finished with all the tests, place the remainder of your product in the collection container in the fume hood.

From the measured mass and volume of your product, calculate its density. Is it in agreement with the reported density of 1-bromobutane (1.27 g/mL)? Are the results of your Beilstein test, NaI/acetone test, and index of refraction measurement also consistent with the product being 1-bromobutane? How confident are you that your product is indeed 1-bromobutane?

EXPERIMENT 10.2 DISPLACEMENT OF BROMIDE BY IODIDE: PREPARATION OF 1-IODOHEXANE BY THE FINKELSTEIN REACTION

Estimated Time:
2.5 hours

Prelab

1. Calculate the mass of 80 mmol of NaI and the volume of 70 mmol of *n*-hexyl bromide.

Special Hazards

Sodium iodide, 1-bromohexane, and 1-iodohexane are all irritants.

Procedure

In a 100-mL round-bottomed flask, place sodium iodide (80 mmol, __ g), *n*-hexyl bromide (70 mmol, __ g, __ mL), and acetone (40 mL). Add a condenser, heating mantle, and variable transformer. To protect the product from light, shield the top portion of the boiling flask with aluminum foil. Do not place foil between the heating mantle and the flask. Heat the mixture to reflux (the NaI will dissolve). Reflux for 45 minutes, during which time the yellow color disappears and a white precipitate of NaBr appears. At the end of the reflux period decant the solution away from the NaBr, add a couple of boiling chips, and remove most of the acetone on a steam bath in the hood.

Meanwhile, set up an apparatus for simple distillation. After most of the acetone has been removed on the steam bath, distill the residue to yield 1-iodohexane (lit. bp 179–180°). Record the boiling range, mass, and refractive index of your product.

EXPERIMENT 10.3 CONVERSION OF ISOBUTYL ALCOHOL TO ISOBUTYL BROMIDE USING PBr₃

Estimated Time:
2.0 hours

Prelab

1. Calculate the volumes of 100 mmol of isobutyl alcohol and of 33 mmol of PBr₃. (density 2.88 g/mL)
2. What is the theoretical yield of isobutyl bromide?

Special Hazards

Phosphorus tribromide is highly corrosive, irritating, and reacts violently with water to form HBr and H_3PO_3.

Procedure

In a 125-mL Erlenmeyer flask place isobutyl alcohol (100 mmol, __ g, __ mL). Add a thermometer clamped securely. In a 10-mL graduated cylinder place PBr₃ (33 mmol, __ g, __ mL). Add the PBr₃ dropwise by Pasteur pipette at such a rate that the temperature of the solution remains below 50°. At the end of the addition allow the mixture to stand for an additional 20 minutes with occasional swirling. To remove the H_3PO_4 formed, slowly and *with caution* pour the mixture into a separatory funnel containing 25 mL of H_2O. Shake and vent. Draw off the lower (organic) layer and discard the aqueous layer into a beaker marked "waste." Dry the organic layer over Na_2SO_4, decant, and distill to obtain isobutyl bromide.

EXPERIMENT 10.4 CONVERSION OF 1-PENTANOL TO 1-IODOPENTANE USING PHOSPHORUS AND IODINE

Estimated Time:
2.5 hours

Prelab

1. What is the mass of 10 mL of 1-pentanol? Calculate the millimoles corresponding to 10 mL of 1-pentanol, 0.90 g of phosphorus, and 12.0 g of iodine.
2. Calculate the theoretical yield of 1-iodopentane.
3. Why should the product be protected from light?

Special Hazards

Red phosphorus is flammable and can be ignited by friction. Use due caution. Do not use white or yellow phosphorus, which are pyrophoric.

Procedure:

In a 50-mL round-bottomed flask in the fume hood, place 1-pentanol (__ mmol, __ g, 10 mL), red phosphorus (__ mmol, 0.90 g), and iodine (__ mmol, 12.0 g). The mixture is refluxed for 1.5 hours. Cool the flask in an ice bath and pour the mixture into a separatory funnel containing 50 mL of water and a pinch of sodium bisulfite (to reduce I_2 to I^-). Shake, vent, and separate the layers. Dry the organic layer over Na_2SO_4 and decant it into a clean 50-mL round-bottomed flask. To increase the yield, rinse the drying agent with a few milliliters of ether and add this to the flask. Add boiling chips, a Claisen head with thermometer, water-cooled condenser, and heating mantle with variable transformer. Cover the upper portion of the flask with foil to insulate it and minimize photochemical decomposition of the iodide. Record the boiling range, mass, and refractive index of the 1-iodopentane obtained. A slight reddish color indicates the presence of a small amount of I_2 formed from decomposition of the organic iodide.

MECHANISMS REVIEW

S_N2: Substitution, Nucleophilic, Bimolecular

A nucleophile displaces a leaving group in a single step, going through a planar intermediate and bonding to carbon with inversion. These reactions require a good nucleophile and a leaving group on a methyl, primary, or secondary carbon to avoid steric hindrance. Beta branching can prevent reaction.

Transition state
with three R
groups planar

S_N1: Substitution, Nucleophilic, Unimolecular

A leaving group breaks off of carbon, taking both bonding electrons with it and leaving a planar carbocation intermediate. The carbocation can then be attacked by a nucleophile from either side with equal probability. If the starting material is chiral at only this carbon, racemic product is formed even if the starting material is optically active. The nucleophile can be weak, and a polar solvent is needed to stabilize the carbocation. The leaving group should be on a tertiary, allylic, benzylic, or (rarely) secondary carbon. If the intermediate carbocation is secondary and can rearrange to give a more stable carbocation via a methide or hydride shift, it will do so.

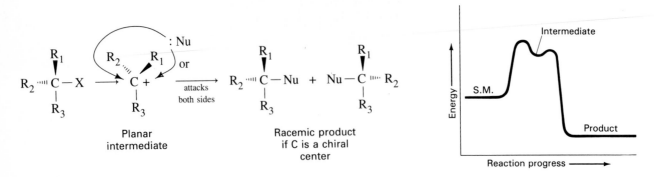

Planar intermediate

Racemic product if C is a chiral center

E2: Elimination, Bimolecular

This reaction competes with S_N2 and involves removal of a hydrogen without its electrons on the carbon next to the leaving group. The hydrogen must be in an anti conformation to the leaving group when they are removed. The major product will be the most substituted alkene (Saytzeff's rule). A very hindered base such as *t*-butoxide can give a large amount of the least substituted alkene (Hofmann product). *Note:* If both carbons are chiral, R_1 and R_3 wind up cis to each other as the molecule becomes an alkene and flattens out.

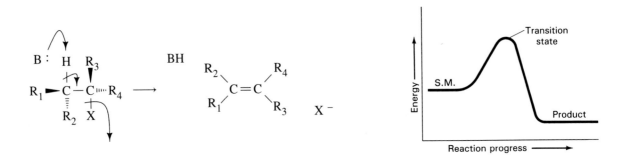

E1: Elimination, Unimolecular

This reaction competes with S_N1 and goes through the same carbocation intermediate, which requires a polar solvent to form. It is favored over S_N1 by high temperatures and over E2 by *weak* bases. *Note:* R_1 and R_3 can be either cis or trans to each other due to free rotation in the carbocation intermediate. Saytzeff's rule still applies.

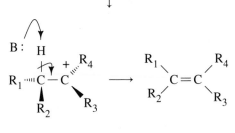

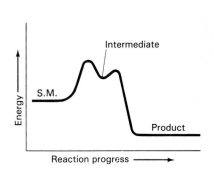

Mechanism Practice

For each of the following reactions, give the most likely (major) product and label the mechanism (S_N1, S_N2, E1, E2, or radical).

$$CH_3-CH_2I \xrightarrow{\text{NaCN}}$$

Mechanism:

(methylcyclopentane) $\xrightarrow[h\nu]{\text{Br}_2}$

Mechanism:

$$\begin{array}{c} CH_3 \\ | \\ CH_3-C-OH \\ | \\ CH_3 \end{array} \xrightarrow{\text{HBr}}$$

Mechanism:

$$\begin{array}{c} Cl \\ | \\ CH_3-C-CH_2CH_3 \\ | \\ CH_3 \end{array} \xrightarrow{\text{CH}_3\text{CH}_2\text{OH}}$$

Mechanism:

$\xrightarrow[\Delta]{\text{OH}^-}$

Mechanism:

SECTION 11

S$_N$1 REACTIONS AND KINETICS

Overview

In Section 10 we discussed several types of S$_N$2 reactions. There is another class of nucleophilic substitution reactions, which are unimolecular (abbreviated S$_N$1). For these reactions the rate-determining (slow) step involves only one molecule. A leaving group departs in the slow step to form a carbocation. Since the stability of carbocations is tertiary > secondary > primary, this mechanism will take place most often on tertiary carbons. The nucleophile need not be a strong one, since the difficult part of the reaction (formation of the carbocation) occurs before it attacks. For example, in Experiment 11.1 an alcohol is converted to a chloride by protonating the oxygen, forming a good leaving group (water). In the slow step, water departs, leaving a tertiary carbocation. The carbocation formed bonds with chlorine in a fast step.

$$CH_3CH_2-\overset{\overset{\displaystyle CH_3}{|}}{\underset{\underset{\displaystyle CH_3}{|}}{C}}-\overset{..}{\underset{..}{O}}H \xrightarrow{H^+} CH_3-\overset{\overset{\displaystyle CH_3}{|}}{\underset{\underset{\displaystyle CH_3}{|}}{C}}\overset{+}{\underset{..}{O}}H_2 \xrightarrow{-H_2O} CH_3CH_2-\overset{\overset{\displaystyle CH_3}{|}}{\underset{\underset{\displaystyle CH_3}{|}}{C}}+ \xleftarrow{Cl^-} CH_3-CH_2-\overset{\overset{\displaystyle CH_3}{|}}{\underset{\underset{\displaystyle CH_3}{|}}{C}}-Cl$$

t-Amyl alcohol
(2-methyl-2-butanol)
MW 88.2, bp 102°
dens. 0.81 g/mL

Protonated Alcohol

3° Carbocation

t-Amyl chloride
(2-chloro-2-methylbutane)
MW 107, bp 85 – 86°
dens. 0.87 g/mL

It is also possible to convert a chloride to an alcohol (the reverse of the reaction above) by a similar mechanism in which chloride departs, leaving a carbocation that is attacked by water. The overall reaction in Experiment 11.2 is the hydrolysis of *t*-butyl chloride by this process.

Overall reaction:

$$CH_3-\overset{\overset{\displaystyle CH_3}{|}}{\underset{\underset{\displaystyle CH_3}{|}}{C}}-Cl + H_2O \longrightarrow CH_3-\overset{\overset{\displaystyle CH_3}{|}}{\underset{\underset{\displaystyle CH_3}{|}}{C}}-OH + HCl$$

t-Butyl chloride
(2-chloro-2-methylpropane)
MW 92.6, bp 51 – 52°
dens. 0.85 g/mL

t-Butyl alcohol
(2-methyl-2-propanol)
MW 74.1, mp 25 – 26
bp 83°, dens. 0.79 g/mL

The mechanism of this reaction is also S$_N$1, involving loss of chloride ion in the slow first step, leading to formation of a tertiary carbocation. Water attacks and bonds to this carbocation in a fast second step, and loss of H$^+$ yields the tertiary alcohol.

Mechanism:

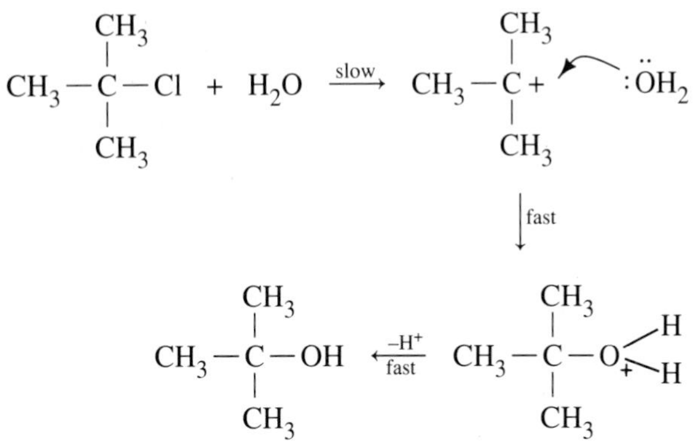

For this reaction you will study the kinetics by monitoring the rate of formation of HCl.

Kinetics reveals a great deal about the mechanisms by which reactions proceed. Kinetic studies tell us, for example, whether one or two molecules (or more) are involved in the rate-determining step of a reaction. Let's review first- and second-order kinetics briefly. If the rate-determining step is unimolecular, the rate depends only on the concentration of the single species undergoing the slow step, or

$$rate = k[A]$$

Where rate is measured in moles per liter per unit time, [A] is measured in moles per liter, and the first-order rate constant k has units of inverse time. If two species, A and B, were involved in the rate-determining step, the rate expression would be

$$rate = [A][B]$$

where k now has units of liters per mole per unit time (or inverse molarity, inverse time).

In a first-order reaction it is also true that the rate of reaction is the rate of disappearance of starting material, or

$$rate = \frac{-d[A]}{dt}$$

Combining this with the first equation gives

$$\frac{d[A]}{dt} = -k[A]$$

Rearranging yields

$$\frac{d[A]}{[A]} = -k\, dt$$

and integrating gives

$$\int \frac{d[A]}{[A]} = \int -k\, dt$$

or

$$\int \frac{dA}{[A]} = -\int k\, dt$$

which becomes

$$\ln[A] = -kt + C$$

Using the initial conditions, when $t = 0$ and $[A] = [A]_0$,

$$\ln[A]_0 = C$$

So replacing C with $\ln[A]_0$ gives

$$\ln[A] = -kt + \ln[A]_0$$

Rearranging, we have

$$\ln[A] - \ln[A]_0 = -kt$$

or

$$\ln\frac{[A]}{[A]_0} = -kt$$

Multiplying through by -1 gives

$$\ln\frac{[A]_0}{[A]} = kt$$

Since in this case $A = RX$ (the alkyl halide), the equation becomes

$$\ln\frac{[RX]_0}{[RX]} = kt$$

This useful equation states that a plot of the natural logarithm of the ratio of the initial concentration of RX to its concentration at time t versus time should give a straight line with a slope equal to k, the rate constant.

In this experiment you will follow the concentration of RX by the production of HCl during the reaction, since each mole of t-butyl chloride hydrolyzed gives rise to 1 mole of HCl. If this is the case, the rate should depend only on the concentration of the starting halide, t-butyl chloride, abbreviated RX. The rate of reaction will equal the rate of disappearance of RX, which is proportional to its concentration.

In Experiment 11.2, aliquots of NaOH are added to a solution of t-butyl chloride in water. As the halide undergoes hydrolysis, HCl is produced and NaOH is consumed. By having an indicator present, the time required for neutralization of the NaOH can be determined.

The ratio of the initial concentration of the halide to concentration remaining at time t is equal to the ratio of the volume of NaOH required to neutralize all the HCl produced (to infinite time) to the difference between the milliliters of NaOH required at infinite time and milliliters of NaOH used to time t.

$$\frac{[RX]_0}{[RX]} = \frac{(\text{mL NaOH})_\infty}{(\text{mL NaOH})_\infty - (\text{mL NaOH})_t}$$

Feeding this into an earlier equation yields

$$\ln \frac{(\text{mL NaOH})_\infty}{(\text{mL NaOH})_\infty - (\text{mL NaOH})_t} = kt$$

Thus a plot of the left-hand expression in the equation above versus t should yield a straight line with a slope of the first-order rate constant k.

QUESTIONS

1. What are the mechanism and kinetics equations for an S_N1 reaction?
2. What would be the product of reaction of t-butyl alcohol with HBr?
3. What would be the product of hydrolysis of 2-chloro-2-methylbutane?
4. Write the rate expression for the reaction in Question 3.
5. Which should occur faster: hydrolysis of t-butyl chloride or of isopropyl chloride, and why?

EXPERIMENT 11.1 CONVERSION OF *t*-AMYL ALCOHOL TO *t*-AMYL CHLORIDE USING HCL

Estimated Time:
2.0 hours

Prelab

1. Calculate the grams and milliliters corresponding to 100 mmol of t-amyl alcohol, and the theoretical yield of t-amyl chloride.

Special Hazards

Concentrated hydrochloric acid is extremely corrosive. Five percent aqueous NaOH is also caustic. Use caution and wash immediately with copious quantities of water if contact occurs. t-Amyl alcohol is toxic on ingestion; t-amyl chloride is an irritant. Both are flammable.

Procedure

In a 125-mL separatory funnel, place 2-methyl-2-butanol (t-amyl alcohol, 100 mmol, ___ g, ___ mL) and cautiously add concentrated (12 M) HCl (300 mmol, 25 mL). With the stopper off, swirl gently for 1 minute. Add the stopper, invert, and vent. Shake and vent intermittently to release pressure for several minutes. Allow the mixture to stand until two distinct clear layers are formed. Draw off the aqueous (lower) layer, wash the remaining organic layer with 5% aqueous NaOH, and dry it over anhydrous Na_2SO_4. Decant into a 50-mL round-bottomed flask, add boiling chips, and distill, collecting the product at 83–85°. Record the mass of t-amyl chloride obtained. To confirm the presence of a halide, carry out the Beilstein test by cleaning a piece of copper wire in a flame (well away from flammable solvents), dipping it in your product, and replacing it in the flame. A green color indicates the presence of a halide. For comparison test the starting material (*t*-amyl alcohol), too.

EXPERIMENT 11.2 KINETIC STUDY OF THE HYDROLYSIS OF *t*-BUTYL CHLORIDE

Estimated Time:
2.5 hours

Prelab

1. What information do you need in order to calculate a first-order rate constant?
2. Describe the effects of errors in timing, concentrations, and volumes.

Special Hazards

t-Butyl chloride is an irritant and both it and acetone are flammable.

Procedure

In a 50-mL Erlenmeyer flask, place 25 mL of 9:1 water–acetone solution. Add a magnetic stir bar and record the temperature of the solution. Clamp the flask securely over the center of a magnetic stirring plate and start stirring.

Now fill a plastic 5-mL syringe with 4.0 mL of 0.010 *M* NaOH solution. As an indicator add 2 drops of bromthymol blue solution to the flask, then add exactly 1.0 mL of 0.010 *M* NaOH from the syringe. The indicator turns blue, showing that the solution is basic. In another syringe place 1.0 mL of a solution of 0.050 *M* *t*-butyl chloride in acetone. Get ready to time the run, then add the *t*-butyl chloride solution to the flask quickly (within 1 second) and note the time. Watch the solution carefully while standing by with the syringe of NaOH solution. The instant the solution turns yellow, showing it to be acidic, record the time and add another 1.0 mL of base. Continue recording the time and adding another 1.0 mL of base each time the solution turns yellow until 4.0 mL of NaOH solution has been added and the solution has turned yellow again.

Repeat the entire procedure twice more to collect additional data and complete Table 11–1.

TABLE 11–1

	Time to turn yellow after addition:			
Run	1	2	3	4
1				
2				
3				
Average				

QUESTIONS

1. How reproducible are your results? Do any of the data points look unreliable? If any differ widely from the other runs, you may wish to exclude them from the average.
2. How many milliliters of the NaOH solution would be required to neutralize the HCl at infinite time?
3. Using the answer from Question 2 as (mL NaOH)$_\infty$, prepare a plot of

$$\ln \frac{(\text{mL NaOH})_\infty}{(\text{mL NaOH})_\infty - (\text{mL NaOH})_t}$$

versus time, using total time elapsed for each point. Calculate the slope k (using a least-squares plot if possible) and report it with the correct units and number of significant figures.

RADICAL HALOGENATION AND GAS CHROMATOGRAPHY

Overview

Radical reactions. Because alkanes are relatively unreactive there are only a few types of reactions commonly performed with them: halogenation, combustion, and thermal cracking or dehydrogenation. In a halogenation reaction a halogen (Cl or Br) replaces a hydrogen. This occurs in a series of steps where a halogen radical removes a hydrogen, leaving an alkyl radical, which then forms a bond with another halogen. To get a radical reaction going, we must have an initiation step. Sometimes this consists of shining light on a halogen molecule, for example:

$$\text{Cl}-\text{Cl} \;+\; h\nu \;\longrightarrow\; 2\text{Cl}\cdot$$

or

$$\text{Br}-\text{Br} \;+\; h\nu \;\longrightarrow\; 2\text{Br}$$

Note that a fishhook arrow used in a radical reaction represents movement of one electron.

Although starting a radical reaction with halogens and light is a useful industrial process, working with chlorine gas or molecular bromine in the laboratory is somewhat awkward and dangerous. There is a more convenient way of generating halogen radicals on a small scale, which we will use to avoid the necessity of handling toxic chlorine gas or liquid bromine. This alternative method employs a radical initiator and a potential radical source such as sulfuryl chloride (SO_2Cl_2) or N-bromosuccinimide (NBS). A radical initiator is any compound that gives radicals easily on heating. The ones used here are 2,2′-azobis(2-methylpropionitrile) [also known as 2,2′-azobis(isobutyronitrile) or AIBN] and benzoyl peroxide. These compounds decompose easily on heating. AIBN decomposes readily because it can give off nitrogen gas and form a resonance-stabilized radical:

$$CH_3-\overset{\overset{\displaystyle CN}{|}}{\underset{\underset{\displaystyle CH_3}{|}}{C}}-N=N-\overset{\overset{\displaystyle CN}{|}}{\underset{\underset{\displaystyle CH_3}{|}}{C}}-CH_3 \longrightarrow \;:N\equiv N: \;+\; 2\left[\overset{\ddot{N}}{\underset{\underset{CH_3 \quad CH_3}{}}{\overset{|||}{\underset{C \cdot}{C}}}} \longleftrightarrow \overset{\dot{\ddot{N}}}{\underset{\underset{CH_3 \quad CH_3}{}}{\overset{||}{\underset{C}{C}}}}\right]$$

AIBN
MW 164

This radical then abstracts a chlorine atom from SO_2Cl_2:

$$CH_3-\overset{\overset{\displaystyle CN}{|}}{\underset{\underset{\displaystyle CH_3}{|}}{C\cdot}} \;+\; Cl-\overset{\overset{\displaystyle O}{||}}{\underset{\underset{\displaystyle O}{||}}{S}}-Cl \longrightarrow CH_3-\overset{\overset{\displaystyle CN}{|}}{\underset{\underset{\displaystyle CH_3}{|}}{C}}-Cl \;+\; \cdot\overset{\overset{\displaystyle O}{||}}{\underset{\underset{\displaystyle O}{||}}{S}}-Cl$$

The new radical decomposes spontaneously to give a chlorine radical and SO_2 gas:

$$\cdot\overset{\overset{\displaystyle O}{||}}{\underset{\underset{\displaystyle O}{||}}{S}}-Cl \longrightarrow \overset{\overset{\displaystyle O}{||}}{\underset{\underset{\displaystyle O}{||}}{S}}: \;+\; Cl\cdot$$

The net result, then, of the use of the initiator AIBN and sulfuryl chloride is the creation of chlorine radicals without the use of Cl_2. The chlorine radicals then proceed through the propagation steps. Here are the propagation steps for a radical chlorination:

$$Cl\cdot \;+\; H-R \longrightarrow Cl-H \;+\; R\cdot$$

$$R\cdot \;+\; Cl-\overset{\overset{\displaystyle O}{||}}{\underset{\underset{\displaystyle O}{||}}{S}}-Cl \longrightarrow R-Cl \;+\; \cdot\overset{\overset{\displaystyle O}{||}}{\underset{\underset{\displaystyle O}{||}}{S}}-Cl \longrightarrow \;:SO_2 \;+\; Cl\cdot$$

In the first propagation step a halogen radical abstracts a hydrogen atom from the alkane. In the second step the alkyl radical abstracts chlorine from sulfuryl chloride, giving the alkyl halide and a radical that decomposes. Note that the chlorine radical is regenerated in the second propagation step, making this a chain reaction, in other words, one that can proceed hundreds or thousands of times until a termination step occurs.

Termination steps occur whenever two radicals combine, for example:

$$R\cdot \ + \ Cl\cdot \ \longrightarrow \ RCl$$

$$Cl\cdot \ + \ Cl\cdot \ \longrightarrow \ Cl_2$$

$$R\cdot \ + \ R\cdot \ \longrightarrow \ R-R$$

Looking at the overall sequence of reactions, you can see that for each mole of 1-chlorobutane that is chlorinated 1 mole of SO_2 gas and 1 mole of HCl gas are produced. Since both gases are very irritating and SO_2 produces sulfurous acid on contact with water (e.g., in eyes or lungs) according to the equation below, we must either work in the fume hood or use an efficient gas trap.

$$SO_2 \ + \ H_2O \ \longrightarrow \ H_2SO_3$$

The overall reaction is therefore

$ClCH_2CH_2CH_2CH_3 \quad \xrightarrow{SO_2Cl_2} \quad SO_2 \ + \ HCl \ +$

$Cl-\overset{\displaystyle H}{\underset{\displaystyle Cl}{C}}-CH_2CH_2CH_3$	1,1-Dichlorobutane, bp 114°
$+ \ ClCH_2\overset{\displaystyle Cl}{CH}-CH_2CH_3$	1,2-Dichlorobutane, bp 124°
$+ \ ClCH_2CH_2\overset{\displaystyle Cl}{CH}-CH_3$	1,3-Dichlorobutane, bp 134°
$+ \ ClCH_2CH_2CH_2CH_2Cl$	1,4-Dichlorobutane, bp 161°

1-Chlorobutane
MW 92.6
dens. 0.886 g/mL

Sulfuryl
chloride
MW 135
dens. 1.35 g/mL

Since we will obtain a mixture of products that are all reasonably volatile liquids, we can analyze the ratios of products obtained using gas chromatography, which will be described after a brief discussion of Experiment 10.2.

Another radical initiator is benzoyl peroxide, which can form two resonance-stabilized benzoyloxy radicals.

Benzoyl peroxide   Benzoyloxy radical   Phenyl radical

Benzoyl peroxide, incidentally, is the active ingredient in such acne-fighting medicines as Oxy-5. This stuff really blasts those little zit-forming bacteria.

Because of its high reactivity, benzoyl peroxide is explosive if overheated. In fact, any peroxide should not be heated on dry glassware. So be sure that there is always plenty of solvent present.

Once the radical formation has been initiated, a bromine radical is removed from NBS, leaving a succinimide radical that is resonance-stabilized.

The succinimide radical can remove a hydrogen radical from a benzylic or allylic position, such as that found on *m*-toluic acid.

The benzyl radical abstracts a bromine from another molecule of NBS to propagate the chain.

Thus the overall reaction for Experiment 12.2 is

m-Toluic acid	*N*-Bromo-succinimide	α-Bromo-*m*-toluic acid	Succinimide
MW 136	MW 178	MW 215	MW 99
mp 108–110°	mp 180–183°	mp 147–149°	mp 123–125°

Gas chromatography.   Gas chromatography (GC), also known as gas-liquid chromatography (GLC), separates compounds using the same principle as that used in thin-layer or column chromatography. That is, each compound spends part of the time attached to a stationary phase and part of the time in a mobile phase. Since different compounds have different affinities for the stationary and mobile phases, they move at different speeds.

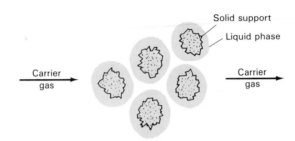

Figure 12-1 Highly Magnified View of GC Packing

In GC the mobile phase is an inert gas such as nitrogen or helium and the stationary phase is a powder of silica or ceramic material (called the support) coated with a thin layer of an oily liquid (called the liquid phase), as shown in Figure 12-1. As the carrier gas flows past the solid support coated with the liquid phase, the molecules spend a certain fraction of their time in the gas and a certain fraction of time dissolved in the liquid, so they move at a certain fraction of the speed of the carrier gas. Increasing the temperature causes the molecules to spend more time in the gas phase and therefore travel faster. A molecule that travels faster has a lower *retention time*, that is, time to pass through the column. Heating a column lowers the retention times for all the molecules.

A schematic diagram of a gas chromatograph appears in Figure 12-2. The sample (about 1 μL) is injected by syringe through a rubber septum into the heated injection port, where it vaporizes. The carrier gas then carries it through the column over the liquid phase, separating the components. The detector signals when a compound comes off the column and sends the signal to a recorder. There are several types of detectors. The most common type is a flame ionization detector (FID). When a compound passes over the flame, it pyrolyzes, causing ions to form. These ions bridge the gap between two electrodes and create a current, which is measured and recorded.

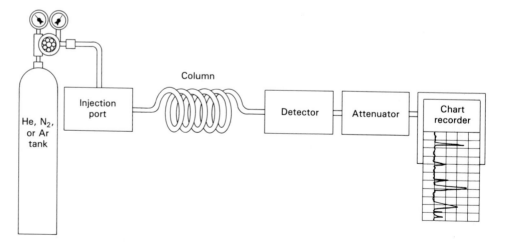

Figure 12-2 Schematic Diagram of Gas Chromatograph

APPLICATION: GAS CHROMATOGRAPHY AND ENVIRONMENTAL PROTECTION

The Environmental Protection Agency is responsible for monitoring environmental pollutants, stopping polluters, and helping clean up contaminated areas. One of the main tools of EPA and other environmental chemists is the gas chromatograph.

Samples from environmental sources (such as river water) often contain organic contaminants, many of which can be analyzed by GC. A GC trace of one sample of contaminated river water is shown below. Note that a capillary column (a very long, thin column with the liquid coating on its inner wall instead of a packing) gives much higher resolution than that of normal packed column (see table on next page).

By following certain well-defined procedures, technicians can use GC to test for the presence of many common pollutants, including volatile organics and pesticides. Careful standardization using calibration standards is essential for accurate quantitative results.

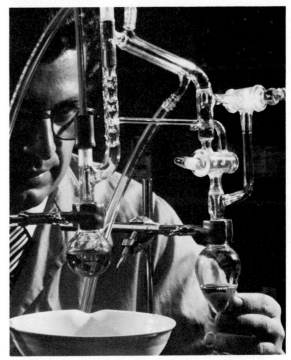

EPA Technician Prepares a Sample for Distillation (courtesy US EPA)

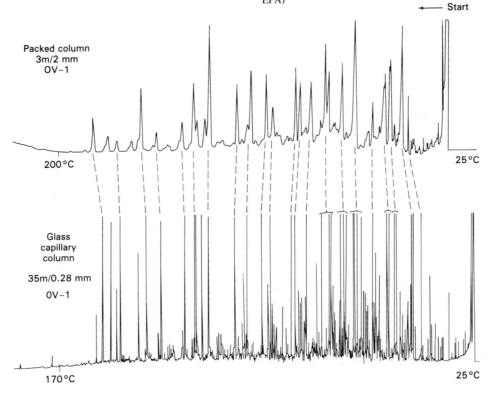

GC Trace of Polluted Water

COLUMN CHARACTERISTICS AND CHROMATOGRAPHIC CONDITIONS

	Packed	Capillary (narrow bore)
Length	3 m	35 m
Inner diameter	2 mm	0.28 mm
Coating	3% OV-1 on Gaschrom Q, 80–100 mesh	OV-1 film 1.0×10^{-5} cm
Liquid-phase load	120 mg	1.5 mg
Carrier gas flow	He, 10–22 mL/min (progr.)	H_2, 2.2 mL/min
Temperature range	50–200°	25–170°
Temperature program rate	2°/min	3.5°/min
Analysis time	90 min	45 min
Sample size	0.15 μL direct	0.6 μL splitting 1:25
Attenuation	× 32	× 4
Number of peaks	118	490

The liquid phase you will be using separates compounds according to their boiling points, with the lowest-boiling components coming off first. This is because we use a nonpolar coating (silicone grease) that does not bond strongly with the compounds. Other columns with more polar liquid phases separate on the basis of polarity, with the most polar components coming off last, since they bind to the column more strongly. Check with your instructor to find out whether the column packing you use is nonpolar or polar.

A typical GC trace appears in Figure 12–3. The area under each peak is proportional to the amount of compound present. Therefore, we can estimate the quantities of each component present by comparing the areas under the peaks. One easy way to find

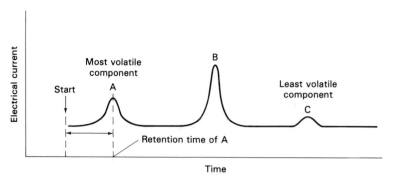

Figure 12-3   A Gas Chromatogram

the area under the peak is to draw a triangle approximating the peak and use the formula *area* = 1/2 *base* × *height*. Another way is to cut out and weigh the peaks. More sophisticated recorders do the integration of area under the peaks electronically.

The START signals the time of injection. The retention time of a compound is measured as the time (or distance) between the start and the top of the compound peak.

A gas chromatograph can be hooked up to a mass spectrometer in the setup known as a GC-MS. In this arrangement as each peak comes out of the GC, part of the material is diverted to the mass spectrometer and its mass spectrum is obtained. This serves to identify each component coming off the column. Many GC-MS instruments allow for computerized matching of mass spectra with an internal library of 40,000 or more known spectra to make the identification easier.

QUESTIONS

1. Write out initiation, propagation, and termination steps for the monobromination of ethane using Br_2.
2. How many different products are formed in the monochlorination of methylcyclopentane? Draw each one.
3. Predict the percentage of each of these products if the relative reactivities for hydrogens are $1°:2°:3° = 1:3.5:5.0$.

Concept Map

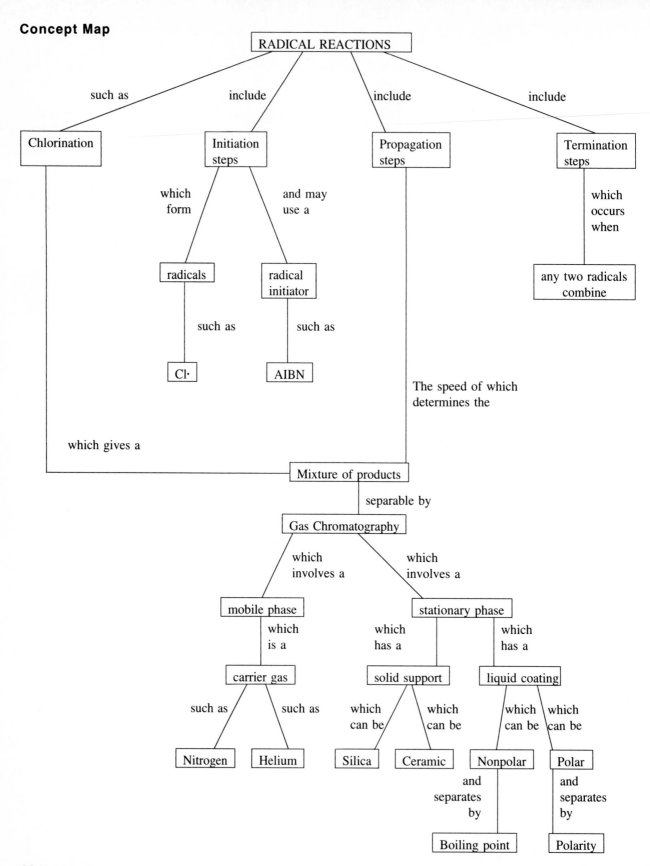

EXPERIMENT 12.1: CHLORINATION OF 1-CHLOROBUTANE

Estimated Time:
2.5 hours

Special Hazards

Sulfuryl chloride is a very irritating and toxic liquid. It should be handled only in the hood, using gloves. On contacting water it reacts violently, forming HCl and H_2SO_4. Sulfur dioxide gas (SO_2), which is produced in this experiment, is very irritating and has an unpleasant, penetrating odor. If you use a trap, make sure that it is airtight and functioning properly before heating the reaction mixture.

Procedure

To a 50-mL round-bottomed flask in the fume hood, add 1-chlorobutane (10 mL, 96 mmol), sulfuryl chloride (4.5 mL, 45 mmol), 2,2'-azobis(2-methylpropionitrile) (0.1 g, 0.6 mmol), and a boiling chip. Add a water-cooled condenser and heating mantle. If there is room for everyone to work in the fume hood, no traps are needed, but if you are working outside the hood you will need to set up a gas trap for the evolved SO_2 as shown in Figure 12–4. If you use a trap, read and follow carefully the directions given in Experiment 8.1 on setting up the trap.

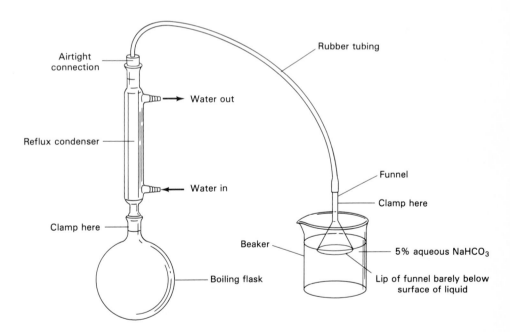

Figure 12-4   Reflux Apparatus with Gas Trap

Heat the mixture to gentle reflux for 30 minutes. Cool the flask in an ice bath, then pour the contents through a funnel into a separatory funnel containing 25 mL of water, shake and vent, and separate the two phases. The organic phase may be either the upper or the lower layer, depending on the relative amounts of starting material (1-chlorobutane, with density less than water) and products (dichlorobutanes, with densities greater than water). The organic layer should have the smaller volume. If in doubt, test by adding a few drops of water and seeing which layer it goes to. Be sure to save all layers until your final product is in hand.

To remove any remaining HCl, wash the organic phase with one 25-mL portion of 5% aqueous $NaHCO_3$. Washing, of course, means adding another solution to dissolve unwanted materials, shaking and venting, then discarding the unwanted layer. In this case, the aqueous layer is unwanted, since the product remains in the organic phase. After separating the layers, dry the organic layer with anhydrous $CaCl_2$. When it appears dry (the cloudiness is gone and the $CaCl_2$ does not all clump together) filter it into a dry tared vial and weigh it.

Gas chromatography. To analyze the product, take it over to your instructor at the gas chromatograph (GC). The temperature of the GC column should be approximately 90–100°, slightly lower than the boiling points of the products you want to separate. Your instructor will inject a 1-μL sample into the machine. Since the GC will be using a nonpolar column which separates according to boiling points, the first peak will represent leftover starting material (1-chlorobutane), followed by the products in order of their boiling points.

The starting material peak should be allowed to run off-scale so that the product peaks are large. The first major peak after the starting material should be 1,1-dichlorobutane (bp 114°), followed by the 1,2 (bp 124°), 1,3 (bp 134°), and finally the 1,4 (bp 161°) isomers. Note that the peaks for 1,1-, 1,2-, and 1,3-dichlorobutanes are very close, while the distance between 1,3- and 1,4- is farther. This is because there is a greater difference in boiling points between 1,3 and 1,4 than between the other isomers.

When the last peak has come out, be sure that all peaks are labeled, cut off the paper, and save it in your notebook for calculating the percentages of each isomer. When finished, dispose of your product mixture in the waste container as instructed.

QUESTIONS

1. For each peak estimate the area by drawing the best triangle you can and using the formula area = 1/2 base × height, with distances measured in millimeters. Or you may find the area by multiplying the *width at half-height* by the height of the peak. Width at half-height should equal 1/2 the base length. Compare the areas of the triangles to get a percentage of the total product for each isomer.

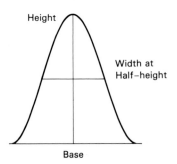

2. Find the relative reactivity for each type of hydrogen by dividing the percentage of each by the statistical factor (the number of hydrogens that could have given that product). For exam-

ple, the number of hydrogens whose replacement leads to 1,1-dichlorobutane is two. Since a normal methyl hydrogen is usually taken as the standard of comparison (with a relative reactivity of 1.0), divide each of your reactivities by the reactivity of the methyl group hydrogens to get relative reactivities.

3. Using these relative reactivities, predict the percentage of each product possible in the radical chlorination of 1-chloropropane.

4. What reaction occurs between SO_2, H_2O, and $NaHCO_3$ in the gas trap?

EXPERIMENT 12.2 BROMINATION OF *m*-TOLUIC ACID USING NBS

Estimated Time:
2.5 hours

Prelab

1. Calculate the masses of 10.0 mmol of *m*-toluic acid and 10.1 mmol of NBS.
2. What is the theoretical yield of α-bromo-*m*-toluic acid?

Special Hazards

N-Bromosuccinimide is an irritant. Benzoyl peroxide is a strong oxidizer and can detonate if overheated on dry glassware. Carbon tetrachloride is highly toxic and a cancer suspect agent; avoid exposure and use only in the fume hood. The product (α-bromo-*m*-toluic acid) is a lachrymator, though not highly volatile.

Procedure

In a 50-mL round-bottomed flask in the fume hood, place *m*-toluic acid (10.0 mmol, __ g) and *N*-bromosuccinimide (10.1 mmol, __ g). To this mixture carefully add benzoyl peroxide (0.10 g) so that none of it remains on the ground-glass joint. Add carbon tetrachloride (20 mL; *use caution*) so as to rinse down any benzoyl peroxide particles on the joint. Add a condenser, heating mantle, and variable transformer, check that water flow through the condenser is moderate, and reflux for 1 hour with occasional swirling. During this time the solution takes on a red-orange color due to the presence of a small quantity of molecular bromine (Br_2) formed by combination of bromine radicals. Cool the flask in an ice bath and collect the precipitate by suction filtration in the fume hood. Wash the product with a little hexane (to remove CCl_4 and Br_2) then with water (to dissolve the succinimide, which has also precipitated).

Recrystallize from 1:1 methanol–water (approximately 5 mL). Avoid prolonged heating to prevent hydrolysis of the benzylic bromide. Record the mass and melting point (lit. mp 147–149°).

REFERENCE

TULEEN, D. L., and B. A. HESS, Jr., ''Free-Radical Bromination of *p*-Toluic Acid,'' *Journal of Chemical Education* (1971) *48*, 476.

INFRARED SPECTROSCOPY

Overview

The basic idea behind infrared (IR) spectroscopy is very simple. As you know, every molecule contains bonds, and every bond acts like a spring. So if we shine the right wavelength of light on a sample some of the energy will be absorbed by the "springy" bonds, which then stretch and vibrate. Actually, there are several ways in which atoms in molecules can move. A bond undergoes various types of stretching and bending motions, as shown below.

Some vibrational modes for the methylene group

Stretching vibrations

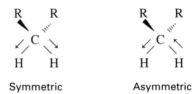

Symmetric   Asymmetric

Bending vibrations

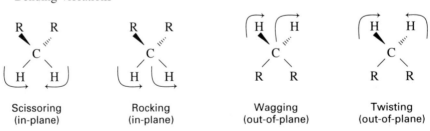

Scissoring   Rocking   Wagging   Twisting
(in-plane)   (in-plane)   (out-of-plane)   (out-of-plane)

Each motion will have a characteristic energy and will absorb a specific wavelength of light. In the IR spectrum we are looking mostly at stretching and bending modes, so you will sometimes see peaks labeled as C—H stretch, C—C bend, and so on.

Since light is an electromagnetic wave, it has the properties that all waves have: wavelength (abbreviated λ, lambda) frequency (abbreviated ν, nu), and speed (c).

Wavelength is measured in meters, centimeters, micrometers (μm, also called microns), or nanometers (nm), and the infrared region ranges from 2.5 to 15 μm. *Frequency* (ν, not v) is the number of cycles per second, measured in hertz (Hz), and 1 Hz = 1 cycle per second (cps). Frequency in the infrared ranges from 2×10^{13} to 1.2×10^{14} Hz.

Another way to describe waves is by *wavenumber*, which tells how many waves there are in 1 cm. The units of wavenumber are cm^{-1} (per centimeter or inverse centimeters, also called wavenumbers). This is the most common method of reporting IR peaks. On an infrared spectrum the wavenumber commonly ranges from 4000 to 667 cm^{-1}. To find the wavenumber corresponding to a certain wavelength light, just take 1 cm and divide by the wavelength in centimeters. For example, if light has a wavelength of 8.60 μm = 8.60×10^{-6} m = 8.60×10^{-4} cm, the wavenumber would be $1/8.60 \times 10^{-4}$ cm = 1160 cm^{-1}.

Getting back to frequency and wavelength, the product of the length of each wave times the number of waves passing per second gives the speed of propagation:

$$\lambda\nu = c$$

where c is the speed of light, approximately 3.00×10^8 meters per second. Therefore, if we know either the wavelength or frequency of light, we can easily calculate the other.

Now the energy of light also depends on its wavelength and frequency. Just like at the ocean, if you have short, choppy waves (low wavelength, high frequency, high wavenumber) they contain more energy than similarly sized gentle swells (high wavelength, low frequency, low wavenumber). The equation relating energy to frequency or wavelength, as discovered by Max Planck, is

$$E = h\nu = \frac{hc}{\lambda} = hc\left(\frac{1}{\lambda}\right)$$

where h is Planck's constant, with a value of 6.63×10^{-34} joule-second. So we can calculate the energy of light from its frequency, wavelength, or wavenumber (which *is* $1/\lambda$; no need to invert). You can see from the formula that the energy of a light wave is directly proportional to its wavenumber; the higher the wavenumber, the higher the energy. A bond absorbing at 1720 cm^{-1} takes higher energy light to stretch it, so it is a stronger bond than one absorbing at 1680 cm^{-1}, assuming that they are stretching in the same mode.

Now for the good news. You do not need any of these calculations to interpret IR spectra. Neither do you need to know how the IR spectrophotometer works. All you need to know is which functional groups absorb at which wavenumbers. But let's talk about the ''guts'' of the machine for a moment for the sake of completeness, before we go on to interpreting spectra. Schematically, the IR spectrometer consists of five parts, as outlined here.

The light source is an inert solid (such as silicon carbide, nichrome wire, or rare-earth oxides) heated to produce ''black-body'' radiation (i.e., a broad range of wavelengths). These light rays pass through a prism or diffraction grating turned at an angle, so that a particular wavelength passes through a slit to the sample. The prism or grating (known as the monochromator) can rotate to vary the wavelength. The sample, if it is a liquid, can be smeared between two salt plates (polished slabs of rock salt, NaCl). This is the easiest way to take an IR spectrum, because salt plates are transparent to infrared (and visible) light. They are expensive, though, and easily damaged by chipping or contact with water from fingerprints or wet samples.

For a solid the two most common ways to prepare a sample are by grinding it up to a fine paste with a minimum of mineral oil (Nujol) to make a Nujol mull, which is then spread on salt plates like a liquid, or by grinding it up with dry KBr and then squeezing the heck out of it in a special press, which melts the KBr and forms a transparent window containing the sample. If the solid is low-melting, it can be melted between two salt plates using a heat lamp. Both solids and liquids can be dissolved and placed in solution cells, which are so expensive they are often not available to students. Solution cells also have the disadvantage that most solvents absorb in the IR region, so we need to stick to solvents that do not absorb except in the fingerprint region, such as chloroform or methylene chloride.

After the light has passed through the sample, the detector compares it electronically with a reference beam that has not gone through the sample. By comparing the two beams the instrument can tell what fraction of the light has been absorbed, and it sends a signal to the recorder, which moves a pen to record the downward-pointing peaks.

A diagram of a Perkin-Elmer model 710 infrared spectrometer appears in Figure 13–1.

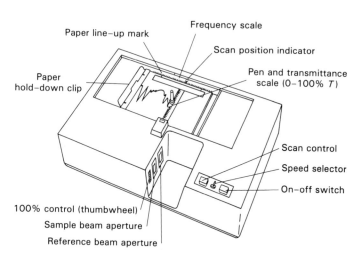

Figure 13-1 Perkin-Elmer Model 710 Infrared Spectrometer (courtesy of Perkin-Elmer Instruments)

Now we are ready to look at some IR spectra. A typical spectrum looks something like the one shown in Figure 13–2. Peaks (coming down from the top) represent light absorbed, so in the figure we would say that we have peaks at 3400, 3100, 2950, and 1640 cm^{-1}, as well as some peaks in the "fingerprint" region below 1400 cm^{-1}.

To interpret the peaks we must either memorize or have at hand a list of where the common functional groups absorb. There are all kinds of lists of IR peaks available, most of which are too detailed and mind-boggling for the beginner. A "quick-and-dirty" list of IR peaks good enough for most purposes is shown in Table 13–1.

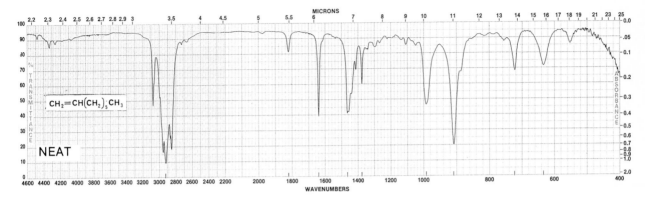

Figure 13-2   IR Spectrum of 1-Octene (all IR and NMR Spectra courtesy of Aldrich Chemical Co.)

TABLE 13–1   INFRARED PEAK POSITIONS OF COMMON FUNCTIONAL GROUPS

Peak position $(cm^{-1})^a$	Group
3300–3600 broad, m to s	OH or NH (if two peaks, NH_2); note that NH peaks are generally smaller and not as broad as OH peaks
2900–3000 w, m, or s	C — H, always present, no information
2100–2250 w to m	$C \equiv C$ or $C \equiv N$
1670–1800 s	$C = O$
1630–40 m	$C = C$
1600–1450 three or four peaks, m to s	Aromatic ring
1370–85 two sharp peaks, m	Isopropyl group
1050–1150 s	C — O
700–900 m to s	Peaks here give substitution pattern on aromatic rings (see Table 13–2)

^{a}w, Weak; m, medium; s, strong.

Table 13-2 lists some characteristic peaks of substituted benzenes.

TABLE 13–2   C — H BENDING ABSORPTION OF SUBSTITUTED BENZENES

Substitution pattern	Appearance	Position of absorption (cm^{-1})
Monosubstituted	Two peaks	690–710, 730–770
o-Disubstituted	One peak^a	735–770
m-Disubstituted	Two peaks	690–710, 810–850
p-Disubstituted	One peak	800–860

^{a}An additional band is often observed at about 680 cm^{-1}.

Ortho, abbreviated *o-*, means a 1,2 relationship (adjacent positions) on a benzene ring. For example, the following aromatic compounds are ortho-disubstituted:

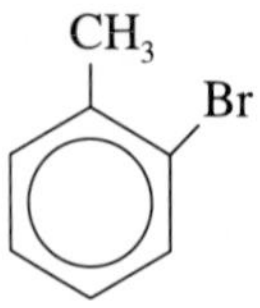

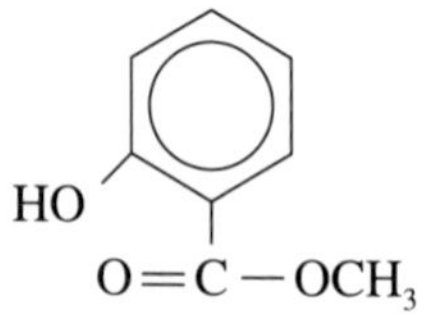

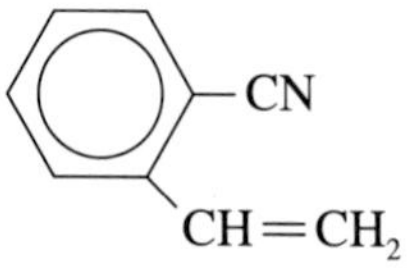

Similarly, *meta* (abbreviated *m-*) means a 1,3 relationship.

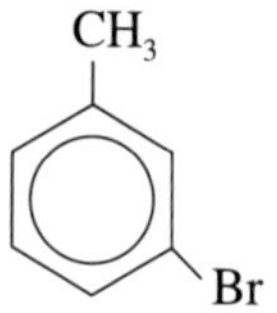

Para, abbreviated *p-*, means a 1,4 relationship. This means that the two substituents occupy opposite positions on the benzene ring:

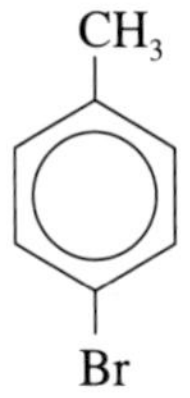

Evidence for aromatic substitution patterns also appears in the region 2000–1667 cm^{-1}. Bands due to C—H bending vibrations appear in this range. Since the exact position and intensity of these bands depends on the number of *adjacent* hydrogen atoms bound to an aromatic nucleus, a characteristic pattern appears for each substitution pattern. Bear in mind that these peaks are small ones; they are enlarged in Figure 13–3 for clarity.

It would be a good idea to memorize the peaks for common functional groups, since it will save a lot of time in searching through tables. There are only eight peaks to learn and you can make flash cards easily if desired. For now let's assume that you have this list available while working problems. As a practical matter, it might be a good idea to find out what kind of table your instructor will give you on the exams, and practice with that list as well.

After the peaks in the first table above have been mastered, you may want to learn the absorptions of different types of carbonyl compounds, since the carbonyl is the most versatile group in organic synthesis and appears frequently. These values are shown in Table 13–3 (p. 144).

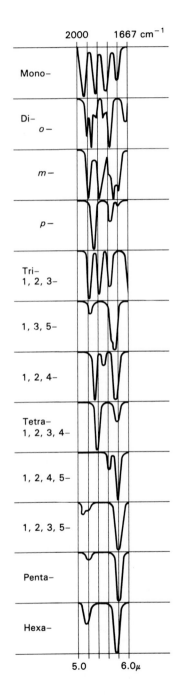

Figure 13-3 Typical Absorption Patterns of Substituted Aromatics in the 2000 to 1670 cm^{-1} Region

TABLE 13–3 INFRARED PEAK POSITIONS FOR CARBONYL GROUPS

Type of carbonyl compound	Peak position (cm^{-1})
Amides	1640–1690
Aldehydes and ketones	1710–1720
Esters	1735
Carboxylic acids	1710–1760
Acid halides	Near 1800
Acid anhydrides, two peaks	1760 and 1820

Note: The values in Table 13–3 refer to unconjugated carbonyl groups. Conjugation lowers the peak wavenumber by approximately 10–30 cm^{-1}. Conjugated means separated from another multiple bond by one single bond. For example, the following carbonyl groups are conjugated:

For comparison, the following carbonyl groups are unconjugated:

There are five common methods of preparing samples for IR spectra, which are described in turn below.

1. *Neat liquid.* This is not neat in the sense of "neat-o-keen-o" (although some spectra can be pretty exciting). Rather, it means a *pure* liquid. This is the easiest sample to prepare, because it just involves placing 1–3 drops of the liquid sample on the face of a salt plate, placing another salt plate on top, and squeezing them together to form a thin, uniform film. They should be squeezed using a Kimwipe or screw-type holder, since fingers should never touch the face of the salt plate (they cloud the surface). Also, remember that water or wet samples should never contact salt plates. When done they should be wiped off carefully, rinsed with anhydrous CH_2Cl_2 or $CHCl_3$ or CH_3OH, and stored inside their metal box in the desiccator. Since this is the easiest method for preparing a liquid sample, this is the one we will use in this introductory experiment.

2. *Melt.* In the case of a solid that melts at a low temperature (below $100°$ or so) a melt can often be prepared by placing some of the solid between two salt plates

and then heating either with a heat lamp or by placing the "sandwich" in a drying oven. This melts the solid to form a thin, even layer which solidifies on cooling but is transparent enough to take an IR spectrum.

3. *Nujol mull.* To prepare a mull, a sample of a solid is ground to a fine powder with a small mortar and pestle. A minimum of paraffin oil (one brand of which is Nujol) is blended in to form a paste, which is then pressed between two salt plates to form a translucent layer. When interpreting the spectra it is necessary to ignore the peaks due to the oil.

4. *KBr pellet.* A common way of taking spectra of solids is to grind them up very finely with *dry* KBr (about 100:1 KBr:compound by weight) and then press the resulting powder into a transparent "window," known as a KBr pellet. The way to do this is either in a small press requiring two wrenches or by using a hydraulic jack. The difficulty with KBr pellets is that it often takes several tries to get a good one that is transparent enough and does not crumble. When using the small pellet maker with two wrenches (or one wrench and a vise), be sure to squeeze hard without jarring and to loosen the bolts gently. The entire press, including the KBr pellet, is then placed in the sample holder of the IR spectrometer.

5. *Solution.* A liquid or solid may be dissolved in a solvent having few peaks in the IR, for example $CHCl_3$ or CCl_4, and placed in a solution cell which consists of two salt plates with a gap in between. Since not very many solvents work in these cells and they are more expensive than the sample holders used in the other methods, we will not use solution cells in this lab.

QUESTIONS

1. Draw the predicted infrared spectra for the following molecules, showing only the most important peaks.

(a)   (b)

(c)   (d)

2. A compound of formula C_4H_9NO shows bands in the IR at 3300–3500 cm^{-1} (two peaks), 1680 cm^{-1}, and 1380 cm^{-1} (two peaks). What is it?

Concept Map

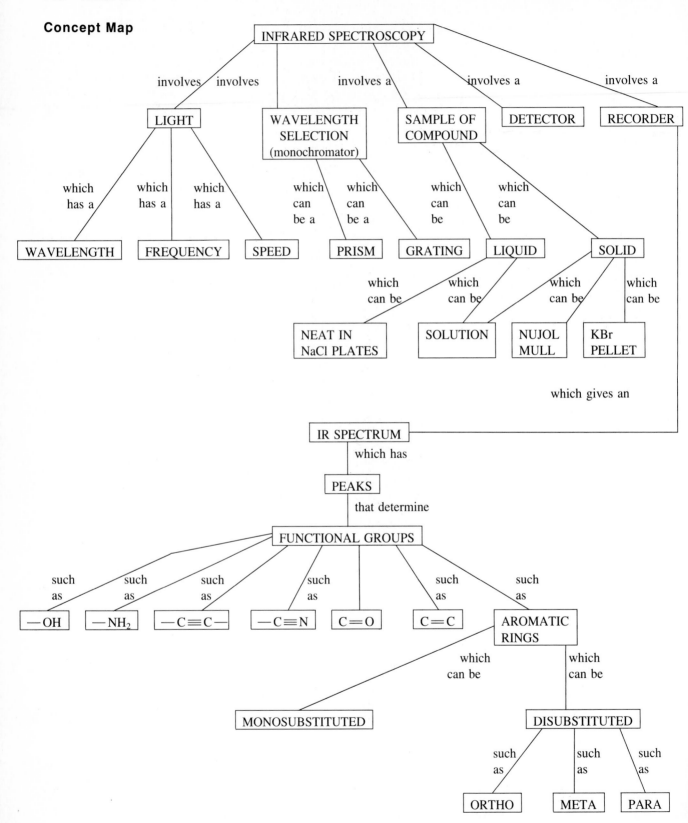

EXPERIMENT 13.1 UNKNOWN DETERMINATION BY IR SPECTROSCOPY

Estimated Time:
1.0 hour

Special Hazards

Probably the main hazard in this lab is the instructor's wrath if you damage the salt plates or the IR spectrometer. As always, treat unknown compounds as toxic and flammable.

Procedure

Pick a liquid unknown, then record its number and the molecular formula given on the bottle. Form groups of four students; each group will share one pair of salt plates. Determine the order in which you will run the samples; then the first person should lay a Kimwipe on the benchtop and set the two salt plates out on it. *Remember never to handle them on the faces or expose them to water*—it clouds the surface and makes them unusable. If you need to touch them, pick them up firmly by the edges. Handle them only over a nearby soft surface, such as a Kimwipe on a lab bench. Salt plates are very brittle and, if dropped, crack easily. Each plate costs more than you might expect, and you may be held responsible for their replacement if damaged.

Leaving the two plates lying on the Kimwipe, using an eyedropper, place 2–3 drops of the unknown on top of one plate. Place the other plate on top and press down gently using another Kimwipe to get a thin uniform film between the plates. Wipe off any excess sample from around the edges. Mount the plates in a screw-type holder, being careful not to tighten too much. When ready, carry your mounted sample carefully over to the IR machine.

Although it takes a long time to describe running the spectrum, the actual process takes about 2 minutes. Preparing the sample takes about 1 minute, as does cleaning the salt plates afterwards. There should be plenty of time for everyone to run a spectrum.

Instructions for running the IR spectrometer

1. If you are the first person to use it, make sure that it is *on* so that it can warm up for at least 5 minutes before running the spectrum.
2. Slide the sample holder into the sample beam. The sample beam is the beam nearest you; the one farther away is the reference beam.
3. Check that a new sheet of the correct paper is on the instrument.
4. Take the pen out of its storage container, check that it works, and place it in the machine.
5. Turn the dial to *reset*, and, if it is a drum-type instrument, rotate the drum gently until the pen is at 4000 cm^{-1}. (*Note:* The drum must never be turned by hand except when the dial is on *reset*. Never "force" the drum; this throws off the calibration. For instruments without drums, reset the pen at 4000 cm^{-1} by moving the sliding bar across.)
6. Adjust the 100% knob and/or the comb in the reference beam to get approximately 80–90% transmission to start with. The purpose of a comb (shown in Figure 13–4) is to cut down the light intensity in the reference beam so that it can better match the intensity of the sample beam.

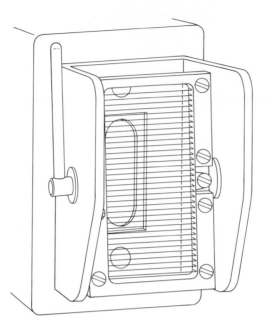

Figure 13-4 Variable Reference Beam Attenuator (Comb)

7. Turn the dial to *Fast*, make sure that the pen is down, and enjoy watching your spectrum being drawn.
8. When the spectrum is finished (at the end of the paper) the machine will stop automatically. Turn the dial to *stop*, take off the paper with your spectrum, and insert a fresh piece of paper for the next user. Turn the dial to *reset* and reset the spectrometer to 4000 cm^{-1}.
9. If someone else will be using the machine within a few minutes, you are done. If you are the last user, remove the pen and place it upright in its storage container. Turn the machine off and pull the dust cover over it. By following these rules you and your co-workers will enjoy an instrument that is reliable, easy to use, and requires little maintenance.

If your section is small, at the instructor's option you may want to calibrate each spectrum using a polystyrene film. Polystyrene has a very sharp peak at 1603 cm^{-1}, so before removing your spectrum from the instrument you can replace your sample with a polystyrene calibration film, set the dial on *reset* and find this large, sharp peak. Once you have found it, run only this one peak on top of your spectrum by lowering the pen as it approaches the peak and raising it immediately afterward. The position of this calibration peak on the paper will tell you whether your peaks are shifted to the left or right on the paper.

When you are through running the spectrum, immediately fill in your name, date, sample number, and phase (neat, NaCl plates). Remove the sample holder, take it back to your desk, carefully remove the salt plates (handling only by the edges), rinse them with CH_2Cl_2 in the frame hood, wipe with a clean Kimwipe, and give them to the next person in your group.

QUESTIONS

1. Make a list of the major peaks in your unknown spectrum and give an interpretation for each.
2. How many degrees of unsaturation does your formula dictate? Have you found them all?
3. Propose one or more structures consistent with the IR spectrum and formula.
4. Look up the *Aldrich Library of FT-IR Spectra* or the *Aldrich Library of Infrared Spectra,* both by C. J. Pouchert, in the reference section of the library. Check the formula index and examine the spectra to see if you can match up the fingerprint region. If you can match the spectrum, photocopy the literature spectrum and staple it to your own.
5. Give the name and structure of your unknown. Write the unknown number and formula next to it so that the instructor can check it easily.

SECTION 14

STRUCTURE DETERMINATION PRACTICE

Overview

Throughout the course of your study of organic chemistry you will frequently encounter questions of structure determination. These can be thought of as questions in a detective game, where you are asked to look at all the clues and to come up with the correct chemical structure.

Students sometimes have difficulty with this part of the course, primarily because they do not get enough practice. The purpose of this learning lab is to improve your skills by having a thorough discussion by the instructor followed by practice with many examples.

It is easy to feel overwhelmed when first presented with spectroscopy because of the large amount of information to be learned in a short time. Therefore, it is important to cut out unnecessary information and take the remainder in small, easily digested segments.

In this section we present a variety of practice problems involving all the common functional groups. It is particularly suitable for one 3–hour laboratory meeting, and can also be used in discussion sections.

The key to success in working spectroscopy questions is practice, practice, and more practice. So it is very important to work a lot of examples, preferably three or four times each. It is a good idea to leave at least an hour before reworking the same problem, because we want the ability and knowledge to go into *long-term* memory, not short-term.

To develop a reasonable ability requires working *at least a hundred* examples, and this is where many textbooks fall short. They may provide only 20 or 30 examples, which are not enough. Spending some additional time with this section can save you a lot of frustration and help make your encounters with spectroscopy pleasant experiences instead of disasters.

The most important rules in working these problems are:
1. Keep all the information *organized*.
2. Proceed logically from the data with the *least* information to that with the *most*.
3. Keep in mind those atoms and functional groups that have been accounted for and those that have not yet been found.

To keep all the information organized, it is essential to *outline* the question. This is particularly important in a ''road map'' or ''cowtrail'' question involving more than one compound. For example if a question states that compound *A*, of known formula, on heating with acid gave compound *B*, also of known formula, and infrared (IR) and hydrogen nuclear magnetic resonance (NMR) spectra are given for *A* and *B*, the outline would look something like this:

$$A \xrightarrow[\text{heat}]{\text{H}^+} B$$

Formula of *A*	Formula of *B*
List of IR peaks of *A*	List of IR peaks of *B*
List of NMR peaks of *A*	List of NMR peaks of *B*

Note that a lot of space needs to be left in the outline, because we are going to go back and put an interpretation next to each piece of data. For example, each peak listed should have enough room next to it and below it to put a partial structure deduced from that peak. Results of chemical tests (for example, Br_2/CCl_4), mass spectral (MS) or ultraviolet (UV) data, should go after the formula and before IR or NMR data.

Sometimes a complete IR or NMR spectrum will be given, in which case you can do your interpretations on or near the printed spectrum. This saves time in case you are rushed, but in general it is still a good idea to write out the list of peaks in numerical form before trying to interpret them.

Both in writing the outline and in going back and interpreting the data, we want to go from the *least* amount of information to the *most*. The approximate order should be:

Formula	(least information)
Ultraviolet spectrum	
Mass spectrum	
Infrared spectrum	
NMR spectrum	(most information)

In this way, by the time you get to the NMR spectrum you have a pretty good idea what the compound is, or at least some of its pieces. This makes the NMR interpretation much easier and allows you to nail down the structure using the NMR.

Let's talk first about the information gained from the formula. If the formula is given or can be deduced from the elemental analysis or mass spectrum, the number of degrees of unsaturation is the first thing to determine. Each degree of unsaturation (abbreviated °unsat) means a multiple bond or ring in the molecule. Another way to look at it is that each degree of unsaturation means that two hydrogens are missing from the saturated formula. Saturated, of course, means open chain with no multiple bonds. It is easy to see that these are really the same definition because if you imagine a straight-chain hydrocarbon and want to form a ring or a double bond, you would have to remove two hydrogens. For example, to start conceptually with *n*-pentane (C_5H_{12}) and make cyclopentane (C_5H_{10}), we would have to remove the two end hydrogens, as shown here.

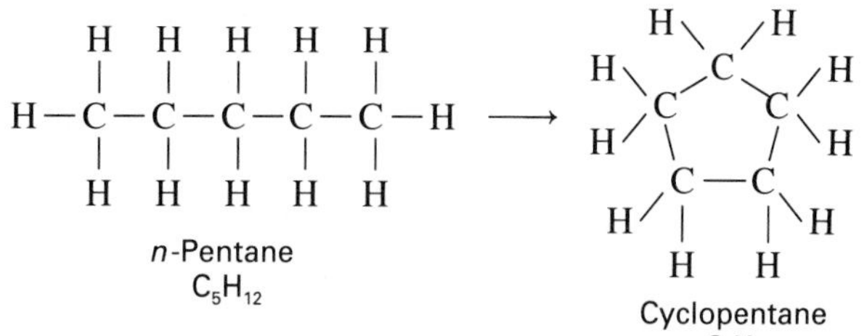

n-Pentane
C_5H_{12}

Cyclopentane
C_5H_{10}

Similarly, if we wanted to make one of the pentenes from *n*-pentane, we would have to remove a pair of adjacent hydrogens.

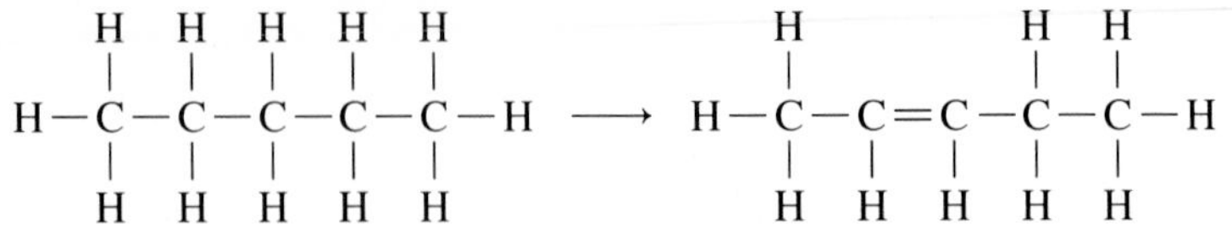

$$
\begin{array}{ccccc}
\text{H} & \text{H} & \text{H} & \text{H} & \text{H} \\
| & | & | & | & | \\
\text{H}-\text{C}-\text{C}-\text{C}-\text{C}-\text{C}-\text{H} \\
| & | & | & | & | \\
\text{H} & \text{H} & \text{H} & \text{H} & \text{H}
\end{array}
\longrightarrow
\begin{array}{ccccc}
\text{H} & & & \text{H} & \text{H} \\
| & & & | & | \\
\text{H}-\text{C}-\text{C}=\text{C}-\text{C}-\text{C}-\text{H} \\
| & | & | & | & | \\
\text{H} & \text{H} & \text{H} & \text{H} & \text{H}
\end{array}
$$

2-Pentene
C_5H_{10}

So to find the degrees of unsaturation from a formula, we need to compare that formula to the *saturated hydrocarbon* with the same number of carbons (formula C_nH_{2n+2}) and see how many pairs of hydrogens are missing. In the example above we would compare the formula C_5H_{10} to the saturated formula for five carbons, C_5H_{12}, and see that both cyclopentane and 2–pentene are missing one pair of hydrogens, for 1° of unsaturation. We could also observe that the formula C_nH_{2n} represents one degree of unsaturation, since it is missing one pair of hydrogens compared to the saturated formula C_nH_{2n+2}. Similarly, the formula C_nH_{2n-2} represents two degrees of unsaturation (missing two pairs of hydrogens), and so on.

So, for example, how many degrees of unsaturation are there in a compound of formula C_5H_8? Answer this yourself before reading the next paragraph. Answer: _____

Since, compared to C_5H_{12}, C_5H_8 is missing four hydrogens and follows the formula C_nH_{2n-2}, we know that it has 2° of unsaturation. This means that it could have two double bonds, two rings, one double bond and one ring, or a triple bond. One isomer for each of these possibilities is shown here:

$$CH_2{=}CH{-}CH{=}CH{-}CH_3$$

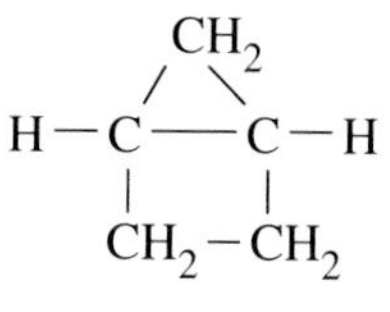

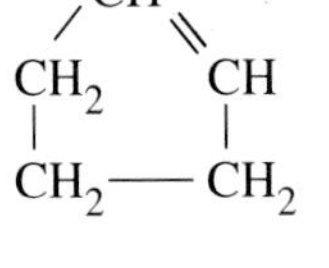

$$
\begin{array}{c}
\text{CH}_2 \\
\diagup \quad \diagdown \\
\text{H}-\text{C}\text{------}\text{C}-\text{H} \\
| \qquad\quad | \\
\text{CH}_2-\text{CH}_2
\end{array}
$$

$$
\begin{array}{c}
\diagup\text{CH}\diagdown\!\!\diagdown \\
\text{CH}_2 \qquad \text{CH} \\
| \qquad\qquad | \\
\text{CH}_2\text{------}\text{CH}_2
\end{array}
$$

$$HC{\equiv}C{-}CH_2CH_2CH_3$$

As the number of degrees of unsaturation goes up, the number of possible isomers usually increases rapidly, and we will need to rely more heavily on spectral data to find the structure. If there are *four* or more degrees of unsaturation there is a good likelihood of having a benzene ring (one ring with three double bonds).

Now, what if there are other elements besides C and H in the molecule? Oxygen is easy, because it does not affect the degrees of unsaturation. Since it makes two bonds, you can think of it as a sort of "spacer" between two other atoms, not changing the number of hydrogens needed. So, for example, $C_5H_{10}O$ has the same number of degrees of unsaturation as C_5H_{10}, namely one. A halide (F, Cl, Br, or I) takes the place of a hydrogen and therefore counts like a hydrogen for our purposes. So, for example,

$C_{10}H_{15}Cl$ counts like $C_{10}H_{16}$, and compared to $C_{10}H_{22}$ is missing six hydrogens (formula C_nH_{2n-4}), so it has 3° of unsaturation. A nitrogen makes three bonds and so acts as a "spacer" requiring one extra hydrogen, so we must subtract one hydrogen from the formula for each N present in order to compare it to the saturated formula. For example, $C_7H_{13}N$ counts like C_7H_{12} (formula C_nH_{2n-2}), and compared to the saturated C_7H_{16}, is missing 4H, so has 2° of unsaturation.

If there is more than one heteroatom present we just adjust the formula for all the heteroatoms present, compare to the saturated formula, and *voilà!*—we have degrees of unsaturation. For example, in the formula $C_{12}H_{14}NOCl_3$ we ignore O, subtract one H for N, and add three H's for the three chlorines, to give for comparison purposes $C_{12}H_{16}$. Compared to the saturated $C_{12}H_{26}$ it is missing 10 hydrogens, for 5° of unsaturation. I always prefer reasoning these out in this way, but for those who prefer using a formula there is a nifty one, which is just a condensed form of the reasoning we were using above. It reads:

$$°\text{unsat} = 1 + (\text{no. carbons}) - 1/2(\text{no. hydrogens} + \text{no. halogens}) + 1/2(\text{no. nitrogens})$$

In words, the degrees of unsaturation is one plus the number of carbons, minus half the sum of hydrogens and halogens, plus half the number of nitrogens. Let's check out this formula for $C_{12}H_{14}NOCl_3$, the same compound we did above:

$$°\text{unsat} = 1 + 12 + 1/2(14 + 3) - 1/2(1) = 1 + 12 + 17/2 - 1/2 = 5$$

Amazing, isn't it? Whichever way we did it, the answer came out the same.

Now here are some examples for you to do to be sure that you have got it down. It might be a good idea to do them using both the "reasoning" and "formula" methods as a double-check.

Question 1

How many degrees of unsaturation do the following molecules have?

(a) C_6H_{10} (b) $C_{11}H_{10}$
(c) $C_{14}H_{30}$ (d) $C_{11}H_{16}O$
(e) C_7H_7N (f) $C_{15}H_{21}Cl$
(g) $C_9H_{13}NBr_2$ (h) $C_{47}H_{91}N_5O_7I_2$

Good work! Now we'll go on to talk about chemical tests.

Often in a question you are given the results of some chemical tests on a compound. For example, you may be told that it decolorized bromine in carbon tetrachloride or that it reacts with Tollens reagent. To interpret these tests it is important to remember the reactions you have learned up to this point. For example, reddish Br_2 in carbon tetrachloride as a solvent adds across carbon-to-carbon double bonds to give a colorless vicinal dibromide. Therefore, the Br_2/CCl_4 test is positive for alkenes. Most of the common tests are listed in Table 14–1. Since some of these reactions are not encountered until late in an organic chemistry course you may not know them all, and probably will not need to know them until after they have been presented in class.

When writing the results of a chemical test on the outline, the name of the test followed by a (+) or (−) is a handy way to indicate a positive or negative result. For example, you might have a compound X, $C_7H_6O_2$, which is soluble in 5% NaOH and gives a negative Br_2/CCl_4 test. It could be outlined as follows:

$$X$$
$$C_7H_6O_2$$
Sol. 5% NaOH
Br_2/CCl_4 $(-)$

We can put IR and NMR peaks on the outline without even knowing much about IR or NMR. If you were told in this example that the compound X has peaks in the IR at 3000–3300 and 1680 inverse centimeters (cm^{-1}), and hydrogen NMR peaks at delta (δ) 7.3–7.8 (multiplet for five hydrogens) and δ 11.5 (singlet for one hydrogen), the outline would now look like:

$$X$$
$$C_7H_6O_2$$

Sol. 5% NaOH
Br_2/CCl_4 $(-)$

IR (cm^{-1}): 3000–3300
1680

NMR: δ 7.3–7.8 m, 5H
11.5 s, 1H

Let's practice outlining the data on a couple of compounds.

TABLE 14–1 SUMMARY OF QUALITATIVE TESTS

Reagent	Name	Positive test is:	Positive test means:
Cu	Beilstein	Green flame	Halide
Br_2/CCl_4	Bromine in carbon tetrachloride	Decolorization	Alkene or alkyne
Br_2/H_2O	Bromine water	Precipitate	Phenol
$Na_2Cr_2O_7/H_2SO_4$	Chromic acid	Blue-green Cr^{+3} ion	1° or 2° alcohol
2,4-Dinitrophenyl-hydrazine	2,4-DNP or 2,4 DNPH	Yellow, orange, or red ppt	Aldehyde or ketone
$FeCl_3$	Ferric chloride	Color change	Phenol
p-Toluenesulfonyl chloride	Hinsberg	Precipitate	1° or 2° amine; 1° if soluble in 5% NaOH, 2° if insoluble in 5% NaOH
$I_2/NaOH$	Iodoform	Yellow ppt of CHI_3	Methyl ketone or methyl carbinol
$ZnCl_2/HCl$	Lucas	Formation of layer of alkyl chloride	2° or 3° alcohol
$KMnO_4$	Permanganate (Baeyer)	Purple vanishes, brown MnO_2 appears	Alkene or alkyne
$AgNO_3/EtOH$	Silver nitrate in ethanol	Ppt of AgX	2° or 3° halide
NaI	Sodium iodide in acetone	Ppt of NaCl or NaBr	1° or 2° halide
$Ag(NH_3)_2^+$	Tollens	Silver mirror	Aldehyde

Question 2

Without trying to interpret any of the data, simply outline the information on the following compounds:

(a) Compound A, of formula C_3H_6O, was soluble in water and gave a positive iodoform test. In the IR it had a peak at 1720 cm^{-1} and in the NMR only one peak, a singlet, at δ 2.3.

(b) Compound B (C_8H_8O) was insoluble in 5% aqueous $NaHCO_3$, gave a yellow precipitate with 2,4–DNPH reagent, and gave a negative Tollens test. The IR showed peaks at 1680, 1600, 1510, 1460, 760, and 700 cm^{-1}. The NMR showed two peaks; at δ 2.2 a singlet for three hydrogens, and from δ 7.3 to 7.9 a multiplet for five hydrogens.

Hydrogen nuclear magnetic resonance (^{1}H NMR, or simply NMR) spectroscopy yields the most information about organic molecules of any type of spectroscopy. It is also the easiest to interpret in moderate detail, so it is the most useful type of spectroscopy for us. Let's talk about what it is like to run an NMR spectrum, the theory behind it, and the type of information obtained.

To take an NMR spectrum, the sample is placed in a long, thin glass tube shaped like a test tube, and a solvent not containing hydrogen is added. If the solvent contained hydrogen, it would swamp out any signals from the sample. The solvent may be naturally without hydrogen, such as CS_2 or CCl_4, or it may be a deuterated version of a normal solvent. Common deuterated solvents are $CDCl_3$ (deutero-chloroform), CD_3COCD_3 (deuterated acetone, or acetone-d$_6$), C_6D_6 (deuterobenzene or benzene-d$_6$), and CD_3SOCD_3 (deuterated dimethyl sulfoxide, or DMSO-d$_6$). Of course, the deuterated solvents are fairly expensive, but only about 1 mL is needed per sample. Once the sample is dissolved in the solvent, one drop of tetramethylsilane (TMS) is added as an internal reference. This is so that we can compare the positions of the other peaks to the positions of the peak from TMS. The sample tube is now capped and lowered into a strong magnetic field, where radio waves are passed through it (Figure 14–1). The tube spins to average out imperfections in the glass.

The whole principle behind NMR is that as the radio waves pass through the sample, some of their energy is absorbed. In other words, each hydrogen will absorb a characteristic frequency, depending on the magnetic field it experiences. Some hydro-

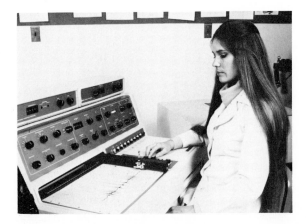

Figure 14-1   Running a Spectrum on a Varian EM-360 NMR Spectrometer (Courtesy Varian Associates, Palo Alto, CA)

TABLE 14–2 HYDROGEN NMR PEAK POSITIONS

Approximate peak position (δ)	Type of H
0.9	CH_3-
1.2	$-CH_2-$
1.5	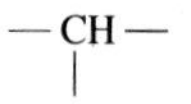 $-\overset{\displaystyle\vert}{C}H-$

| 2.1–2.5 | $-\overset{\overset{\textstyle H}{\vert}}{\underset{\underset{\textstyle}{\vert}}{C}}-\overset{\overset{\textstyle O}{\|}}{C}-$, $\quad-\overset{\overset{\textstyle H}{\vert}}{\underset{\underset{\textstyle}{\vert}}{C}}-\overset{\overset{\textstyle H}{\vert}}{C}=C\diagdown^{\diagup}$, $\quad-\overset{\overset{\textstyle H}{\vert}}{\underset{\underset{\textstyle}{\vert}}{C}}-\bigcirc\!\!\!\!\!\bigcirc$ |

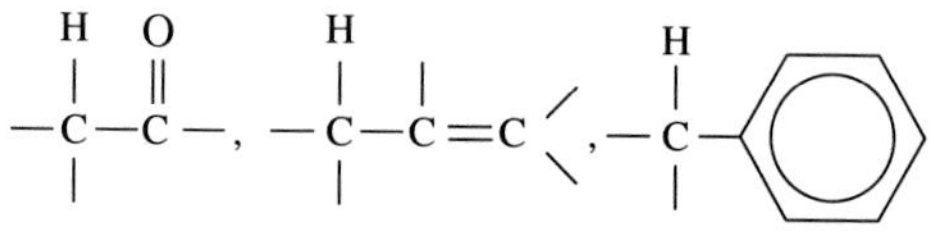

(H on C next to sp^2 carbon)

| 2.5–4.0 | 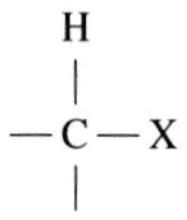 $-\overset{\overset{\textstyle H}{\vert}}{\underset{\underset{\textstyle}{\vert}}{C}}-X \qquad X = Cl, Br, I$ |

| 3.6–4.0 | 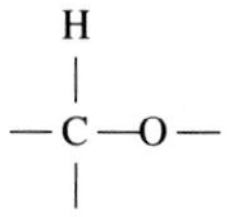 $-\overset{\overset{\textstyle H}{\vert}}{\underset{\underset{\textstyle}{\vert}}{C}}-O-$ |

| 5–6.5 | 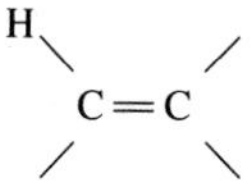 $\overset{H}{\diagdown}C=C\diagup^{\diagdown}_{\diagup}$ |

| 6.5–8 | 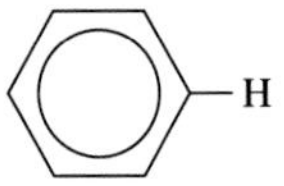 $\bigcirc\!\!\!\!\!\bigcirc-H$ |

| 9.5 | $H-\overset{\overset{\textstyle O}{\|}}{C}-$ |

| 11 | $-\overset{\overset{\textstyle O}{\|}}{C}-OH$ |

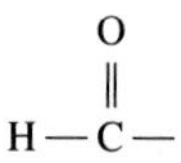

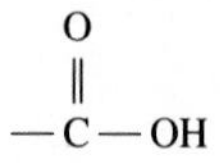

Variable peaks:

Alcohols   OH   1–6 $\left.\vphantom{\begin{matrix}1\\2\end{matrix}}\right\}$ usually broad, unsplit
Amines   NH   1–5

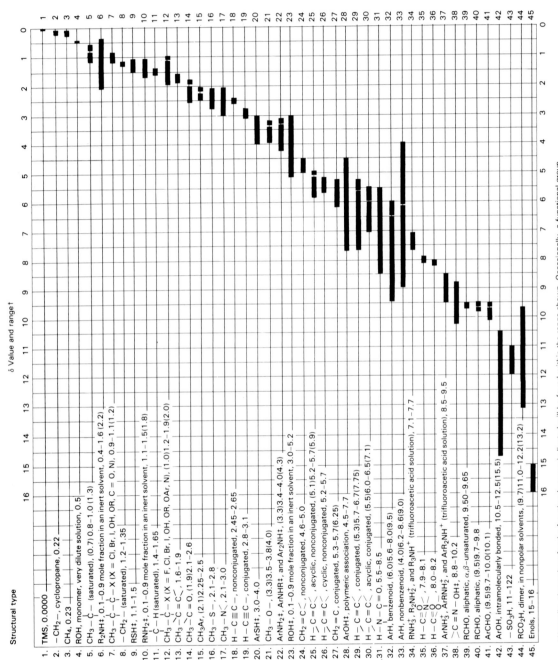

Figure 14-2 Detailed Hydrogen NMR Correlation Chart

†Normally, absorptions for the functional groups indicated will be found within the range shown. Occasionally, a functional group will absorb outside this range. Approximate limits for this are indicated by absorption values in parentheses and by shading in the figure.

‡The absorption positions of these groups are concentration-dependent and are shifted to lower δ values in more dilute solutions.

157

gens that are unaffected by neighboring atoms will experience exactly the magnetic field that is applied, called H_0. Other hydrogens will experience less magnetic field than the applied field; in other words, they are shielded by their neighbors. Still others will experience more magnetic field than H_0, being deshielded by their neighbors. So different environments affect the magnetic fields of the hydrogens, causing them to absorb at different frequencies. So if we look at a chart of absorbed energy versus frequency, we will get peaks for each type of hydrogen present.

Why does the sample absorb radio waves? Well, a proton (such as the one in the hydrogen nucleus) creates its own magnetic field, and this field can be aligned either with or against the applied external magnetic field, called H_0. If the field of the proton were at some angle to H_0, it would experience a torque and rotate to one of these stable positions. Now if each of these states has a definite energy, there is a certain amount of energy needed to ''flip'' the proton's magnetic field between these states. This energy happens to fall in the radio-frequency range.

There are three types of information obtained from NMR spectra. The *position* (δ value) of the peaks tells about the *environment* of the hydrogens; in other words, what they are attached to. We can look up these positions on a table such as Table 14–2. It is a good idea to memorize the most common peak positions, for example, δ 0.9 for a ''normal'' methyl group (CH_3–attached to a saturated carbon such as a $-CH_2-$) and δ 3.6–4.0 for a hydrogen on a carbon singly bonded to an oxygen

$$(H-\overset{\textstyle |}{\underset{\textstyle |}{C}}-O-).$$

In Figure 14–2 you'll find a more detailed listing of hydrogen NMR peak positions for various functional groups.

The *splitting* of the peak gives the number of *neighboring* hydrogens. To get the number of neighbors, subtract one from the number of peaks. For example, a singlet (one peak) means no neighbors. A doublet has one, a triplet has two, and a quartet has three neighbors. Figure 14–3 illustrates these patterns. Any peak split by more than three neighbors is usually called a multiplet, although on a very well resolved spectrum you can sometimes make out a quintet, sextet, or septet.

Why are the peaks split by neighboring hydrogens? Each neighboring hydrogen aligns itself either with or against H_0, which changes the magnetic field felt by the

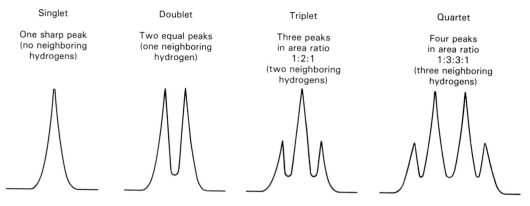

Singlet	Doublet	Triplet	Quartet
One sharp peak (no neighboring hydrogens)	Two equal peaks (one neighboring hydrogen)	Three peaks in area ratio 1:2:1 (two neighboring hydrogens)	Four peaks in area ratio 1:3:3:1 (three neighboring hydrogens)

Figure 14-3 Splitting Patterns in Hydrogen NMR

proton we are looking at. For example, if there were one neighboring hydrogen, it could be aligned with or against the applied magnetic field. So the proton we are looking at experiences either $H_0 + B$ or $H_0 - B$, where B is the contribution to the magnetic field by the neighbor. Since half of the protons we are looking at experience $H_0 + B$ and half experience $H_0 - B$, half will absorb radio waves at one frequency and half at another, leading to a split peak (a doublet). Similarly, if a hydrogen has two neighbors, they can both be aligned with H_0, one with and one against (which cancels), or both opposing H_0. Note that there are two ways in which one neighbor is with and one against the applied field, making this arrangement twice as likely as either of the others. Let's do the same analysis for three neighbors. These can all be up, or two up and one down (three ways to do this) or one up and two down (three ways) or all down. So this leads to four peaks, in the ratio of areas $1:3:3:1$, as shown in Figure 14–4.

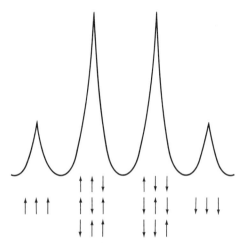

Figure 14-4 Splitting and Spin Multiplicity for a Quartet

Question 3

Predict the ratio of areas in a quintet (from a hydrogen with four neighbors).

You may notice that these area ratios are coming out in a familiar pattern from algebra called Pascal's triangle. This gives the coefficients of binomial, trinomial, and so on, expansions, and is easily drawn by making a triangle of ones, then adding the two upper numbers to get the lower one. If the areas are not in the correct ratios, you are dealing with more than one peak, rather than a split peak.

The total area under the peaks gives the relative numbers of identical hydrogens of each type. This is called the integration. After running an NMR spectrum, the operator pushes a button marked "integrate" and the machine electronically adds up the areas under the peaks. A typical integrated spectrum (for ethanol) is shown in Figure 14–5.

The relative heights of the integration curve ($20:10:30$ or $2:1:3$) gives the ratios of the three types of hydrogens. Integration alone does not tell us the number of hydrogens; we do not know if there are $2:1:3$ or $4:2:6$ or $6:3:9$. This we must get by knowing the total number of hydrogens from the formula.

Now let's interpret the NMR spectrum in Figure 14–5 together. First we have to write the peaks in a compact way. The standard format for writing an NMR peak is

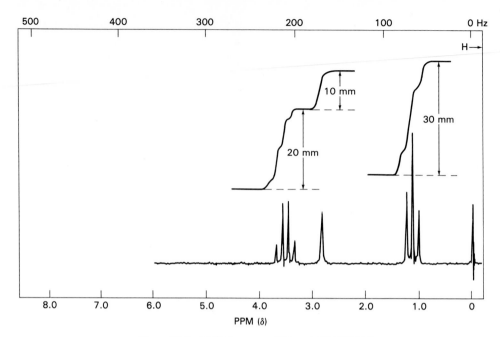

Figure 14-5 NMR Spectrum of Ethanol (CH₃CH₂OH)

to list the δ value, then the multiplicity (splitting), then the relative number of hydrogens. In writing multiplicities, s = singlet, d = doublet, t = triplet, q = quartet, and m = multiplet. Let's write out the peaks for the example of ethanol above. Starting from the right (upfield, near TMS) to left (downfield) we get

$$\delta\ 1.2\ t,\ 3H$$
$$2.9\ s,\ 1H$$
$$3.6\ q,\ 2H$$

To interpret these, we would say that the first peak has three identical protons, so it must be a CH_3—. The fact that it comes at δ 1.2 means that it is almost a normal alkyl methyl group but has been moved slightly downfield. The fact that it is a triplet means it must have two equivalent neighboring hydrogens, so it must be next to a —CH_2—. That —CH_2— must be at least a quartet (split by the CH_3 and maybe something else on the other side). So we look over the list of peaks for a —CH_2— and aha! We find it in the third peak. Two hydrogens, a quartet, so no neighbors other than the CH_3, and at δ 3.6, which means next to a singly bonded oxygen. So now we have

$$CH_3—CH_2—O—$$

The remaining peak must be the OH. Hydrogens on oxygen (e.g., alcohols) are usually not split by their neighbors and often give broad peaks because of hydrogen bonding, since their environment is constantly changing. Also because of hydrogen bonding, —OH and —NH peaks are variable, coming anywhere over a range of several δ units, depending on concentration and temperature. The value observed here (2.9) is well within the normal range of 1–6 for alcohols.

The fact that hydrogens on oxygen or nitrogen are hydrogen bonded and easily exchangeable leads to the common practice in taking NMRs of adding a drop of D_2O and rerunning the spectrum after the first run. What this does is to exchange the hydrogen on any —OH or —NH for a deuterium, so that peak disappears from the new spectrum. A new peak for H—O—D appears at approximately δ 4.8, but that does not tell us anything. The important thing is that if a peak vanishes on addition of D_2O, it is called D_2O exchangeable and means it is an —OH or —NH.

Nuclear magnetic resonance can be used with nuclei other than hydrogen as long as they possess a magnetic moment. Some of these nuclei are ^{13}C, ^{15}N, ^{19}F, and ^{31}P. After hydrogen NMR, ^{13}C NMR is the most widely used by organic chemists.

In carbon-13 NMR (also called ^{13}C NMR or CMR), it is the *carbon* nuclei, rather than the hydrogen, that are absorbing radio waves. Therefore, a peak is observed for each type of carbon. Normally, in a so-called *coupled* spectrum, the hydrogens bonded to a particular carbon atom will split the peak for that carbon, in the same patterns as hydrogens on adjacent carbons split each other's peaks in hydrogen NMR. The splitting follows the n + 1 rule: For n hydrogens, there will be *n* + 1 peaks.

For example, the peak for the carbon methyl group, CH_3—, will appear as a quartet (three hydrogens give a quartet or n = 3, n + 1 = 4). Similarly, a methylene group (—CH_2—) will appear as a triplet, and a methine group (—CH—) as a doublet. A carbon with no attached hydrogens appears as a singlet.

Peaks in ^{13}C NMR appear over a much wider range of δ (ppm) values than peaks in hydrogen NMR. For hydrogen, the common range of peak positions is approximately 0–12 ppm, while for carbon the range spans 0–230 ppm. A correlation chart of functional group and ^{13}C NMR peak positions appears in Figure 14–6. You can see that, similarly to hydrogen NMR, the more electron-poor (deshielded) the carbon, the farther downfield (higher δ value) it is. The chemical shift of each carbon atom indicates its environment (functional group and neighboring groups).

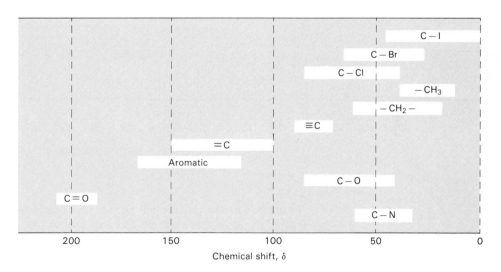

Figure 14-6 Chemical Shift Correlations for ^{13}C NMR

^{13}C NMR spectra are more time consuming, difficult, and expensive to obtain than hydrogen spectra, primarily because of the low natural abundance (1.08%) of the isotope carbon-13. Since the nucleus of carbon-12, the most abundant isotope of carbon, does not possess net spin, it cannot be observed by nuclear magnetic resonance.

The magnetic moment (μ) of ^{13}C is also quite low. Because of these two factors, ^{13}C resonances are approximately 6000 times weaker than hydrogen resonances. Using computerized equipment with Fourier-transform (FT) techniques and averaging the results of many scans over several hours, high-quality spectra can be obtained.

If a sample is irradiated in a broad spectrum of frequencies in the proper range, the spins of the hydrogen atoms are randomized and the splitting of the carbon peaks by hydrogen is eliminated. Each type of carbon appears as a singlet no matter how many hydrogen atoms are on it. Examples of uncoupled and coupled spectra (of 3–methyl–2–butanone) are shown in Figures 14–7 and 14–8. Integration is unreliable in ^{13}C NMR, so the relative number of carbons of each type is not available accurately, although a larger peak may indicate more carbons of that type. Splitting is normally not observed between carbons because the probability of two adjacent ^{13}C atoms is so low (1.08% x 1.08%).

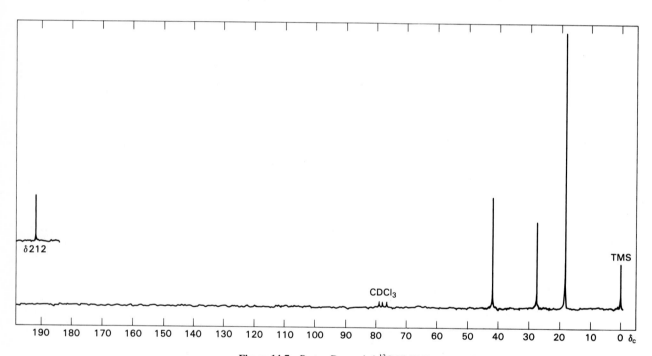

Figure 14-7   Proton-Decoupled ^{13}C NMR Spectrum of 3-Methyl-2-Butanone

In mass spectroscopy, a small sample of a compound is placed in a vacuum so that some of the molecules vaporize. Once in a vapor phase, the material is bombarded with high-energy electrons. Impact with one of these electrons knocks an electron out of the molecule to form the molecular or parent ion, M^+.

$$M + e^- \rightarrow M^+ + 2e^-$$

These molecular (parent) ions have almost the same mass as the original compound and possess a single positive charge. Once formed, these ions are then accelerated toward a negatively charged plate. Some strike the plate, while others pass through a focusing slit into a region of high magnetic field. As you may recall from physics, a charged particle moving in a magnetic field experiences a force perpendicular to the

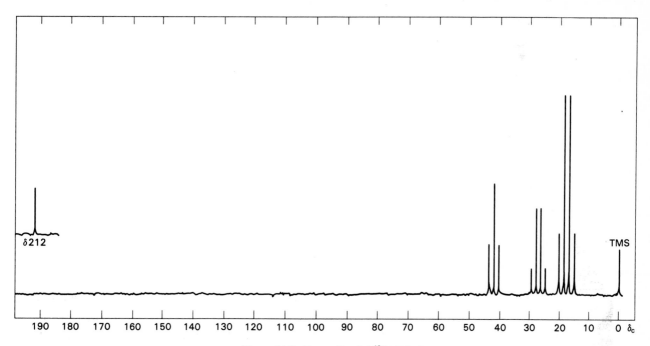

δ212

TMS

| 190 | 180 | 170 | 160 | 150 | 140 | 130 | 120 | 110 | 100 | 90 | 80 | 70 | 60 | 50 | 40 | 30 | 20 | 10 | 0 δ_c |

Figure 14-8   Proton-Coupled ^{13}C NMR Spectrum of 3-Methyl-2-Butanone

direction of motion. The direction of this force vector **F** is given by the right-hand rule in the equation.

$$\mathbf{F} = q\,\mathbf{V} \times \mathbf{B}$$

where q is charge, **V** is the velocity vector, and **B** is the magnetic field vector. The particle is pulled into a circular orbit of radius

$$R = \frac{mV}{Bq} = \frac{m}{q}\frac{V}{B}$$

where m is the mass of the particle. In other words, the radius of curvature is directly proportional to the mass-to-charge ratio (m/q, also called m/e or m/z) of the particle, since V and B are held constant. We will assume that we are dealing only with singly charged ions ($q = +1$). Thus heavier particles (with larger m) have larger radii and strike the collection screen farther away from the slit than lighter particles, as illustrated in Figure 14–9.

When the parent ion strikes the collector, a parent peak (or molecular ion) is recorded. This ion, which is generally the most massive on the spectrum, gives the molecular mass of the compound. After initial ionization, parent ions may undergo cleavage to form smaller ions. For example, fragments may break off of alkyl chains.

$$CH_3CH_2CH_3 \xrightarrow[-2e^-]{+e^-} CH_3CH_2CH_3^{\,\cdot\,+} \xrightarrow{\text{fragmentation}} CH_3CH_2^+ + CH_3\cdot$$

Parent ion

A long-chain hydrocarbon will give a series of peaks due to loss of methyl (mass 15) and successive methylene (—CH_2—, mass 14) groups. This may be observed in the mass spectrum of *n*-octane shown in Figure 14–10.

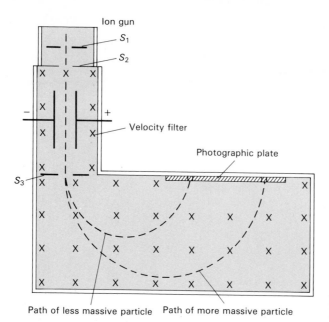

Figure 14-9 Schematic of a Bainbridge Mass Spectrometer

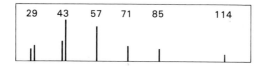

Figure 14-10 Mass Spectrum of *n*-Octane

As another example, if a compound yields a parent ion at $m/e = 158$ and has another peak at $m/e = 143$, it is reasonable to assume that it contained a methyl group. If an additional peak is present at $m/e = 129$ (loss of another 14 units or $-CH_2-$), it would indicate a probable ethyl group.

A table of fragments commonly lost appears in Table 14–3. Since heteroatoms (O, N, S, Cl, Br, etc.) have unshared electrons, relatively high electron densities, and low ionization potential, on impact an electron is often lost from the heteroatom. In this case, a carbon-to-heteroatom bond may break to yield major fragments, as illustrated below for chlorobenzene:

$$Cl-C_6H_5 \xrightarrow[-2e^-]{+e^-} {}^+Cl-C_6H_5 \longrightarrow Cl\cdot + C_6H_5{}^+$$

In a technique called gas chromatography–mass spectrometry (GC-MS), a mass spectrometer is hooked up so as to sample compounds as they elute from a GC column. A computer may also be attached to compare the mass spectra obtained with an internal library of spectra to identify the best match(es). This system is an excellent way to purify and identify small quantities of volatile compounds rapidly. A photograph of one such setup appears in Figure 14–11.

Once we have all the data down we can go back and start interpreting, from the formula (least information) to the NMR (most information). Interpreting IR peaks is explained in Section 13; for now getting back to our first example, let's assume we know that in the IR a broad peak from 3000 to 3300 cm^{-1} is an $-OH$ group, and a

TABLE 14–3 SOME FRAGMENTS COMMONLY LOST IN MASS SPECTROMETRY

Mass	Fragment lost
1	H·
15	CH_3·
17	HO·
18	H_2O
26	$HC \equiv CH$, ·$C \equiv N$
27	$CH_2 = CH$·, $HC \equiv N$
28	$CH_2 = CH_2$, CO
29	CH_3CH_2·, ·CHO
30	NH_2CH_2·, CH_2O, NO
31	·OCH_3, ·CH_2OH, CH_3NH_2
32	CH_3OH, S
35	Cl·
43	C_3H_7·, $CH_3\overset{\overset{\displaystyle O}{\|}}{C}$·
57	C_4H_9·
71	C_5H_{11}·
77	C_6H_5·

Figure 14-11 Photograph of a Hewlett Packard GC-MS System

peak at 1680 cm^{-1} is a carbonyl. In the NMR let's assume we know that only aromatic hydrogens come between δ7.3 and 7.8, and that a peak at δ11.5 has to be a carboxylic acid. The outline now looks as follows:

$$X$$
$$C_7H_6O_2 \qquad 5° \text{ unsat}$$

Sol 5% NaOH	acid or phenol
Br_2/CCl_4 (−)	not alkene

IR: 3000–3300 cm^{-1} (broad)   —OH
       1680   C=O

IR peaks taken together suggest possible —COOH

NMR: δ 7.3–7.8 m, 5H   monosubstituted aromatic ring
     11.5 s, 1H   —COOH

Now if we check the formula we see we have found all seven carbons (a benzene ring and a COOH), the two oxygens (the COOH), the six hydrogens, and all 5° of unsaturation (four for the benzene ring, one for the carbonyl). Furthermore, all the data are in agreement. What bliss! We now simply connect up the largest pieces we have found, to get the structure

$$
\text{(benzene ring)} - \; + \; -\overset{\overset{\displaystyle O}{\|}}{C}-OH \;\Rightarrow\; \text{(benzene ring)} - \overset{\overset{\displaystyle O}{\|}}{C} - O - H
$$

As a final step we go back and check that this structure is in agreement with all the data, which it is.

Note that in writing the pieces deduced from the IR and NMR data it is a good idea to draw the vacant bonds on the pieces. This helps a lot in putting them together at the end, and is a good habit to get into. I like to think of pieces as being either "end" pieces (only one free bond) or "middle" pieces (two or more free bonds). Then we can hook them together into a complete structure. For example, let's say that you deduced the following pieces for a molecule of formula $C_4H_8O_2$:

$$
-\overset{\overset{\displaystyle O}{\|}}{C}- \qquad CH_3CH_2- \qquad CH_3O-
$$

Now since there is only one middle piece and there are two end pieces, there is only one way to put these together:

$$
CH_3CH_2\overset{\overset{\displaystyle O}{\|}}{C}-OCH_3, \qquad \text{same as} \qquad CH_3O\overset{\overset{\displaystyle O}{\|}}{C}CH_2CH_3
$$

If there is more than one middle piece, it gets a little more involved; that is why we want to deduce as large pieces as we can, usually from the NMR. For example, let's say that we have deduced the following pieces for a compound of formula $C_8H_{17}NO$:

$$
-\overset{\overset{\displaystyle O}{\|}}{C}- \qquad -CH_2- \qquad CH_3-\overset{\overset{\displaystyle CH_3}{|}}{\underset{\underset{\displaystyle CH_3}{|}}{C}}- \qquad \overset{CH_3}{\underset{CH_3}{{>}N-}}
$$

Since there are two middle pieces there are two ways to put the molecule together:

$$
CH_3-\overset{\overset{\displaystyle CH_3}{|}}{\underset{\underset{\displaystyle CH_3}{|}}{C}}-\overset{\overset{\displaystyle O}{\|}}{C}-CH_2-N{\overset{\displaystyle CH_3}{\underset{\displaystyle CH_3}{<}}} \qquad \text{or} \qquad CH_3-\overset{\overset{\displaystyle CH_3}{|}}{\underset{\underset{\displaystyle CH_3}{|}}{C}}-CH_2-\overset{\overset{\displaystyle O}{\|}}{C}-N{\overset{\displaystyle CH_3}{\underset{\displaystyle CH_3}{<}}}
$$

To tell these apart, we would need to go back and examine the data carefully for additional information. In this case the exact position of the carbonyl peak in the IR would tell us whether we are dealing with a ketone or an amide. Also, the exact position of the $-CH_2-$ group in the NMR would tell us whether it was next to both a carbonyl and a nitrogen, or just a carbonyl. Before we get into these details, let's practice putting some pieces together.

Question 4

If you have deduced the following partial structures, what are all the possibilities for the complete structure?

1. $-CH_2-$ $-CH_3$ $H-\overset{\displaystyle |}{C}=O$

2. $-\overset{\displaystyle O}{\overset{\displaystyle \|}{C}}NH_2$ $CH_3-\overset{\displaystyle \overset{\textstyle H}{|}}{\underset{\displaystyle \underset{\textstyle CH_3}{|}}{C}}-O-$ $-CH_2CH_2-$

3. CH_3- CH_3- $-\overset{\displaystyle |}{\underset{\displaystyle |}{C}}-H$ $-CH_2OH$

4. ⬡$-\overset{\displaystyle O}{\overset{\displaystyle \|}{C}}OH$ $-CH_2-$ $-Br$

5. $-\overset{\displaystyle \overset{\textstyle H}{|}}{\underset{\displaystyle \underset{\textstyle CH_3}{|}}{C}}-$ $-\overset{\displaystyle O}{\overset{\displaystyle \|}{C}}OH$ ⬡$-$ $-CH_2Cl$

6. $CH_3CH_2-N\begin{smallmatrix}\diagup\\[4pt]\diagdown\end{smallmatrix}$ $\underset{\displaystyle H}{\overset{\displaystyle H}{\diagdown}}C=C\begin{smallmatrix}\diagup\\[4pt]\diagdown\end{smallmatrix}$ CH_3-CH_3-

7. $CH_3-\overset{\displaystyle \overset{\textstyle H}{|}}{\underset{\displaystyle \underset{\textstyle CH_3}{|}}{C}}-$ $-\overset{\displaystyle O}{\overset{\displaystyle \|}{C}}OCH_3$ $-CH_2-$ CH_3- $-\overset{\displaystyle |}{\underset{\displaystyle |}{C}}-H$

Now that you know how to outline a question, find degrees of unsaturation, interpret spectra, and put the pieces together at the end, you are ready to do some structure determination. To summarize the process of structure determination, then:

1. Outline the question, leaving space for interpretations.
2. From the molecular formula (which may be obtained from mass spectral or analytical data) calculate the degrees of unsaturation (multiple bonds or rings) present. The formula also indicates the heteroatoms present, which places limits on the number and types of functional groups.
3. Interpret any chemical test data available to suggest possible functional groups. For example, if a compound is soluble in aqueous base, it suggests a carboxylic acid or phenol.

4. Interpret the spectra in the following order: UV, IR, MS, NMR. For each one available, write down the list of peaks, then go through and write the interpretation next to each peak listed. If a peak can have two interpretations, put both of them down. Also jot down the functional groups that are obviously absent based on the spectrum. For example, if an IR spectrum has no large peak between 1640 and 1810 cm^{-1}, you can state "no C=O."

5. Make sure that you have found all the pieces by checking the molecular formula, and examine the data for consistency.

6. Put the pieces together; there may be more than one way to do this.

7. Double-check all the information to see if it is consistent with the structure(s) obtained, and if any structures can be eliminated.

8. The stereochemistry of the molecule often is not fully determined by the spectra, and additional tests may be needed if it is desired to determine it.

Procedure

First your instructor will take a few minutes to review the major concepts of IR spectroscopy. After the discussion you will take 5 minutes to work the first problem involving IR spectroscopy on your own.

It is important to learn to work these problems quickly, since you want to develop proficiency and you will probably encounter many of them on exams. For each question, be sure to follow the strategies outlined in the overview in order to stay organized. At the end of this time the instructor will go over the problem with you and answer any questions. This procedure is repeated for the next question, and so on.

After finishing the IR questions, take a 10–minute break to relax and refresh yourself and reward yourself for the good work. Now the instructor will give a review of NMR spectroscopy, answer any questions, and you will work the NMR questions in the same way. You should try to do as much as possible on your own, but if stuck you may ask classmates or the instructor for advice.

Infrared spectroscopy

1. Identify the functional groups in each compound from the infrared spectra.

   (a)

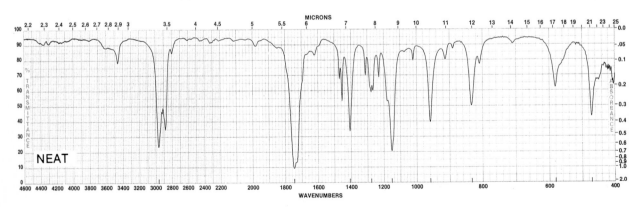

(b)

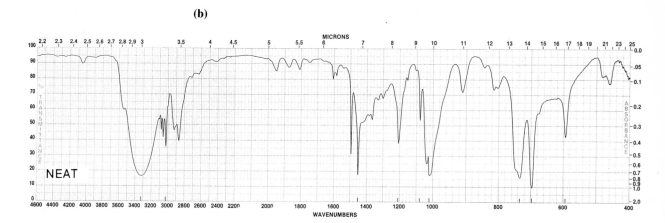

(c)

(d)

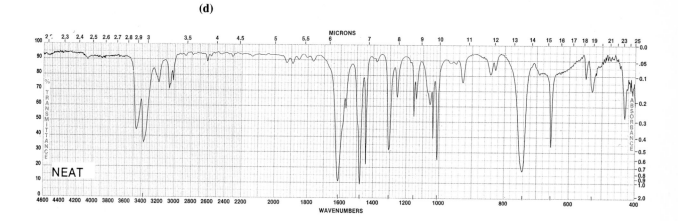

2. Identify the following compounds from their formulas and IR spectra.

(a) C$_7$H$_9$N

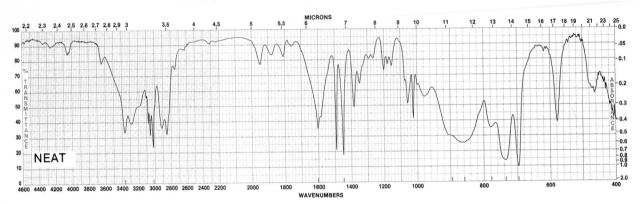

(b) C$_9$H$_{12}$

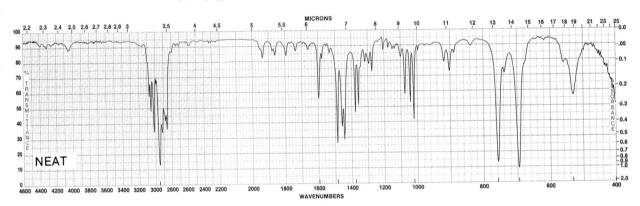

(c) C$_7$H$_4$BrN

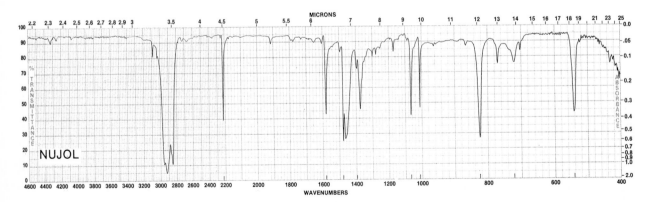

Hydrogen NMR spectroscopy

1. Fill in the blank spaces in the following statement. In the NMR spectrum of ethyl iodide, the peaks for methyl hydrogens appear at $\delta = 1.9$ ppm and the peaks for the methylene hydrogens appear at $\delta = 3.2$ ppm with $J = 8$ Hz. The number of peaks given by the methyl hydrogens is _____ with the approximate area ratio: _____ . These peaks are separated by _____ Hz. The number of peaks given by the methylene hydrogens is _____ with the approximate area ratio: _____ . These peaks are separated by _____ Hz. The ratio of the total area of the methyl peaks to the methylene peaks is _____ . Of these two groups of peaks, the _____ peaks are farther downfield. The chemical shift difference between these peaks of 1.3 ppm corresponds in a 60–MHz instrument to _____ Hz.

2. Draw the predicted NMR spectra for the following compounds.
   (a) *t*-Butyl alcohol
   (b) Isobutyl bromide
   (c) Ethylbenzene
   (d) Methyl butyrate

3. The NMR spectra for some isomers of $C_5H_{10}Br_2$ are given below. Find the structure corresponding to each spectrum.
   (a) δ 1.0 (s, 6H); 3.4 (s, 4H)
   (b) δ 1.0 (t, 6H); 2.4 (quart, 4H)
   (c) δ 0.9 (d, 6H); 1.5 (m, 1H); 1.85 (t, 2H); 5.3 (t, 1H)
   (d) δ 1.0 (s, 9H); 5.3 (s, 1H)
   (e) δ 1.0 (d, 6H); 1.75 (m, 1H); 3.95 (d, 2H); 4.7 (doublet of triplets, 1H)
   (f) δ 1.3 (m, 2H); 1.85 (m, 4H); 3.35 (t, 4H)

4. Give a structure for each of the following from the formula and NMR data.
   (a) $C_4H_8Br_2$
      (1) δ 1.7 (d, 6H)
      (2) 4.4 (quart, 2H)
   (b) C_4H_9Br
      (1) δ 1.04 (d, 6H)
      (2) 1.95 (m, 1H)
      (3) 3.33 (d, 2H)
   (c) $C_{10}H_{14}$
      (1) δ 1.30 (s, 9H)
      (2) 7.28 (s, 5H)
   (d) $C_{10}H_{14}$
      (1) δ 0.88 (d, 6H)
      (2) 1.86 (m, 1H)
      (3) 2.45 (d, 2H)
      (4) 7.12 (s, 5H)
   (e) $C_{10}H_{13}Cl$
      (1) δ 1.57 (s, 6H)
      (2) 3.07 (s, 2H)
      (3) 7.24 (s, 5H)
   (f) $C_{10}H_{12}$
      (1) δ 0.65 (m, 2H)
      (2) 0.81 (m, 2H)
      (3) 1.35 (s, 3H)
      (4) 7.17 (s, 5H)
   (g) $C_9H_{11}Br$
      (1) δ 2.15 (quintet, 2H)
      (2) 2.75 (t, 2H)
      (3) 3.38 (t, 2H)
      (4) 7.22 (s, 5H)

5. Identify the following compounds from their formulas and hydrogen NMR spectra.
   (a) C_9H_{10}

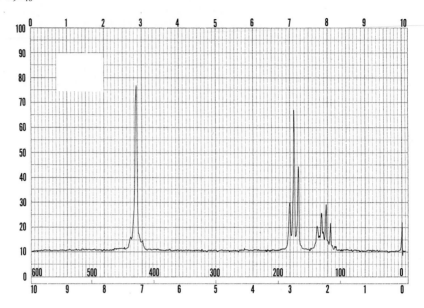

   (b) $C_6H_{13}NO_2$

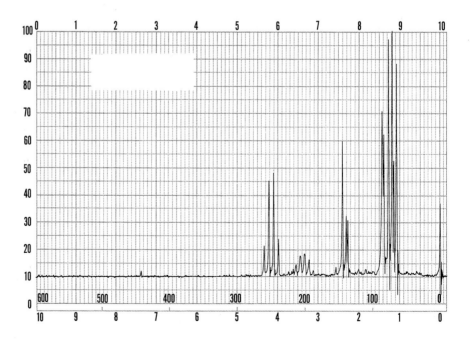

6. Identify the following compounds from their formulas and IR and NMR spectra.
    (a) $C_4H_7ClO_2$

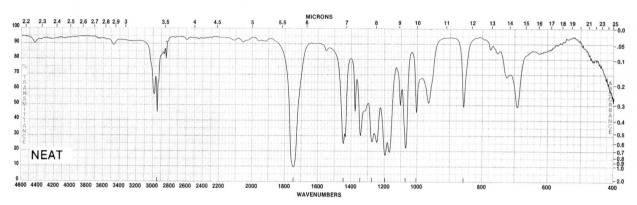

NEAT

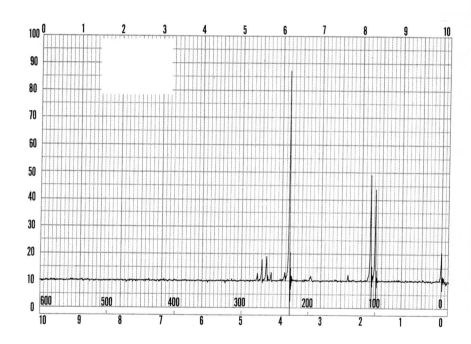

(b) C_4H_6BrN

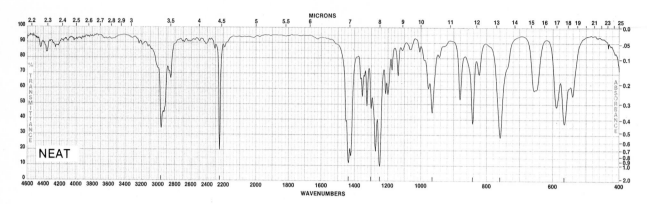

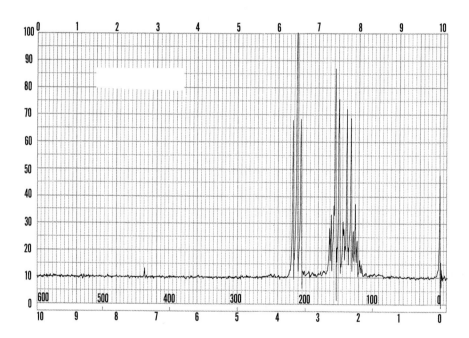

(c) $C_8H_{11}NO$

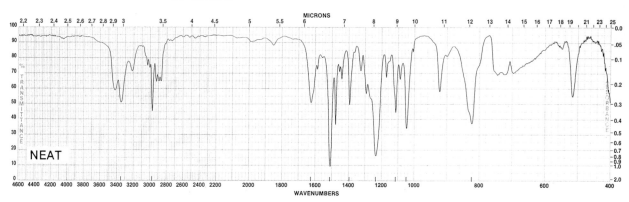

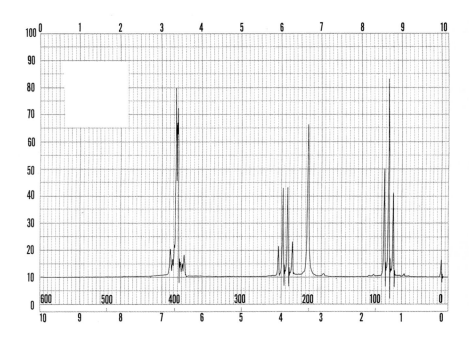

7. Word problems.

(a) Compound A was found to have the following formula, IR and NMR spectra. What is it?

$C_{10}H_{12}O$ IR (cm^{-1}): 1720, 1600, 1580, 1500, 1450, 760, 700

NMR (δ values): 2.2 s, 3H; 2.5 t, 2H; 2.6 t, 2H; 7.4 s, 5H

(b) Compound B (C_5H_8, IR peak at 2180 cm^{-1}) was treated with H_2 and a Pd/BaSO$_4$/quinoline catalyst (Lindlar's) to give compound C (C_5H_{10}). Compound C had an IR peak at 1630 cm^{-1} and the following NMR:

1.1 t, 3H; 2.3 d, 3H; 2.5 d of q, 2H; 5.8 d of q, 1H; 6.0 d of t, 1H

What are the structures of B and C?

(c) Compound D (C_4H_6), which showed a peak in the IR at 2200 cm^{-1}, was treated with sodium amide (NaNH$_2$) followed by isopropyl bromide to give compound E (C_7H_{12}), which still had an IR peak near 2150 cm^1 and the following NMR: δ 1.0 t, 3H; 1.1 d, 6H; 2.2 q, 2H; 2.4 m, 1H. What are the structures of D and E?

(d) A compound F (C_8H_{14}) was treated with ozone and zinc dust (reductive workup) to give G ($C_8H_{14}O_2$, IR: 1710 cm^{-1}, NMR: δ 0.9 d, rel. area 3, 2.3 d 2, 1.5 m 1, 9.5 s l). Compound G was treated with excess methylene triphenylphosphine ($\phi_3P=CH_2$) to give H ($C_{10}H_{18}$). What are the structures of F, G, and H?

(e) Compound I ($C_8H_{14}O$, IR: 1720, 1375 cm^{-1}, NMR: δ 1.0 s, 6H; 1.4–1.6 m, 6H; 2.4 t, 2H) upon treatment with methylene triphenylphosphine gave compound J (C_9H_{16}, 1R: 1630, 1375 cm^{-1}) NMR: δ 1.2 s, 6H; 1.4–1.6 m, 6H; 2.2 t, 2H; 5.5 s, 1H; 5.75 s, 1H). What are I and J?

(f) Compound K (C_7H_7Br, 1R, 1600, 1580, 1460, 750 cm^{-1} NMR: δ 2.5 s, 3H; 7.4–7.8 m, 4H) was treated with magnesium in ether followed by compound L ($C_{12}H_{10}O$, IR: 1680, 1600, 1510, 1470, NMR: δ 2.3 s, 3H, 7.8–8.2 m, 7H) to yield (after acidification) compound M ($C_{19}H_{18}O$, IR: 3200–3400 broad, 1600, 1570, 1500, 1440, 760; NMR: δ 1.4 s, 3H; 2.3 s, 3H; 2.5 broad s, 1H vanishes on D$_2$O exchange; δ 7.2–8.0 m, 11H). What are K, L, and M?

(g) Compound N ($C_{11}H_{16}$, IR: 1600, 1570, 1500, 830 cm^{-1}, NMR: δ 1.1 s, 9H; 2.3 s, 3H; 7.2–7.6 d of d, 4H) on treatment with NBS followed by NaCN/DMSO, then H$^+$/H$_2$O followed by CH$_3$OH/H$^+$ gave compound O ($C_{13}H_{18}O_2$, IR: 1710, 1590, 1500, 1470, 850 cm^{-1}, NMR: δ 1.1 s, 9H; 2.9 s, 2H; 4.0 s, 3H; 7.3–7.7 d of d, 4H). What are N and O?

(h) Compound P ($C_5H_{11}N$, IR 3100–3200, NMR 1.2 d, 3H, 1.5 m, 4H, 2.0 broad s, 1H, 3.0 t, 2H, 3.1 m, 1H) was treated with excess methyl iodide, followed by silver oxide and heat to give Q ($C_7H_{15}N$). When the same process was repeated on Q, a compound R (C_5H_8) was formed, which had a peak in the IR at 1630 cm^{-1}.

(i) When acetophenone (structure below) was treated with nitric and sulfuric acids, then Fe/HCl, then base, compound S was formed (C_8H_9NO, IR + NMR below). When S was treated with acetyl chloride, T was formed ($C_{10}H_{11}NO_2$, IR below). When T was treated with LiAlH$_4$ then H$_2$O, U was formed ($C_{10}H_{15}NO$, 1R and NMR below).

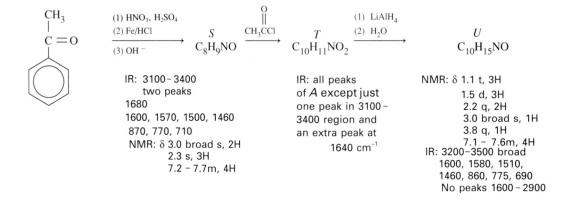

^{13}C NMR spectroscopy

1. How many different types of carbons are found in the following molecules?
   (a) Isobutyl chloride
   (b) Methylisopropylamine
   (c) t-Butyl acetate
   (d) 4–Methylbenzoic acid
2. Draw the predicted proton-decoupled ^{13}C NMR spectra for the compounds in Question 1.
3. Same as Question 2, but proton coupled.

4. Determine the structures of the following compounds from their formulas and proton-decoupled CMR spectra.

(a) $C_4H_{10}O$

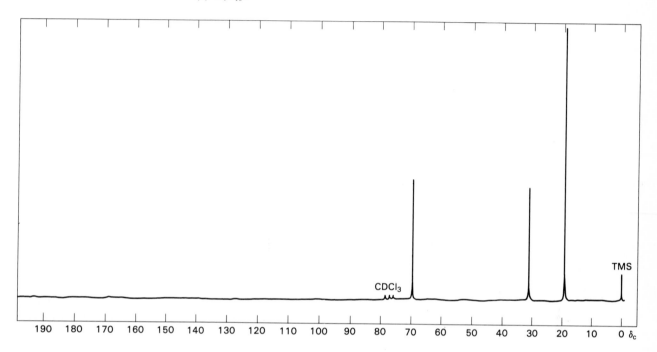

(b) C_4H_7NO

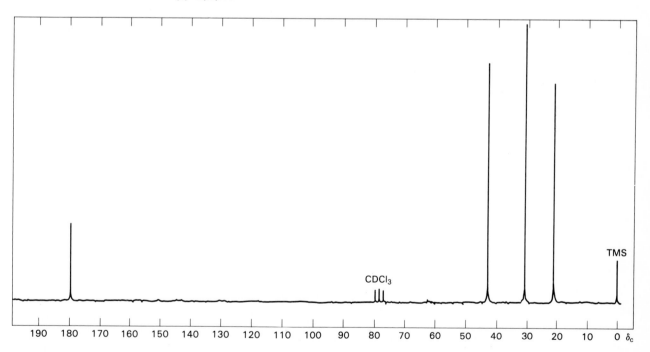

Mass spectrometry

1. What would be the masses of the molecular ion and major fragments from the following compounds?

   (a) Butane

   (b) 2–Butanone

   (c) *p*-Chlorotoluene

2. Identify the compound from its mass spectrum.

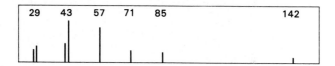

DIELS-ALDER REACTIONS AND INSECTICIDES

Overview

A Diels-Alder reaction is also known as a [4 + 2] cycloaddition, since one molecule (the diene) supplies four pi electrons while the other (the dienophile) supplies two. The electrons move around in a circle forming two new sigma bonds and one new pi bond, forming a cyclohexene ring. The new pi bond ends up *between* where the two double bonds were on the diene. This is shown below for the Diels-Alder reaction of cyclopentadiene and *cis*-1,2-dicyanoethene.

It would be well worth taking a couple of minutes to run through this reaction using your molecular model kit. This may help you visualize the six electrons swinging around in a circle, forming two new single bonds and one new double bond as the three original double bonds are broken.

An unusual feature of this reaction is that substituents on the dienophile (e.g., the cyano groups) end up pointing toward the larger, two-carbon bridge rather than the smaller, one-carbon bridge. This so-called *endo* product is formed in preference to the *exo* product, except in very polar solvents.

Endo (substituents toward larger bridge)

Exo (substituents toward smaller bridge)

For a long time researchers were puzzled as to why the more hindered endo product was formed until it was explained by R. B. Woodward, R. Hoffmann, and K. Fukui in the 1960s that overlap of the **p** orbitals on the substituents with **p** orbitals on the diene is favorable, helping to bring the two molecules together one on top of the other, as shown in Figure 15–1. These explanations of molecular orbitals, along with their other contributions, are so useful to chemists that all three of these men received Nobel prizes.

You will probably learn about their description of molecular orbitals later. For now all you need to do is to look at Figure 15–1 and notice the favorable overlap (matching light or dark lobes) of the diene and the substituent on the alkene. In this way the double bond of the dienophile lies underneath the end carbons of the diene, as it must to form the new sigma bonds, and the substituents are attracted to the middle two carbons of the diene.

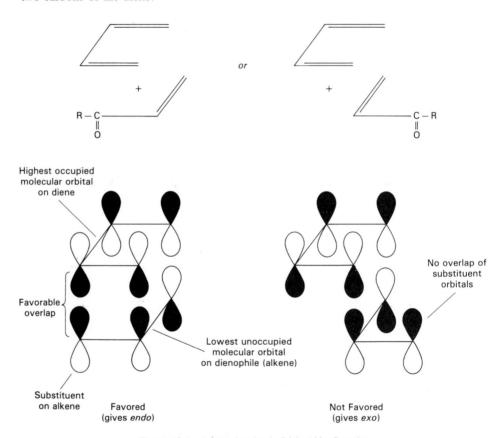

Figure 15-1 Orbital Overlap in Diels-Alder Reactions

On standing for a few days at room temperature, cyclopentadiene undergoes a Diels-Alder reaction with itself to form dicyclopentadiene.

Dicyclopentadiene

This reaction can be reversed by heating dicyclopentadiene and distilling off the cyclo-pentadiene as it is formed.

Since cyclopentadiene cannot be stored for long without dimerizing, we distill it freshly before use by "cracking" dicyclopentadiene in a *reverse* or *retro* Diels-Alder reaction caused by heating. Dicyclopentadiene is inexpensive and comes from distilling coal tar or from the by-products of ethylene production. The freshly cracked cyclopen-tadiene is then allowed to react with maleic anhydride to form the adduct.

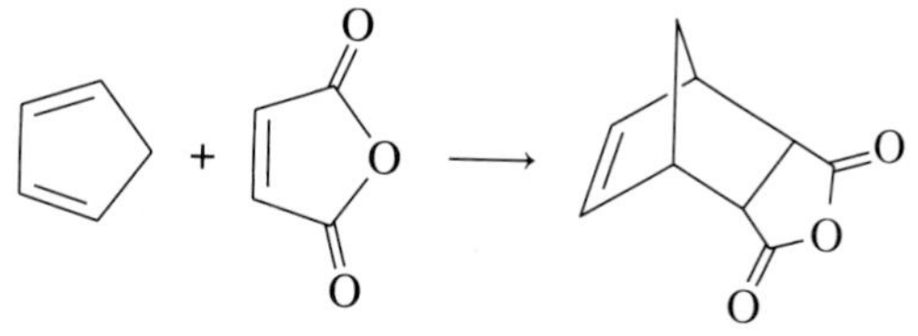

Once you obtain the Diels-Alder adduct of cyclopentadiene and maleic anhydride, you will hydrolyze the acid anhydride to a dicarboxylic acid.

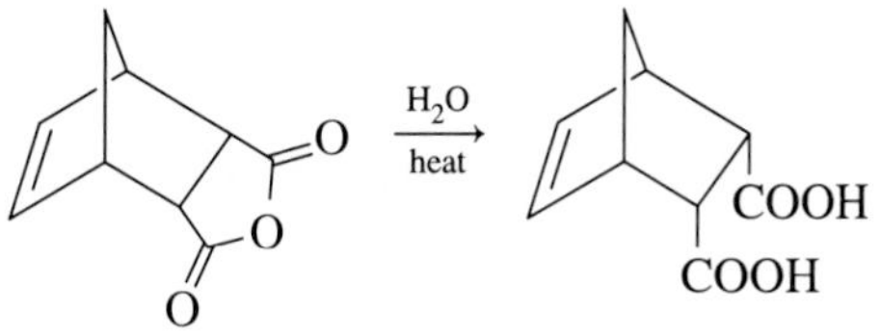

Most anhydrides form carboxylic acids spontaneously on contact with water. For ex-ample, if acetic anhydride touches water, it reacts within a few minutes at room tem-perature to give acetic acid.

The mechanism of this reaction is that the electron-rich (slightly negative) oxygen of water attacks the electron-poor (slightly positive) carbon of the carbonyl group and, after transferring a proton, displaces a carboxyl group, as shown here.

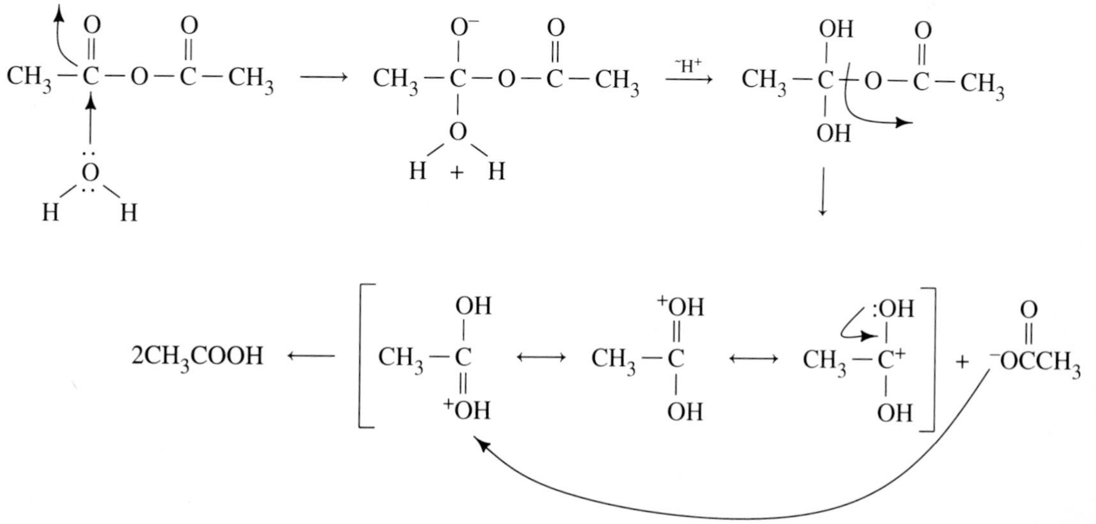

In our molecule the carboxyl groups are held very closely together by the shape of the molecule, and this makes it much harder to hydrolyze the anhydride. As you know, five- and six-membered rings are the easiest to form and the hardest to break. In this case we are breaking up a five-membered ring anhydride, and it requires boiling water to do it.

Overall reaction:

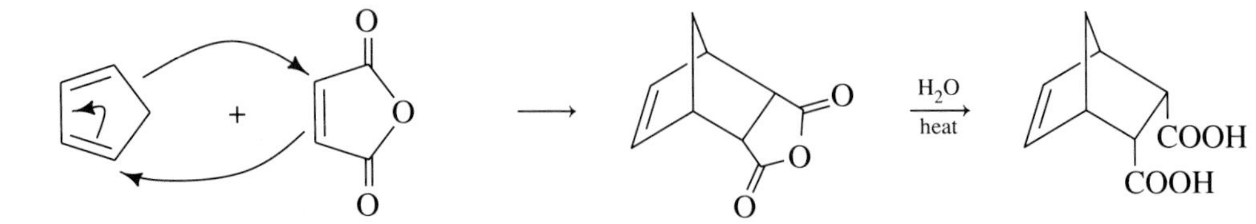

Cyclopentadiene	Maleic anhydride	cis-5-Norbornene-2,3-endo-dicarboxylic anhydride	cis-5-Norbornene-2,3-endo-dicarboxylic acid MW 182
MW 66.1, bp 41° dens. 0.80 g/mL	MW 98.1, mp 53°	MW 164, mp 165°	mp 180–185°d

Several strong insecticides have been prepared using Diels-Alder reactions. Dieldrin and aldrin (named after Diels and Alder), chlordane, and heptachlor are all prepared using hexachlorocyclopentadiene as the diene, as shown in Figure 15–2.

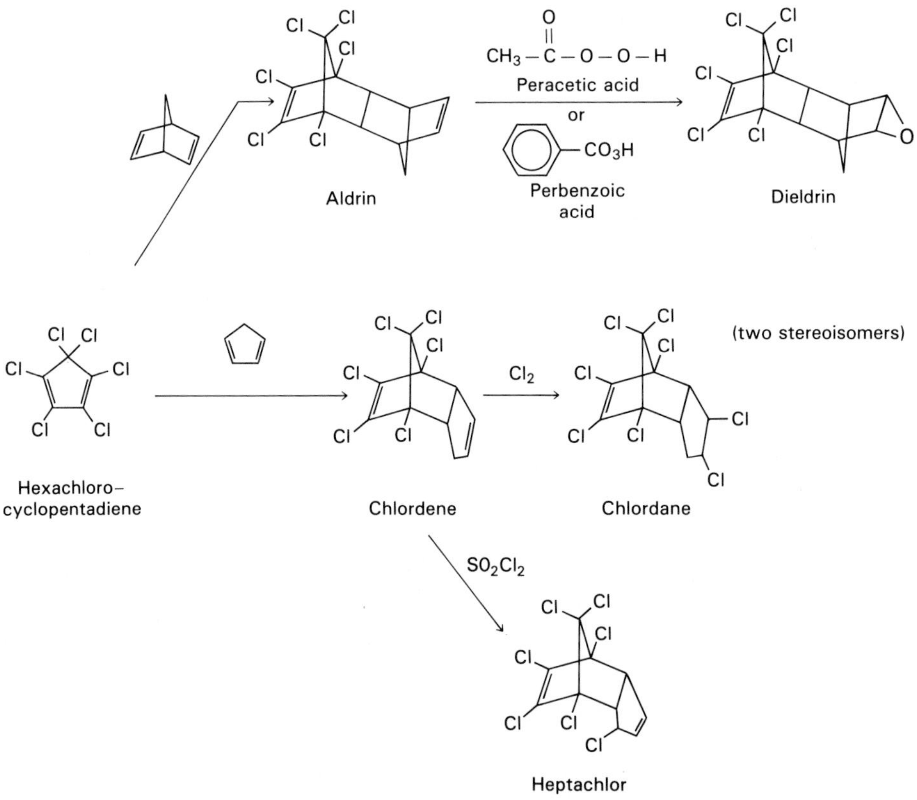

Figure 15-2 Synthesis of Some Chlorinated Pesticides Involving Diels-Alder Reactions

Aldrin and heptachlor have been widely used as soil insecticides and for control of grasshoppers and pests on fruit trees and cotton plants. Chlordane is used to control cockroaches, ants, termites, and some pests on vegetable and field crops. Dieldrin has a longer environmental half-life and has been used to treat houses to eliminate malaria-carrying mosquitoes and for mothproofing. Like most other effective contact insecticides, these compounds are efficiently absorbed by insect cuticle and have high toxicity to insects. Aldrin, for example, has an LD_{50} value (lethal dose for 50% of the sample) for cockroaches of 1.9 mg/kg topically applied. Because of the high lipophilicity of these cyclodiene insecticides, their environmental persistence, high toxicity to fish, and the increase of resistant strains of insects, use of aldrin and dieldrin was banned in the United States by the Environmental Protection Agency in 1974. In 1978, most uses of heptachlor and chlordane were banned as well.

Because of the billions of dollars of damage caused annually worldwide by insect pests, and the problems associated with use of chlorocarbon insecticides, chemists are developing new insecticides which are more selective and less persistent. Organophosphorus insecticides, some of which are shown below, do not persist in the environment and therefore do not build up in the food chain. They are highly toxic to people, though, and extreme care must be used in their application.

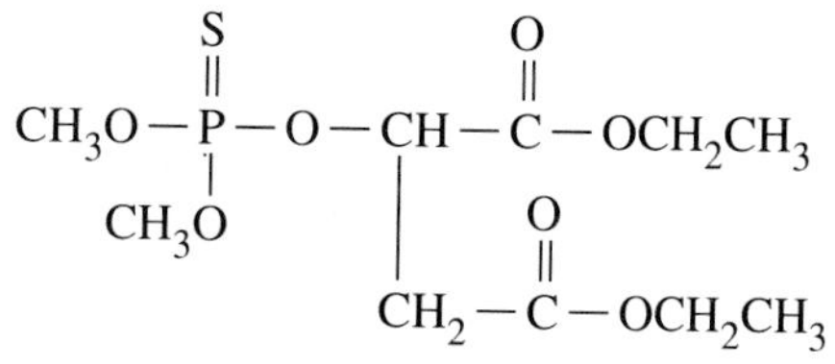

$$CH_3O-\overset{\overset{S}{\|}}{\underset{\underset{CH_3O}{|}}{P}}-O-\overset{\overset{}{\underset{\underset{\overset{O}{\underset{\|}{CH_2-C-OCH_2CH_3}}}{|}}{}}{CH}-\overset{\overset{O}{\|}}{C}-OCH_2CH_3$$

Malathion

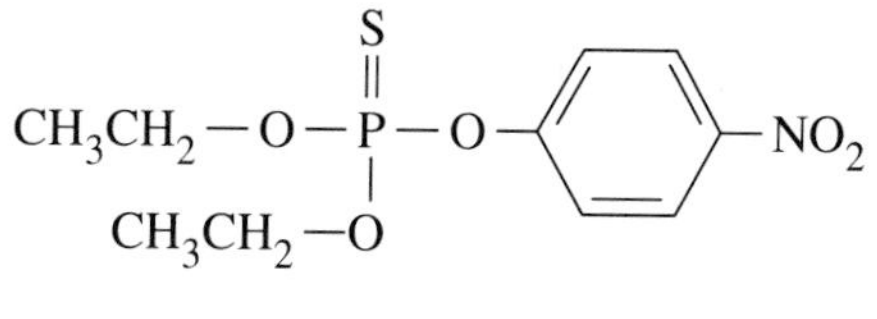

$$CH_3CH_2-O-\overset{\overset{S}{\|}}{\underset{\underset{CH_3CH_2-O}{|}}{P}}-O-\!\!\!\bigcirc\!\!\!-NO_2$$

Parathion

$$CH_3O-\overset{\overset{O}{\|}}{\underset{\underset{CH_3O}{|}}{P}}-O-CH=C\overset{\diagup Cl}{\diagdown Cl}$$

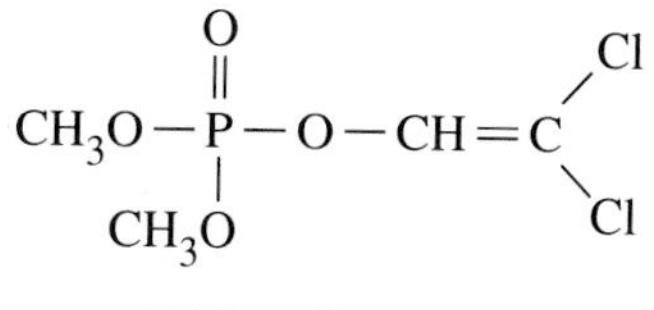

DDVP or Dichlorvos

An attractive alternative to the use of these compounds is to use the types of chemicals produced by insects themselves for communication or regulation of growth. Pheromones (sex attractants), many of which are relatively simple (and easily biodegradable) molecules, can be used to attract insects to traps or confuse them so they cannot find a mate. Table 15–1 shows the structures of a few sex pheromones used in insect control.

TABLE 15–1   SOME SEX PHEROMONES USED IN INSECT CONTROL

Compound	Structure	Species
cis-7,8-Epoxy-2-methyloctadecane	$CH_3(CH_2)_9CH —\!\!\overset{\displaystyle O}{\overset{\displaystyle /\backslash}{}}\!\!— CH(CH_2)_4CH(CH_3)_2$	Gypsy moth, *Porthetria dispar*
trans-8,*trans*-10-Dodecadienol	$CH_3CH = CHCH = CH(CH_2)_6CH_2OH$	Codling moth, *Laspeyresia pomonella*
cis-11-Hexadecenal	$CH_3(CH_2)_3CH = CH(CH_2)_9CHO$	Corn earworm, *Heliothis zea*
cis-3,*cis*-13-Octadecadienyl acetate	$CH_3(CH_2)_3CH = CH(CH_2)_8CH = CH(CH_2)_2\overset{\displaystyle O}{\overset{\displaystyle \|}{O}}CCH_3$	Peach-tree borer, *Sanninoidea exitiosa*
cis-9,*trans*-12-Tetradecadienyl acetate	$CH_3CH = CHCH_2CH = CH(CH_2)_8\overset{\displaystyle O}{\overset{\displaystyle \|}{O}}CCH_3$	Almond moth, *Cadra cautella* Fig moth, *Cadra figulilella* Mediterranean flour moth, *Anagasta kuehniella*

The discovery of a class of insect juvenile hormones, which regulate growth and development, has made possible the preparation of thousands of hormone analogs which may arrest development and cause death. Methoprene, for example, has high toxicity toward mosquitoes and houseflies, and very low toxicity to mammals.

Methoprene

Plants possess many natural insect-repellant and insecticidal compounds, and sometimes these can be used to human advantage. For example, *pyrethrins*, a class of compounds found in chrysanthemum flowers, may be sprayed on crops or livestock to protect them. A mixture of diatomaceous earth and pyrethrins, sold under the trade name Diacide, has proven very effective for household use against cockroaches. Pyrethrins possess the advantages of rapid knockdown of pests, low toxicity to mammals, and biodegradability. The structures of pyrethrins I and II are

Pyrethrin R
I : R = CH_3
II : R = $COOCH_3$

QUESTIONS

1. Show arrows for electron movement in each of the Diels–Alder reactions below, and show the product.

(a)

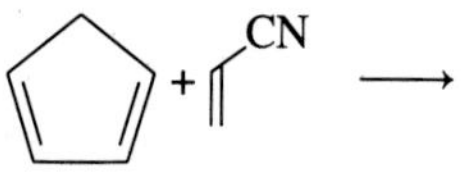

(b)

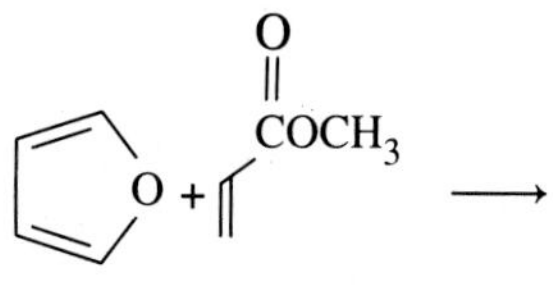

(c)

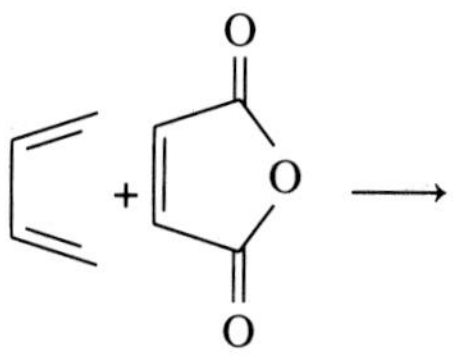

(d)

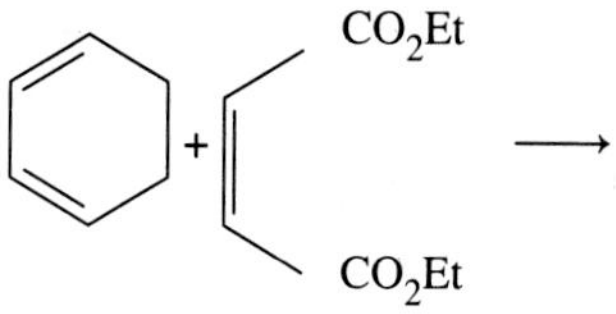

(e)

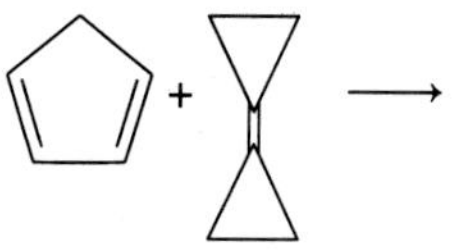

(f)

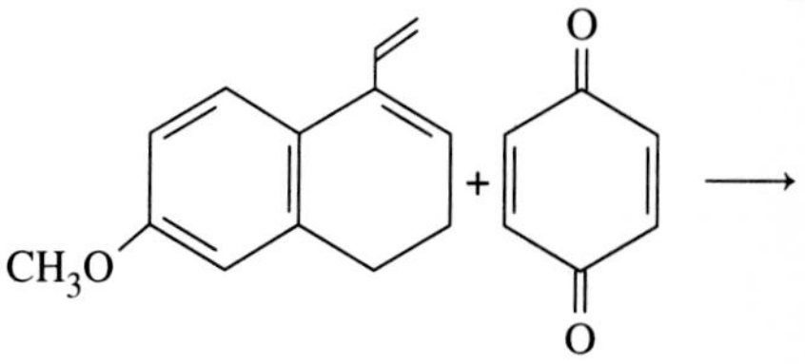

excess

2. Show how to make each of the following. (*Hint:* Draw arrows first to reverse a Diels-Alder reaction; always look for the cyclohexene that any Diels-Alder reaction formed from an alkene and a diene.)

(a)

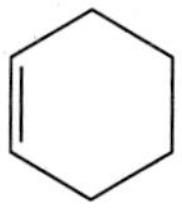

(b)

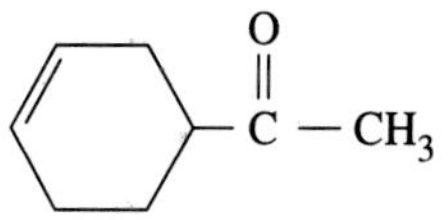

(c)

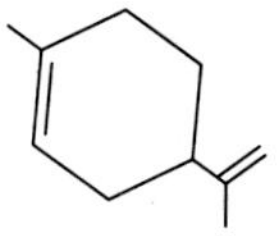

(d)

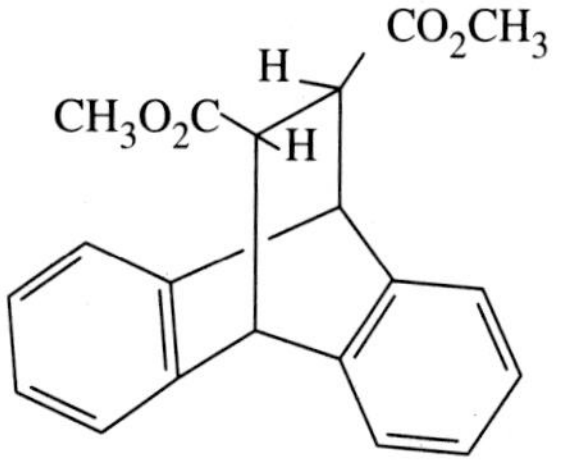

(e)

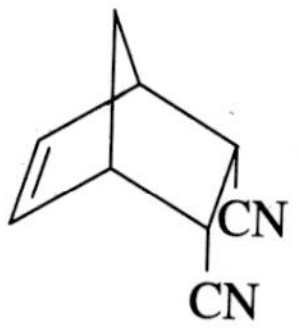

REFERENCE

METCALF, R., ''Insect Control Technology'' in H. F. Mark et al., eds., *Kirk-Othmer Encyclopedia of Chemical Technology*, 3rd ed., Vol. 13, Wiley-Interscience, New York, 1981, pp. 413–485.

Concept Map

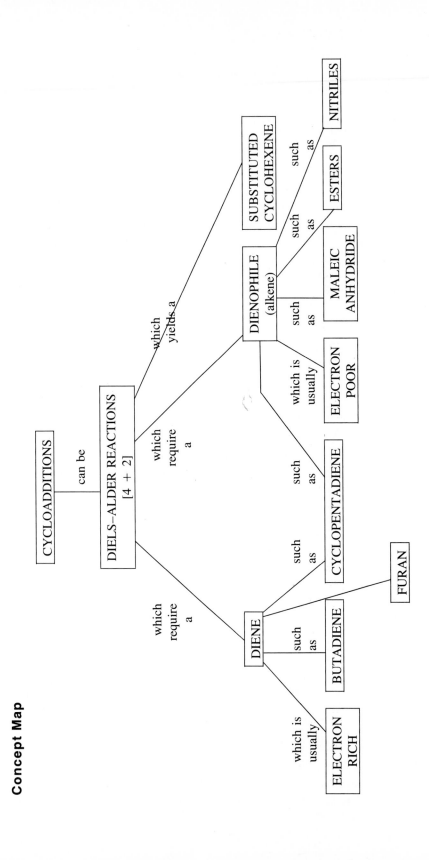

EXPERIMENT 15.1 CYCLOADDITION OF CYCLOPENTADIENE AND MALEIC ANHYDRIDE

Estimated Time:
4.0 hours for procedures a and b

Prelab

1. In your notebook fill out a reagent table and fill in the number of millimoles in the procedure.
2. Calculate the theoretical yield for each step.

Special Hazards

Cyclopentadiene has a very unpleasant odor, so the room should be well ventilated and the cyclopentadiene should be kept in the fume hood until after it is added to the solution of maleic anhydride.

Procedure 15.1a: *cis*-5-Norbornene-*endo*-2,3-dicarboxylic anhydride

Estimated time:
2.5 hours

Place maleic anhydride (5 g, _____ mmol) in a 125–mL Erlenmeyer flask. If the maleic anhydride is in the form of large briquets, it may be necessary to grind it up using a mortar and pestle before weighing. Add ethyl acetate (20 mL) and heat on a hot plate until the maleic anhydride is dissolved. Add ligroin (bp 60–80°, also called petroleum ether, 20 mL) as a cosolvent, cool the solution thoroughly in an ice bath, and leave it in the ice bath. Take the ice-cold solution of maleic anhydride over to the fume hood and add cyclopentadiene (5 mL, _____ mmol) directly from the burette provided. Before the lab your instructor has freshly distilled (cracked) the cyclopentadiene from dicyclopentadiene, dried it over $CaCl_2$, and placed some in a burette in the fume hood for easy dispensing.

   Take the flask back to your bench and swirl the solution in the ice bath until the exothermic reaction subsides and the adduct precipitates as a white solid. To recrystallize the product from this solution, heat on a hot plate until all the solid dissolves, then allow it to cool to room temperature, and you will be rewarded with beautiful needles of the product. Collect the product, air dry, and record its mass and melting point.

Procedure 15.1b: Hydrolysis to the endo, *cis*-diacid

Estimated time:
1.5 hours

To form the diacid from the anhydride, place the bicyclic anhydride (3.0 g, _____ mmol) in a 100–mL beaker, add water (40 mL), and bring to a boil on a hot plate. Continue heating until all the material is dissolved. Remove the beaker from the hot plate and allow it to stand undisturbed to let the product crystallize in long needles. If necessary, scratch the walls to induce crystallization. After the solution has cooled to room temperature, place it in an ice bath to drive out more product. Collect the diacid by suction filtration, air dry, weigh, and record the melting (decomposition) range. Note that this compound does not truly melt in the usual sense but undergoes a chemical change over a temperature range that varies with the rate of heating. This may be easily shown by letting the melted sample cool, then taking the melting point again. What do you think the new compound is? Save your diacid product (*cis*-5-norbornene-2,3-*endo*-dicarboxylic acid) for use as the starting material in the next experiment.

Catalytic Hydrogenation

This section completes a three-step sequence started in Section 15 with the Diels–Alder reaction.

Overview

Conceptually, catalytic hydrogenation is very simple: adding two hydrogens across a carbon-to-carbon double bond.

$$\text{C}=\text{C} \xrightarrow[\text{catalyst}]{\text{H}_2} -\overset{\overset{\displaystyle \text{H}}{|}}{\text{C}}-\overset{\overset{\displaystyle \text{H}}{|}}{\text{C}}-$$

Several metals catalyze this reaction, the most common being platinum, palladium, and nickel.

The mechanism of the reaction is still under investigation. It appears that the hydrogen molecules dissociate into atoms on the surface of the metal. The alkene comes down and bonds to the metal, breaking its pi bond. Hydrogen atoms migrate over to the carbon atoms and form C—H bonds. The driving force for this reaction is the formation of stronger bonds (two C—H bonds, approximately 90–100 kcal/mol each) from a weak C=C pi bond (approximately 70 kcal/mol) and an H—H bond (104 kcal/mole).

Once all carbons have bonded to hydrogens, there is no more attraction to the metal and the molecule is released from the surface, freeing the metal catalyst for further reaction. This scheme is shown next

Since reaction occurs only at the *surface* of a metal, we want our particles of metal to be as small as possible (instead of in chunks) to get maximum surface area, particularly if we are using an expensive metal such as platinum. One effective way to do this is to deposit the metal on some kind of support, such as fine particles of charcoal. In this experiment we deposit platinum metal on charcoal by reducing platinum(IV) to metallic platinum using sodium borohydride. Sodium borohydride (NaBH$_4$) is an extremely useful reducing agent and can be thought of as a source of hydride ions, H$^-$. So the redox reaction to produce the catalyst is

$$Pt^{+4} + 4H^- \longrightarrow Pt(s) + 2H_2$$

where four electrons are transferred from hydrogen to platinum.

Once the catalyst is ready we can also generate hydrogen *in situ* by mixing sodium borohydride and hydrochloric acid, for the overall result

$$H^- + H^+ \longrightarrow H_2(g)$$

or, in more complete form,

$$NaBH_4 + 4HCl \longrightarrow 4H_2(g) \uparrow + NaCl + BCl_3$$

Since platinum is a very good catalyst, if we hold the hydrogen gas over the catalyst and the substrate (starting material) for a few minutes, the reaction will be complete. One nice feature of catalytic hydrogenations is that they usually either go l00% or 0% (in which case you can recover the starting material). Whether they go or not depends on whether the catalyst is clean or has been ''poisoned'' by nitrogen- or sulfur-containing impurities that get stuck on the surface of the metal and prevent reaction.

In this experiment our overall reaction is

cis-5-Norbornene-endo-
2, 3-dicarboxylic acid
MW 182, mp 180 – 190°d

cis-Norbornane-endo-
2, 3-dicarboxylic acid
MW 184, mp 170 – 175°d

APPLICATION: CATALYTIC HYDROGENATION OF VEGETABLE OIL

How many times have you eaten a candy bar containing partially hydrogenated vegetable oil? By appropriate partial or complete hydrogenation, a wide range of liquid or partially solid fats can be converted into solids such as margarine or shortening.

Many oils may be used interchangeably as raw materials, enabling a producer to use whichever are most readily available and lowest in cost. The table shows some common fatty acids which may be present in oils as mono, di-, or triglycerides (esters of glycerol).

SOME COMMON FATTY ACIDS

Name	Carbons	No. of double bonds	Structure
Saturated			
Lauric	12	0	$CH_3(CH_2)_{10}COOH$
Myristic	14	0	$CH_3(CH_2)_{12}COOH$
Palmitic	16	0	$CH_3(CH_2)_{14}COOH$
Stearic	18	0	$CH_3(CH_2)_{16}COOH$
Unsaturated			
Palmitoleic	16	1	$CH_3(CH_2)_5CH = CH(CH_2)_7COOH$ (cis)
Oleic	18	1	$CH_3(CH_2)_7CH = CH(CH_2)_7COOH$ (cis)
Linoleic	18	2	$CH_3(CH_2)_4CH = CHCH_2CH = CH(CH_2)_7COOH$ (cis,cis)
Linolenic	18	3	$CH_3CH_2CH = CHCH_2CH = CHCH_2CH = CH(CH_2)_7COOH$ (cis,cis,cis)
Arachidonic	20	4	$CH_3(CH_2)_4(CH = CHCH_2)_4CH_2CH_2COOH$ (all cis)

Hydrogenation decreases the degree of unsaturation and makes fats less susceptible to oxidation and rancidity. During storage, the carbon-to-carbon double bonds in unsaturated fats may be attacked by molecular oxygen in the air, resulting in cleavage to lower-molecular-weight carboxylic acids.

Alkene group in oil

It is these small carboxylic acids that impart a rancid odor and flavor.

Commercially, hydrogenation is carried out in batch reactors (also called converters; see figure below) with capacities of 5–30 tons. These reactors include systems for handling gas and for introducing and removing catalysts, gas-dispersing agitators, heating and cooling coils, plus safety devices and controls.

To carry out the hydrogenation, a refined and bleached liquid fat is pumped in, then subjected to vacuum to remove any air present. The catalyst (usually 0.02–0.15% nickel on an inert support) is added and hydrogen gas is introduced with stirring, normally at an operating pressure of 1–6 atm (200–700 kPa). The hydrogenation reaction is exothermic, so the system must usually be cooled to maintain the desired temperature.

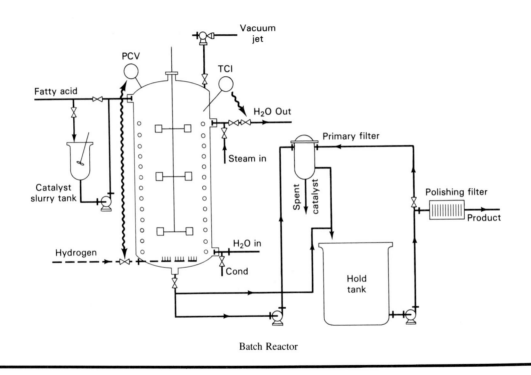

Batch Reactor

Concept Map

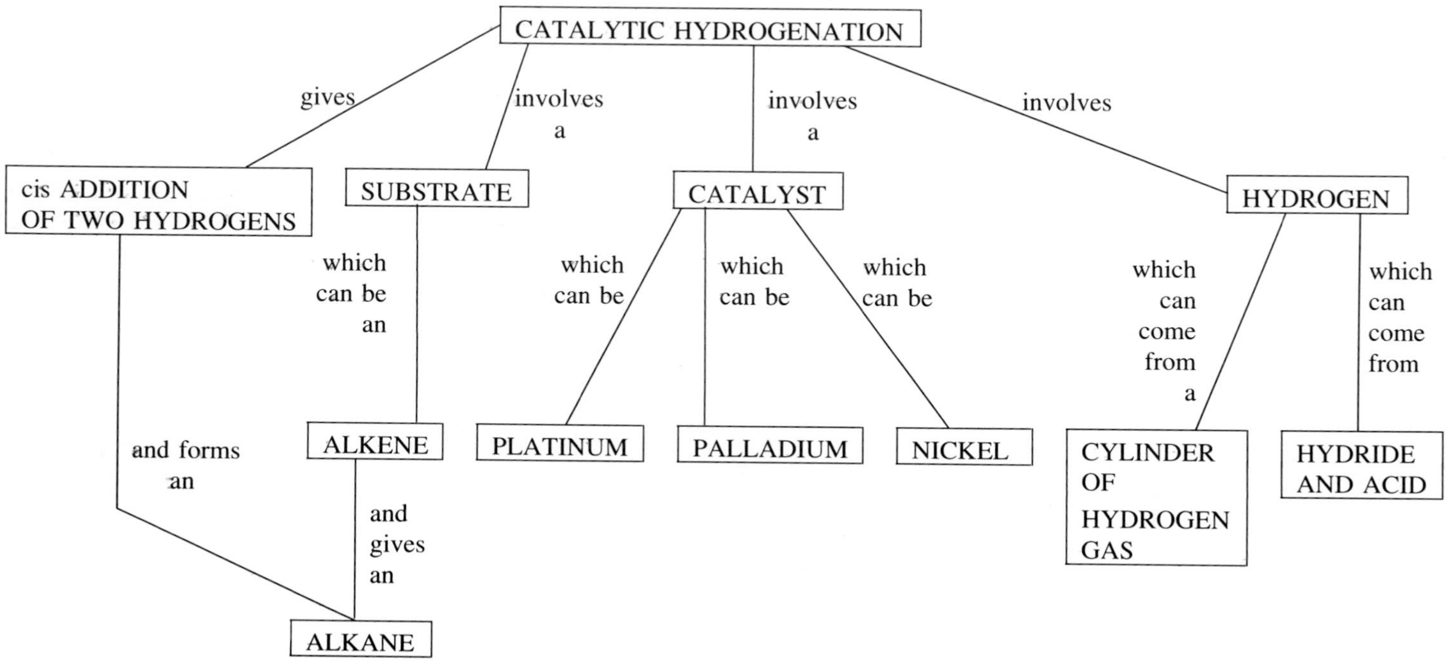
CATALYTIC HYDROGENATION
gives
involves a
involves a
involves
cis ADDITION OF TWO HYDROGENS
SUBSTRATE
CATALYST
HYDROGEN
which can be an
which can be
which can be
which can be
which can come from a
which can come from
and forms an
ALKENE
PLATINUM
PALLADIUM
NICKEL
CYLINDER OF HYDROGEN GAS
HYDRIDE AND ACID
and gives an
ALKANE

EXPERIMENT 16.1 HYDROGENATION OF *CIS*-5-NORBORNENE-*ENDO*-2,3-DICARBOXYLIC ACID USING Pt/H₂

Estimated Time:
2.5 hours

Prelab

1. Calculate the millimoles of *cis*-5-norbornene-*endo*-2,3-dicarboxylic acid contained in 1.00 g and the theoretical yield of hydrogenated product.
2. How many millimoles of hydrogen gas are generated in this experiment after the catalyst is formed? How many millimoles of excess hydrogen are formed over the amount needed to reduce the double bond?

Special Hazards

The hydrogen gas generated in this experiment is flammable and explosive in air; avoid flames. In general, hydrogenation catalysts can be pyrophoric (spontaneously flammable) in air on occasion, especially when hydrogen-rich such as after a reduction. The risks are minor as long as the catalysts are kept wet.

Procedure

Wire a balloon securely onto the sidearm of a 125–mL filter flask using copper wire and pliers. Be sure to wrap the wire around the sidearm twice to ensure an airtight seal, then twist the two ends together using pliers, cut off the excess, and fold the protruding wire out of the way. In the flask place water (10 mL), powdered charcoal (0.5 g), and 5% aqueous platinum(IV) chloride solution (0.5 mL, by syringe). In a different syringe take up 3 mL of 1 *M* sodium borohydride solution, then add it to the flask while swirling. Allow 5 minutes for the platinum(IV) to be reduced to platinum metal. While waiting for the catalyst to form, dissolve *cis*-5-norbornene-2,3-*endo*-dicarboxylic acid (the product from Experiment 15.1, 1.0 g, 5.5 mmol) in hot water (10 mL).

Pour concentrated HCl (4 mL) into the reaction flask followed by the hot solution of the unsaturated diacid. Cap the flask with a serum stopper and wire it on securely with pliers. With a syringe take up 1.5 mL of sodium borohydride solution and thrust the needle through the serum stopper. Add the solution dropwise while swirling. The balloon may not inflate because the hydrogen is reacting with the alkene so quickly. Add a second 1.5–mL portion of NaBH₄ solution, then swirl the flask for 10 minutes. Release the pressure by removing the plunger from a syringe and poking the needle through the serum cap.

Set up a Büchner funnel for suction filtration. Wet the filter paper and cover it with a layer of Celite (a filter aid consisting of diatomaceous earth) to help catch the fine particles of charcoal. Wet the Celite with a few drops of water, then filter the solution through it. With suction still going, wash ether (80 mL) through the filter cake to dissolve any product adsorbed on the carbon or the catalyst. Disconnect the vacuum just as the last of the ether passes through the filter cake. Put the filter cake in the container marked "Platinum Catalyst Recovery," and place the filtrate (both water and ether) in a separatory funnel. If too much of the ether has evaporated to give a good layer, add some more. Shake the mixture and separate the layers. If in doubt about which layer is which, review the extraction lab, look up the densities of ether and water, or add a drop of water, and see which layer it goes to.

Caution. In any extraction save all unwanted layers until the final product is in hand. If you are in doubt about which layer contains your product, think about the fact that both it and its alkene precursor, which have similar structures, can be recrystallized from water, so its solubility in cold water is low. Also, even though the molecule contains two carboxylic acid groups, it has a large nonpolar portion.

Separate the layers, saving the ether layer (which contains your product) and putting the aqueous layer back in the separatory funnel. Extract with one more 30–mL portion of ether to get out any of the product that remained in the aqueous phase. Combine the ether extract with the first one and wash them with saturated aqueous NaCl to remove most of the water which has dissolved in the ether phase. Discard the aqueous phase, then dry the ether over Na_2SO_4, filter, and evaporate to dryness on a steam bath in the hood to obtain your crude product.

To recrystallize the crude product, place it in a 25–mL Erlenmeyer flask, add water (3 mL) and bring it to a boil. Add 10 drops of concentrated HCl to help decrease the solubility of the diacid by the common-ion effect. Remove the flask from the heat and let it cool to room temperature, then cool it in an ice bath. Collect the product by suction filtration, rinse with cold water, air dry, weigh, and record the melting point.

QUESTIONS

1. In the reaction of platinum (IV) ion with hydride ion to form platinum metal and hydrogen gas, what is oxidized and what is reduced? What is the oxidizing agent? The reducing agent?

2. Can you think of two advantages of generating platinum metal in the flask compared to simply adding a pinch of powdered platinum metal? What are two disadvantages of forming the platinum catalyst *in situ*?

3. For an industrial hydrogenation of vegetable oil, for example, why might you ever choose platinum instead of nickel, which is cheaper? In other words, what are some factors to look for in a catalyst besides price?

Isolation of Natural Flavors and Fragrances

Overview

Have you ever wondered what makes a flavor or fragrance? Or what are the differences between natural and artificial flavors?

While the tongue can register only four basic tastes (sweet, sour, salty, and bitter) the nose has a much wider range of responses. It was postulated by the early Greek atomist Lucretius that odor arose from tiny ''atoms'' traveling through the air and arriving at suitably shaped pores in the nose. Recent theories agree with this idea and postulate seven basic ''primary'' odors. The odor of a molecule depends on its shape and dipole, which dictate which receptor site it fits (Figure 17–1). Putrid and pungent receptor sites appear not to correspond to a particular shape but rather to the dipole of the molecule.

If the receptor sites are chiral, it might be expected that two enantiomers might smell different. This is in fact the case in many instances; for example, compare the odors of spearmint [of which a major component is (+)-carvone] and caraway [containing largely (−)-carvone].

A flavor or fragrance consists of a combination of volatile compounds which contact the olfactory bulb simultaneously. Many natural flavors or fragrances, such as strawberry, vanilla, coffee, rose, or jasmine, contain a hundred or more volatile components. Other flavors, such as banana, peach, lemon, or clove, may contain only a handful of important volatile components.

A simple artificial flavor or fragrance reproduces a small number of these components; a high-quality one recreates many of the major components. Economics dictates that many artificial flavors lack the richness and depth of the natural sources, although theoretically it is possible to analyze and match a natural flavor or fragrance almost precisely. Often a natural oil or extract may be enhanced by adding a small number of synthetic ingredients to improve the total impact and to replace volatile components lost during the extraction or distillation of the natural essential oil.

The history of flavors and fragrances makes a fascinating story. Early trade in spices and aromatic herbs is reported to have taken place in China, where about 2700 B.C. Shen Nung, the founder of Chinese medicine, discovered the curative powers of over 100 plants. Confucius in the fifth century B.C. described the use of ginger and cassia (Chinese cinnamon).

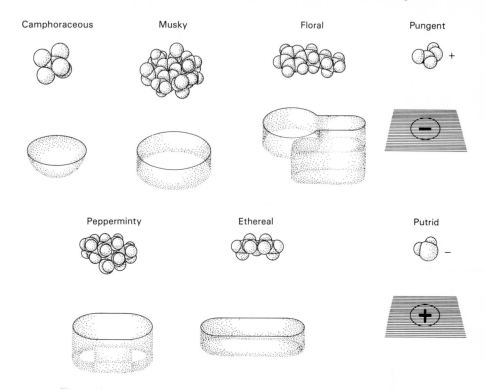

Figure 17-1 Odor Receptor Sites (from "the Stereochemical Theory of Odor," by J. E. Amoore, J. W. Johnston, Jr., and M. Rubin, copyright © 1964 by Scientific American, Inc. All rights reserved.)

An Egyptian medical scroll from approximately 1515 B.C. described almost 800 drugs, herbs, and spices, many of which were only available from China and other regions of Asia, implying that trade was occurring. The Old Testament in the book of Genesis describes a land called Hevila, through which a branch of the river from the Garden of Eden flowed, in which were found fragrant resins, including myrrh.

In the fourth century B.C. Alexander the Great established Greek settlements and trading posts along the spice trading routes to Asia. The Romans expanded the spice trade, making use of pepper, clove, cinnamon, and nutmeg in food and wine. Spice trading with Europe diminished with the fall of the Roman Empire, but in the late thirteenth century Marco Polo's travels led to the establishment of direct trade between Europe and Asia.

Much of the subsequent world exploration by Europeans was driven by the desire to expand the spice trade. The Italians, Portuguese, Spanish, Dutch, French, and British competed fiercely for control of spice-producing regions such as India, Ceylon (now Sri Lanka), Singapore, Malaysia, Sumatra, and the Spice Islands off the east coast of Africa.

Throughout much of history the total markup between producers and retailers of spices has been on the order of 100:1. Now the ratio is closer to 2:1, testimony to more efficient transportation plus a freer market. Today the international market for spices and essential oils is a multibillion dollar industry. Table 17–1 shows 1986 statistics on U.S. import tonnage and dollar value of some popular flavoring materials.

TABLE 17–1   IMPORTS OF SOME FLAVORING MATERIALS TO THE UNITED STATES IN 1986

Flavoring material	Metric tons	Value ($millions)
Caraway	3,700	4.2
Cassia	12,000	15.2
Cloves	1,000	4.3
Cumin	3,600	4.0
Ginger	5,600	5.8
Mustard	47,100	23.2
Nutmeg	1,800	6.9
Origanum	5,400	8.0
Paprika	7,200	10.3
Pepper, black	37,800	152
Pepper, white	3,500	17.8
Red peppers	8,200	8.8
Sage	2,100	5.6
Sesame seed	34,500	25.0
Vanilla beans	1,000	58.6

Source: U.S. Department of Commerce.

In recent times the compositions of many flavors and fragrances have yielded to gas chromatographic analysis and mass spectroscopy. By extracting a natural material and injecting the extract into a gas chromatograph which is attached to a mass spectrometer (a method known as GC-MS), many volatile components can be identified. As an example, a gas chromatogram for a steam distillate of Brazilian coffee, showing approximately 70 peaks and shoulders, appears in Figure 17–2.

For coffee flavor, over 500 components have been identified. These are listed by classes in Table 17–2. Often the aromatic components of a natural material can be concentrated in an essential oil, which may be used in perfumery or cooking.

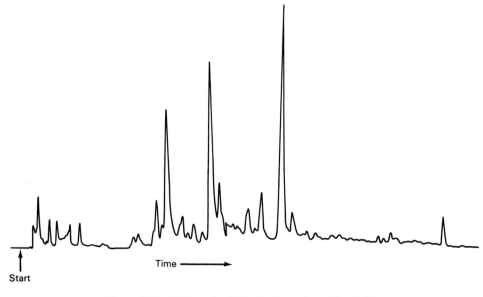

Figure 17-2   GC Trace for Steam Distillate of Brazilian Coffee

TABLE 17–2   COMPONENTS OF COFFEE FLAVOR (By Class)

Acids	22	Phenols	21
Alcohols	19	Pyrazines	67
Aldehydes	24	Pyridines	7
Amines	4	Pyrroles	25
Esters	29	Quinolines	2
Hydrocarbons	49	Quinoxalines	11
Indoles	3	Sulfides	17
Ketones	83	Thiazoles	28
Lactones	7	Thiols	5
Oxazoles	25	Thiophenes	26

For fragrances where some of the components would be destroyed by heating, essential oils are obtained by squeezing in presses (e.g., rose and orange oils) or by grinding and extracting with ethanol (used for jasmine and vanilla). Clove oil, as well as smelling good and acting as a preservative, has been used as a temporary anesthetic for dental pain for thousands of years. Clove cigarettes imported from Indonesia are said to be mildly hallucinogenic, as well as highly carcinogenic.

The main component of clove oil is eugenol, which is accompanied by a small amount of acetyleugenol. Clove oil makes up about 14–20% of dry whole cloves by weight.

Eugenol
(4-allyl-2-methoxyphenol)
MW 164, bp 225°

Acetyleugenol
(4-allyl-2-methoxyphenol acetate)
MW 206

Eugenol is also found in allspice, cinnamon leaf, and West Indian bay. In Experiment 17.1 clove oil is isolated by steam distillation of freshly ground whole cloves. Steam distillation, used in the isolation of many essential oils, relies on the fact that the vapor pressures of two immiscible liquids *add up* in opposing atmospheric pressure.

If two liquids are *miscible*, the mixture obeys Raoult's law. This means that if we have a liquid that is 70% (by moles) compound *A* and 30% (by moles) compound *B*, the vapor pressure above the liquid will be 70% of the vapor pressure of pure *A* plus 30% of the vapor pressure of pure *B*.

On the other hand, if two liquids are *immiscible*, their full vapor pressures add up to oppose atmospheric pressure. Therefore, in the case of eugenol, which would normally boil only at 255° but has an appreciable vapor pressure at 100°, if we have water present at its boiling point, it creates a high vapor pressure to push back the atmosphere and allow both water and eugenol to distill over.

In the collection flask the eugenol and acetyleugenol will form visible droplets in the water, so you can follow the progress of the distillation visually. After you have isolated your product you will take an infrared spectrum to see if it matches a literature spectrum of eugenol.

APPLICATION: COMMERCIAL DISTILLATION OF ESSENTIAL OILS

All spices consist mostly of cellulose, which contributes nothing to the aroma and flavor. By extracting the essential oil the flavor and aroma can be concentrated greatly. The advantages and disadvantages of essential oils relative to spices are outlined in the first table.

ADVANTAGES AND DISADVANTAGES OF SPICE ESSENTIAL OILS

Advantages
   Hygienic, free from all microorganisms
   Flavoring strength within acceptable limits
   Flavor quality consistent with source of raw material
   No color imparted to the end product
   Free from enzymes
   Free from tannins
   Stable in storage under good conditions
Disadvantages
   Flavor good but often incomplete
   Flavor often unbalanced
   Some readily oxidize
   No natural antioxidants
   Readily adulterated
   Very concentrated, so difficult to handle and weigh accurately
   Not readily dispersible, particularly in dry products

Most spices contain 0.5% or more of a volatile oil (see second table) which can be collected by distillation. Care must be taken not to overheat the spice, which could destroy some of the desired components and might result in the loss of the most volatile constituents (the "top notes") of the fragrance. Other methods of obtaining essential oils include *pressing* or extracting with an organic solvent, which is then evaporated to yield an *oleoresin*.

VOLATILE OIL CONTENT OF SPICES

Spice	Volatile oil content (% v/v)
Allspice (pimento)	3.0–4.5
Anise	1.5–3.5
Anise, China star	About 30
Capsicum	Nil
Caraway	2.7–7.5
Cardamom seed	4.0–10.0
Cassia (Chinese cinnamon)	0.5–4.0
Celery	2.5–3.0
Cinnamon	0.5–1.0
Clove	15–20
Coriander	0.4–1.0
Cumin	2.5–4.5
Dill	2.0–4.0
Fennel	4.0–6.0
Fenugreek	Trace
Ginger	1.0–3.0
Mace	12–15
Nutmeg	6.5–15
Paprika	Nil
Pepper, black	2.0–4.5
Pepper, white	1.5–2.5
Turmeric	4.0–5.0

For water distillation, the plant material is loaded into a still which contains a slow-speed paddle stirrer. Water is added and the mixture is brought to a boil by submerged steam coils. Distillation is continued (and condensed water continuously returned to the still) until all the essential oil has been collected (see figure below).

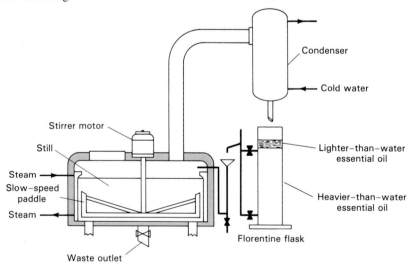

Still for Extracting Essential Oils

Experiment 17.2 is the isolation of limonene, the main component of orange oil, also by steam distillation.

Limonene
MW 136
bp 175 –177°
n_D^{20} = 1.4715
dens. 0.84 g/mL

Commercial orange oil is not prepared this way because other volatile components are lost in the process, and some of the top notes of the fragrance disappear. The process of steam distillation results in attaining limonene of approximately 97% purity. Limonene is an example of a terpene, a 10-carbon molecule whose skeleton consists of two linked isoprene units.

Isoprene

Hundreds of terpenes are known, as are numerous sesquiterpenes (three isoprene units, 15 carbons), diterpenes (four units, 20 C), and triterpenes (30 C). Rubber is a very long chain of isoprene units hooked head to tail. In Figure 17–3 are shown some typical mono-, sesqui-, di-, and triterpenes, as well as rubber. Terpenoid compounds arise biogentically from oligomerization of isopentenyl pyrophosphate in the cell (Figure 17–4). Enzymes may then catalyze oxidation, addition of water, or other reactions to produce functional groups on the carbon skeleton.

There is also some politics involved in the essential oil trade. Madagascar, for example, raises a large proportion of the world's vanilla bean crop, which is extracted with ethanol to give the essential oil. When Coca-Cola changed its formula in 1986, the government of Madagascar complained loudly, since old Coke uses real vanilla flavor, whereas new Coke presumably uses artificial vanilla. Of course, the folks at Coke will not divulge either formula, but there is no doubt that their competitors at Pepsi have extracted and analyzed Coke syrup quite thoroughly.

The main component of vanilla flavor, vanillin, has a structure related to eugenol.

Vanillin

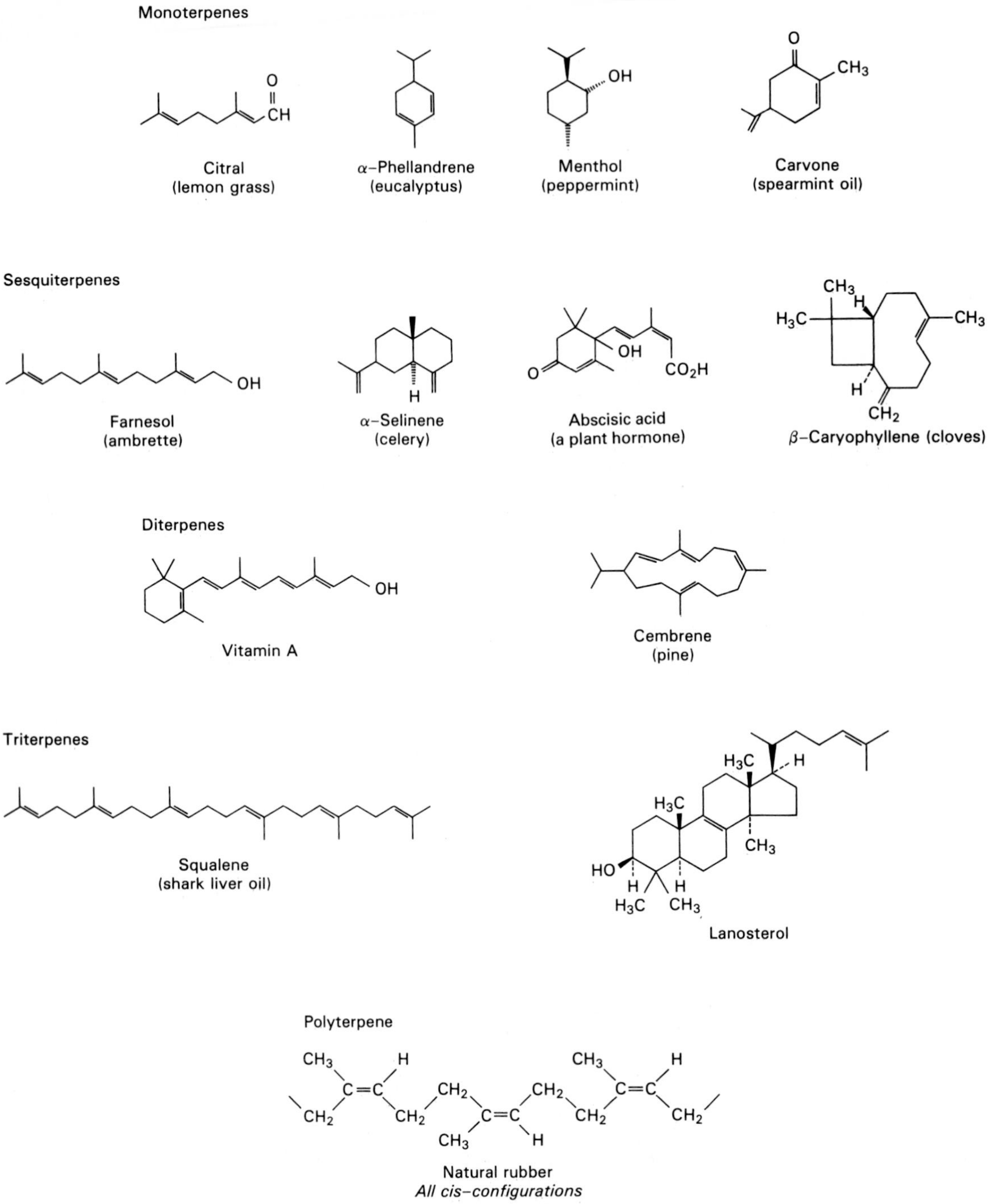

Monoterpenes

Citral
(lemon grass)

α–Phellandrene
(eucalyptus)

Menthol
(peppermint)

Carvone
(spearmint oil)

Sesquiterpenes

Farnesol
(ambrette)

α–Selinene
(celery)

Abscisic acid
(a plant hormone)

β–Caryophyllene (cloves)

Diterpenes

Vitamin A

Cembrene
(pine)

Triterpenes

Squalene
(shark liver oil)

Lanosterol

Polyterpene

Natural rubber
All cis-configurations

Figure 17-3   Some Terpenoid Compounds

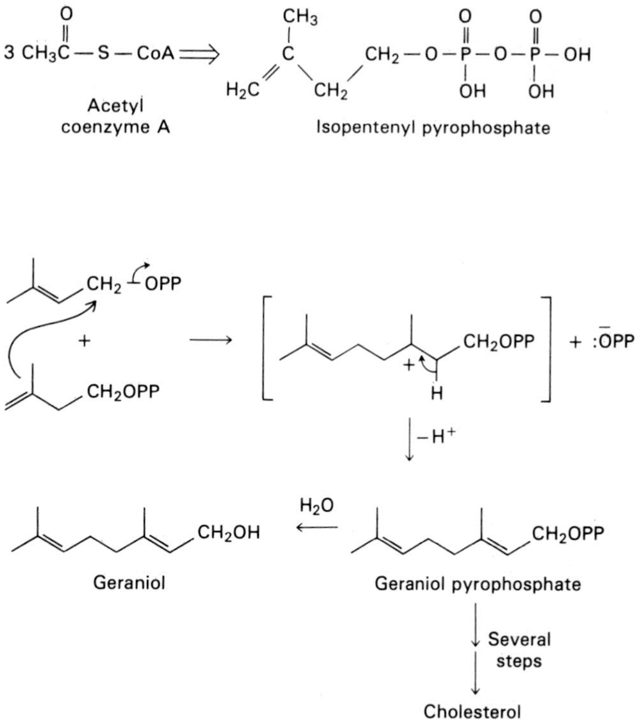

Figure 17-4   Biosynthesis of Terpenoid Compounds

In fact, vanillin has been made commercially from eugenol. Vanillin is also made very inexpensively by digesting wood pulp with strong acid. Both vanillin and eugenol are classified as *lignans* because of their structural similarity to *lignin*, a major component of wood for which the complete structure is not known. Lignin exists as a polymer. Its monomer is a phenol with a methoxy group *ortho* and a three-carbon chain *para* to the hydroxyl:

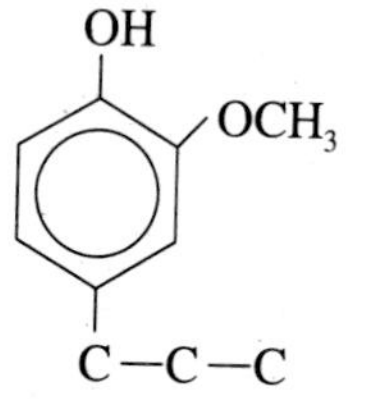

Lignin monomer
(complete polymeric
structure unknown)

   If you can discover important new uses for cheap, renewable starting materials such as lignin or cellulose, you could become a millionaire. Cellulose, which is a polymer of glucose found in plants, accounts for an estimated 50% of all the carbon on the earth. That is certainly food for thought (as well as for cows).

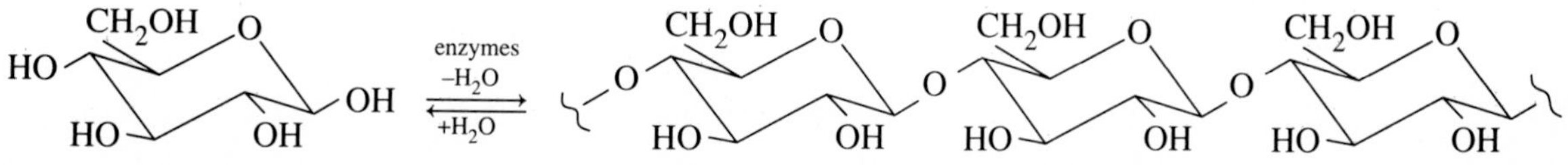

APPLICATION: VANILLA FLAVOR FROM WOOD

Paper Plant (Courtesy of Georgia-Pacific Corp.)

A story is told of a guy working in a paper mill who one day noticed some white crystals forming on a cooling pipe. On sniffing them, he discovered that they smelled like vanilla. The crystals turned out to be vanillin.

Nowadays large quantities of vanillin are made by alkaline air oxidation of fermented spent-waste liquor from sulfite paper mills. Sodium hydroxide is added to the crude liquor and it is heated at about 160–175° for 2 hours at a pressure of approximately 10 atm air. This process oxidizes lignin to vanillin, which is then extracted with an organic solvent and purified by vacuum distillation and recrystallization.

Vanillin can also be made from guaiacol (*o*-methoxyphenol), which is obtained from wood tar or coal tar formed during the destructive distillation of wood. From guaiacol there are several routes to vanillin. The Reimer–Tiemann reaction can be used to introduce an aldehyde group onto an aromatic ring.

$$\text{Guaiacol} \xrightarrow[\text{KOH}]{\text{CHCl}_3} \text{Vanillin}$$

Guaiacol   Vanillin

The Gattermann synthesis using HCN has also been used (with great caution, we hope), although several isomeric products are formed.

$$\text{Guaiacol} \xrightarrow[\text{HCl}]{\text{HCN}}$$

Guaiacol

Vanillin + Isovanillin + *o*-Vanillin

REFERENCES

AMMORE, J. E., J. W. JOHNSTON, JR., and M. RUBIN, "The Stereochemical Theory of Odor," *Scientific American* (Feb. 1964) 210.

HEATH, H. B., and G. REIN, *Flavor Chemistry and Technology*, AVI, Westport, Conn., 1986.

THEIMER, E. T., ed., *Fragrance Chemistry*, Academic Press, New York, 1982.

Concept Map

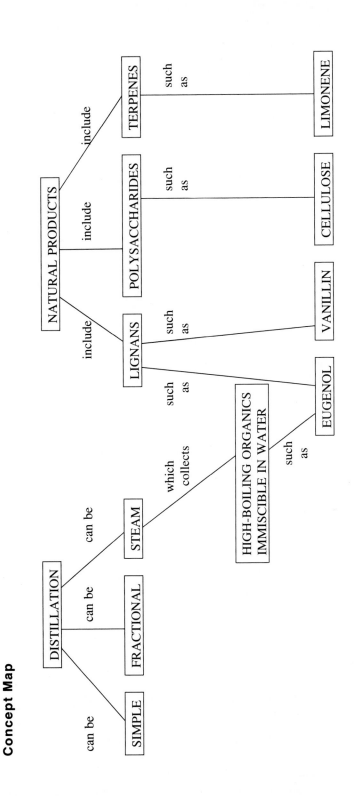

EXPERIMENT 17.1 STEAM DISTILLATION OF EUGENOL FROM CLOVES

Estimated time:
2.0 hours

Special Hazards

Dichloromethane is irritating and toxic. Use in the fume hood and avoid exposure. Dispose of any dichloromethane-containing wastes in a specially marked container in the fume hood; do not put down the drain.

Procedure

In a 250–mL round-bottomed flask place freshly ground cloves (10.0 g) and water (80 mL). Set up for simple distillation, heat to boiling, and collect the distillate until you no longer see droplets of organic liquid coming out of the condenser (about 40 mL). Place the distillate in a separatory funnel and extract with three 10–mL portions of dichloromethane. Combine the dichloromethane extracts and dry them over Na_2SO_4.

Now you will examine the purity of the product using thin-layer chromatography (TLC). Using a clean filed-off-flat syringe needle, spot a small droplet of the liquid about 1 cm from the end of a thin-layer chromatography slide, develop with dichloromethane, then visualize by UV light. Circle the spots in pencil, then place the slide in an iodine chamber for a few minutes, examine it, and circle any new spots. Finally, dip the slide quickly in a solution of 2% phosphomolybdic acid in 95% ethanol, wipe off the back, and place it on a hot plate set on *low*. The phosphomolybdic acid (PMA) will cause any spots to turn dark green, orange, or brown. Refer to Section 4 for details.

While the TLC slide is developing decant the combined CH_2Cl_2 extracts into a tared beaker and evaporate by heating on a steam bath on the hood. Weigh the product and calculate the percent oil based on the 10 g of cloves used. Enjoy the smell of the product, which is much the same as clove oil sold in drugstores as a dental anesthetic. Use of your product for this purpose is not recommended, though, because it still contains a trace of dichloromethane. Combine your product with that of several others and take an infrared spectrum. If necessary, refer to Section 13 for details.

QUESTIONS

1. Interpret the most important peaks in the IR spectrum of your product.
2. How closely does your spectrum match the literature spectrum of eugenol shown in Figure 17–5? What differences are there, and how do you explain them?

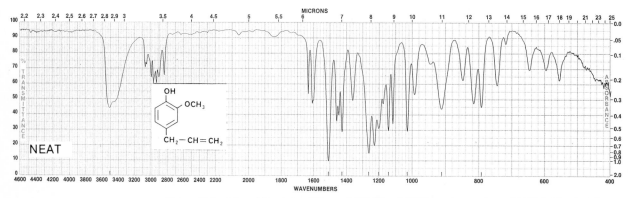

Figure 17-5   Infrared Spectrum of Eugenol (All IR and NMR spectra courtesy of Aldrich Chemical Co.)

EXPERIMENT 17.2 STEAM DISTILLATION OF ORANGE OIL

Estimated Time:
2.0 hours

Special Hazards

Ether is highly flammable; avoid exposure to heat or flames. If the orange from which the peel is taken is to be eaten, it must be done outside the lab.

Procedure

Cut or tear one orange peel into pieces less than 1 cm on a side, so that they can be removed from the round-bottomed flask after the experiment. Be aware that they will swell during the heating. Place these pieces in a 250–mL round-bottomed flask with 100 mL of water. Set up for simple distillation by adding a Claisen head, thermometer with adapter, and condenser, plus a collection container (a beaker or round-bottomed flask with a bent adapter) as well as a heating mantle and variable transformer. Distill and collect 30–50 mL of distillate. Swirl the collection flask and observe the droplets of limonene on the surface of the water. Smell the product and describe the odor. Place the distillate in a separatory funnel. Rinse the collection flask with a 20–mL portion of ether and add this to the separatory funnel. Shake and vent, then remove the aqueous phase and dry the ether extract over Na_2SO_4, decant into a beaker, add a boiling chip, and remove the ether on a steam bath in the hood. Note the odor and record the mass of the product (0.2–0.5 g). Combine your product with that of several others and take an IR spectrum and refractive index. Dispose of the orange rind in a trashcan; do not put it in the sink because it may clog the drain.

QUESTIONS

1. Interpret the most important peaks in the IR spectrum you obtained.
2. How well does your IR spectrum match the literature spectrum in Figure 17–6? Explain any differences.

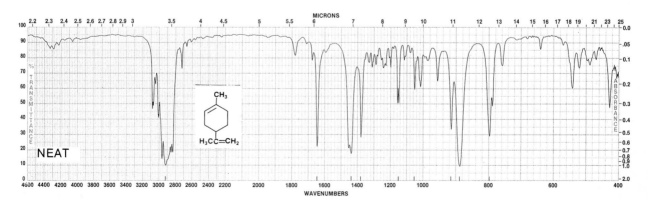

Figure 17-6   Infrared Spectrum of Limonene

Electrophilic Aromatic Substitution

Overview

There are not very many reliable ways to put functional groups on aromatic rings because they are so stable and unreactive. One of the best ways found so far is through *electrophilic aromatic substitution*. Using this method we generate a positive species strong enough to pull electrons out of the ring to bond to it. When the ring loses H^+ to rearomatize, voilà!—we have a substituted ring.

In this way we can nitrate, sulfonate, halogenate, alkylate, and acylate aromatic rings, by generating (respectively) NO_2^+, HSO_3^+, Cl^+ or Br^+, R^+, and $R-\overset{+}{C}=O$. The latter two methods are so valuable they are named after their discoverers, Charles Friedel (a Frenchman) and James Crafts (an American).

In Experiment 18.1 you will be doing a nitration by generating nitronium ion, NO_2^+, in a solution of mixed nitric and sulfuric acids. Since sulfuric acid is a stronger acid than nitric acid, it protonates the nitric acid, causing it to lose water, leaving NO_2^+:

This nitronium ion is a strong electrophile and attacks benzene rings:

If there is a substituent on the benzene ring, it directs the incoming nitro group ortho-para or meta to it by stabilizing or destabilizing resonance structures of the intermediate. An electron-donating group ($-OR$, $-NR_2$) activates the ring (makes reaction faster) and directs ortho-para. An electron-withdrawing group ($-\overset{\displaystyle O}{\overset{\displaystyle \|}{C}}-R$, $-\overset{\displaystyle O}{\overset{\displaystyle \|}{C}}-O-R$, $-NO_2$, $-\overset{+}{N}R_3$) deactivates the ring (makes reaction slower) and directs meta. In Experiment 18.1, we are working with a meta director, the carbomethoxy group on methyl benzoate. In Experiment 18.1 methyl benzoate is nitrated in the meta position. The temperature is kept low so that only one nitro group is introduced.

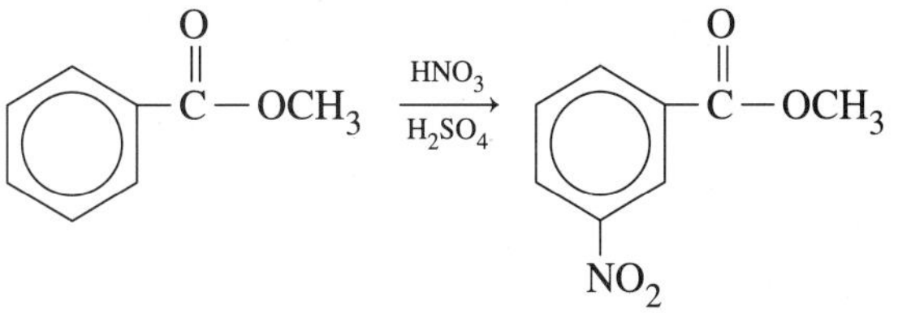

Methyl benzoate
MW 136, bp 200°
density 1.09 g/mL

Methyl *m*-nitrobenzoate
MW 181, mp 78°

Experiment 18.2 is a similar example of an aromatic dinitration; it illustrates the effect of a higher temperature.

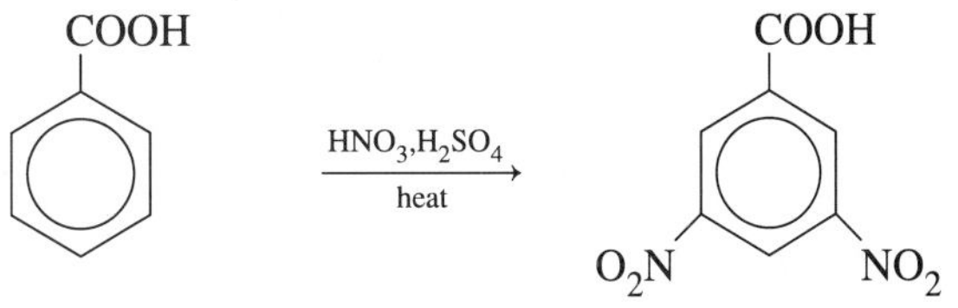

Benzoic acid
MW 122, mp 122–123°

3,5-Dinitrobenzoic acid
MW 212, mp 205–207°

Many explosives contain several nitro groups because nitro-containing compounds give off a lot of energy when burned and supply some of their own oxygen. Here are some examples:

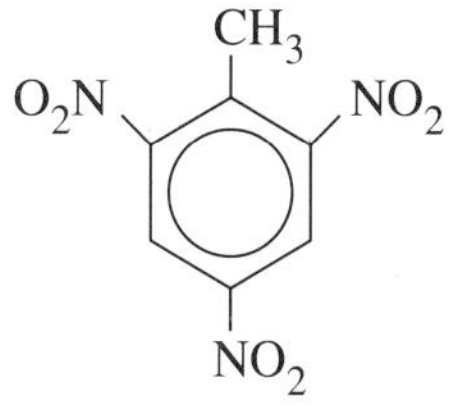

Trinitrotoluene
(TNT)

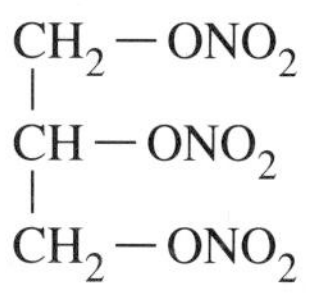

Nitroglycerin

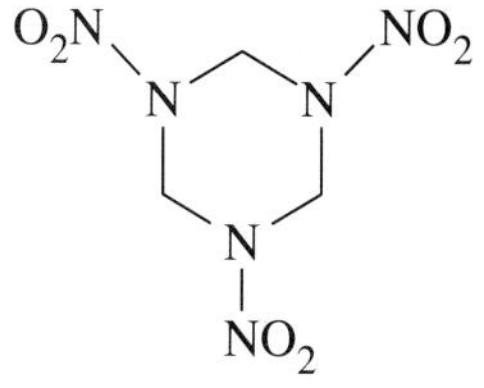

Cyclonite

So-called "nitro-burning" race cars and some model airplanes use nitromethane (CH_3NO_2) as fuel, for the same reasons as above. Luckily, the compounds we are making are not explosive or especially flammable.

Experiment 18.3 is an example of an aromatic bromination reaction, in which the bromine attacks the ring on the most electron-rich and unhindered site.

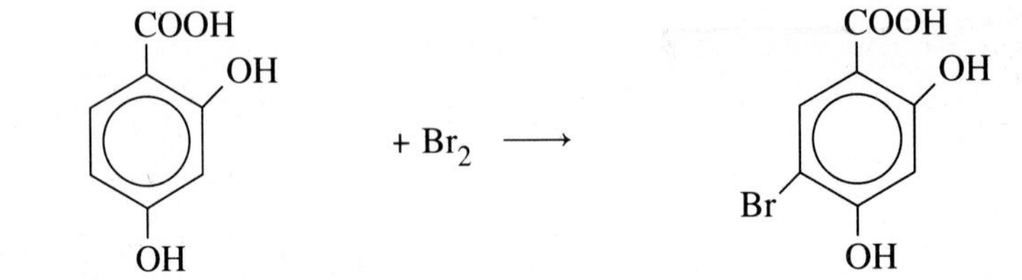

Resorcylic acid
(2, 4-dihydroxybenzoic acid)
MW 154, mp 225–227° d

5-Bromoresorcylic acid
(5-bromo-2, 4-dihydroxybenzoic acid)
MW 233, mp 194–200°

Experiment 18.4 is an example of a Friedel–Crafts alkylation. The Friedel–Crafts reaction is one of the most reliable methods for putting an alkyl or acyl group on an aromatic ring. The idea is to form a carbocation that attacks the ring, after which the ring loses H^+ to rearomatize.

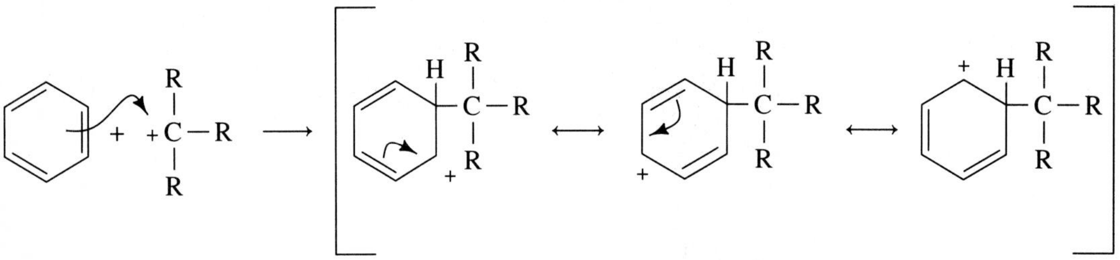

Aromatic ring   Carbo-cation

Intermediate

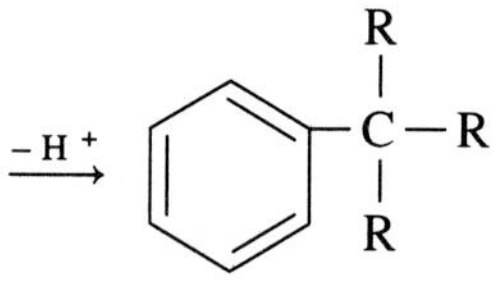

An alkylbenzene

Similarly,

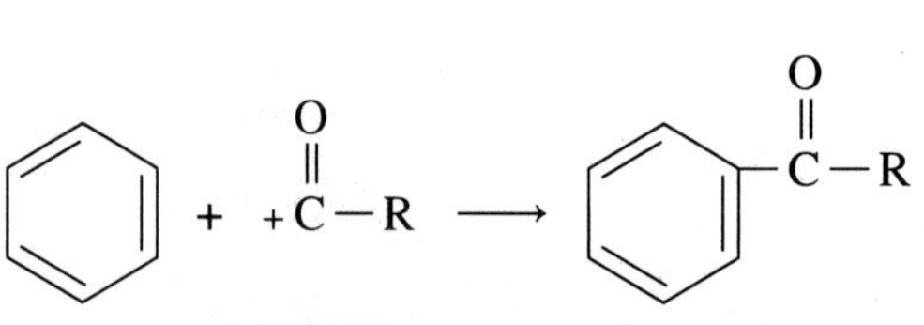

Acylium ion

An acylbenzene

Carbocations and acylium ions are formed by removing a negative leaving group using a Lewis acid, for example,

$$CH_3-\underset{\underset{\displaystyle CH_3}{|}}{\overset{\overset{\displaystyle CH_3}{|}}{C}}-Cl + AlCl_3 \longrightarrow CH_3-\underset{\underset{\displaystyle CH_3}{|}}{\overset{\overset{\displaystyle CH_3}{|}}{C^+}} + AlCl_4^-$$

or

$$CH_3-\underset{\underset{\displaystyle CH_3}{|}}{\overset{\overset{\displaystyle CH_3}{|}}{C}}-OH + H^+ \longrightarrow CH_3-\underset{\underset{\displaystyle CH_3}{|}}{\overset{\overset{\displaystyle CH_3}{|}}{C^+}} + H_2O$$

or

$$CH_3\overset{\overset{\displaystyle O}{\|}}{C}-Cl + AlCl_3 \longrightarrow CH_3\overset{\overset{\displaystyle O}{\|}}{C^+} + AlCl_4^-$$

If there are substituents already on the ring, they direct the electrophilic attack of the carbocation or acylium ion so that the new group goes ortho or para to electron-donating groups. This occurs because the substituents stabilize the resonance forms of the intermediate leading to ortho- or para-substituted products. In the case of electron-withdrawing groups which ordinarily would direct an incoming electrophilic meta, Friedel–Crafts reactions do not work at all because the ring is too deactivated.

One problem with alkylation is that it is difficult to stop at monoalkylation since the product is more electron-rich than the starting material. Since alkyl groups are electron donating, the monoalkylated material reacts faster with carbocations than the starting material. For this reason we will do a dialkylation, which can be done cleanly.

Overall reaction:

| p-Dimethoxybenzene (hydroquinone dimethyl ether) MW 138, mp 57° | t-Butyl alcohol (2-methyl-2-propanol) MW 74, dens. 0.79 g/mL mp 25°, bp 83° | 1,4-Di-t-butyl-2,5-dimethoxybenzene MW 250, mp 104–105° |

The product does not alkylate further because of steric hindrance* and forms crystals easily because of its symmetry.

QUESTION

1. Starting with a different monosubstituted benzene for each, give examples of nitration, sulfonation, halogenation, and Friedel–Crafts alkylation and acylation reactions.

*During moments of panic on tests (and even when calm) it may be helpful to remember that almost any reaction in organic chemistry can be explained by using the Big Three: (1) steric hindrance, (2) inductive effects, and (3) resonance and/or molecular orbitals.

Concept Map

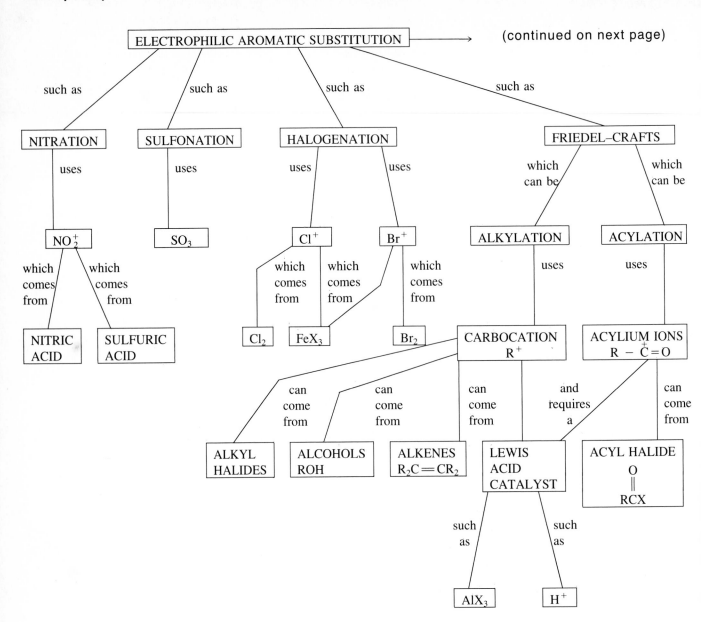

(continued on next page)

Concept Map

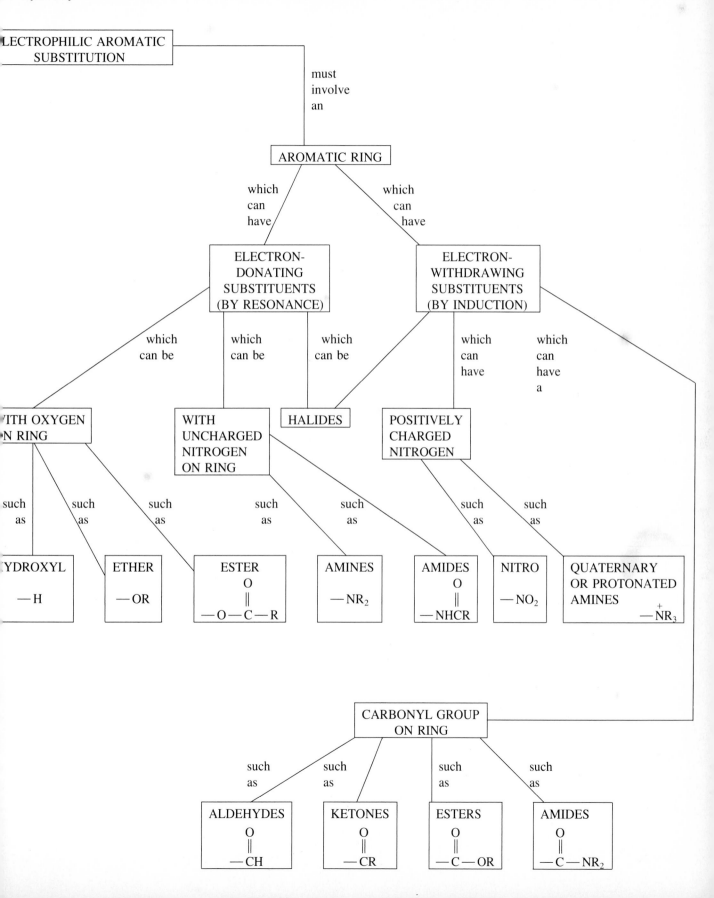

EXPERIMENT 18.1 NITRATION OF METHYL BENZOATE

Estimated Time:
2.0 hours

Prelab

1. What are the mass and volume of 12 mmol of methyl benzoate? Show your calculation and fill in the volume on the second line of the procedure.
2. Calculate the theoretical yield.

Special Hazards

Concentrated nitric and sulfuric acids are highly corrosive; use adequate precautions. Methyl benzoate is an irritant.

$$V = \frac{1.632}{1.09} = 1.497$$

Procedure

$$12 \times .136 = 1.632$$

In a 25–mL Erlenmeyer flask cool concentrated sulfuric acid (3 mL) to 0–3°, then add methyl benzoate (__ mL, __ g, 12 mmol). In a 10–mL graduated cylinder, cool a mixture of concentrated H_2SO_4 (1 mL) and concentrated HNO_3 (1 mL). Now, using a Pasteur pipette, add the solution of mixed HNO_3/H_2SO_4 dropwise to the solution of methyl benzoate. During the addition swirl the flask often and maintain the temperature in the range 5–20°. Be sure to swirl the flask vigorously enough to mix the reagents thoroughly.

After addition of all the nitric acid solution, allow the flask to warm up to room temperature and stand for 15 minutes. To cause the product to precipitate, pour it onto ice (approximately 10 g) in a 50–mL beaker. To increase the recovery, rinse out the reaction flask with a little water and add it to the product in the beaker. Collect the product by suction filtration, and take a melting point of the crude product.

Recrystallize from a minimum of methanol (about 5–10 mL), collect the product, rinse with a little ice-cold methanol, air dry, weigh, and record the melting point (lit. mp 78°).

QUESTIONS

1. Show intermediate resonance structures leading to both methyl *meta-* and *para*-nitrobenzoates. Based on these resonance structures, which product should be favored, and why?
2. If methyl *meta*-nitrobenzoate were nitrated again, what would the product be, and why?
3. What by-products would you expect from the nitration of methyl benzoate?
4. What would be the product in each case? (*Hint:* Mononitration often occurs near 0°, dinitration near 60°, and trinitration near 90°.)

(a)

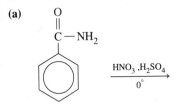

(b)

OCH$_3$

$$\xrightarrow[0°]{\text{HNO}_3 ,\text{H}_2\text{SO}_4}$$

(c)

CH$_3$ CH$_3$
N

C=O
CH$_3$

$$\xrightarrow[0°]{\text{HNO}_3 ,\text{H}_2\text{SO}_4}$$

(d)

CH$_3$

$$\xrightarrow[90°]{\text{HNO}_3 ,\text{H}_2\text{SO}_4}$$

EXPERIMENT 18.2 DINITRATION OF BENZOIC ACID

Estimated Time:
1.5 hours

Prelab

1. Calculate the millimoles corresponding to 0.61 g of benzoic acid and 5 mL of concentrated H_2SO_4.
2. Calculate the theoretical yield.

Special Hazards

Concentrated H_2SO_4 and fuming HNO_3 are extremely corrosive. Use caution and wash with water immediately if contact occurs. NO_2 is a highly toxic gas; generate only in the fume hood.

Procedure

In a small beaker in the fume hood, place benzoic acid (___ mmol, 0.61 g), then add 5 mL of concentrated H_2SO_4 and 2 mL of fuming HNO_3 cautiously by Pasteur pipette. The mixture becomes hot. After about 5 minutes the beaker is placed on a steam bath or hot plate on a low setting in the hood and warmed for 30 minutes. Brown fumes of NO_2 (*caution*: avoid breathing fumes) are evolved. If using a hot plate, be careful not to allow the solution to boil over. Cool the mixture in an ice bath, cautiously add ice and water, collect the yellow precipitate by suction filtration, wash with water, and air dry. Record the mass and melting point.

REFERENCE

BREWSTER, R. Q., B. WILLIAMS, and R. PHILLIPS, *Organic Syntheses Collective Volume 3* (1955) 337.

EXPERIMENT 18.3 MICROSCALE BROMINATION OF RESORCYLIC ACID

Estimated Time:
1.0 hour

Procedure

In a 15 × 125 mm test tube place 100 mg of resorcylic acid (2,4–dihydroxybenzoic acid, weighed on an analytical balance) and 0.80 mL of glacial acetic acid (by pipette or burette). While adding the acetic acid, rinse down any solid resorcylic acid remaining on the walls of the tube. Warm the tube in a steam bath until the solid dissolves. Using a burette or pipette in the fume hood, add 0.65 mL of a solution of 1 M Br_2 in acetic acid. After the addition, warm the solution for another 5 minutes in the fume hood. The red color of Br_2 should diminish as reaction occurs. Cool the tube in an ice bath and add ice-cold water (10 mL). Cool the tube in an ice bath. Cool and stir for a few minutes until precipitation appears complete and the precipitate has coagulated into sizable particles from the milky solution. Collect the product by suction filtration, wash with water, and allow to air dry. Record the mass and melting range (lit. mp 194–200°).

REFERENCE

SANDIN, R. B., and R. A. McKEE, *Organic Syntheses Collective Volume 2* (1943) 100.

EXPERIMENT 18.4 MICROSCALE FRIEDEL-CRAFTS ALKYLATION OF *p*-DIMETHOXYBENZENE

Estimated Time:
1.0 hour

Prelab

1. Fill out the data table in your notebook and copy the number of millimoles of reagents into the procedures.
2. Calculate the theoretical yield.
3. If you have two plastic bottles of *t*-butyl alcohol (mp 25°) at room temperature and one is in solid form while the other is liquid, which is probably purer? How could you liquefy the solid easily for pouring without melting its plastic container?

Procedure

MW 138

In a 15 × 125 mm test tube, place *p*-dimethoxybenzene (100 mg, ___ mmol), *t*-butyl alcohol (0.20 mL, ___ mmol, by pipette, or approximately 19 drops), and acetic acid (0.30 mL, ___ mmol, or approximately 14 drops), and cool the tube in a beaker containing ice and water. In another test tube place concentrated sulfuric acid (0.40 mL or approximately 34 drops, ___ mmol). Cool this tube in the ice bath as well. Add the cold sulfuric acid dropwise by Pasteur pipette to the tube containing the *p*-dimethoxybenzene at the rate of 1 drop per second or so with occasional swirling. An exothermic reaction should occur, but on this scale the heat evolved may not be noticeable. At the end of the addition some solid product should be visible in the yellow mixture. Remove the tube from the ice bath and let it stand at room temperature for about 5 minutes with occasional swirling.

To collect the product, add a little ice to the tube to dilute the sulfuric acid and to absorb the heat of dilution, fill the tube about three-fourths of the way to the top with water. Stir to suspend the white product and collect it by suction filtration (using a small Hirsch funnel if available; otherwise a Büchner funnel). Wash the crystals with water.

Recrystallize by dissolving the product in a minimum volume of methanol (about 2–3 mL) in a test tube heated on a steam bath. When the tube has cooled to room temperature, cool it in an ice bath, collect the product, and wash the product (1,4-di-*t*-butyl-2,5-dimethoxybenzene) with a few drops of ice-cold methanol. Air dry, weigh, and record the melting point (lit. mp 104–105°).

QUESTIONS

1. Draw the predicted NMR spectrum for the product.
2. What would be the expected products of the following reactions?

(a)

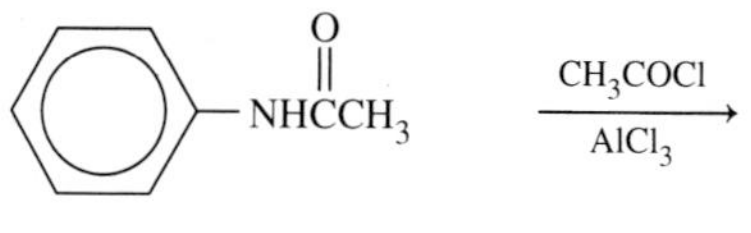

$$\underset{AlCl_3}{\xrightarrow{CH_3COCl}}$$

(b)

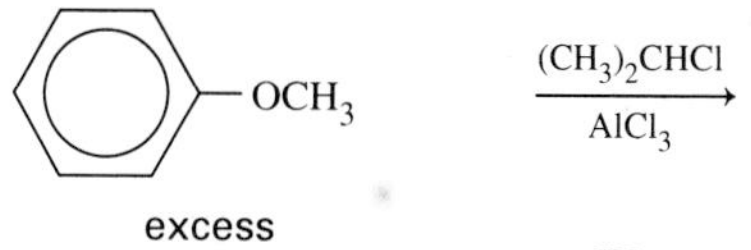

excess

$$\underset{AlCl_3}{\xrightarrow{(CH_3)_2CHCl}}$$

(c)

$$\underset{H^+}{\xrightarrow{}}$$

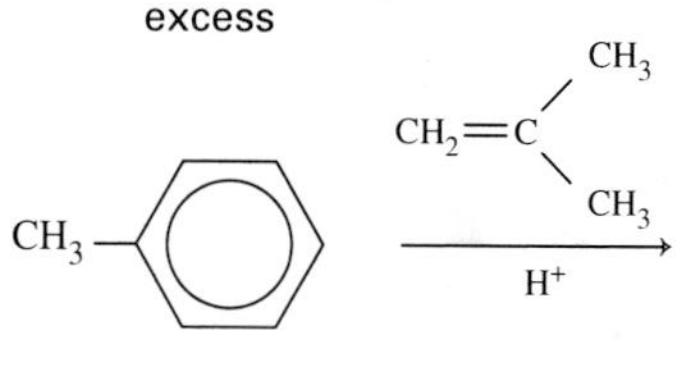

(d)

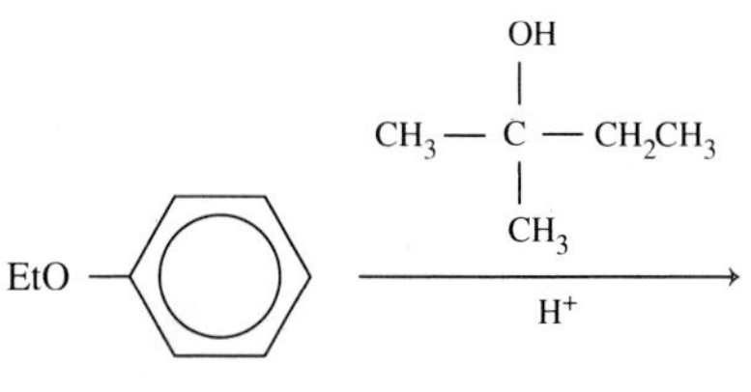

excess

$$\underset{H^+}{\xrightarrow{}}$$

3. What would be another way to dry and recrystallize the crude product of this experiment?

4. Two common preservatives used in breakfast cereals, crackers, and other foods are BHA and BHT, which stand for butylated hydroxyanisole and butylated hydroxytoluene, respectively. These, of course, are not systematic names.

BHA   BHT

These molecules are called radical scavengers. In the presence of a radical they lose H· to form a phenoxy radical, which is stabilized by resonance and by the steric bulk of the *t*-butyl groups. Since they form stable radicals they stop chain reactions that can destroy the flavor and freshness of foods. How would you make each one using a Friedel–Crafts reaction?

5. A carbocation produced during a Friedel–Crafts alkylation may undergo rearrangement, like any other carbocation, if it is favorable to do so. What rearranged product might you expect from the reaction below?

Oxidation of a Side Chain on an Aromatic Ring

Overview

Oxidations are often used in organic synthesis to convert one functional group into another. You may recall that in organic terms, oxidation usually amounts to the gain of oxygen or loss of a pair of hydrogens by the molecule. Oxidations you may have encountered so far include formation of a ketone from a secondary alcohol and conversion of an alkene to a vicinal *cis*-diol, both of which are shown below.

$$R-\underset{\underset{H}{|}}{\overset{\overset{OH}{|}}{C}}-R' \xrightarrow[H_2SO_4]{Na_2Cr_2O_7} R-\overset{\overset{O}{||}}{C}-R'$$

$$\underset{/}{\overset{\backslash}{C}}=\underset{\backslash}{\overset{/}{C}} \xrightarrow[\substack{(1)\ Os_{4,}\ O_4 \\ (2)\ NaHSO_3}]{\substack{KMnO_4,\ at\ 0° \\ or}} -\underset{|}{\overset{\overset{OH}{|}}{C}}-\underset{|}{\overset{\overset{OH}{|}}{C}}-$$

Also notice that many common oxidizing agents, such as $Na_2Cr_2O_7$, $KMnO_4$, and OsO_4, contain a metal in a highly oxidized state; this metal atom will remove electrons from the organic molecule, reducing the metal and oxidizing the organic species.

Under strongly oxidizing conditions it is possible to oxidize the side chain on an aromatic ring to a carboxylic acid.

$$\text{⬡}-R \xrightarrow{[O]} \text{⬡}-COOH$$

This reaction will work when R is an alkyl or acyl group. The reaction conditions are fairly harsh; only a few unreactive or highly oxidized functional groups (such as halogens or nitro or sulfonate groups) can survive intact on the ring. While the exact mechanism is not well understood, the process appears to proceed through a benzylic radical. One piece of evidence is that a benzylic hydrogen must be present initially for the reaction to succeed, as shown in the following examples.

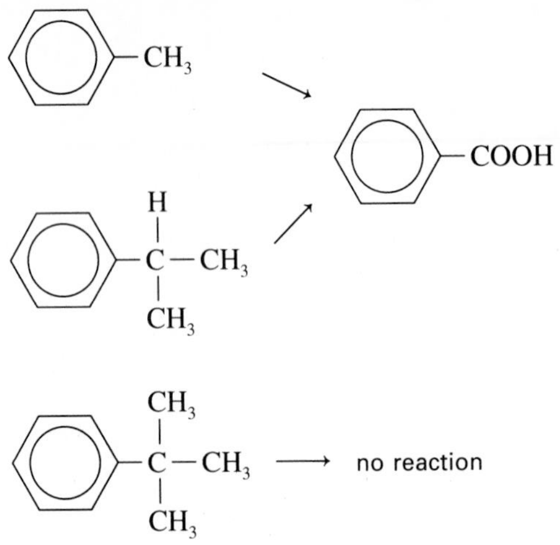

Experiment 19.1 is the oxidation of *o*-chlorotoluene to *o*-chlorobenzoic acid using KMnO$_4$, which illustrates the unreactivity of aromatic halides toward oxidation.

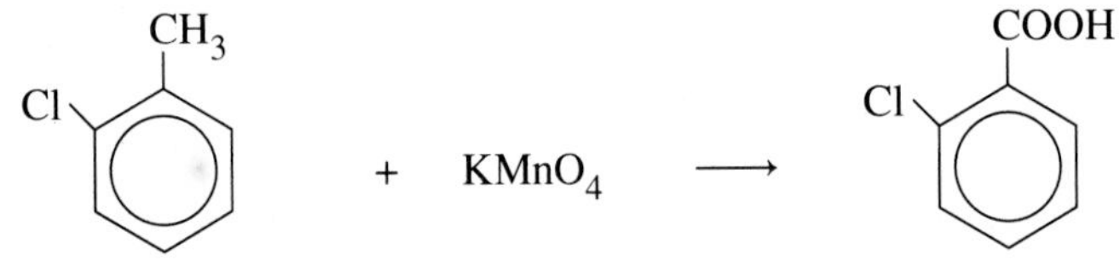

o-Chlorotoluene
MW 126, bp 157–159°
dens. 1.08 g/mL

Potassium
permanganate
MW 158

o-Chlorobenzoic acid
MW 157, mp 138–140°

Experiment 19.2 uses a different oxidant system, sodium dichromate and sulfuric acid, to oxidize *p*-nitrotoluene to *p*-nitrobenzoic acid.

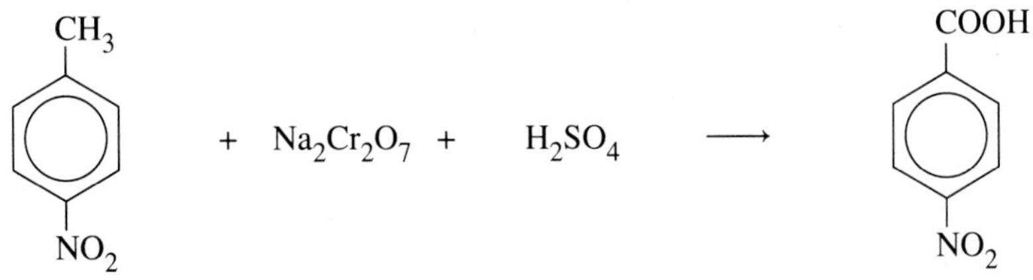

p-Nitrotoluene
MW 137, mp 52–54°

Sodium
dichromate
(as dihydrate)
MW 298

Sulfuric
acid
MW 98
dens. 1.84 g/mL

p-Nitrobenzoic acid
MW 167, mp 239–241°

Experiment 19.3 illustrates the fact that this reaction works well on side chains of heterocyclic aromatic compounds such as pyridines as well. This experiment also provides the experience of working with an amino acid.

2-Methylpyridine
(2-picoline
α-picoline)
MW 93.1, bp 144°
dens. 0.97 g/mL

KMnO$_4$
MW 158

Potassium salt
of 2-pyridine-
carboxylic acid
(2-picolinic acid)

HCl

2-Picolinic
acid
hydrochloride
MW 160
mp 210–212°

· HCl

QUESTIONS

1. Predict the products of the following reactions.

(a) Cl—⟨benzene ring⟩—CH$_2$CH$_3$ $\xrightarrow{\text{KMnO}_4}$

(b) ⟨furan ring with CH$_3$⟩ $\xrightarrow[\text{H}_2\text{SO}_4]{\text{K}_2\text{Cr}_2\text{O}_7}$

(c) ⟨naphthalene with NO$_2$ and C(=O)—CH$_3$ groups⟩ $\xrightarrow{\text{H}_2\text{CrO}_4}$

2. If KMnO$_4$ and NaOH are used to oxidize a side chain, why is it necessary to acidify the final mixture?

3. Phthalic acid is a high-volume industrial chemical used in the manufacture of plastics and as an intermediate for synthesis of many other compounds. Show two compounds it could be prepared from using dichromate oxidation.

EXPERIMENT 19.1 OXIDATION OF *o*-CHLOROTOLUENE TO *o*-CHLOROBENZOIC ACID USING KMnO$_4$

Estimated Time:
3.0 hours

Prelab

1. Calculate the millimoles contained in 3.00 g of KMnO$_4$ and in 1.00 g of *o*-chlorotoluene. Also calculate the volume of 1.00 g of *o*-chlorotoluene.

2. Calculate the theoretical yield of *o*-chlorobenzoic acid.

Special Hazards

Potassium permanganate is a strong oxidant; avoid contact. If it stains hands or clothing, rinse with aqueous NaHSO$_3$. Toluene is flammable.

Procedure

In a 100–mL round-bottomed flask place KMnO$_4$ (_____ mmol, 3.00 g), *o*-chlorotoluene (_____ mmol, 1.00 g, _____ mL), and water (35 mL). Heat to reflux for 1.5 hours, during which time most of the purple color of MnO$_4^-$ disappears and a mirror-like coating of MnO$_2$ appears on the inside walls of the flask. Filter the hot solution using suction to remove MnO$_2$ and cool the filtrate in an ice bath. Slowly add a little solid sodium bisulfite to destroy any remaining permanganate. (*Caution:* exothermic reaction.) Filter again if necessary. Continue cooling the solution in an ice bath and acidify cautiously (dropwise) with 2.5 mL of concentrated HCl. Collect the product by suction filtration and recrystallize from toluene. Record the mass and melting range.

REFERENCE

Clarke, H. T., and E. R. Taylor, *Organic Syntheses Collective Volume 2* (1943) 135.

EXPERIMENT 19.2 OXIDATION OF *p*-NITROTOLUENE TO *p*-NITROBENZOIC ACID USING Na$_2$Cr$_2$O$_7$/H$_2$SO$_4$

Estimated Time:
1.5 hours

Prelab

Calculate the number of millimoles in 3.4 g of Na$_2$Cr$_2$O$_7$, 4.5 mL of concentrated H$_2$SO$_4$ (96%, 18M, dens. 1.84 g/mL), and 1.15 g of *p*-nitrotoluene.

Special Hazards

Chromium-containing compounds are often carcinogenic. *p*-Nitrotoluene is an irritant and highly toxic. Avoid contact and dispose of wastes properly.

Procedure:

In a 50-mL round-bottomed flask place Na$_2$Cr$_2$O$_7$ (_____ mmol, 3.4 g) and *p*-nitrotoluene (_____ mmol, 1.15 g). Cool concentrated H$_2$SO$_4$ (4.6 mL, 8.5 g, _____ mmol) in an ice bath and add it to the flask dropwise over the course of 5 minutes. *Caution:* Too rapid addition will result in a violent reaction. At the end of the addition set up the flask for reflux and boil the contents for 15 minutes. Cool in an ice bath, add water (20 mL), collect the precipitate by suction filtration, and wash it with water. To purify the product, recrystallize from water to which enough ethanol is added at the boiling point to dissolve the product. Filter the hot solution to remove a small amount of insoluble chromium salts and dispose of them in a properly marked waste container. Cool the resulting solution in an ice bath. Collect the precipitate by suction filtration, wash with water, and air dry. Record the mass and melting range.

REFERENCE

KAMM, O., and A. O. MATTHEWS, *Organic Syntheses Collective Volume 1* (1941) 392.

EXPERIMENT 19.3 OXIDATION OF 2-METHYLPYRIDINE TO 2-PYRIDINECARBOXYLIC ACID USING KMnO₄

Estimated Time:
2.5 hours

Prelab

1. Calculate the millimoles and volume of 2-methylpyridine corresponding to 1.00 g. Calculate the millimoles of KMnO₄ in 2.00 g.
2. Calculate the theoretical yield of 2-pyridinecarboxylic acid.
3. Would 2-pyridinecarboxylic acid be soluble in 5% aqueous HCl? In 5% aqueous NaOH? Explain your reasoning.

Special Hazards

Potassium permanganate is a strong oxidizing agent. 2-Methylpyridine is flammable and toxic. Ether is highly volatile and flammable. Exercise due care and avoid exposure or contact with flames.

Procedure:

In a 50-mL round-bottomed flask, place 2-methylpyridine (α-picoline, _____ mmol, 1.00 g, _____ mL), KMnO₄ (_____ mmol, 3.6 g), and water (15 mL). Reflux the mixture for 45 minutes. At this time almost all of the purple color of the KMnO₄ has vanished. Cool and filter to remove MnO₂. Cool the filtrate in an ice bath and cautiously add a little NaHSO₃ to destroy any remaining permanganate (*caution:* heat is evolved). Refilter if necessary, continue cooling and acidify with 3 mL of concentrated HCl. To remove the water, place the solution in a 100-mL round-bottomed flask, add boiling chips, and boil off almost all of the water. Be sure to watch the flask closely toward the end so that it does not dry out completely and scorch the contents. Cool the flask in an ice bath to obtain the crude product. To remove residual KBr, add ethanol (20 mL) and boil to dissolve the picolinic acid hydrochloride. Dry the warm solution over MgSO₄, then gravity filter. Cool the filtrate in an ice bath (be sure that it is below 30°C), then add ether (30 mL), and cool the solution in an ice bath. Collect the crystallized product by suction filtration, wash with a little ether, and air dry. Record the mass and melting range (lit. mp 210–212° with decomposition, if carried out very slowly).

REFERENCE

SINGER, A. W., and S. M. MCELVAIN, *Organic Syntheses Collective Volume 3* (1955) 740.

Ether Synthesis

Overview

A Williamson ether synthesis consists of using an alkoxide ion (RO^-) to attack a halide ($R'X$), displacing the halide in a back-side (S_N2) attack. The abbreviation S_N2, of course, stands for substitution, nucleophilic, bimolecular. This reaction is a substitution because RO^- replaces X^-, it is nucleophilic because one nucleophile (RO^-) replaces another (X^-), and it is bimolecular because the slow step (and therefore overall rate) involves two species, RO^- and $R'X$. Here is the rate-determining step of the reaction:

$$RO^- + R' {-} X \longrightarrow ROR' + X^-$$

Alkoxide	Alkyl halide	Ether

In Experiment 20.1 the reaction is

$$\underset{\substack{\text{Phenoxide ion from}\\ \text{phenol}\\ \text{MW 94.1, mp }41^\circ\\ \text{dens. 1.07 g/mL}}}{\boxed{\bigcirc}{-}O^-} \;+\; \underset{\substack{\text{1-Iodoethane}\\ \text{(ethyl iodide)}\\ \text{MW 156, bp 69}-73^\circ\\ \text{dens. 1.95 g/mL}}}{CH_3CH_2{-}I} \longrightarrow \underset{\substack{\text{Ethyl phenyl ether}\\ \text{(ethoxybenzene, phenetole)}\\ \text{MW 122, bp }170^\circ\\ \text{dens. 0.966 g/mL}}}{\boxed{\bigcirc}{-}OCH_2CH_3} \;+\; I^-$$

A special feature in this case is that the reaction takes place in two phases, an aqueous and an organic phase. This is because ethyl iodide is soluble in organic solvents, not water, while hydroxide (the base we use to make the phenoxide from phenol) is soluble in water, not in organic solvents. We will use a phase-transfer catalyst (PTC) to move anions between the aqueous and organic phases. A PTC is not a true catalyst in the sense of lowering the activation energy of the reaction; its purpose is to bring the reagents into physical contact in a two-phase system. The PTC looks like this:

$$CH_3(CH_2)_3$$
$$|$$
$$CH_3(CH_2)_3 - N^+ - (CH_2)_3CH_3 \quad Br^-$$
$$|$$
$$CH_3(CH_2)_3$$

Tetra-*n*-butylammonium bromide
(abbreviated $R_4N^+Br^-$)
MW 322, mp 103–104°

When all the reagents and solvents are mixed, the hydroxide ion deprotonates the phenol, forming phenoxide ion which dissolves in the aqueous layer. So to begin with, the flask looks as shown in Figure 20-1. Then the R_4N^+ complexes with the PhO^- and carries it into the organic phase (the ethyl iodide), where the phenoxide ion reacts with EtI in an S_N2 displacement.

$$H_2O \quad R_4N^+ \quad Pho^-$$

$$E_tI$$

Figure 20-1 Aqueous and Organic Phases in a Round-bottom Flask

Even though $R_4N^+PhO^-$ (tetra-*n*-butylammonium phenoxide) is ionic, it is still soluble in ethyl iodide because the long "arms" of the PTC wrap around the charges, making it nonpolar on the outside:

Once the phenoxide has reacted with ethyl iodide, the R_4N^+ carries the displaced I^- back to the aqueous phase, where the R_4N^+ picks up another PhO^- to repeat the cycle.

As you might expect from the fact that the PTC must travel back and forth between the aqueous and organic phases, good mixing of layers is essential for this reaction to work. To ensure this, you will keep the magnetic stir bar stirring as vigorously as possible.

Experiment 20.2 is a similar example but does not require the use of a phase-transfer catalyst, since acetone is used as a solvent and the ions involved are slightly soluble in acetone,

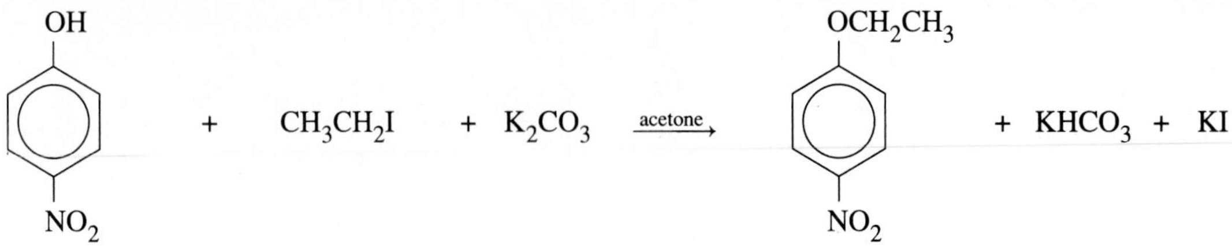

p-Nitrophenol MW 139, mp 113 − 115°	Iodoethane (ethyl iodide) MW 156, bp 69−73° dens. 1.95 g/mL	Potassium carbonate MW 138	p-Ethoxynitrobenzene (p-nitrophenetole) MW 167, mp 60°

In a similar reaction, Experiment 20.3 illustrates the conversion of one analgesic drug, acetaminophen, into another, phenacetin.

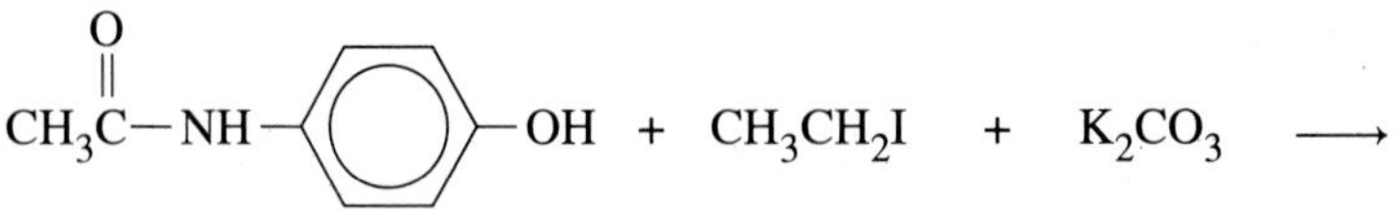

Acetaminophen (4-acetamidophenol) MW 151, mp 169 − 172°	Iodoethane (ethyl iodide)	Potassium carbonate MW 138

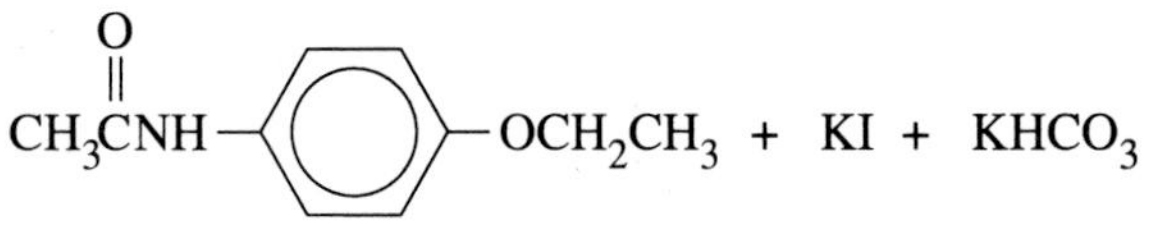

Phenacetin
(p-4-acetamidoethoxybenzene)
MW 179, mp 134−136°

Acetaminophen is the only active ingredient in Tylenol and Datril, while phenacetin is an important component of Excedrin, Vanquish, and Empirin (which also contain other physiologically active ingredients). In the human body the ether group of phenacetin is cleaved metabolically in the reverse of the synthetic equation above to form acetaminophen, which is the active analgesic.

Ether formation need not always proceed by the Williamson (S_N2) mechanism. An S_N1 ether formation is illustrated in Experiment 20.4.

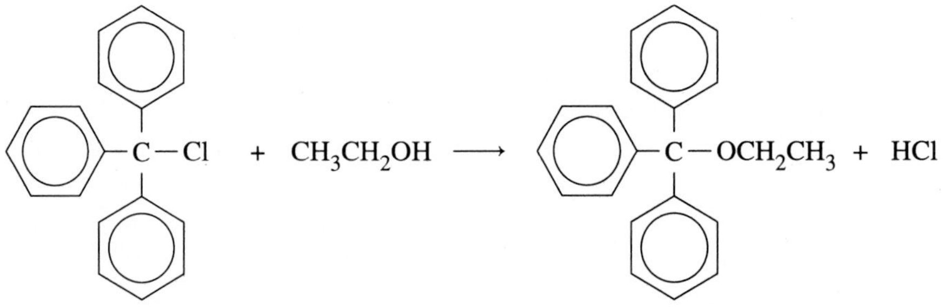

Trityl chloride (triphenylmethyl chloride) MW 279, mp 110 −112°	Ethanol MW 46, bp 78° dens. 0.78 g/mL	Ethyl trityl ether (ethyl triphenylmethyl ether) MW 288, mp 82 − 83°

This reaction cannot proceed by an S_N2 mechanism for two reasons; the carbon with the halide is tertiary and much too hindered, and ethanol is a poor nucleophile. However, this setup is excellent for an S_N1 reaction. Triphenylmethyl ("trityl") chloride can form a carbocation very easily because of the extensive (tribenzylic) resonance possibilities. Once the planar carbocation has been formed, attack by ethanol proceeds easily.

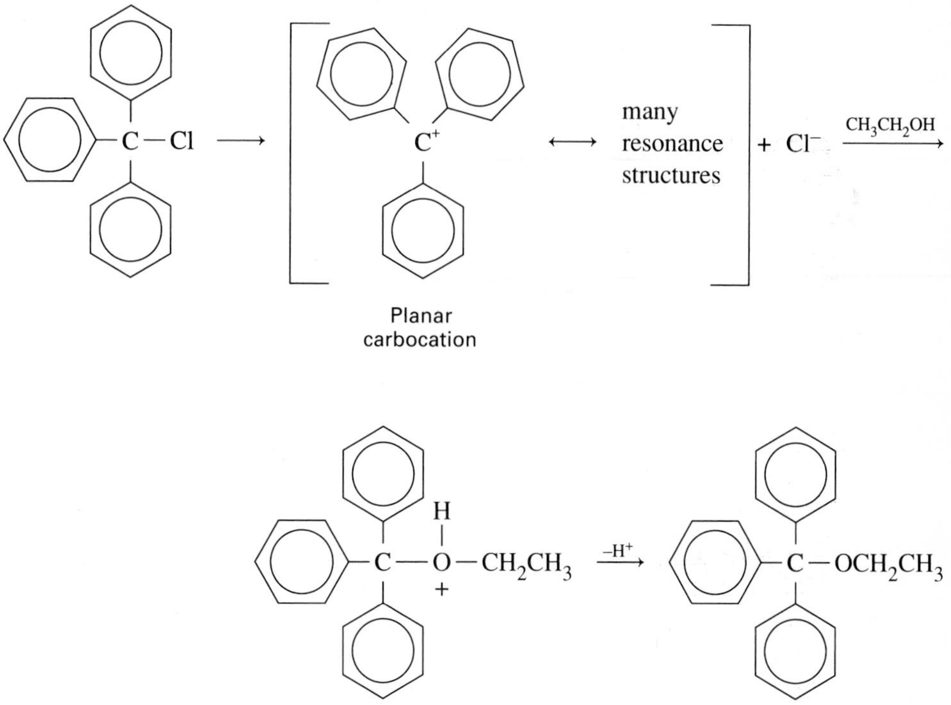

Planar
carbocation

QUESTIONS

1. What are two advantages of using ethyl iodide instead of ethyl bromide (bp 43°) in a Williamson ether synthesis?

2. How would you prepare each of the following compounds?

(a)

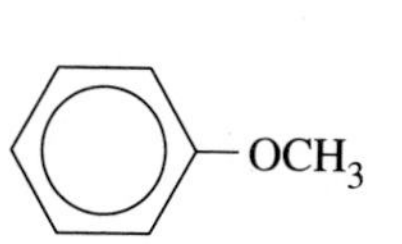

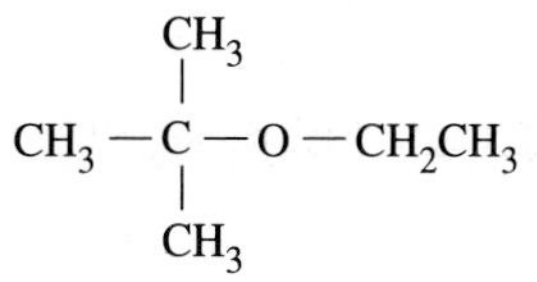

(b)

(c)

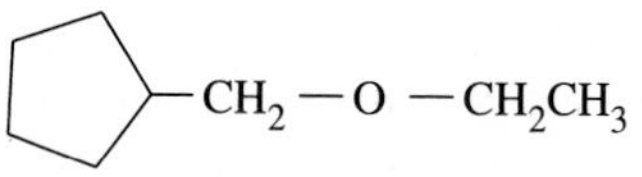

(d)

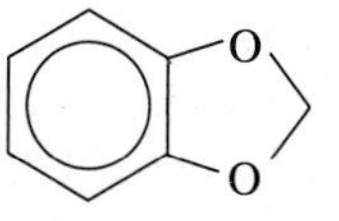

3. Draw a concept map for this experiment.

EXPERIMENT 20.1 WILLIAMSON SYNTHESIS OF ETHYL PHENYL ETHER USING PHASE-TRANSFER CATALYSIS

Estimated Time:
2.5 hours

Prelab

1. Calculate the number of millimoles in 4.5 mL of iodoethane and in 0.5 g of tetra-*n*-butylammonium bromide, and write these into the procedure.
2. Calculate the theoretical yield.
3. After the reflux period in the procedure, what substances will be present in the organic phase? The aqueous phase? Why will the density of the organic phase be lower than that of the aqueous phase?
4. At what thermometer readings do you expect to collect the product, ethyl phenyl ether, during the final distillation?

Special Hazards

Phenol is caustic to skin and can cause severe chemical burns if not washed off immediately. It is also poisonous on ingestion. Ethyl iodide is corrosive and moisture sensitive.

Procedure

In a 50-mL round-bottomed flask, place a magnetic stirring bar, sodium hydroxide (2.0 g, 50 mmol), and water (10 mL). Stir to dissolve the sodium hydroxide; considerable heat will be evolved. To the hot solution add phenol (4.7 g, 50 mmol). It should dissolve, so you now have a concentrated solution of aqueous sodium phenoxide.

Add the phase-transfer catalyst tetra-*n*-butylammonium bromide (0.5 g, _____ mmol). Now add ethyl iodide (iodoethane, 4.5 mL, _____ g, _____ mmol), which will form a separate layer. Place a heating mantle (or steam bath) under the flask and a condenser on top to complete the reflux apparatus. *Before* starting to heat the flask, be sure that the magnetic stirrer is mixing vigorously and that water is flowing through the condenser at a slow constant rate.

While maintaining effective stirring, heat the contents of the flask to a gentle reflux, keeping the condensing vapor in the lower half of the condenser. While the solution is refluxing, assemble the equipment needed for an air-cooled semimicro distillation (shown in Figure 20-2), which will be needed later.

After refluxing for 30 minutes, cool the flask briefly in an ice bath, remove the condenser, and pour the reaction mixture into a separatory funnel. Be careful not to drop the stir bar into the separatory funnel; if it does fall in, retrieve it with a stir-bar retriever, which consists of a magnet on a long handle.

Take off the aqueous (lower) layer, label it, and set it aside. Wash the remaining organic layer, which contains your product, with one 10-mL portion of water to remove any phase-transfer catalyst remaining in the organic phase. Draw off the organic (lower) layer into a small Erlenmeyer and add some drying agent such as Na_2SO_4, $CaCl_2$, or $MgSO_4$ to absorb any residual water. Decant the clear solution into a 50-mL round-bottomed flask. To improve the recovery rinse the drying agent with ether (5 mL) and add this to your product.

Add a couple of boiling chips and set up a semimicro air-cooled distillation apparatus as shown in Figure 20-2. An ice bath for the collection flask is not necessary in this case because of the high boiling point of the product.

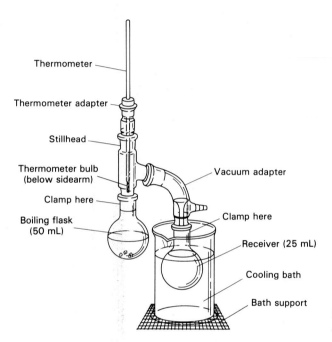

Figure 20-2   Semimicro Distillation Apparatus

Collect the forerun, at thermometer readings up to 5° below the expected boiling point for the product. Because of the high boiling point of the product (170° at 760 torr) it may be helpful to insulate the top of the distilling flask to help drive the product over. Collect the ethyl phenyl ether at its expected boiling point. Be sure not to distill completely to dryness.

Record the mass and boiling point of the product obtained. Pool your product with that of several other members of the class and take an infrared spectrum, which can be photocopied for each person. Be sure to fill in the date, names, source of product, and phase for the spectrum.

QUESTIONS

1. Interpret the most important peaks in your IR spectrum.
2. Draw the predicted NMR spectrum for ethyl phenyl ether, and show which structural features give rise to each peak.

EXPERIMENT 20.2 WILLIAMSON SYNTHESIS OF *p*-ETHOXYNITROBENZENE

Estimated Time:
2.5 hours

Prelab

1. Calculate the volume of 24 mmol of ethyl iodide and the theoretical yield of *p*-ethoxynitrobenzene.

Special Hazards

p-Nitrophenol is an irritant and toxic on ingestion. Ethyl iodide is corrosive. Avoid contact or breathing fumes.

Procedure

In a 50-mL round-bottomed flask in the fume hood, place *p*-nitrophenol (10 mmol, 1.4 g), anhydrous K_2CO_3 (10 mmol, 1.4 g), acetone (10 mL), and ethyl iodide (24 mmol, _3.74_ g, _1.92_ mL). Reflux the mixture for 1 1/4 hours, then remove the condenser and allow almost all the acetone and excess EtI to evaporate. Watch carefully and be sure to remove the heat source before complete dryness. Cautiously add cold water (20 mL) and cool the flask in an ice bath. Add ether (20 mL), stir and swirl thoroughly, and pour the mixture into a separatory funnel. Draw off the aqueous (lower) layer, dry the ether extract over anhydrous K_2CO_3, filter, add two to three boiling chips, and remove the ether on a steam bath in the hood. Recrystallize the residue from aqueous ethanol by dissolving the product in 5 mL of boiling ethanol, adding water dropwise to approach the cloud point, then cooling first to near room temperature, then in an ice bath. Collect the product by suction filtration, wash with water and air dry. Record the mass and melting range.

REFERENCE

ALLEN, C., and J. GATES, JR., *Organic Syntheses Collective Volume 3* (1955) 140.

EXPERIMENT 20.3 CONVERSION OF ACETAMINOPHEN INTO PHENACETIN

Estimated Time:
2.0 hours

Prelab

1. Calculate the millimoles contained in 1.30 g of acetaminophen and in 2.50 g of K_2CO_3. What volume of ethyl iodide corresponds to 1.8 g?
2. Why is acetaminophen soluble in 5% aqueous NaOH, while phenacetin is not?
3. Calculate the theoretical yield of phenacetin.

Special Hazards

Methyl ethyl ketone is highly flammable. Iodoethane is corrosive.

Procedure

In a 50 mL round-bottomed flask, place acetaminophen (1.30 g or four crushed tablets of Tylenol, Datril, or generic acetaminophen containing 325 mg each, _____ mmol). Add powdered anhydrous K_2CO_3 (2.50 g, _____ mmol), 15 mL of methyl ethyl ketone (MEK, 2-butanone) as solvent, and ethyl iodide (12 mmol, 1.8 g, _____ mL). Set up for reflux and boil for 1 hour. At the end of the reflux period, cool the flask in an ice bath and gravity filter the contents into a separatory funnel. Rinse some ether through the residue on the filter paper to carry additional product into the separatory funnel. Wash the organic phase with 20 mL of 5% NaOH to remove any unreacted acetaminophen. Dry the organic layer over anhydrous Na_2SO_4, decant, and evaporate the MEK and ether on a steam bath in the hood to obtain the crude product. Record the melting point of this material, then recrystallize it from water. It may be slow to crystallize; use an ice bath and scratch to induce precipitation. Collect the phenacetin by suction filtration, wash with water, and air dry. Record the mass and melting range of this purified phenacetin.

REFERENCE

VOLKER, E. J., E. PRIDE, and C. HOUGH, *Journal of Chemical Education* (1979) *56*, 831.

EXPERIMENT 20.4 MICROSCALE PREPARATION OF TRITYL ETHYL ETHER

**Estimated Time:
1.0 hour**

Prelab

1. Calculate the millimoles of triphenylmethyl chloride in 100 mg and the theoretical yield of ethyl trityl ether.

Special Hazards

Triphenylmethyl chloride is corrosive and a lachrymator. Ethanol is flammable.

Procedure

In a 50-mL round-bottomed flask, place triphenylmethyl chloride (100 mg, _____ mmol), and 5 mL of anhydrous ethanol. A condenser, heating mantle, and variable transformer are added. The flask is brought to reflux for 1 hour. A piece of moist indicator paper held over the top of the condenser during this time indicates that HCl is evolved.

At the end of the heating period, remove the condenser, add a boiling chip, and allow the ethanol to boil down to a volume of 0.5–1 mL, then cool the flask in an ice bath. A precipitate of ethyl phenyl ether will appear. Add methanol (5 mL) to dissolve impurities and cause the product to solidify. Keep cooling in the ice bath, collect the precipitate, wash with a little cold methanol, allow the ethyl trityl ether to air dry, and record the mass and melting range (lit. mp 82–83°).

REFERENCE

Nixon, A., and G. Branch, "The Rates of Alcoholyses of Triarylmethyl Chlorides," *Journal of the American Chemical Society* (1936) *58*, 492.

Epoxidation of Alkenes

Overview

Alkenes may be converted to epoxides by hydrogen peroxide or peracids. Peracids (RCO_3H) such as peracetic, perbenzoic, and *m*-chloroperbenzoic acid (MCPBA) have been found to be very effective reagents for this transformation. *m*-Chloroperbenzoic acid is particularly convenient to use because it is a relatively stable crystalline solid.

The mechanism of this reaction is an electrophilic attack by the ''extra'' oxygen atom of the peracid on the alkene, accompanied by displacement of a carboxylate.

Epoxides are versatile intermediates because they can be hydrolyzed in either acid or base to yield the corresponding glycols (1,2 diols) or opened with nucleophiles on the less hindered side.

233

Epoxides have also found numerous industrial and household applications. Have you ever repaired an object with epoxy glue? This process usually involves mixing two substances, a resin and a curing agent. Epoxy resins have gained widespread use because of their excellent adhesion, toughness, and chemical resistance as well as high electrical resistance. There are many types of epoxy resins and of curing agents: one example is shown below. Two equivalents of epichlorohydrin react with one equivalent of bisphenol A to yield a liquid epoxy resin. When this resin is mixed with a polyamine curing agent, polymerization occurs to yield a strong, hard, inert material.

Epoxy Glue: Resin and Hardener (Courtesy Atlas Minerals & Chemicals, Inc.)

$$2\ CH_2\!-\!CH\!-\!CH_2Cl + HO\!-\!\bigcirc\!-\!\overset{\overset{\textstyle CH_3}{|}}{\underset{\underset{\textstyle CH_3}{|}}{C}}\!-\!\bigcirc\!-\!OH$$

Epichlorohydrin   Bisphenol A

$$\downarrow\ 2HCl$$

$$CH_2\!-\!CH\!-\!CH_2\!-\!O\!-\!\bigcirc\!-\!\overset{\overset{\textstyle CH_3}{|}}{\underset{\underset{\textstyle CH_3}{|}}{C}}\!-\!\bigcirc\!-\!OCH_2\!-\!CH\!-\!CH_2$$

Epoxy resin

$$\downarrow\ \overset{\textstyle H_2NCH_2CH_2NHCH_2CH_2NH_2}{}$$
Diethylaminetriamine
(curing agent)

$$\Big(\!-\!NHCH_2CH_2NHCH_2CH_2NH\!-\!CH_2\!-\!\overset{\overset{\textstyle OH}{|}}{CH}\!-\!CH_2\!-\!O\!-\!\bigcirc\!-\!\overset{\overset{\textstyle CH_3}{|}}{\underset{\underset{\textstyle CH_3}{|}}{C}}\!-\!\bigcirc\!-\!O\!-\!CH_2\!-\!\overset{\overset{\textstyle OH}{|}}{CH}\!-\!CH_2\!-\!\Big)_n$$

Polymer
(will cross-link on nitrogens)

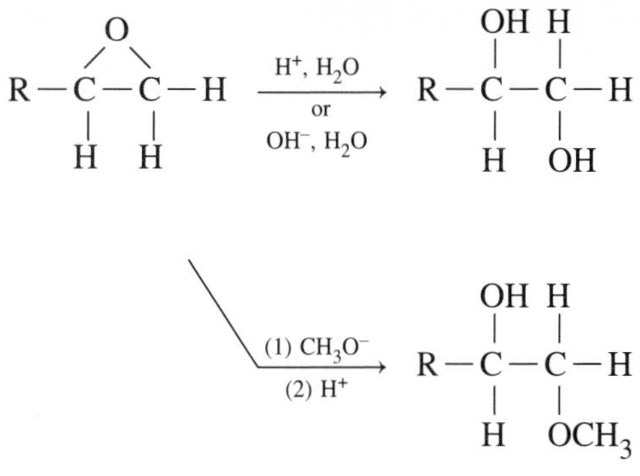

$$R - \underset{\underset{H}{|}}{\overset{\overset{O}{\diagup\,\diagdown}}{C}} - \underset{\underset{H}{|}}{\overset{}{C}} - H \quad \xrightarrow[\substack{\text{or} \\ OH^-,\, H_2O}]{H^+,\, H_2O} \quad R - \underset{\underset{H}{|}}{\overset{\overset{OH}{|}}{C}} - \underset{\underset{OH}{|}}{\overset{\overset{H}{|}}{C}} - H$$

$$\xrightarrow[\text{(2) } H^+]{\text{(1) } CH_3O^-} \quad R - \underset{\underset{H}{|}}{\overset{\overset{OH}{|}}{C}} - \underset{\underset{OCH_3}{|}}{\overset{\overset{H}{|}}{C}} - H$$

Certain epoxides have been implicated in the mechanism of carcinogenesis of some polynuclear aromatic hydrocarbons. For example, benzo[*a*]pyrene is a potent carcinogen found in tobacco smoke, chimney soot, and as a contaminant on foods cooked on barbecue grills. It has been shown to be oxidized in the liver to the diol epoxide, which induces mutations leading to cancer.

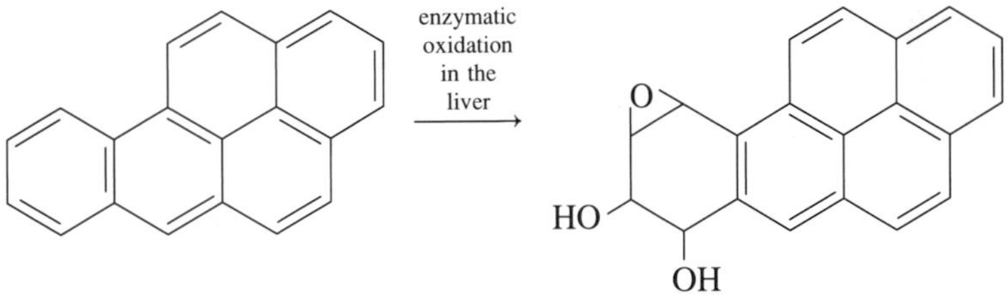

Benzo[*a*]pyrene → (enzymatic oxidation in the liver) → 7,8-Dihydroxy-9,10-epoxy-7,8,9,10-tetrahydrobenzo[*a*]pyrene

QUESTIONS

1. Supply the missing starting material, reagent, or product for each of the following reactions.

(a)

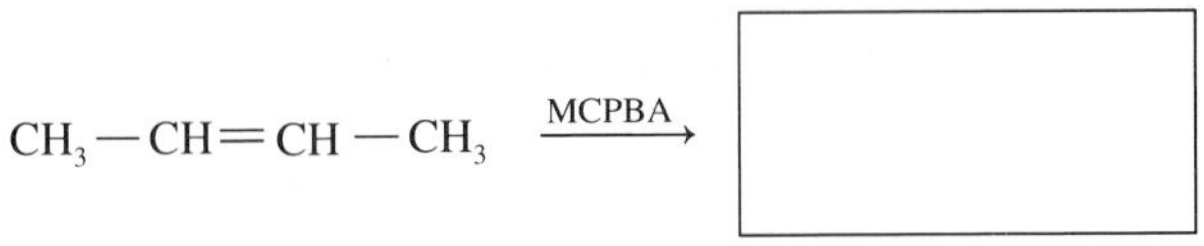

$$CH_3 - CH = CH - CH_3 \quad \xrightarrow{MCPBA} \quad \boxed{}$$

(b)

$$\boxed{} \quad \xrightarrow{CH_3CO_3H} \quad \text{(furan)} - CH - CH_2 \overset{O}{\diagup\,\diagdown} \quad \xrightarrow[\text{(2) } H^+]{\text{(1) } CH_3MgBr} \quad \boxed{}$$

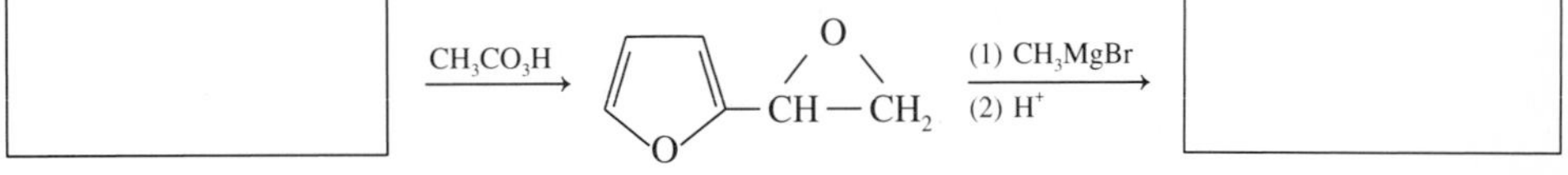

2. Which epoxy resin and curing agent would yield the following polymer?

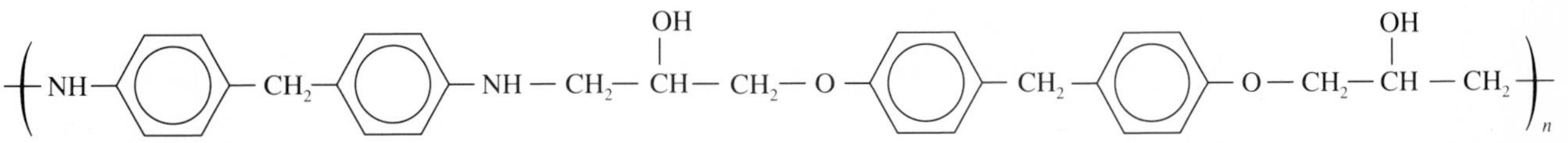

$$\left[-NH - \bigcirc - CH_2 - \bigcirc - NH - CH_2 - \underset{\underset{OH}{|}}{\overset{}{CH}} - CH_2 - O - \bigcirc - CH_2 - \bigcirc - O - CH_2 - \underset{\underset{OH}{|}}{\overset{}{CH}} - CH_2 - \right]_n$$

EXPERIMENT 21.1 PREPARATION OF STILBENE OXIDE USING MCPBA

Estimated Time:
2.5 hours

Prelab

1. Write the reaction between *trans*-stilbene (*trans*-1,2-diphenylethene) and *m*-chloroperbenzoic acid and show the product formed. Consult the *Aldrich Catalog* or the *CRC Handbook of Chemistry and Physics* to determine the MWs and melting points of these compounds.
2. Calculate the millimoles of stilbene in 0.90 g and of MCPBA in 1.15 g of 85% purity.
3. Calculate the theoretical yield of *trans*-stilbene oxide.

Special Hazards

m-Chloroperbenzoic acid is a strong oxidizing agent and irritant.

Procedure

In a 50-mL round-bottomed flask, place stilbene (0.90 g, _____ mmol, use the product of Experiment 30.1 if available). Add *m*-chloroperbenzoic acid (technical grade, 85% pure, 1.15 g, _____ mmol) and dichloromethane (20 mL). Reflux the mixture for 1 hour. Since CH_2Cl_2 has such a low boiling point, be careful not to overheat and lose the solvent through the top of the condenser.

Meanwhile prepare a separatory funnel containing 25 mL of 5% NaOH solution. At the end of the reaction period cool the flask slightly in an ice bath (some *m*-chlorobenzoic acid by-product may precipitate), and pour the reaction mixture into the separatory funnel. Shake and vent several times to extract the *m*-chlorobenzoic acid by-product into the aqueous phase. Draw off the organic (lower) layer, dry it over a drying agent, decant it, add a couple of boiling chips, and remove the dichloromethane on a steam bath in the hood. Recrystallize the residual oil from a small volume of methanol and cool in an ice bath. Collect the precipitated *trans*-stilbene oxide by suction filtration, wash with 3–5 mL of ice-cold methanol, and air dry. Record the mass and melting point.

REFERENCE

AUGUSTINE, R. L., *Oxidation: Techniques and Applications in Organic Synthesis*, Vol. 1, Marcel Dekker, New York, 1969, p. 226.

CASSADY, J. M., and J. D. DOUROS, eds., *Anticancer Agents Based on Natural Product Models*, Academic Press, New York, 1980.

Carbenes

Overview

A carbene is a carbon with two unshared electrons and no charge, the simplest example being carbene itself, :CH$_2$. It is a highly reactive intermediate with a short lifetime. Carbenes can be formed by removing the elements of HCl from a single carbon atom, called alpha-elimination (alpha because both atoms are removed from the same carbon). For example, in Experiment 22.1 you will be forming dichlorocarbene from chloroform:

Carbenes have the unusual property that they are both nucleophilic and electrophilic, since they have both an unshared pair of electrons and a vacant orbital, as shown here:

full *sp*^{2} orbital

vacant *p* orbital

For this reason carbenes can both donate and accept electrons at the same time. This is why they insert themselves into the double bonds of alkenes to form cyclopropanes:

237

To bring the OH⁻ (which is water soluble) into contact with the chloroform (which is not water soluble) we need to use a *phase-transfer catalyst*. A phase-transfer catalyst (PTC) carries ions back and forth between the aqueous and organic phases and lets us use inexpensive bases such as sodium hydroxide instead of more expensive organic-soluble bases, such as *n*-butyllithium (*n*-BuLi), lithium diisopropylamide (LDA), or potassium *t*-butoxide (KO*t*-Bu).

$$CH_3CH_2CH_2CH_2^- \ Li^+ \qquad\qquad Li^+N \qquad\qquad K^{+-}O-\underset{\underset{CH_3}{|}}{\overset{\overset{CH_3}{|}}{C}}-CH_3$$

n-Butyllithium (*n*-BuLi)	Lithium diisopropylamide (LDA)	Potassium *t*-butoxide (KO*t*-Bu)

The PTC we will use is a quaternary ammonium salt, tetra-*n*-butylammonium bromide.

$$CH_3CH_2CH_2CH_2-\underset{\underset{CH_2CH_2CH_2CH_3}{|}}{\overset{\overset{CH_2CH_2CH_2CH_3}{|}}{N^+}}-CH_2CH_2CH_2CH_3 \qquad Br^-$$

Tetra-*n*-butylammonium bromide
MW 322, mp 103–104°

When we add hydroxide to this it replaces the bromide and the PTC carries the hydroxide into the organic phase by wrapping it up in its long hydrophobic arms, holding the hydroxide tightly near the positively charged nitrogen.

$$N^+ \ OH^-$$

Once in the organic phase the hydrophobic arms unfold and release the hydroxide, which reacts quickly with the chloroform to form dichlorocarbene, which in turn reacts with cyclohexene to give the product. The PTC then returns to the aqueous phase and picks up another hydroxide ion to repeat the cycle. Since the phase-transfer catalyst must travel back and forth between the aqueous and organic phases it is essential to keep a large surface area of contact between the phases. This means maintaining *vigorous* stirring during the reaction.

Overall reaction:

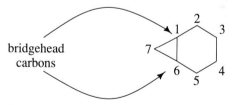

Cyclohexene	Chloroform				7,7-Dichloro-bicyclo [4.1.0] heptane (7,7-dichloronorcarane)
MW 82.2, bp 83°	MW 119, bp 62°				MW 165, bp 197°
dens. 0.810 g/mL	dens. 1.48 g/mL				dens. 1.21 g/mL

To name a bicyclic system such as the one in our product, count up the total number of carbons involved in the two rings. In this case it is seven, so it is a bicyclo*heptane*.

Bicyclo[3.2.0]heptane

3-Amino-bicyclo[4.4.1]undecane

Bicyclo[5.3.0]decan-2-one

Now going from one bridgehead carbon to the other, count the number of carbons encountered going each way and list them from the most to the least. In this case, going to the right there are 4, to the left there is 1, and in the middle there are 0 (direct connection). So this is named [4.1.0]bicycloheptane. The numbering goes from the bridgehead carbon (1) around the larger ring, then on to the smaller ring. So the compound we make in this experiment is called 7,7-dichloro-bicyclo[4.1.0]heptane.

Some other bicyclic compounds with their names are given below:

bridgehead carbons

Bicyclo[4.1.0]heptane

In Experiment 22.2, dibromocarbene is generated from bromoform (CHBr$_3$) and the base potassium *t*-butoxide. The dibromocarbene adds to 1,1-diphenylethylene to give 1,1-dibromo-2,2-diphenylcyclopropane.

$$CHBr_3 \quad + \quad KOt\text{-}Bu \quad + \quad (C_6H_5)_2C=CH_2 \quad \longrightarrow \quad (C_6H_5)_2C\text{——}CH_2 \quad + \quad t\text{-}BuOH$$

Bromoform	Potassium *t*-butoxide	1,1-Diphenylethylene	1,1-Dibromo-2,2-diphenylcyclopropane
MW 253, mp 8°	MW 112	MW 180, mp 6°	MW 352
bp 150–151°		bp 270–271°	mp 154–156°
dens. 2.89 g/mL		dens. 1.02 g/mL	

This reaction is carried out on a micro (100 mg) scale. A phase-transfer catalyst is not necessary in this case because all the reagents are soluble in pentane.

QUESTIONS

1. How would you name the following compounds?

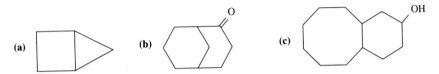

(a)   (b)   (c)

2. What would be the products of the following reactions?

(a) $\xrightarrow[n\text{-Bu}_4\text{N}^+\text{Br}^-]{\text{CHCl}_3, \text{OH}^-}$ (b) $\xrightarrow[\text{KO}t\text{-Bu}]{\text{CHBr}_3}$

3. The Simmons–Smith reaction involves generating carbene (:CH$_2$) by using CH$_2$I$_2$ and a zinc–copper couple to remove the iodines in a redox reaction. What would you expect to be the product of the following reaction?

$\xrightarrow[\text{Zn/Cu}]{\text{CH}_2\text{I}_2}$

4. How would you make the following? Show only the step forming the cyclopropane ring.

(a)   (b)   (c)

Concept Map

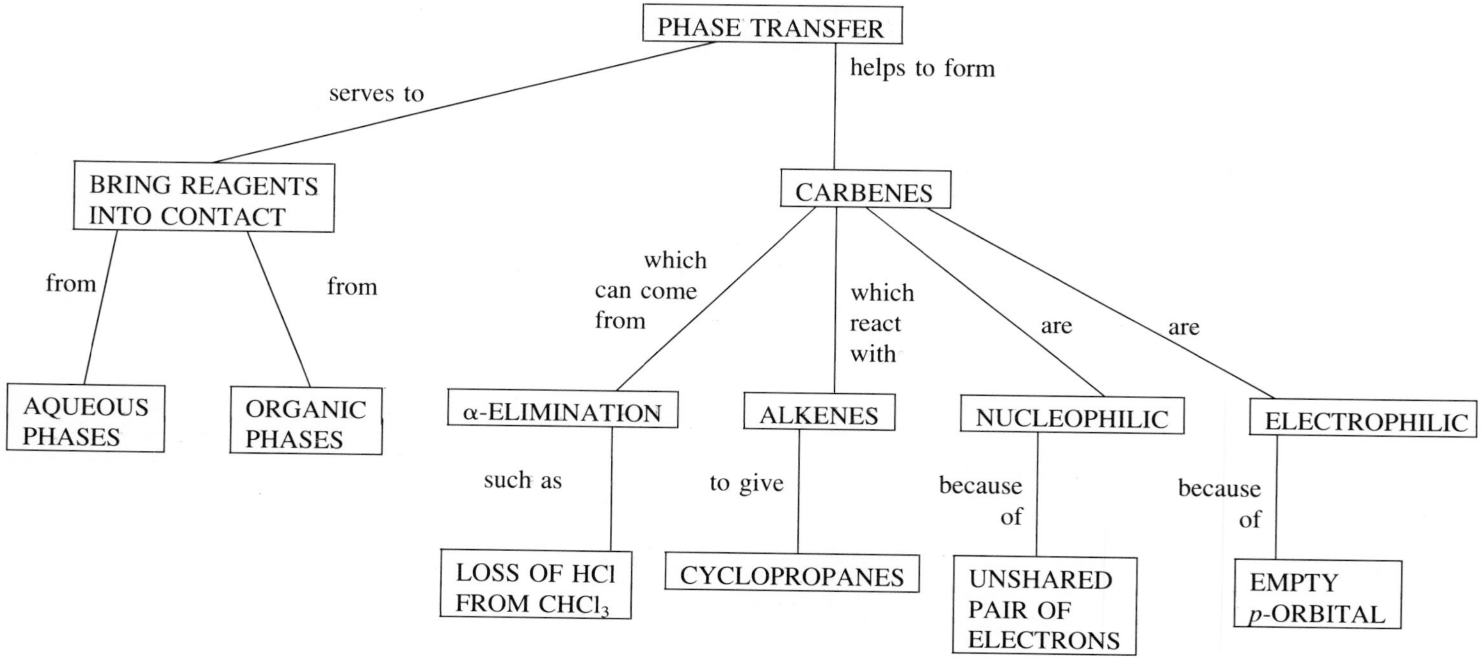

PHASE TRANSFER

— helps to form → CARBENES

— serves to → BRING REAGENTS INTO CONTACT
- from → ORGANIC PHASES
- from → AQUEOUS PHASES

CARBENES
- are → ELECTROPHILIC — because of → EMPTY p-ORBITAL
- are → NUCLEOPHILIC — because of → UNSHARED PAIR OF ELECTRONS
- which react with → ALKENES — to give → CYCLOPROPANES
- which can come from → α-ELIMINATION — such as → LOSS OF HCl FROM $CHCl_3$

EXPERIMENT 22.1 ADDITION OF DICHLOROCARBENE TO CYCLOHEXENE

Estimated Time:
2.5 hours

Prelab

1. Calculate the volumes of 200 millimoles of chloroform and of 70 mmol of cyclohexene.
2. Calculate the theoretical yield of the product.
3. If a flask contains an aqueous layer of 10 M NaOH (density 1.33 g/mL), an organic layer consisting of about 17 mL of chloroform (density 1.49 g/mL), and about 7 mL of 7,7-dichloro[4.1.0]bicycloheptane (density 1.21 g/mL), which will be the lower layer, and why?
4. At what boiling point do you expect to collect your product during the distillation? In view of this boiling point do you need to be concerned about losing any of the product while evaporating off the chloroform in the fume hood before distillation?
5. What is the expected bp range of the forerun during the distillation?

Special Hazards

Chloroform has been found to be a carcinogen on ingestion by mammals. Therefore, it is advisable to use adequate ventilation and avoid breathing the vapors as much as possible. Another reason to ensure good ventilation is that there may be a small amount of carbon monoxide produced from further reaction of dichlorocarbene with hydroxide ion. A concentrated sodium hydroxide solution such as 10 M is very corrosive and must be rinsed off immediately and thoroughly if it touches the skin.

Procedure

In this experiment there are two procedures to be carried out in one lab period: a reflux and a semimicro distillation. There is plenty of time for both, but you will need to be well organized and use your time efficiently.

In a 250-mL round-bottomed flask, place chloroform (200 mmol, _____ mL), cyclohexene (70 mmol, _____ mL), tetra-n-butylammonium bromide (phase-transfer catalyst, 0.5 g, 1.5 mmol), 10 M sodium hydroxide solution (50 mL), and a magnetic stirring bar.

Set up for reflux by adding a condenser and heating mantle. A stir plate should be used underneath the heating mantle instead of an iron ring. The stir plate contains a magnet that spins, driving the Teflon-coated magnetic stirring bar in your flask. For a 250-mL round-bottomed flask, 1-inch stir bars seem to work well. Smaller ones may not stir adequately, while larger ones may or may not work. You may wish to experiment with the stir bars available to you; some are egg-shaped or disk-shaped, especially to fit round-bottomed flasks. The flask must be *centered* over the stir plate for the stirrer to work well, since your magnetic stir bar should be directly over the spinning magnet at the center of the stir plate. The setup is shown in Figure 22-1.

Center your flask on the stirring plate, then turn it on (stirring only, not heat) and gradually get the stir bar in the flask spinning as fast as possible. If it "spins out," in

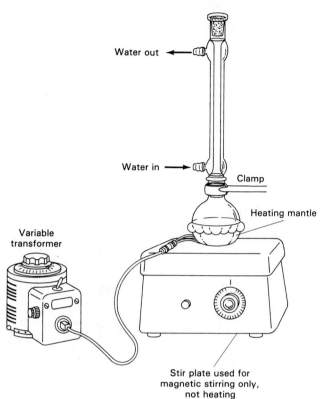

Figure 22-1   Apparatus for Reflux with Magnetic Stirring

other words, if the stir bar stops spinning and starts wiggling or jumping around, shut off the stirrer and start again, turning it up slowly to the fastest sustainable stirring rate. Be sure that the phases are mixing well in a vortex and you cannot see two distinct layers. Once it is stirring well and water is flowing through the condenser, turn on the Variac to the heating mantle and heat until the chloroform refluxes gently. While the mixture is refluxing, assemble the apparatus for semimicro distillation shown in Figure 22-2. After 45 min of refluxing, cool the mixture in an ice bath, pour it into a separatory funnel, and separate the layers. If they do not separate cleanly, add ether (20 mL) to decrease the density of the organic layer, and shake gently. If an emulsion forms during the extraction, it may be caused by the presence of tetra-*n*-butylammonium hydroxide, n-Bu$_4$N$^+$OH$^-$. If you have trouble with an emulsion, acidify the aqueous phase with HCl and the phases will separate cleanly.

Dry the organic layer over anhydrous MgSO$_4$, decant, add some boiling chips, and evaporate off most of the chloroform on a steam bath in the hood. Purify the product by semimicro distillation using the apparatus in Figure 22-2, using a 50-mL distilling pot and the correct size heating mantle for the pot. A water-cooled condenser is unnecessary because of the high boiling point of the product (it recondenses easily). Discard the forerun, consisting of residual chloroform and any unreacted cyclohexene, then collect the product in a small tared round-bottomed flask, clamped securely and in an ice bath. Because of the high boiling point of the product, the variable transformer can be on a fairly high setting, but remember not to distill completely to dryness. Record the boiling range (lit. bp 197°) and mass of the product.

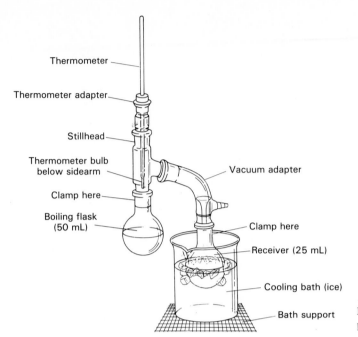

Thermometer

Thermometer adapter

Stillhead

Thermometer bulb
below sidearm

Clamp here

Boiling flask
(50 mL)

Vacuum adapter

Clamp here

Receiver (25 mL)

Cooling bath (ice)

Bath support

Figure 22-2 Semimicro Distillation Apparatus (air-cooled)

REFERENCES

DOERING, W. V. E. and A. K. HOFFMAN, ''The Addition of Dichlorocarbene to Olefins,'' *Journal of the American Chemical Society* (1954) *76*, 6162.

GOKEL, G. W., and W. P. WEBER, ''Phase Transfer Catalysis,'' *Journal of Chemical Education* (1978) *55*, 350, 429.

MAKOSZA, M., and M. WAWRZYNIEWICZ, ''Catalytic Method for Preparation of Dichlorocyclopropane Derivatives in Aqueous Medium,'' *Tetrahedron Letters* (1969) 4659.

WEBER, W. P., and G. W. GOKEL, *Phase Transfer Catalysis in Organic Synthesis*, Springer-Verlag, Berlin, 1977.

EXPERIMENT 22.2 MICROSCALE ADDITION OF DIBROMOCARBENE TO 1,1-DIPHENYLETHYLENE

Estimated Time:
1.5 hours

Prelab

1. Calculate the millimoles corresponding to 100 mg of 1,1-diphenylethylene, 110 mg of potassium *t*-butoxide, and 260 mg of bromoform. What is the volume of 260 mg of bromoform?

2. What is the theoretical yield of 1,1-dibromo-2,2-diphenylcyclopropane?

Special Hazards

Potassium *t*-butoxide is a strong base and highly corrosive. Bromoform is toxic and a lachrymator. Handle only in the fume hood. Pentane is highly flammable.

Procedure

In a dry 15 × 125 mm test tube in the fume hood place 1,1-diphenylethylene (100 mg, _____ mmol) and potassium *t*-butoxide (110 mg, _____ mmol). Add dry hexane (1 mL) and cool the tube in an ice bath. By pipette add bromoform (260 mg, _____ mmol, _____ mL). Agitate the tube to mix the reagents and observe the formation of a yellow precipitate of 1,1-dibromo-2,2-diphenylcyclopropane. After 20 minutes, add ice water (about 2 mL) to dissolve the KBr. Push a cotton ball to the bottom of the tube; the crude product will stick to it. Using a Pasteur pipette, remove both liquid phases. To recrystallize the crude product on the cotton ball, add 2 mL of either 2-propanol or methanol and heat the tube in a steam bath or hot water bath. When a clear solution results, remove the hot solution by Pasteur pipette to a clean test tube. Allow the tube to cool to room temperature then further cool it in an ice bath to give long needles of 1,1-dibromo-2,2-diphenylcyclopropane. Collect the purified product, air dry, and record the mass and melting range. A second crop may be obtained by reheating the mother liquor to boiling, adding water to the cloud point, and cooling.

REFERENCE

SANDLER, S. R., "Reactions of *gem*-Dihalocyclopropanes with Electrophilic Reagents," *Journal of Organic Chemistry* (1967) *32*, 3876.

SECTION 23

Carbocation Rearrangements

Overview

You may recall that the order of stability of carbocations (also known as carbonium ions) obeys the following trend:

next to O or N > tertiary, allylic, or benzylic > secondary > primary > methyl

In fact, it is extremely difficult to form a primary carbocation and virtually impossible to form a methyl carbocation in solution. If a secondary carbocation can rearrange to form a tertiary carbocation by shifting an alkyl, aryl, or hydride group from a neighboring carbon, it is likely to do so, as in the following example:

$$CH_3-\underset{\underset{CH_3}{|}}{\overset{\overset{CH_3}{|}}{C}}-\underset{\overset{|}{H}}{\overset{+}{C}}-CH_3 \xrightarrow[\text{shift}]{\text{1,2 methide}} CH_3-\underset{\underset{CH_3}{|}}{\overset{+}{C}}-\underset{\underset{H}{|}}{\overset{\overset{CH_3}{|}}{C}}-CH_3$$

2° Carbocation   3° Carbocation

This particular example is described as a 1,2-methide shift since the shift is to an adjacent carbon and the methyl group moves *with* its bonding pair of electrons (like an anion, hence meth*ide*). This type of shift would also occur if the new site of the carbocation would be next to an oxygen or nitrogen atom, which stabilize adjacent positive charges by resonance, leading to carbocations that are even more stable than tertiary ones. Exactly this type of shift occurs in a famous reaction called the pinacol rearrangement. In this reaction, protonation of an alcohol (actually a diol, pinacol) and loss of water lead to formation of a tertiary carbocation, which rearranges via a 1,2-methide shift to form a carbocation that is stabilized by an adjacent oxygen atom.

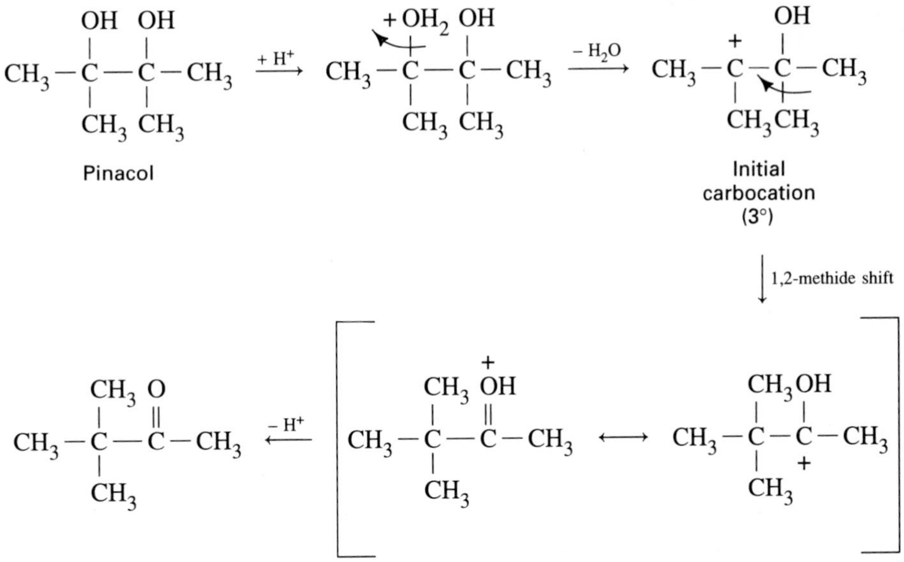

Pinacol

Initial carbocation (3°)

Pinacolone

Rearranged carbocation

When this new carbocation, which in one resonance form contains a carbon-to-oxygen double bond, loses a proton, the acid catalyst is regenerated and the product ketone (pinacolone) is formed. The overall reaction, then, is

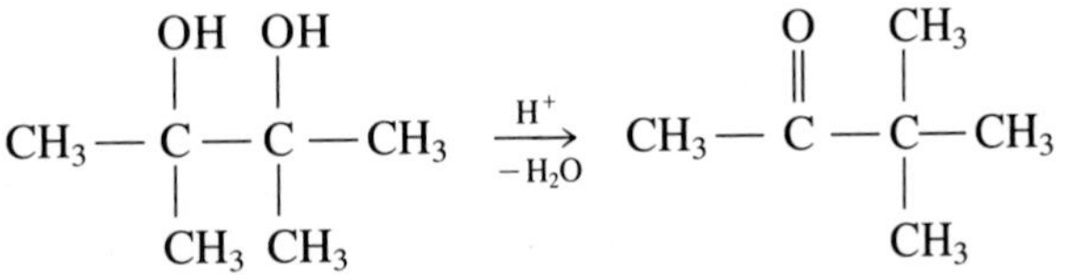

Pinacol
(2,3-dimethyl-2,3-butanediol)
MW 118, mp 40–43°

Pinacolone
(*t*-butyl methyl ketone)
MW 100, bp 106°
dens. 0.80 g/mL

The procedure for carrying out this experiment is described in Experiment 21.1. A similar rearrangement occurs (with a 1,2-phenyl shift) in benzopinacol, described in Experiment 21.2.

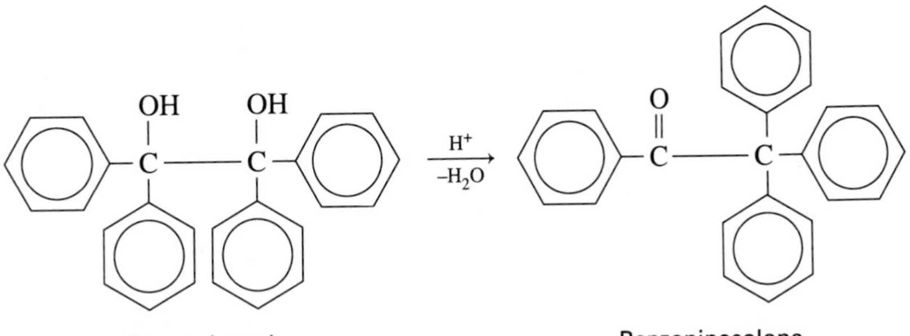

Benzopinacol
(1,1,2,2-tetraphenyl-1,2-ethanediol)
MW 366, mp 184–186°

Benzopinacolone
(2,2,2-triphenylacetophenone)
MW 348, mp 182–184°

Carbocation rearrangements are also often observed in ring systems. When a carbocation can rearrange to form a less strained system, it will often do so, as in the expansion from a cyclobutyl-substituted cation to a cyclopentyl cation:

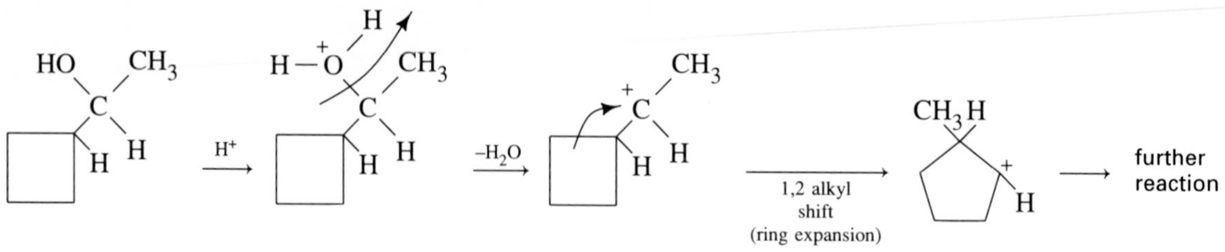

Similarly, a strained bicyclic (two-ring) system will rearrange if given the chance, and that is the key in our multistep synthesis of camphor from camphene. The mechanism of this rearrangement is as follows:

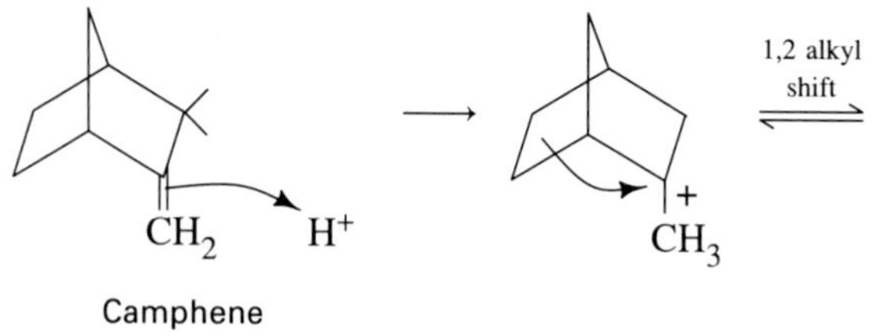

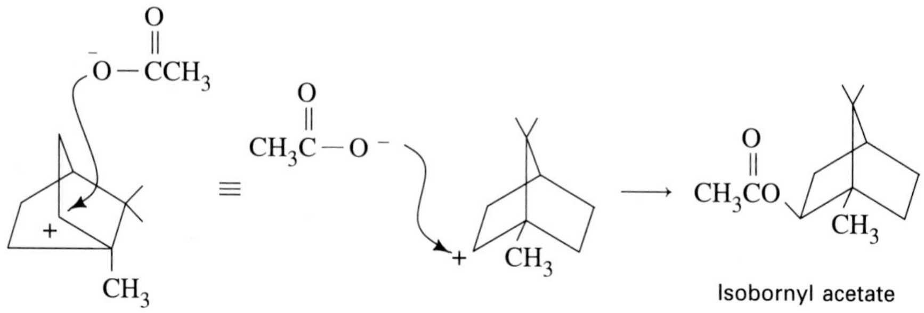

Camphene

Isobornyl acetate

Note that acetate must attack from the direction shown because the alkyl group shifting back and forth is blocking the other side. This is called a nonclassical carbocation and is similar to the bromonium ion you have seen before.

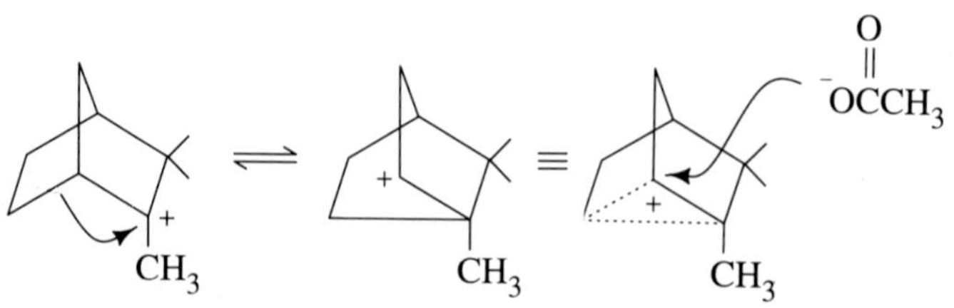

The overall sequence, then, is the rearrangement of camphene to isobornyl acetate, followed by hydrolysis of the ester and oxidation of the isoborneol to form camphor.

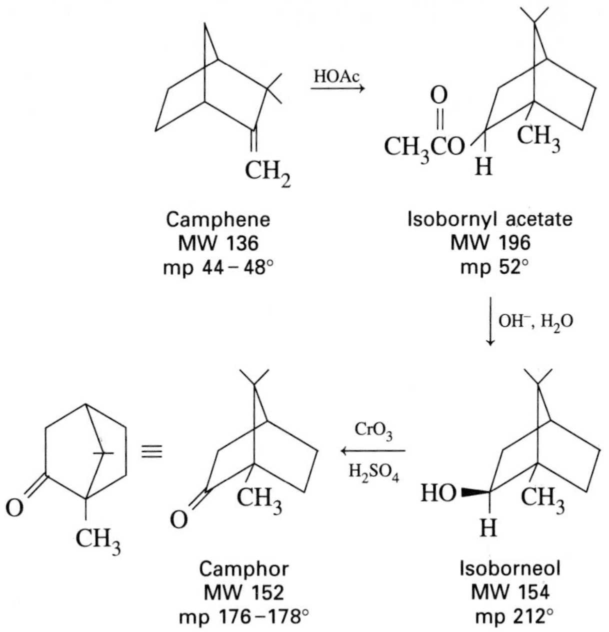
HOAc
CH2
CH3CO
CH3
H
Camphene
MW 136
mp 44–48°
Isobornyl acetate
MW 196
mp 52°
OH−, H2O
CrO3
H2SO4
O
CH3
HO
Camphor
MW 152
mp 176–178°
Isoborneol
MW 154
mp 212°

Some of the rearrangements of bicyclic compounds are difficult to visualize on paper; you are strongly encouraged to build models to help you picture the processes occurring.

Camphor has been used for hundreds of years in lacquer and varnishes, moth repellents, explosives, and embalming fluid. A description of it appears in the text *Alchymia* by Libavias in 1595. It occurs naturally in the Chinese camphor tree *Cinnamomum camphora*, which is native to Taiwan and has also been grown successfully in California and Florida. Because this tree is the only natural source of camphor in large quantities, and it is expensive to collect and extract the naturally occurring camphor, a synthesis from readily available inexpensive starting materials is desirable. Such a synthesis was developed by the French chemist, Berthelot, in 1869. He succeeded in converting camphene into camphor by chromic acid oxidation, although the yield was very low. Camphene is an inexpensive starting material, being made from the pinenes, which comprise a large percentage of turpentine.

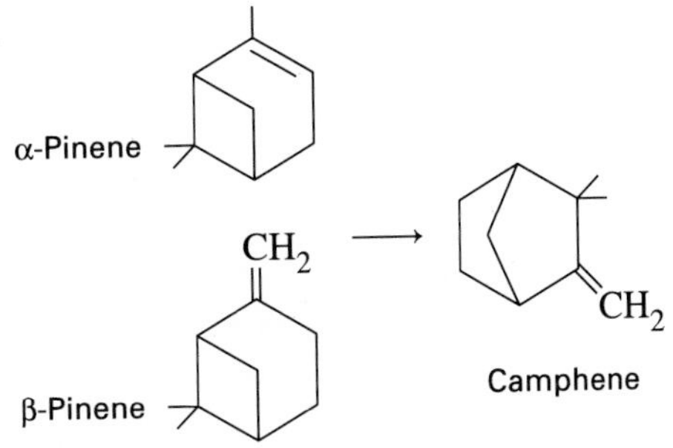
α-Pinene
CH2
β-Pinene
CH2
Camphene

Both pinene and camphene fall into the category of *terpenes*, naturally occurring compounds whose carbon skeletons consist of two five-carbon isoprene units.

Camphor possesses a very pungent odor, and it is believed that the olfactory bulb in human beings possesses receptor sites that fit its nonpolar hemispherical shape. This hypothesis is supported by the fact that these other relatively nonpolar molecules, possessing hemispherical portions, also exhibit a camphoraceous odor.

Cyclooctane	Hexachloroethane (perchloroethane)	Dimethyl-pentamethylene silicon	2-Nitroso-2-methyl-propane	Thiophosphoric acid dichloride ethylamide

Molecules possessing a camphoraceous odor

QUESTIONS

1. Draw the mechanism for the acid-catalyzed rearrangement of benzopinacol.

2. Interpret the NMR spectra of pinacol and pinacolone (Figures 23-1 and 23-2).

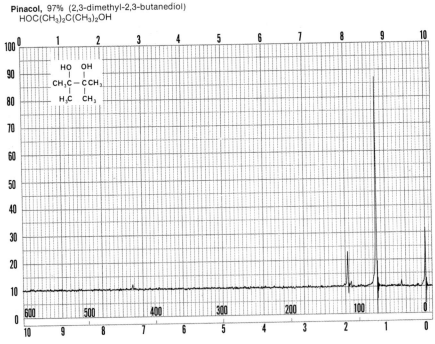

Pinacol, 97% (2,3-dimethyl-2,3-butanediol)
HOC(CH$_3$)$_2$C(CH$_3$)$_2$OH

Figure 23-1   NMR Spectrum of Pinacol

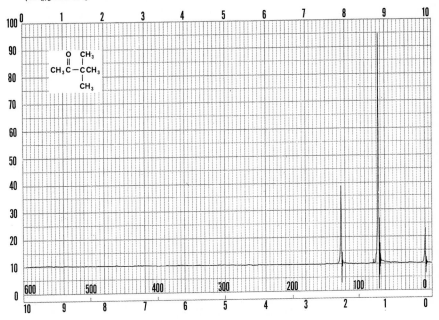

Figure 23-2 NMR Spectrum of Pinacolone

EXPERIMENT 23.1: THE PINACOL REARRANGEMENT

Estimated Time:
2.0 hours

Prelab

1. Calculate the number of millimoles of anhydrous pinacol in 6.0 g and the mass of the equivalent number of millimoles of pinacol hexahydrate.
2. Calculate the theoretical yield of pinacolone.

Special Hazards

Pinacol and pinacolone are irritants and flammable. Sulfuric acid is highly corrosive; avoid contact.

Procedure

In a 100-mL round-bottomed flask, place anhydrous pinacol (6.0 g, _____ mmol, or the equivalent amount of pinacol hexahydrate) and 30 mL of 6 N sulfuric acid. Set up for simple distillation and collect the distillate, which separates into an aqueous (lower) and pinacolone (upper) phase. When the volume of the upper phase stops increasing (after about 20 minutes), stop the distillation, pour the collected material into a separatory funnel, and draw off the aqueous phase. Dry the crude pinacolone over Na_2SO_4, $CaCl_2$, or $MgSO_4$, decant, and distill the product, collecting the fraction boiling at 103–107°. Record the yield and refractive index of the product.

REFERENCE

HILL, G. A., and E. W. FLOSDORF, *Organic Syntheses Collective Volume 1* (1941) 462.

EXPERIMENT 23.2 REARRANGEMENT OF BENZOPINACOL

Estimated time:
1.5 hours

Procedure

In a 25-mL Erlenmeyer flask, place iodine (0.10 g, _____ mmol) and acetic acid (5 mL). Add benzopinacol (1.0 g, _____ mmol) and a boiling chip, and heat the mixture to boiling on a hot plate in the fume hood. Maintain boiling for 5 minutes to obtain a dark red solution. Cool the flask in an ice bath; crystals of benzopinacolone precipitate as a solid mass. Add water (20 mL) and stir. Collect the grayish product by suction filtration, then wash with water followed by a little ethanol. If desired, this crude material may be recrystallized from ethanol (approximately 100 mL). Allow the product to air dry. Record the mass and melting range.

Since the melting ranges of the starting material and product are close, further confirmation of the identity of the product is required. Prepare a Nujol mull of the product and take its infrared spectrum to compare with a literature spectrum.

REFERENCE

BACHMANN, W. E., *Organic Syntheses Collective Volume 2* (1943) 71.

EXPERIMENT 23.3 THREE-STEP SYNTHESIS OF CAMPHOR FROM CAMPHENE

Estimated Time:
5.0 hours for parts a, b, and c

Prelab

1. Build a model of camphene and follow the reaction through the formation of isobornyl acetate. Draw each of the structures involved, using a three-dimensional perspective different from those given in this book.
2. Calculate the mass of 80 mmol of camphene and the volume of 400 mmol of acetic acid.

3. In the hydrolysis of the ester isobornyl acetate to yield isoborneol (step b of the procedure), how many milliliters of 2.5 M KOH in 75% ethanol would you use if you have 8.4 g of isobornyl acetate?

4. In the oxidation of isoborneol to camphor (step c of the procedure), if you had 5.7 g of isoborneol, how many milliliters of Jones reagent (which is 2.67 M in Cr^{6+}) would you add?

Special Hazards

Acetic acid and potassium hydroxide can cause serious chemical burns, especially to eyes. Jones reagent is strongly acidic and a powerful oxidizing agent. The chromium trioxide it contains has been found to cause cancer in laboratory animals. Camphor is an irritant, can damage eyes, and is toxic on ingestion. Avoid contact with all these materials and wash immediately and thoroughly if contact occurs.

Procedure 23.3a: Rearrangement of Camphene to Form Isobornyl Acetate

Estimated Time:
1.0 hour

$$\text{Camphene} \xrightarrow{\text{CH}_3\text{COOH}} \text{Isobornyl acetate}$$

Camphene   Isobornyl acetate

In a 100-mL beaker in the fume hood, place glacial acetic acid (400 mmol, _____ g, _____ mL) and camphene (80 mmol, _____ g). Heat the mixture for 15–20 minutes on a steam bath or hot plate, keeping the temperature within the range 90–100°. Add 15 mL of cold water and cool the beaker in an ice bath. Place the mixture in a separatory funnel and remove the lower (aqueous) layer, keeping the upper layer of isobornyl acetate in the funnel. Wash with 10 mL of water, then with 10 mL of 10% aqueous sodium carbonate to remove any residual acetic acid. Dry the organic layer over anhydrous $MgSO_4$, then filter and weigh the product.

Procedure 23.3b: Hydrolysis of Isobornyl Acetate to Isoborneol

Estimated Time:
1.5 hours

Set up a 100-mL round-bottomed flask with a water-cooled condenser, heating mantle, and variable transformer. In the flask, place your isobornyl acetate from step a. Now, based on the mass of your product, calculate and add 2.5 mL of 2.5 M KOH in 75% ethanol/25% water *for every gram of isobornyl acetate used.*

Reflux the mixture for 45–60 minutes. After the reflux period, cool the flask to room temperature using an ice bath, and pour it slowly with swirling into a beaker containing 50 mL of a mixture of ice and water. The isoborneol precipitates; collect this solid by suction filtration and wash it with cold water. Leave the crude solid in an open beaker in the drawer to air dry until your next lab period.

Procedure 23.3c: Oxidation of Isoborneol to Camphor

Estimated Time:
2.5 hours

For each gram of crude isoborneol obtained in part b, place into a beaker 1.75 mL of Jones reagent. Jones reagent consists of CrO_3 in aqueous H_2SO_4, which forms the strong oxidizing agent chromic acid, H_2CrO_4. The solution is 2.67 *M* in Cr^{6+}. In a separate Erlenmeyer flask dissolve the isoborneol in 10 mL of acetone. Over the course of 10 minutes, add the Jones reagent dropwise to the isoborneol solution with swirling. Let the mixture stand for 30 minutes with occasional stirring.

At the end of this time, pour the solution onto 250 mL of an ice–water mixture in a large beaker. Stir well, then vacuum filter to collect the precipitated camphor. Allow the product to air dry at room temperature by placing it on a large filter paper on a watchglass.

While the camphor is drying, assemble the apparatus for sublimation using a filter flask as described in Section 3. Sublime the camphor carefully, then collect and weigh this purified product. To take its melting point, first place a sample in a capillary tube, then seal the open end of the tube in a flame. Now you have a sealed tube, which will prevent the camphor from subliming out the top of the tube when you take its melting point. Record the melting point (lit. mp 176–178°) and place your product in the collection container provided.

Hydride Reductions

Overview

If you carried out the catalytic hydrogenation in Experiment 16.1, you have already used sodium borohydride (NaBH$_4$) as a source of hydride ions. In that experiment the hydride served two purposes; to reduce Pt^{4+} to Pt° in the reaction

$$Pt^{4+} + 4H^- \longrightarrow Pt(s) + 2H_2(g)$$

and to generate hydrogen gas in the presence of acid, according to the equation

$$H^+ + H^- \longrightarrow H_2(g)$$

Hydride ion also functions as a good nucleophile because of its negative charge. It can attack a positive center such as the carbon of a carbonyl group or a carbon with a halogen or other good leaving group on it:

$$R-\overset{\overset{\displaystyle O}{\|}}{C}-R' \;+\; \longrightarrow \;R-\overset{\overset{\displaystyle O^-}{|}}{\underset{\underset{\displaystyle H}{|}}{C}}-R'$$

$$H^-$$

$$R-CH_2{-}X \;+\; \longrightarrow \;R-CH_3 \;+\; X^-$$

$$H^-$$

In the first example, where hydride ion attacks a carbonyl group, if the final product is acidified (in other words, H$^+$ is added), the alkoxide anion is protonated to give an alcohol:

$$R-\overset{\overset{\displaystyle O}{\|}}{C}-R' \;\xrightarrow[\text{(2) H}^+]{\text{(1) NaBH}_4}\; R-\overset{\overset{\displaystyle OH}{|}}{\underset{\underset{\displaystyle H}{|}}{C}}-R'$$

255

This reaction works equally well on aldehydes

$$R-\overset{\overset{\displaystyle O}{\|}}{C}-H \xrightarrow[(2)\ H^+]{(1)\ NaBH_4} R-CH_2OH$$

The acid must, of course, be added in a *second step* after the hydride addition because if a hydride and acid are mixed, they react with each other to form hydrogen gas.

$$H^- + H^+ \longrightarrow H_2$$

This reaction would destroy any hydride present, making its attack on the carbonyl impossible.

There are many hydride reagents available today. The development of various selective hydride reagents by Herbert C. Brown and others since the late 1940s represents a fascinating saga in the history of organic chemistry. You will use sodium borohydride because it is an inexpensive and effective reagent for converting ketones and aldehydes to alcohols.

Lithium aluminum hydride (LiAlH$_4$, sometimes abbreviated LAH) is a stronger reducing agent than NaBH$_4$. In addition to the foregoing reactions, LiAlH$_4$ can, for example, reduce esters to alcohols in a two-step attack:

$$
\begin{array}{ccc}
R-\overset{\overset{\displaystyle O}{\|}}{C}-O-R' & \longrightarrow & R-\overset{O^-}{\underset{H}{C}}\!\!\!|\,O-R' \\[6pt]
\quad\ H^- & & \\
\text{An ester plus hydride} & & \text{ion (from LiAlH}_4)
\end{array}
$$

$$
R-\overset{OH}{\underset{H}{\overset{|}{\underset{|}{C}}}}-H \xleftarrow[\text{(acidification)}]{H^+} R-\overset{O^-}{\underset{H}{\overset{|}{\underset{|}{C}}}}-H \xleftarrow[\substack{\text{(second}\\ \text{attack)}}]{} R-\overset{\overset{\displaystyle O}{\|}}{C}-H + R'O^-
$$

Actually, two alcohols are produced from an ester, since the alkoxide eliminated after the first hydride attack (R'O$^-$) will be protonated during the final acidification step to give R'OH.

There are many hydride reagents that have been "fine-tuned" to a certain strength so that they can carry out a particular reduction without reducing other functional groups present. These reagents are very useful for syntheses of complex molecules containing numerous functional groups.

Many of these reagents have organic groups on them, such as lithium aluminum tri-*t*-butoxyhydride [LiAlH(O*t*-Bu)$_3$]. Compared with plain LiAlH$_4$, adding an electron-rich group such as an alkoxy group on the aluminum decreases the positive charge on the Al and therefore decreases the strength of the hydride ion. As the difference in electronegativity between the Al and H is decreased, the negative character of the H$^-$ is decreased.

So this particular reagent, LiAlH(Ot-Bu)$_3$, with three oxygens on the aluminum, is a fairly weak hydride source and will only reduce the most easily reduced groups.

A few common hydride reagents and the groups they can reduce appear in Table 24-1. This table shows that LiAlH$_4$, a strong reducing agent, will reduce many functional groups, whereas the much weaker NaBH$_4$ will only react with a few.

TABLE 24–1 SOME COMMON HYDRIDE REDUCING AGENTS[a]

		Weakest reducing agent					Strongest reducing agent
		NaBH$_4$	LiAl(O-tBu)$_3$H	LiBH$_4$	B$_2$H$_6$	AlH$_3$	LiAlH$_4$
Easiest to	Aldehyde	×	×	×	×	×	×
reduce	Ketone	×	×	×	×	×	×
	Acid chloride	×	×	×		×	×
	Ester			×	×	×	×
	Carboxylic acid				×	×	×
	Nitrile				×	×	×
	Nitro					×	×
Hardest to reduce	Alkene				×		

[a] × indicates complete reduction in 1 hour.

The products of hydride reductions of acid chlorides, esters, carboxylic acids, nitriles, nitro groups, and alkenes are shown in the following reactions:

$$
\underset{\text{O}}{\overset{\text{O}}{\|}}
$$

$$\text{RC Cl} \longrightarrow \text{RCH}_2\text{OH} + \text{Cl}^-$$

$$\text{RC OR}' \longrightarrow \text{RCH}_2\text{OH} + \text{R}'\text{OH}$$

$$\text{RC OH} \longrightarrow \text{RCH}_2\text{OH}$$

$$\text{R}-\text{C}\equiv\text{N} \longrightarrow \text{R}-\text{CH}_2\text{NH}_2$$

$$\text{R}-\text{NO}_2 \longrightarrow \text{R}-\text{NH}_2$$

$$\text{R}-\text{CH}=\text{CH}-\text{R} \longrightarrow \text{R}-\text{CH}_2\text{CH}_2-\text{R}$$

Note that many of these reactions involve attack of two equivalents of hydride on a somewhat positive center such as a carbonyl, nitrile carbon, or the nitrogen of a nitro group.

The reaction in Experiment 24.1 is reduction of a ketone, cyclohexanone.

Overall reaction:

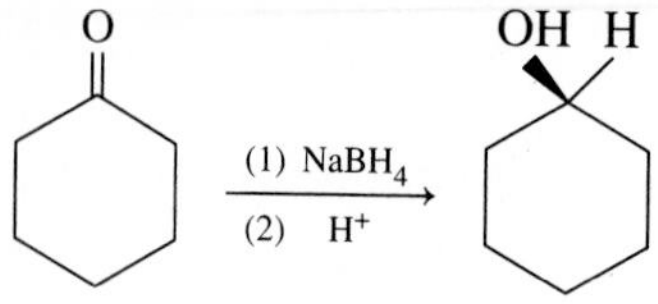

The mechanism is very simple: attack of the hydride ion on the carbonyl carbon to form the alkoxide followed by protonation to form the alcohol.

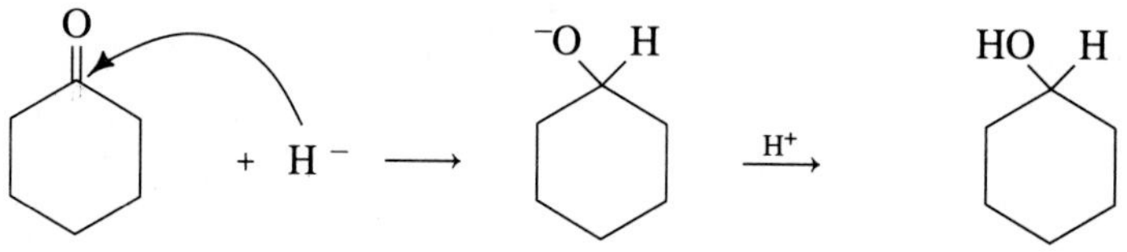

Cyclohexanone
MW 98, bp 155°
dens. 0.95 g/mL

Alkoxide
intermediate

Cyclohexanol
MW 100, bp 160 – 161°
dens. 0.96 g/mL

In Experiment 24.2 an aldehyde is reduced to a primary alcohol.

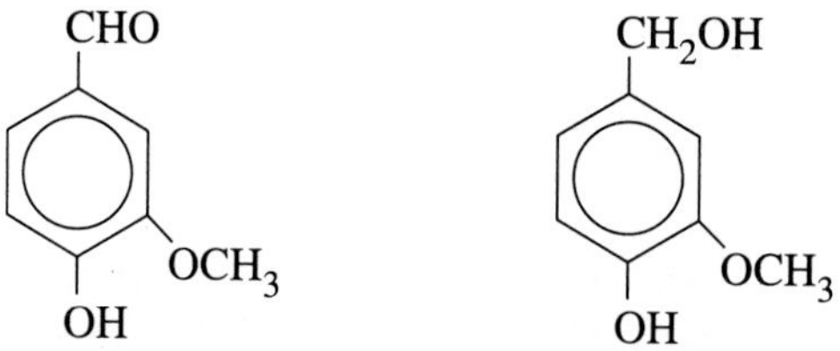

Vanillin
MW 152, mp 81 – 83°

Vanillyl alcohol
(α, 4-dihydroxy-3-methoxytoluene)
MW 154, mp 115°

Experiment 24.3 illustrates the stereochemical requirements of $NaBH_4$ by looking at the ratios of the cis and trans products formed in the reaction with 4-t-butylcyclo-hexanone. This molecule is "locked" in a chair conformation, and equatorial attack is less hindered than axial attack.

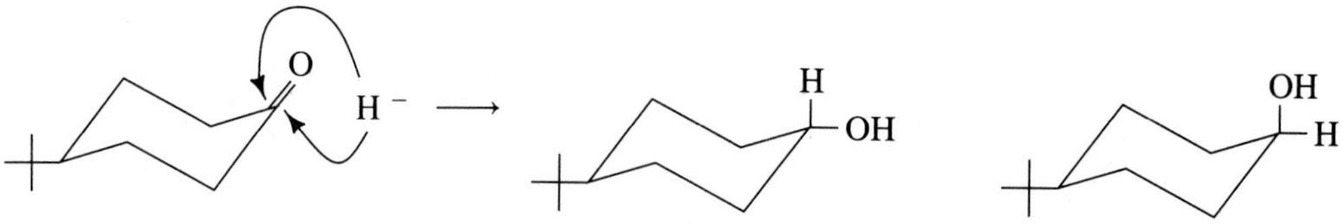

4-t-Butylcyclohexanone
MW 154, mp 47 – 50°

trans and/or *cis*
4-t-Butylcyclohexanol
MW 156

By examining the ratios of products formed we can see how selective $NaBH_4$ is.

Biochemically, hydride ion is often stored as NADH (nicotinamide adenine dinu-cleotide hydride) or NADPH (a phosphorylated version):

X = H NADH
X = OPO$_3^{2-}$ NADPH

NADH or NADPH

Schematically, the loss of hydride from NADH or NADPH can be represented as

NADH or NADPH   NAD$^+$ or NADP$^+$

The system is finely balanced so that the resonance energy of the aromatic ring formed helps make release of the hydride easier. These reducing agents are used by enzymes as cofactors in a multitude of biochemical processes. One example is in the conversion of pyruvic acid to lactic acid.

| Pyruvic acid | Reduced form of coenzyme | | (S)-(+)-Lactic acid | Oxidized form of coenzyme |

As is usually the case in enzyme-catalyzed reactions, the process is completely stereoselective, resulting in formation of only (S)-(+)-lactic acid. The process is also reversible, since the same enzyme can convert lactic acid to pyruvic acid with concomitant formation of NADH from NAD$^+$.

QUESTIONS

1. What is a hydride ion in terms of protons, neutrons, and electrons? Is it electrophilic or nucleophilic, and why?

2. Considering the electronegativities of boron and hydrogen, why is the hydrogen in NaBH$_4$ considered a hydride?

3. Besides reductions of ketones, give specific examples of two other types of reactions that can be conducted using hydride ions.

4. Supply the missing product or reagents for each of the following reactions.

(a) 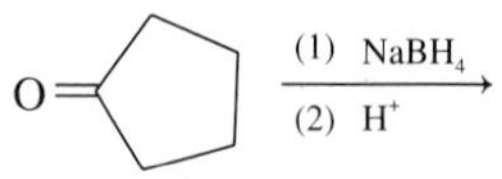 $O=\!\!\!\bigcirc$ $\xrightarrow[\text{(2) }H^+]{\text{(1) NaBH}_4}$

(b) 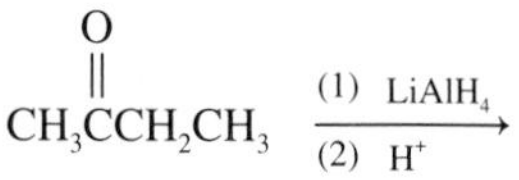 $CH_3CCH_2CH_3$ (with O double bonded to C) 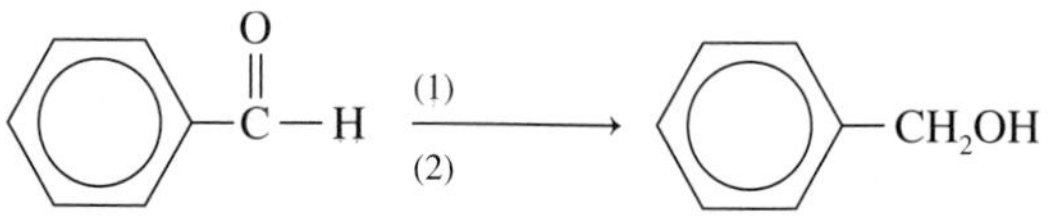 $\xrightarrow[\text{(2) }H^+]{\text{(1) LiAlH}_4}$

(c) $\bigcirc\!\!-\!\!\overset{\overset{\displaystyle O}{\|}}{C}\!\!-\!\!H$ $\xrightarrow[\text{(2)}]{\text{(1)}}$ $\bigcirc\!\!-\!\!CH_2OH$

(d) 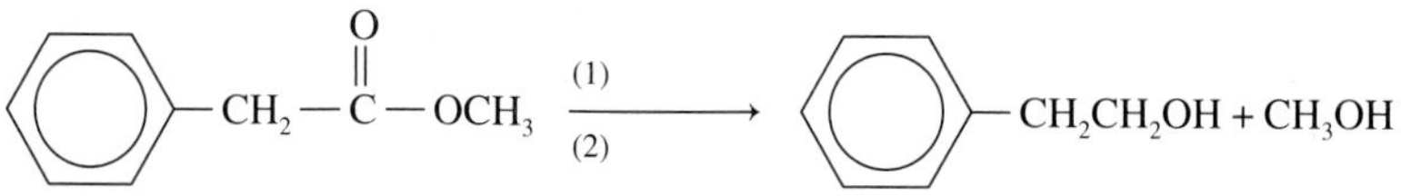 $\bigcirc\!\!-\!\!CH_2\!\!-\!\!\overset{\overset{\displaystyle O}{\|}}{C}\!\!-\!\!OCH_3$ $\xrightarrow[\text{(2)}]{\text{(1)}}$ $\bigcirc\!\!-\!\!CH_2CH_2OH + CH_3OH$

(e) 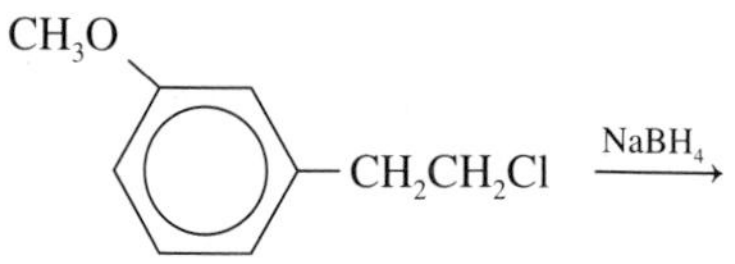 $CH_3O\!\!-\!\!\bigcirc\!\!-\!\!CH_2CH_2Cl$ $\xrightarrow{\text{NaBH}_4}$

5. What reagent could you use to reduce an ester in the presence of a nitrile group, for example in the transformation

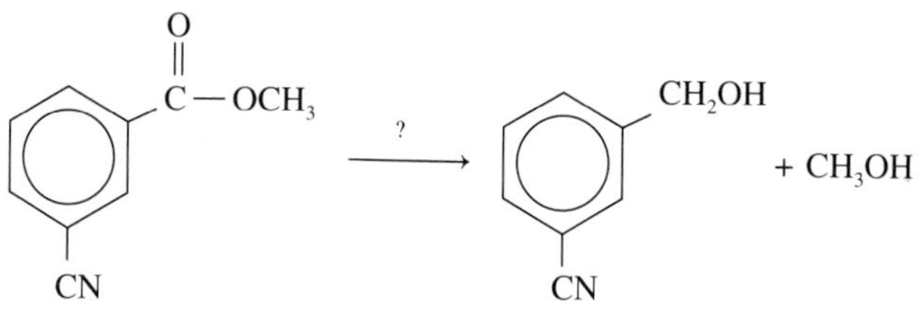

$\bigcirc\!\!-\!\!\overset{\overset{\displaystyle O}{\|}}{C}\!\!-\!\!OCH_3$ (with CN substituent) $\xrightarrow{?}$ $\bigcirc\!\!-\!\!CH_2OH$ (with CN substituent) $+ CH_3OH$

REFERENCE

BROWN, H. C., "Hydride Reductions: A 40-Year Revolution in Organic Chemistry," *Chemical and Engineering News* (Mar. 5, 1979) 24–29.

EXPERIMENT 24.1 REDUCTION OF CYCLOHEXANONE TO CYCLOHEXANOL USING NaBH₄

Estimated Time:
2.5 hours

Prelab

1. Calculate the volume of 90 mmol of cyclohexanone, the mass of 24 mmol of NaBH₄, and the theoretical yield of cyclohexanol.

Special Hazards

Sodium borohydride is a corrosive and flammable solid. Sodium methoxide is a strong base and reacts with water to form NaOH. Cyclohexanone and methanol are flammable. Exercise due care.

Procedure

In a 125-mL Erlenmeyer flask, place cyclohexanone (90 mmol, _____ g, _____ mL) and 20 mL of methanol. In a separate 50-mL beaker place methanol (15 mL), 25% methanolic sodium methoxide (1.6 mL), and sodium borohydride (24 mmol, _____ g). Pour the $NaBH_4$ solution into the flask containing the cyclohexanone solution and swirl occasionally for 5 minutes. Meanwhile, prepare a 250-mL beaker containing 200 mL of a mixture of ice and water to which 10 mL of 5% aqueous HCl has been added. After the 5-minute reaction period, pour the reduction mixture into the ice–water. Place the mixture in a separatory funnel and extract with one 50-mL portion of ether. Wash the ether extract with water (25 mL) and brine (25 mL), then dry it over Na_2SO_4 or $MgSO_4$. Decant into a 100-mL beaker, add boiling chips, and remove the ether on a steam bath in the hood. Purify the crude product by simple distillation. Record the boiling range, mass, and refractive index of your product. At the instructor's option you may combine your product with those of other students and take an IR spectrum.

EXPERIMENT 24.2 REDUCTION OF VANILLIN USING NaBH₄

Estimated Time:
2.0 hours

Prelab

1. Calculate the millimoles in 0.13 g of $NaBH_4$ and the theoretical yield of vanillyl alcohol.

Special Hazards

$NaBH_4$ is flammable and corrosive. Avoid contact of the irritating $NaBH_4$ powder with skin or eyes and flush with a lot of water if contact occurs.

Procedure

In a 50-mL beaker, dissolve vanillin (8.0 mmol, 1.0 g) in 1 M aqueous NaOH (7 mL). Add $NaBH_4$ (0.13g, _____ mmol) and swirl the flask occasionally for the next 20–30 minutes. Cool the solution in an ice bath, and add 7 mL of 5% aqueous HCl to reprotonate the phenol group and drive the product out of solution. If necessary, scratch with a stirring rod to induce precipitation. Collect the product by suction filtration, wash with water, and air dry. Recrystallize the vanillyl alcohol from ethyl acetate, using a ratio of 5 mL per gram.

EXPERIMENT 24.3 STEREOCHEMISTRY OF THE REDUCTION OF 4-*t*-BUTYLCYCLOHEXANONE

Estimated Time:
2.5 hours

Prelab

1. Calculate the volume of 90 mmol of 4-*t*-butylcyclohexanone, the mass of 24 mmol of $NaBH_4$, and the theoretical yield of 4-*t*-butylcyclohexanol.

Special Hazards

See Experiment 24.1.

Procedure

Follow the procedure for the reduction of cyclohexanone (24.1), substituting 4-*t*-butylcyclohexanone for the cyclohexanone. It is not necessary to distill the crude product after drying and removal of ether. Simply inject a sample onto a nonpolar gas chromatographic column at about 150° for analysis. The cis isomer is lower boiling and thus has a shorter retention time. Discuss the relative amounts of cis and trans 4-*t*-butylcyclohexanol obtained and possible mechanistic reasons for this result.

SECTION 25

Oxidation of Alcohols and Aldehydes

Overview

Aldehydes are among the easiest functional groups to oxidize. Aldehydes should be protected from unnecessary exposure to air on storage, since even the oxygen in the air will oxidize them to carboxylic acids over a long time. They can be converted to carboxylic acids by a wide variety of oxidizing agents.

$$RCHO \xrightarrow{[O]} RCOOH$$

These oxidizing agents include the chromium(VI) reagents CrO_3, H_2CrO_4, and $Na_2Cr_2O_7/H_2SO_4$ (Jones reagent), the manganese(VII) reagent $KMnO_4$, and the nitrogen(V) reagent HNO_3. Note that all these oxidizing agents contain an atom in a highly oxidized state, which can accept electrons and become reduced as another species is oxidized. Also, note that all three chromium reagents amount to the same thing because they differ only in the moles of water contained in the formula.

$$CrO_3 + H_2O \longrightarrow H_2CrO_4$$

$$Na_2Cr_2O_7 + 2H^+ + 2H_2O \longrightarrow 2H_2CrO_4 + 2Na^+$$

Since these reagents form H_2CrO_4, they can also be called chromic acid reagents.

Chromic acid oxidation can also be used to oxidize secondary alcohols to ketones. The overall balanced equation is

$$3R_2CHOH + Na_2Cr_2O_7 + 4H_2SO_4 \longrightarrow 3R_2CO + Na_2SO_4 + Cr_2(SO_4)_3 + 7H_2O$$

In Experiment 25.1, 2-methylcyclohexanol is oxidized to 2-methylcyclohexanone using sodium dichromate and sulfuric acid.

$$\underset{\substack{\text{2-Methylcyclohexanol} \\ \text{MW 114, bp 167°} \\ \text{dens. 0.93 g/mL}}}{\overset{\text{OH}}{\text{CH}_3}} \quad \xrightarrow[\text{H}_2\text{SO}_4]{\text{Na}_2\text{Cr}_2\text{O}_7} \quad \underset{\substack{\text{2-Methylcyclohexanone} \\ \text{MW 112, bp 166°} \\ \text{dnes. 0.92 g/mL}}}{\overset{\text{O}}{\text{CH}_3}}$$

This reaction is known as a Jones oxidation. It is possible to oxidize a primary alcohol to an aldehyde this way as well, but unless the aldehyde is distilled out of the pot as it is formed, it will generally be further oxidized to the carboxylic acid.

$$R-CH_2OH \xrightarrow{[O]} R-\overset{\displaystyle O}{\overset{\|}{C}}-H \xrightarrow{[O]} R-\overset{\displaystyle O}{\overset{\|}{C}}-OH$$

During the course of a chromic acid oxidation reaction the chromium normally becomes reduced to the green $+3$ state, forming chromium hydroxide $Cr(OH)_3$. Special care must be taken in using chromium reagents and in disposing of the wastes, since chromium has been found to be carcinogenic on ingestion by mammals.

Permanganate oxidations also involve color changes, beginning with the deep purple $KMnO_4$ in solution and ending up with a brown MnO_2 precipitate, which looks like mud. Experiment 25.2 is the oxidization of furfural to furoic acid using chromic acid, and in Experiment 25.3 heptanal is oxidized to heptanoic acid using potassium permanganate.

Furfural is common and inexpensive, being obtained from distillation of corncobs. Furfural and furoic acid also provide examples of heterocyclic aromatic compounds, since they contain furan rings.

Furfural
(2-furaldehyde)
MW 96, bp 162°
dens. 1.16 g/mL

2-Furoic acid
MW 112, mp 129–130°

$$CH_3(CH_2)_5CHO \xrightarrow[H_2SO_4]{KMnO_4} CH_3(CH_2)_5COOH$$

Heptanal
(heptaldehyde)
MW 114, bp 153°
dens. 0.85 g/mL

Heptanoic acid
MW 130, bp 223°
160° at 100 torr
dens. 0.92 g/mL

Furan is aromatic (and has the corresponding extra stability) because one unshared pair of electrons on the oxygen atom participates in resonance with the two double bonds. The ring therefore involves six electrons and follows Hückel's $4n + 2$ rule for aromaticity. Heptanal is also readily available since it is obtained by distilling castor oil under reduced pressure.

The oxidation of alcohols has forensic applications as well. The National Institute on Alcohol Abuse and Alcoholism estimates that, in purely economic terms, alcohol abuse costs Americans over $100 billion annually. Reduced productivity accounts for approximately half of this staggering figure, while increased medical expenses and mortality account for most of the remainder. Heavy drinkers suffer increased incidence of gastric lesions, liver disease, and heart disease. According to the U.S. Department of Transportation, over 20,000 motor vehicle fatalities are caused by drunk drivers every year. These statistics do not even reflect the extreme emotional suffering by family members and victims of those who abuse alcohol.

One important tool in combating drunk driving is the breathalyzer test for estimating blood alcohol levels. The reaction of a chromium(VI) species (in this case, potassium dichromate) with a primary alcohol (ethanol) forms the basis for this test.

$$3CH_3CH_2OH + 2K_2Cr_2O_7 + 8H_2SO_4 \longrightarrow 3CH_3COOH + 2Cr_2(SO_4)_3 + 2K_2SO_4 + 11H_2O$$

Since alcohol in the air in the lungs is in equilibrium with alcohol in the blood, measuring breath alcohol levels yields reasonable estimates of blood alcohol concentrations.

A simple breath analyzer as used by police officers for initial screening of suspected drunk drivers consists of a glass ampoule containing silica gel impregnated with a potassium dichromate–sulfuric acid reagent. To perform the test, the ends of the ampoule are broken off, a mouthpiece is fitted on one end, and an empty plastic bag is fitted on the other end (Figure 25-1). As the person being tested blows through the mouthpiece to inflate the bag, any alcohol present in the breath reacts with the reddish-orange potassium dichromate to form green chromium sulfate.

Breath alcohol screening device.

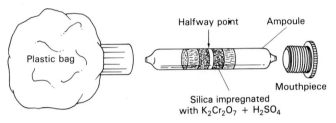

Figure 25–1   Breath Alcohol Screening Device

The concentration of dichromate and volume of the bag are designed so that if the blood alcohol is above the legal limit of approximately 0.10% (equivalent to about six drinks), the green color will extend past the halfway point in the ampoule.

Since this method is necessarily somewhat inaccurate, it is only used as an initial screening. If the results indicate a high blood alcohol count, the person is tested using a more accurate instrument (Figure 25-2). Air blown in the mouthpiece (A) fills a com-

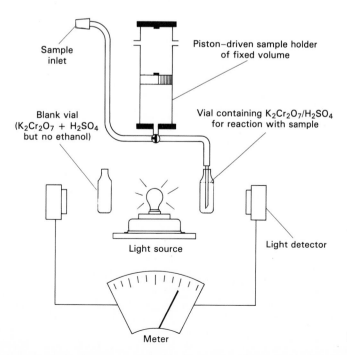

Figure 25–2   Breath Analyzer

partment of known volume (B) causing a piston to rise. The valve is turned and, as the piston descends, this air sample is passed through a bottle containing potassium dichromate and sulfuric acid in water (C). As the alcohol reacts, green Cr^{3+} ions are produced. Filtered light of 600 nm wavelength (green) is passed through this cell and into a spectrometer. The light absorbance is compared electronically with that of a blank cell containing $K_2Cr_2O_7$ and H_2SO_4 but no ethanol. By reading the absorbance and using Beer's law ($A = \epsilon C\ell$, or absorbance A is directly proportional to concentration C) the quantity of alcohol originally present can be determined with an uncertainty of approximately 5%.

REFERENCES

CASEMENT, M. R., "Getting on Top of the Bottom Line: Economics in the Sixth Special Report on Alcohol and Health," *Alcohol Health and Research World*, DHHS Publication (ADM) 88-151, Fall 1987, p. 4.

TIMMER, W. C., "An Experiment in Forensic Chemistry: The Breathalyzer," *Journal of Chemical Education* (1986) *63*, 897.

TREPTOW, R. S., "Determination of Alcohol in Breath for Law Enforcement," *Journal of Chemical Education* (1974) *51*, 651.

KRITSBERG, W., "The Adult Children of Alcoholics Syndrome," Bantam Books, New York, 1988.

EXPERIMENT 25.1 JONES OXIDATION OF 2-METHYLCYCLOHEXANOL

Estimated Time:
2.5 hours

Prelab

1. Calculate the millimoles in 12.0 g of sodium dichromate dihydrate and in 9.0 mL of concentrated H_2SO_4.
2. Calculate the theoretical yield of 2-methylcyclohexanone.

Special Hazards

Chromium salts are suspected carcinogens. Handle with care and avoid contact. Dispose of all chromium-containing wastes in properly marked containers. Concentrated H_2SO_4 is highly corrosive, and 2-methylcyclohexanol and ether are flammable.

Procedure

First prepare the chromic acid solution in a 150-mL beaker by dissolving sodium dichromate dihydrate (12.0 g, _____ mmol) in 50 mL of water. Add concentrated H_2SO_4 (9.0 mL, _____ mmol). Swirl to obtain a homogeneous solution (heat is evolved). This is the Jones reagent.

In a 250-mL Erlenmeyer flask, place a Teflon-coated magnetic stirring bar, 50 mL of ether, and 2-methylcyclohexanol (60 mmol, _____ g, _____ mL). Position the flask in an ice bath atop a magnetic stirring plate. Stir vigorously and cool the solution while placing an addition funnel in an iron ring over the flask. Place the Jones reagent in the addition funnel. Allow the chromic acid solution to drip into the flask at a rate of approximately 3 drops per second so that the entire addition takes about 10 minutes. When the addition is complete, maintain stirring in the ice bath for 20 minutes more, then pour the mixture into a separatory funnel. The lower (aqueous) layer is drawn off and discarded into a container marked for chromium-containing wastes. Wash the remaining ether layer with 20 mL of 5% aqueous Na_2CO_3, then dry it over anhydrous Na_2SO_4. Decant into a beaker, add boiling chips, and remove the ether on a steam bath in the hood. Distill the residue to obtain pure 2-methylcyclohexanone (lit. bp 165°). Record the boiling range and mass of the product obtained.

REFERENCE

KRISHNAMURTHY, S., T. W. NYLUND, N. RAVINDRANATHAN, and K. L. THOMPSON, "Aqueous Chromic Acid Oxidation of Secondary Alcohols in Diethyl Ether," *Journal of Chemical Education* (1979) *56*, 203.

EXPERIMENT 25.2 OXIDATION OF FURFURAL TO FUROIC ACID USING $K_2Cr_2O_7/H_2SO_4$

Estimated Time:
2.0 hours

Prelab

1. Calculate the millimoles corresponding to 1.00 g of furfural, 1.00 g of $K_2Cr_2O_7$, and 2.00 g of concentrated H_2SO_4.
2. Calculate the volume of 1.00 g of furfural and the theoretical yield of 2-furoic acid.

Special Hazards

Furfural is corrosive and toxic. Potassium dichromate is a strong oxidant and suspected carcinogen. Exercise standard precautions with concentrated H_2SO_4.

Procedure

In a small beaker, place furfural (1.00 g, _____ mmol, _____ mL), $K_2Cr_2O_7$ (1.00 g, _____ mmol), water (1 mL), and concentrated H_2SO_4 (25 drops, approx. 1.2 mL, 2.0 g, _____ mmol). The mixture is swirled and heated on a steam bath to initiate the reaction. Once the exothermic reaction has begun the steam is shut off and the flask swirled occasionally for 15 minutes. At the end of the period the solution is cooled in an ice bath and made basic by addition of 15 mL of 6 N NaOH. The solution now contains the dissolved carboxylate salt of 2-furoic acid. Suction filter and dispose of the filter paper (which holds chromium wastes) in the appropriate waste container.

Cautiously acidify the filtrate using concentrated HCl (about 2 mL). Collect the crude precipitate of 2-furoic acid by suction filtration, wash it with a little water, then recrystallize from water to obtain pure 2-furoic acid.

REFERENCE

SANDLER, S. R., and W. KARO, *Organic Functional Group Preparations*, 2nd ed., Academic Press, New York, 1983, Vol. 1, p. 241.

EXPERIMENT 25.3 OXIDATION OF HEPTANAL TO HEPTANOIC ACID USING KMnO₄

Estimated Time:
2.0 hours

Prelab

1. Since heptanoic acid has a polar end and a nonpolar end, something like a soap molecule, what problem may be encountered if distillation is attempted?
2. Balance the redox reaction for the permanganate oxidation of heptanal.

_____$CH_3(CH_2)_5CHO$ + _____$KMnO_4$ + _____H_2SO_4 $\longrightarrow$
  Heptanal

_____$CH_3(CH_2)_5COOH$ + _____K_2SO_4 + _____MnO_2 + H_2O
  Heptanoic acid

(Hint: Find the atoms that are oxidized and reduced, and by how many electrons. Balance these, then balance the spectator ions.)

Special Hazards

Heptanal is flammable and irritating. Potassium permanganate is a strong oxidizing agent and will stain clothes and hands. To remove these stains, rinse with dilute aqueous sodium bisulfite solution. Sulfuric acid is highly corrosive; use due caution. Heptanoic acid has an unpleasant odor.

Procedure

In a 250-mL beaker, place 50 mL of water and add 7.0 mL of concentrated H_2SO_4. Add a thermometer, cool the solution in ice to 10–20, then add heptanal (8.0 mL). Weigh out $KMnO_4$ (6.80 g) and add this in small portions with swirling over 15 minutes at a rate such that the temperature remains below 25°. During this time a brown precipitate of MnO_2 should form. When the addition is complete, keep swirling intermittently for another 5 minutes, then cautiously add enough solid sodium bisulfite ($NaHSO_3$) to destroy any remaining $KMnO_4$ (exothermic reaction). Filter to remove any remaining solids. Pour the mixture into a separatory funnel, add 40 mL of ether, and draw off the aqueous (lower) layer. Extract the remaining organic layer with 50 mL of 5% aqueous NaOH. This aqueous layer now contains the product as heptanoic carboxylate ion. To obtain the product, cool the aqueous extract in an ice bath and carefully acidify using concentrated HCl (about 8–10 mL are needed). Pour the two-phase mixture into a separatory funnel and remove the aqueous (lower) layer. Dry the crude product over $MgSO_4$ or Na_2SO_4, then decant into a sample vial.

Record the mass of your product and measure its index of refraction. Combine your product with that of several others and take an infrared spectrum.

QUESTION

1. In your write-up, interpret your IR spectrum, and find and photocopy a literature spectrum of heptanoic acid. Staple the literature spectrum to yours. How do the two spectra compare?

REFERENCE

RUHOFF, J. R., *Organic Syntheses Collective Volume 2* (1943) 315.

SECTION 26

Reactions of Aldehydes and Ketones with Amines

Overview

In the presence of an acid catalyst, an aldehyde or ketone can condense with a primary amine, forming a carbon-to-nitrogen double bond with the elimination of water.

$$\overset{R}{\underset{R'}{\diagdown}}C{=}O \ + \ H_2N{-}R'' \ \xrightarrow{\ H^+\ } \ \overset{R}{\underset{R'}{\diagdown}}C{=}N\overset{R''}{\diagup} \ + \ H_2O$$

Note that, overall, the oxygen of the carbonyl group and the two hydrogens on the two hydrogens on the amine are lost in the process, forming water. In the case of a primary amine the product is an imine, also known as a Schiff's base. The mechanism of this reaction involves protonation of the carbonyl group followed by nucleophilic attack by the nitrogen on the carbonyl carbon. Subsequent protonation of the resulting OH group results in loss of water and formation of a carbon-to-nitrogen double bond.

Imines can be hydrolyzed back to the corresponding carbonyl compounds and amines by heating with acid and water. They can also be reduced by such reagents as $NaBH_4$ to form secondary amines.

$$\underset{R'}{\overset{R}{>}}C=N-R'' \xrightarrow[H^+]{H_2O} \underset{R'}{\overset{R}{>}}C=O + H_2N-R''$$

$$\xrightarrow[(2)\ H^+]{(1)\ NaBH_4} R-\underset{R'}{\overset{H}{\underset{|}{\overset{|}{C}}}}-\overset{H}{\underset{|}{N}}-R''$$

Several special types of imines are known. When hydrazine is used in this reaction, the product is called a hydrazone.

$$\underset{R'}{\overset{R}{>}}C=O + H_2N-NH_2 \xrightarrow{H^+} \underset{R'}{\overset{R}{>}}C=N-NH_2$$

An aldehyde   Hydrazine
or ketone   A hydrazone

Hydrazine is much more nucleophilic than a normal amine because both nitrogens possess unshared pairs of electrons. These electron pairs repel each other, making both more nucleophilic. This high nucleophilicity is also shared by hydroxylamine, where a nitrogen and an oxygen atom (with two unshared pairs) are adjacent. The products of reaction of aldehydes or ketones and hydroxylamine are called oximes.

$$\underset{R'}{\overset{R}{>}}C=O\ +\ NH_2OH \xrightarrow{H^+} \underset{R'}{\overset{R}{>}}C=N\overset{OH}{\diagup}$$

An aldehyde   Hydroxylamine
or ketone   An oxime

Because of the unshared electron pair on nitrogen, the geometry around nitrogen is *bent*. Therefore, if the two R groups are different, the possibility of two geometric isomers arises, with the —OH group pointing toward one R group or the other. These isomers are called syn and anti (or Z and E), depending on whether the —OH is toward or away from the larger group.

$$\underset{H}{\overset{CH_3}{>}}C=N\overset{OH}{\diagup} \qquad\qquad \underset{H}{\overset{CH_3}{>}}C=N\underset{OH}{\diagdown}$$

syn   anti
Oximes of acetaldehyde

It is sometimes difficult to separate these isomers since their solubility and other physical properties may be similar.

In this experiment you will prepare the oxime of benzophenone, which is symmetrical, so this difficulty will not arise.

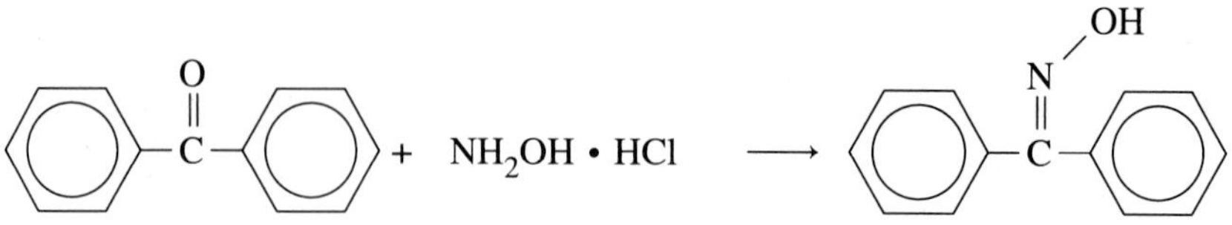

Benzophenone
MW 182, mp 49–51°

Hydroxylamine
hydrochloride
MW 69.5, mp 159°d

Benzophenone oxime
MW 197, mp 144°

Hydroxylamine is most conveniently handled as its hydrochloride. Most of the HCl is removed in solution by adding NaOH; the remaining HCl provides the needed acid catalysis.

Reactions of ketones with amines are essential biochemical processes as well. Excess amino acids for protein construction are often stored by the body in the form of α-keto acids. When needed, these may be reconverted to amino acids by the enzyme-mediated process of transamination. In this process an enzyme known as a transaminase or aminotransferase transfers the amino group off of one amino acid onto an α-keto acid, creating a new amino acid and leaving the former amino acid as a keto acid.

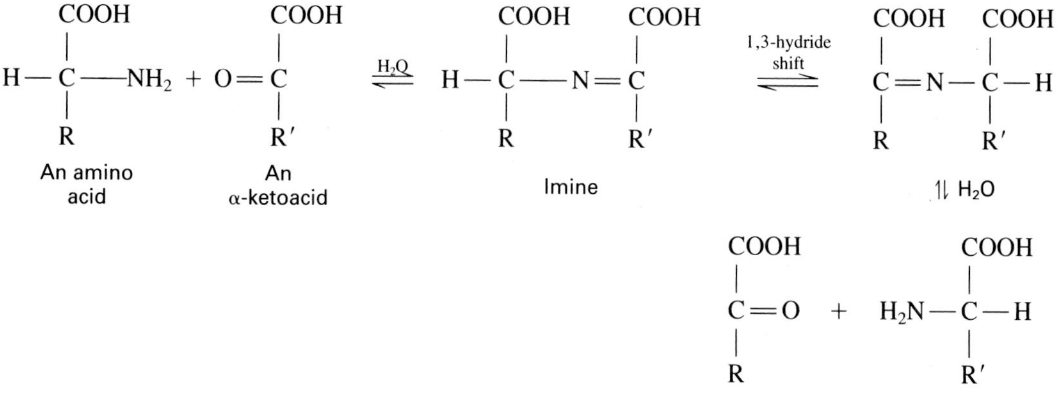

An amino
acid

An
α-ketoacid

Imine

New keto acid

New amino
acid

QUESTION

1. Show the products of the following reactions.

(a)

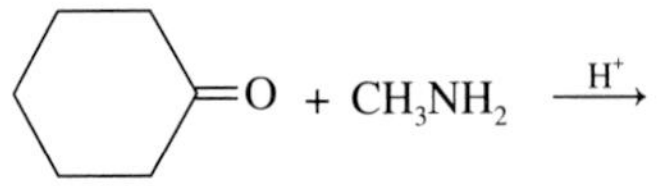

(b)

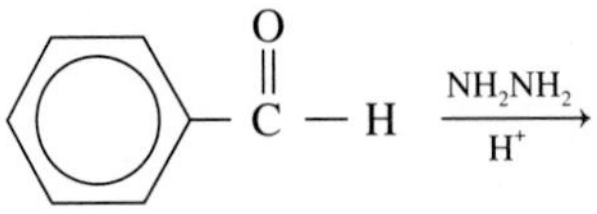

(c) $CH_3CH_2CCH_3$ $\xrightarrow[\text{H}^+]{\text{NH}_2\text{OH}}$

(d) Androsterone (male sex hormone) $+$ 2,4 - dinitrophenyl-hydrazine $\xrightarrow{\text{H}^+}$

EXPERIMENT 26.1 MICROSCALE SYNTHESIS OF BENZOPHENONE OXIME

Estimated Time:
1.0 hour

Prelab

1. Calculate the millimoles of benzophenone in 100 mg and those of hydroxylamine hydrochloride in 60 mg.
2. Calculate the theoretical yield of benzophenone oxime.

Special Hazards

Both NaOH and hydroxylamine hydrochloride are corrosive. Ethanol and methanol are flammable.

Procedure

In a 15 × 125 mm test tube, place benzophenone (100 mg, _____ mmol), hydroxyl-amine hydrochloride (60 mg, _____ mmol), ethanol (2 mL), and water (6 drops). Swirl and add NaOH (110 mg, _____ mmol). It may be necessary to grind NaOH pellets to a powder to weigh out the correct amount. Heat the tube in a steam bath and boil for 5 minutes with occasional swirling. If there is a small amount of gummy white residue, add a little more ethanol. Cool the tube in an ice bath and add 2 mL of 5% aqueous HCl. Swirl to coagulate the off-white precipitate and keep cooling for 5 minutes. Collect the precipitate by suction filtration, wash it with water, then with a few drops of ice-cold methanol, and allow it to air dry. Record the mass and melting range of the crude product (lit. mp 144°). Place it in a clean test tube and recrystallize from methanol/water (approximately 2 mL of methanol plus 15 drops water). Record the melting point of the recrystallized material.

REFERENCE

Lachman, A., *Organic Syntheses Collective Volume 2* (1943) 70.

The Beckmann Rearrangement

Overview

Under acidic conditions oximes can rearrange to form amides. The mechanism of this reaction, known as the Beckmann rearrangement, is as follows:

Protonation of the oxime oxygen leads to loss of water accompanied by a 1,2 alkyl migration. The reaction occurs stereospecifically, with the alkyl group anti to the OH group migrating. This reaction provides a method for overall conversion of a ketone to an amide via the oxime.

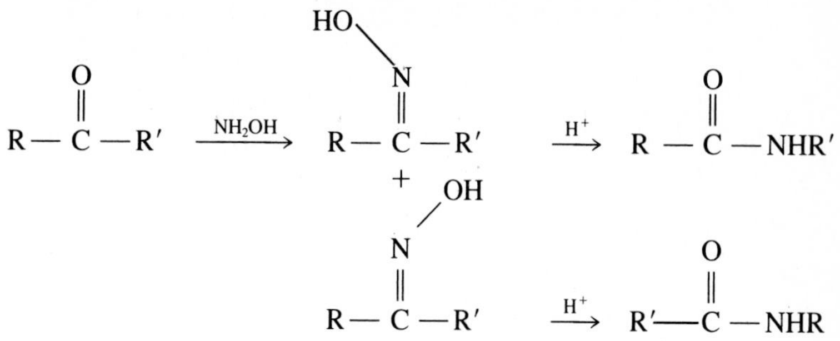

In Experiment 27.1 you will be converting cyclohexanone oxime into caprolactam.

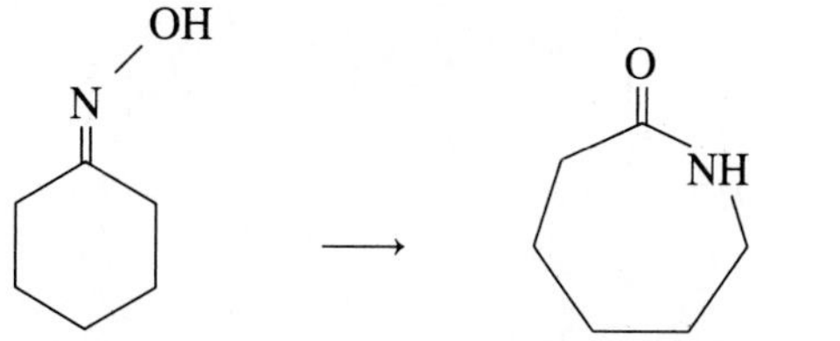

Cyclohexanone oxime
MW 113, mp 89 – 91°

Caprolactam
(6-aminohexanoic acid
lactam)
MW 113, mp 70 – 72°

Over 500 million pounds of caprolactam are produced annually, since it is the monomer used for preparation of nylon 6, a major component of tire cords.

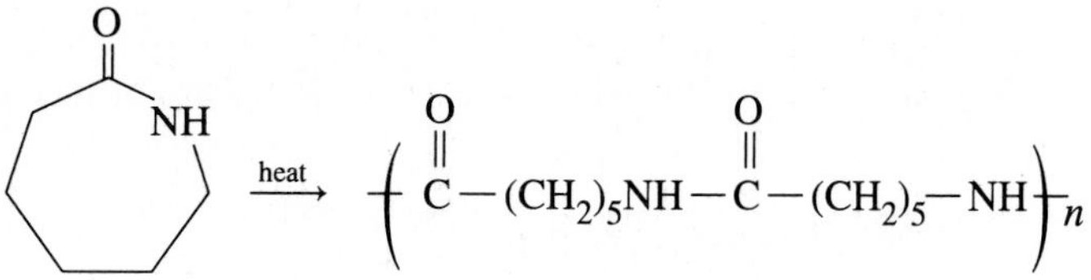

Nylon 6

QUESTIONS

1. Show the mechanism and product of Beckmann rearrangement of benzophenone oxime.

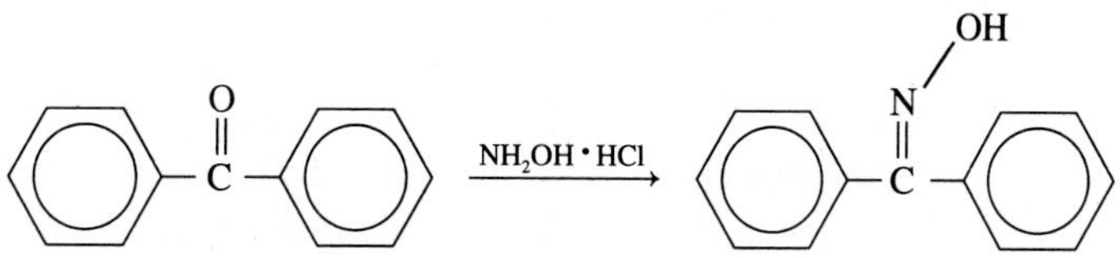

Benzophenone

Benzophenone oxime

2. Fill in the missing compounds.

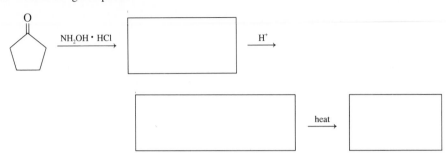

EXPERIMENT 27.1 MICROSCALE CONVERSION OF CYCLOHEXANONE OXIME TO CAPROLACTAM

**Estimated Time:
2.0 hours**

Prelab

1. Calculate the mass of 10 mmol of cyclohexanone oxime and the theoretical yield of caprolactam.

Special Hazards

Caprolactam is an irritant. Exercise standard precautions with 3 M H_2SO_4. While using chloroform, work in the fume hood at all times, since chloroform is a suspected carcinogen. Because of the vigorous nature of this reaction, it should not be performed on a larger scale.

Procedure

In a small beaker, place cyclohexanone oxime (180 mg, _____ mmol). Add 2 drops of water and 5 drops of concentrated H_2SO_4 (*caution*). Be sure to drop the sulfuric acid directly on top of the cyclohexanone oxime. Place the beaker on a hot plate in the fume hood and heat until the solid melts and a sudden exothermic reaction occurs. The mixture turns brown and a small cloud of smoke is produced.

Remove the beaker from the heat and cool it in an ice bath. Add 5% aqueous NaOH until the solution tests slightly basic to litmus (about 3 mL). In the fume hood, add chloroform (3 mL) and swirl thoroughly to extract the caprolactam. Remove the organic (lower) layer by Pasteur pipette and place it in another small flask. Dry this chloroform extract over anydrous sodium sulfate, decant, and evaporate on a steam bath in the fume hood. Recrystallize the syrupy residue from a small volume of ligroin (bp 90–110°). The caprolactam is slow to crystallize. Collect the product by suction filtration, air dry, and record the mass and melting point.

REFERENCE

CARRAHER, C. E., JR., ''Synthesis of Caprolactam and Nylon 6,'' *Journal of Chemical Education* (1978) *55*, 51.

SECTION 28

Malonic and Acetoacetic Ester Syntheses

Overview

The malonic and related acetoacetic ester syntheses have proven to be versatile methods for building the desired carbon skeletons of carbonyl-containing molecules. Both methods rely on the alkylation of an enolate stabilized by two neighboring carbonyl groups, followed by hydrolysis of one or two ester groups and decarboxylation of the resulting carboxylic acid.

$$\underset{\substack{\text{A malonic ester}\\(R = Me \text{ or } Et)}}{ROC\underset{\overset{\displaystyle\|}{O}}{CH_2}\,C\underset{\overset{\displaystyle\|}{O}}{OR}} \xrightarrow{RO^-Na^+}$$

$$\left[\; RO\underset{\overset{\displaystyle\|}{O}}{C}-\underset{\overset{|}{H}}{\bar{C}}-C\underset{\overset{\displaystyle\|}{O}}{OR} \;\longleftrightarrow\; RO\underset{\overset{\displaystyle\|}{O^-}}{C}=\underset{\overset{|}{H}}{C}-C\underset{\overset{\displaystyle\|}{O}}{OR} \;\longleftrightarrow\; RO\underset{\overset{\displaystyle\|}{O}}{C}-\underset{\overset{|}{H}}{C}=C\underset{\overset{\displaystyle\|}{O^-}}{OR} \;\right] \xrightarrow{R'X}$$

$$\underset{\text{Alkylated product}}{ROC\underset{\overset{\displaystyle\|}{O}}{}-\underset{\substack{\overset{\displaystyle R'}{|}\\ \overset{|}{H}}}{C}-C\underset{\overset{\displaystyle\|}{O}}{OR}} \xrightarrow{H^+,\,H_2O} HOC\underset{\overset{\displaystyle\|}{O}}{}-\underset{\substack{\overset{\displaystyle R'}{|}\\ \overset{|}{H}}}{C}-C\underset{\overset{\displaystyle\|}{O}}{OH} \xrightarrow[-CO_2]{\text{heat}} \underset{\substack{\text{A substituted}\\\text{acetic acid}}}{R'CH_2COOH}$$

A carboxylic acid with a carbonyl group in the β position can undergo decarboxylation on heating, through a cyclic transition state. This process forms the enol, which tautomerizes.

Enol

The net result is that a —CO_2H group β to a carbonyl is replaced with a hydrogen. Much of the carbon dioxide we breathe out is produced in just this type of reaction. Decarboxylation of a β-keto acid is one of the essential steps of the tricarboxylic acid (TCA or Krebs) cycle of energy metabolism. When the alcohol group of isocitrate is oxidized to a ketone, the β-carboxyl group is then lost as carbon dioxide, leaving α-ketoglutarate.

Isocitrate   a-Ketoglutarate

Returning to the malonic ester synthesis, if we consider the fact that the R group was put on by alkylation, while the remainder of the molecule came from the original malonic ester, it becomes clear that this is a method for preparing *substituted acetic acids* (RCH_2COOH or $RR'CHCOOH$). Of course, with R being a carbon-containing group, the name will not be an acetic acid.

The acetoacetic ester synthesis is very similar to the malonic ester procedure; the differences are that the starting material is acetoacetic ester (a ketoester) and the products are substituted acetones.

$$\underset{\substack{\text{An acetoacetic ester}\\(R = \text{Me or Et})}}{CH_3\overset{O}{\overset{\|}{C}}CH_2\overset{O}{\overset{\|}{C}}-OR} \xrightarrow{RO^-Na^+}$$

$$CH_3\overset{O}{\overset{\|}{C}}-\underset{H}{\overset{|}{\overset{-}{C}}}-COR \longleftrightarrow CH_3\overset{O^-}{\overset{|}{C}}=\underset{H}{\overset{|}{C}}-COR \longleftrightarrow CH_3\overset{O}{\overset{\|}{C}}-\underset{H}{\overset{|}{C}}=\overset{O^-}{\overset{|}{C}}-OR \xrightarrow{R'X}$$

$$\underset{\substack{\text{Alkylated acetoacetic ester}\\(\text{repetition of above steps}\\\text{results in dialkylation})}}{CH_3\overset{O}{\overset{\|}{C}}-\underset{H}{\overset{R'}{\overset{|}{C}}}-\overset{O}{\overset{\|}{C}}-OR} \xrightarrow{H^+,\,H_2O} CH_3\overset{O}{\overset{\|}{C}}-\underset{H}{\overset{R'}{\overset{|}{C}}}-\overset{O}{\overset{\|}{C}}OH \xrightarrow[-CO_2]{\text{heat}} \underset{\substack{\text{A}\\\text{substituted}\\\text{acetone}}}{CH_3\overset{O}{\overset{\|}{C}}-CH_2R'}$$

Usually, the base used to form the enolate is the alkoxide of the alcohol found in the ester group (methanol or ethanol). Since transesterification is occurring, this prevents changing the identity of the alcohol groups on the ester. While transesterification would not harm the overall reaction, it could create mixtures that are harder to purify or have more complex NMR spectra.

Experiment 28.1 is a malonic ester synthesis involving dialkylation, with the twist that the alkyl halide used is 1,3-dibromopropane ($BrCH_2CH_2CH_2Br$) so that sequential alkylation on both ends of the chain results in formation of a cyclobutane ring.

$$\underset{\substack{\text{Dimethyl malonate}\\\text{MW 132, bp 180}-181°\\\text{dens. 1.16 g/mL}}}{CH_3O\overset{O}{\overset{\|}{C}}CH_2\overset{O}{\overset{\|}{C}}OCH_3} \xrightarrow{CH_3O^-Na^+} \underset{\substack{\text{Sodium}\\\text{methoxide}\\\text{MW 54}}}{} \underset{\text{Enolate}}{CH_3O\overset{O}{\overset{\|}{C}}-\underset{H}{\overset{-}{\overset{|}{C}}}-\overset{O}{\overset{\|}{C}}OCH_3} \xrightarrow{BrCH_2CH_2CH_2Br} \underset{\substack{\text{1,3 - Dibromo-}\\\text{propane}\\\text{MW 202, bp 167°}\\\text{dens. 1.99 g/mL}}}{CH_3O_2C-\underset{H}{\overset{CH_2CH_2CH_2Br}{\overset{|}{\underset{|}{C}}}}-CO_2CH_3} \xrightarrow{CH_3O^-Na^+}$$

Enolate

Dimethyl cyclobutane-
1,1 - dicarboxylate
MW 172, bp 116–117° at 20 torr
174–176° at 165 torr

$\xrightarrow[\text{(2) HCl}]{\text{(1) KOH}}$

Cyclobutane-
1,1-dicarboxylic acid
MW 144, mp 1.58°

$\xrightarrow[-CO_2]{\text{heat}}$

Cyclobutane-
carboxylic acid
MW 100, bp 195 – 196°
dens. 1.05 g/mL
n_D^{20}=1.4413

Experiment 28.2 is an example of an acetoacetic ester synthesis using *n*-butyl bromide to form 2-heptanone.

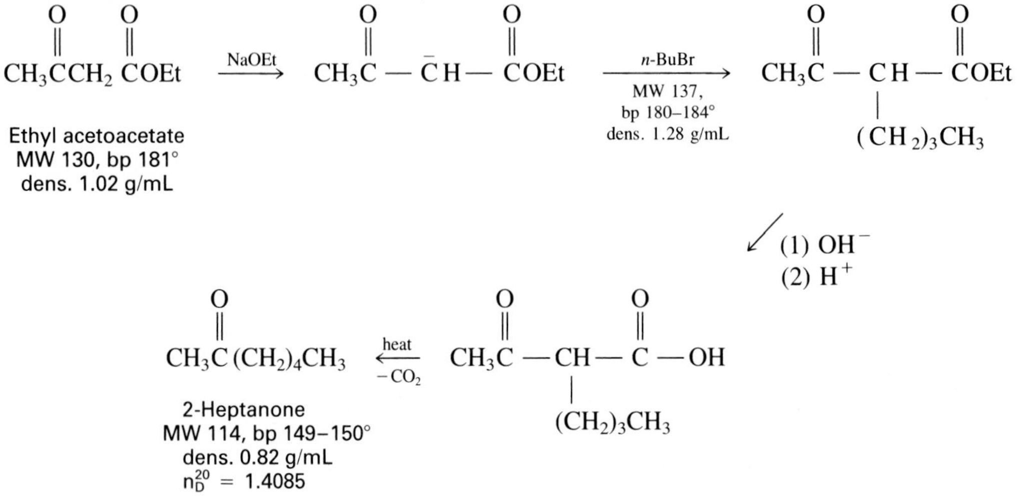

$$\underset{\substack{\text{Ethyl acetoacetate}\\ \text{MW 130, bp 181°}\\ \text{dens. 1.02 g/mL}}}{CH_3\overset{O}{\overset{||}{C}}CH_2\,\overset{O}{\overset{||}{C}}OEt} \xrightarrow{\text{NaOEt}} CH_3\overset{O}{\overset{||}{C}}-\overset{-}{C}H-\overset{O}{\overset{||}{C}}OEt \xrightarrow[\substack{\text{MW 137,}\\ \text{bp 180-184°}\\ \text{dens. 1.28 g/mL}}]{n\text{-BuBr}} \underset{(CH_2)_3CH_3}{CH_3\overset{O}{\overset{||}{C}}-CH-\overset{O}{\overset{||}{C}}OEt}$$

(1) OH⁻
(2) H⁺

$$\underset{\substack{\text{2-Heptanone}\\ \text{MW 114, bp 149-150°}\\ \text{dens. 0.82 g/mL}\\ n_D^{20} = 1.4085}}{CH_3\overset{O}{\overset{||}{C}}(CH_2)_4CH_3} \xleftarrow[-CO_2]{\text{heat}} \underset{(CH_2)_3CH_3}{CH_3\overset{O}{\overset{||}{C}}-CH-\overset{O}{\overset{||}{C}}-OH}$$

QUESTIONS

1. What types of compounds can be made by the malonic ester synthesis?
2. What types can be made by the acetoacetic ester synthesis?
3. What are the differences in starting materials and products between the malonic and acetoacetic ester syntheses?
4. What would be the products of the following reactions?

(a) $CH_3O_2CCH_2CO_2CH_3$
$$\xrightarrow[\substack{(3)\ H^+,\ H_2O\\(4)\ \text{heat}}]{\substack{(1)\ \text{NaOMe}\\(2)\ CH_3I}}$$

(b) $CH_3\overset{O}{\overset{||}{C}}CH_2CO_2CH_2CH_3$
$$\xrightarrow[\substack{(3)\ H^+,\ H_2O\\(4)\ \text{heat}}]{\substack{(1)\ \text{NaOEt}\\(2)\ PhCH_2CH_2Br}}$$

(c) $CH_2(CO_2Et)_2$
$$\xrightarrow[\substack{(3)\ \text{NaOEt}\\(4)\ CH_3I\\(5)\ H^+,\ H_2O\\(6)\ \text{heat}}]{\substack{(1)\ \text{NaOEt}\\(2)\ p-CH_3O-C_6H_4-CH_2Cl}}$$

5. Show how to prepare each of the following, using either a malonic or acetoacetic ester synthesis.

(a)

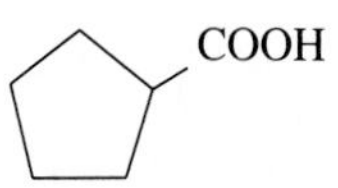

COOH

(b) — CH$_2$CH$_2$COOH

(c) CH$_3$ $\overset{\overset{\displaystyle O}{\|}}{C}CH_2CH_2$ — CH = CH$_2$

(d) CH$_3$CH$_2$ — $\overset{\overset{\displaystyle H}{|}}{\underset{\underset{\displaystyle CH_3}{|}}{C}}$ — $\overset{\overset{\displaystyle O}{\|}}{C}$ — CH$_3$

EXPERIMENT 28.1 MALONIC ESTER SYNTHESIS OF CYCLOBUTANECARBOXYLIC ACID

Estimated time:
6.5 hours for parts a, b, and c

Prelab

1. Calculate the mass of 140 mmol of sodium methoxide and the volumes of 70 mmol each of dimethyl malonate and 1,3-dibromopropane.
2. What will happen if water is present during attempted formation of an enolate?

Special Hazards

Sodium methoxide, either solid or in solution, is strongly basic and corrosive. Avoid contact with skin and eyes and rinse copiously with water immediately if contact should occur.

In a vacuum distillation never use glassware that is cracked or damaged, since the apparatus could .implode with the danger of cuts from broken glass and burns, both physical and chemical.

Procedure 28.1a: Preparation of Dimethyl Cyclobutane-1,1-dicarboxylate

Assemble a 250-mL round-bottomed flask (containing two or three boiling chips) with a Claisen head, condenser, and addition funnel. Place a CaCl$_2$ drying tube on top of the condenser. This setup is shown in Figure 28-1.

Make sure that all flammable chemicals are at least 15 feet away and flame dry the apparatus with a Bunsen burner for about 3 minutes. Move the burner around the round-bottomed flask, condenser, and addition funnel to dry the glass as thoroughly as possible. Heating the glass above the boiling point of water is sufficient. Keep the heat away from Teflon stopcocks or rubber stoppers to avoid melting them.

As the apparatus is cooling, check that no flames are still in use, then place 40 mL of anhydrous methanol (that has been dried over 3-Å molecular sieves) into a dry 125-mL Erlenmeyer flask. Stopper the flask and cool it in an ice bath. To avoid absorbing moisture from the air, quickly weigh out anhydrous sodium methoxide (140 mmol,

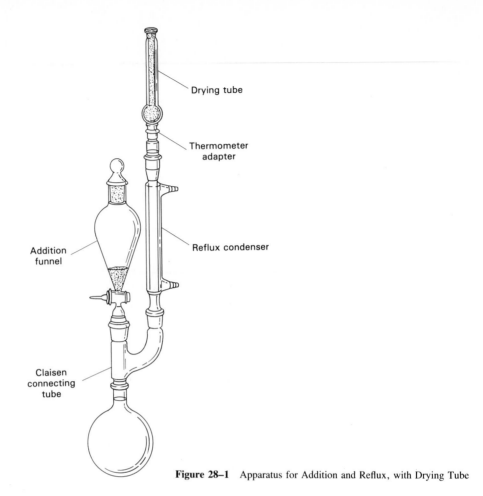

Figure 28–1 Apparatus for Addition and Reflux, with Drying Tube

_____ g), and add this in several portions over the course of 2–5 minutes (not all at once) to the cold methanol in the flask. The heat of solution is high; keep cooling in the ice bath. When all of the sodium methoxide has dissolved, add dimethyl malonate (which has also been dried over 3-Å molecular sieve, 70 mmol, _____ mL) to the flask. Now you have produced the monosodium salt of dimethyl malonate (dimethyl sodiomalonate). This solution also contains the second equivalent of base needed to complete the dialkylation. In a separate dry 50-mL Erlenmeyer flask dissolve 1,3-dibromopropane (70 mmol, _____ g, _____ mL) in anhydrous methanol (15 mL).

Check that the flame-dried setup has cooled to close to room temperature, then open the stopcock and remove the drying tube from the addition funnel. Pour the solution of 1,3-dibromopropane into the top of the addition funnel and allow it to run into the round-bottomed flask. Close the stopcock, then add the solution of dimethyl sodiomalonate to the addition funnel. Replace the drying tube. Add a heating mantle and variable transformer, establish water flow through the condenser, and bring the solution to reflux. After the solution boils, turn the heat down and begin dropwise addition of the methanolic solution of dimethyl sodiomalonate at the rate of about 1 drop per second, so that the addition is complete in about 20–30 minutes. It may be necessary to loosen the stopper on top of the addition funnel to prevent a vacuum from developing and holding up the addition. The reaction is somewhat exothermic, so be cautious to maintain heating at a low level and not to add the reagent too fast. Be ready to shut off the heat and lower the heating mantle if the boiling becomes too vigorous. Continue refluxing the solution for 1 hour, then cool the flask in an ice bath. Stopper it securely with a rubber or cork stopper, cover the stopper with Parafilm, and store the flask upright in your drawer if you are not continuing further during this lab period. If you are continuing with part B, rearrange for simple distillation and remove most of the methanol; skip the initial boiling in the fume hood to remove methanol described in part B.

Procedure 28.1b: Purification of Dimethyl Cyclobutane-1,1-dicarboxylate

Estimated Time:
1.5 hours
Pour the contents of the 250-mL round-bottomed flask into a 400-mL beaker, add some fresh boiling chips, and place it on a steam bath or hot plate in the fume hood. Boil it until most of the methanol is gone. *Caution:* Bumping may occur. The volume should decrease by about 65 mL. Add ice-cold water (50 mL) and swirl to dissolve the salts. Transfer the solution to a separatory funnel and remove the lower (aqueous) layer. Wash the remaining organic layer with water (10 mL) and saturated aqueous NaCl (brine, 10 mL), then dry it over $CaCl_2$, $MgSO_4$, or Na_2SO_4. When dry (about 5–10 minutes), decant the product into a 50-mL round-bottomed flask, add boiling chips, and set up for vacuum distillation using an aspirator. Collect the product (dimethyl cyclobutane-1,1-dicarboxylate, reported bp 116–117° at 20 torr and 174–176° at 165 torr) and record its mass. If possible, take an IR spectrum or refractive index to verify its identity. The residue remaining in the flask is mainly the tetraester resulting from attack of two molecules of dimethyl sodiomalonate on one molecule of 1,3-dibromopropane. The boiling point of this product is much higher: 185–189° at 2 torr.

Estimated Time:
2.5 hours
The following procedure hydrolyzes the two ester groups on dimethyl cyclobutanedicarboxylate and monodecarboxylates the diacid to yield cyclobutanecarboxylic acid. In a 50-mL round-bottomed flask place KOH (85% pure by weight, 8.0 g, _____ mmol), ethanol (15 mL), and dimethyl cyclobutanedicarboxylate (the product from distillation in part B, 25 mmol, _____ mL). Add some boiling chips and set up for reflux by adding a condenser, heating mantle, and variable transformer. Check water flow through the condenser and heat the mixture to reflux for 1 hour. Swirl periodically and check to make sure that the heating rate and water flow are satisfactory.

At the end of this period set up for simple distillation. Remove most of the ethanol, being careful not to distill to complete dryness. Add water (10 mL), stir, and heat if necessary to dissolve the residue, then cool the flask in an ice bath until the solution is below 15°. Cautiously acidify with concentrated HCl (about 7 mL), then pour the contents into a separatory funnel and extract with three 25-mL portions of ether. Dry the combined ether extracts over $CaCl_2$, filter, and evaporate to dryness on a steam bath in the hood. Recrystallize the crude 1,1-cyclobutanedicarboxylic acid from a minimum of ethyl acetate, then record the mass and melting range (lit. mp 156–158°).

To decarboxylate the diacid, place this product in a 50-mL round-bottomed flask set up for simple distillation. Heat strongly and observe the evolution of CO_2. Collect the resulting cyclobutanecarboxylic acid in the boiling range 191–196°. Record the mass and, if possible, take an IR spectrum or refractive index of this product to confirm its identity.

QUESTIONS

1. What can you say about the overall yield and purity of your product? If you took IR spectra, interpret the important peaks and compare them with literature spectra.
2. What by-products may have been produced during this synthesis?

REFERENCE

HEISIG, G. B., and F. H. STODOLA, *Organic Syntheses Collective Volume 3* (1955) 213.

EXPERIMENT 28.2 ACETOACETIC ESTER SYNTHESIS OF 2-HEPTANONE

**Estimated Time:
2.5 hours**

Prelab

1. Calculate the volumes of 50 mmol of ethyl acetoacetate and of 55 mmol of *n*-butyl bromide.
2. Calculate the volume of 21% (by weight) ethanolic NaOEt (dens. 0.87 g/mL) containing 50 mmol of NaOEt.
3. Calculate the theoretical yield of 2-heptanone.
4. What are two ways to tell when most of the ethanol has been removed at the end of part A?

Special Hazards

Sodium ethoxide and sodium hydroxide are strong bases and corrosive. Sulfuric acid is also highly corrosive. Wash immediately and thoroughly with water if contact occurs. Ethyl acetoacetate and *n*-butyl bromide are flammable and irritants.

Procedure

Note: This procedure involves two refluxes and two or three distillations. Careful planning is essential in part A. Care must also be taken to avoid exposure of the reagents to moisture.

A. Preparation of ethyl *n*-butylacetoacetate. In a 250-mL round-bottomed flask place 50 mL of anhydrous ethanol, a 21% ethanolic solution of sodium ethoxide (50 mmol, _____ mL), and ethyl acetoacetate (50 mmol, _____ g, _____ mL). Add a boiling chip, Claisen head, condenser, and addition funnel. In the addition funnel place *n*-butyl bromide (55 mmol, _____ g, _____ mL), then add it at such a rate that the solution becomes warm but does not reflux violently (about 1 drop per second). After the addition is complete and the reaction has subsided, add a heating mantle and variable transformer and reflux for 45 minutes. Shut off the heat briefly, rearrange for simple distillation, and remove almost all of the ethanol.

B. Saponification. Cool the flask in an ice bath, add 50 mL of 5% aqueous NaOH, and bring to reflux for 30 minutes to saponify the ester.

C. Decarboxylation. Cool the flask in an ice bath, then slowly and cautiously acidify by adding 10% aqueous H_2SO_4 (50 mL). The solution should bubble as CO_2 is evolved. When this bubbling diminishes, set up for simple distillation and distill about half the total volume of the flask, or until no more droplets of organic liquid are visible on the surface of the mixture in the distilling pot. This amounts to a steam distillation of 2-heptanone. One way to tell if the distillation is finished is to collect a few drops coming off the condenser in a test tube and to check whether they are cloudy due to the presence of an organic liquid (2-heptanone). If so, continue. At the completion of the distillation, cool the distillate, pour it into a separatory funnel, and allow the layers to separate. Drain off the aqueous (lower) layer and dry the organic layer (2-heptanone, lit. bp 149–150°) over Na_2SO_4. Decant and distill the product to purify it. Record the mass and boiling point observed, and note the distinctive cheesy odor. If possible, take a refractive index and/or IR spectrum.

SECTION 29

Alkylation and Acylation of Carbonyl Compounds

Overview

Alkylating or acylating the carbon next to a carbonyl group (the α carbon) represent extremely useful methods for building the desired carbon skeleton in a molecule. The overall process is as follows:

$$
\underset{\substack{\text{Carbonyl compound} \\ \text{with an } \alpha\text{-hydrogen}}}{-\overset{\overset{\displaystyle O}{\|}}{C}-\overset{\overset{\displaystyle H}{|}}{\underset{|}{C}}-}
\quad\nearrow\quad
-\overset{\overset{\displaystyle O}{\|}}{C}-\overset{\overset{\displaystyle R}{|}}{\underset{|}{C}}-\qquad\text{α-Alkylated product}
$$

$$
\searrow\quad
-\overset{\overset{\displaystyle O}{\|}}{C}-\overset{\overset{\displaystyle O \atop \| \atop C-R}{}}{\underset{|}{C}}-\qquad\text{α-Acylated product}
$$

Note that the hydrogen on the α-carbon has been replaced by an alkyl or acyl group. There are two major approaches to performing this type of reaction. One involves removal of the acidic α-hydrogen (pK_a approximately 20–25) using a strong base such as *n*-butyllithium (*n*-BuLi) or lithium diisopropylamide (LDA). On reaction with a strong base the carbonyl compound loses its α-hydrogen to form a resonance-stabilized enolate ion.

$$
-\overset{\overset{\displaystyle O}{\|}}{C}-\overset{\overset{\displaystyle H}{|}}{\underset{|}{C}}-
\xrightarrow[\text{or LDA}]{n\text{-BuLi}}
\left[\; -\overset{\overset{\displaystyle O}{\|}}{C}-\overset{}{\underset{|}{C}}- \longleftrightarrow -\overset{\overset{\displaystyle O^-}{|}}{C}=C{\Big\langle} \;\right]
$$

Enolate

Treatment of the enolate ion with a primary or secondary alkyl halide results in alkylation. Similarly, treatment of the enolate with an acyl halide or anhydride gives an acylated product.

These methods have found widespread application in synthesis. Their major limitation is the necessity of using strong base. If there are base-sensitive functional groups in the molecule, this alkylation or acylation must be performed in another way. One effective alternative method has been the enamine synthesis, developed by Gilbert Stork and co-workers in 1954. This method accomplishes the same overall result as use of an enolate but without the necessity of strong base.

In an enamine synthesis, the aldehyde or ketone to be alkylated or acylated is first allowed to react with a secondary amine in the presence of an acid catalyst. The nitrogen of the amine attacks the carbon of the protonated carbonyl group. A proton transfer from N to O followed by elimination of water yields the enamine.

The name "enamine" arises from the fact that there is a double bond ("ene") with an amine directly attached to it. Enamines have similar properties to enols, in that they are nucleophilic due to the presence of an unshared electron pair.

In the presence of an electrophile, such as a primary alkyl halide or acyl halide, the enamine will react on carbon.

The resultant iminium ions hydrolyze easily to give back the original carbonyl group.

So the overall scheme consists of formation of an enamine, alkylation or acylation, then hydrolysis. Common secondary amines used include pyrrolidine, piperidine, and morpholine.

Pyrrolidine   Piperidine   Morpholine

The particular reaction you will be performing is the synthesis of 2-acetylcyclohexanone using the morpholine enamine and acetic anhydride.

Cyclohexanone
MW 98, bp 155°
dens. 0.95 g/mL

Morpholine
MW 87.1, bp 129°
dens. 1.00 g/mL

Morpholine
enamine of
cyclohexanone
[1-morpholino-
1-cyclohexene,
4-(-cyclohexene-
1-yl)-morpholine]
MW 167
bp 118 – 120° at
10 torr
dens. 0.995 g/mL

Acetic
anhydride
MW 102
bp 138 – 140°
dens. 1.08
g/mL

2-Acetyl-
cyclohexanone
MW 140
bp 111 – 112°
at 18 torr
dens. 1.08
g/mL

QUESTIONS

1. What is an enamine, and what is it used for?

2. List the three steps of an enamine synthesis and give an example of each.

EXPERIMENT 29.1 ACYLATION OF CYCLOHEXANONE USING THE MORPHOLINE ENAMINE

Estimated Time:
4.0 hours for parts a and b

Prelab

1. Calculate the volumes of 80 mmol of cyclohexanone (dens. 0.95 g/mL), 100 mmol of morpholine (dens. 1.00 g/mL), 100 mmol of triethylamine (dens. 0.73 g/mL), and 90 mmol of acetic anhydride (dens. 1.08 g/mL).

Special Hazards

Cyclohexanone, morpholine, *p*-toluenesulfonic acid, acetic anhydride, and triethylamine are all irritating to eyes, the respiratory system, and skin. Avoid contact and pour these compounds only in the fume hood. Cyclohexanone and triethylamine are highly flammable—keep away from flames and exercise due caution.

Procedure 29.1a: Preparation of 1-Morpholinocyclohexene and Acylation with Acetic Anhydride

Into a 100-mL round-bottomed flask place dry toluene (40 mL), cyclohexanone (80 mmol, _____ g, 8.25 mL), morpholine (100 mmol, _____ g, 8.7 mL), *p*-toluenesulfonic acid (0.10 g), and some boiling chips. Set up for simple distillation by adding a distilling head, thermometer, condenser, adapter, and collecting flask. Add a heating mantle and Variac, and heat strongly to bring the solution to a boil. Distill until you have collected approximately 20 mL. There should be some water visible at the bottom of the collection flask.

Cool the distilling flask, which now contains the enamine, in ice and add triethylamine (100 mmol, _____ g, _____ mL) and acetic anhydride (90 mmol, _____ g, _____ mL). Swirl the solution, stopper it securely, cover the stopper with Parafilm, and store the flask in your drawer until the next laboratory period.

Procedure 29.1b: Hydrolysis of the Iminium Ion and Purification of 2-Acetylcyclohexanone

To the flask add water (30 mL) and reflux the mixture for 15 minutes to hydrolyze the the acylated enamine. Place the contents of the flask in a separatory funnel, remove and discard the aqueous (lower) layer, and dry the organic layer over anhydrous $CaCl_2$, $MgSO_4$, or Na_2SO_4. Decant the liquid back into the round-bottomed flask and set up for reduced-pressure distillation using an aspirator. After removing the toluene and morpholine, collect the product (2-acetylcyclohexanone, bp 125° at 20 torr).

REFERENCES

COOK, A. G., ed., *Enamines: Synthesis, Structure, and Reactions*, Marcel Dekker, New York, 1969.

DYKE, S. F., *The Chemistry of Enamines*, Cambridge University Press, Cambridge, 1973.

STORK, G., A. BRIZZOLARA, H. LANDESMAN, J. SZMUSZKOVICZ, and R. TERRELL, "The Enamine Alkylation and Acylation of Organic Compounds," *Journal of the American Chemical Society* (1963) *85*, 207.

SECTION 30

The Wittig Reaction

Overview

Since its reported discovery in 1953 the Wittig reaction has proven extremely useful for the synthesis of alkenes. The overall process converts a ketone or aldehyde to an alkene, as shown below.

$$
\begin{array}{c}
R_1 \\
 \\
C=O \\
 \\
R_2
\end{array}
\quad + \quad
\phi_3P=C
\begin{array}{c}
R_3 \\
 \\
 \\
R_4
\end{array}
\quad \longrightarrow \quad
\begin{array}{c}
R_1 R_3 \\
 \\
C=C \\
 \\
R_2 R_4
\end{array}
\quad + \quad
\phi_3P=O
$$

An aldehyde   A   An alkene   Triphenylphosphine
or ketone   phosphonium   oxide
                    ylid

The net result, then, is that the phosphorus of the phosphonium ylid (pronounced ilid) removes the oxygen of the carbonyl group, while the two carbons from the carbonyl group and the ylid are hooked together. The driving force for this reaction is the formation of the extremely strong phosphorus-to-oxygen double bond (approximately 510 kJ/mol). In fact, quite a few reactions involve removal of oxygen by phosphorus or by silicon, which also forms strong bonds with oxygen.

To discuss the mechanism, we must consider the fact that the phosphonium ylid has a significant resonance form possessing a positive charge on phosphorus and a negative charge on carbon.

$$
\phi_3P=C
\begin{array}{c}
R \\
 \\
R
\end{array}
\quad \longleftrightarrow \quad
\phi_3\overset{+}{P}-\overset{-}{C}
\begin{array}{c}
R \\
 \\
R
\end{array}
$$

Such a molecule, possessing adjacent positive and negative charges, is called an ylid. The charges arise because of the difference in electronegativity between phosphorus (2.2 Pauling units) and carbon (2.5 units).

An ylid is produced by the attack of a phosphine on an alkyl halide, followed by addition of base.

A phosphine   An alkyl halide   Ylid

Because the ylid possesses a substantial negative charge on carbon it is nucleophilic and can attack the (positive) carbon of a ketone or aldehyde.

A betaine

A phosphine oxide   An alkene   An oxaphosphetane

While the formation of an intermediate containing a four-membered ring is somewhat unusual in organic chemistry due to the 90° angles required, the attraction of the positive and negative charges on the betaine favors formation of the oxaphosphetane.

One drawback to the Wittig reaction is the formation of triphenylphosphine oxide, which is nonpolar and insoluble in water, and therefore is not removed from the organic phase by simple washing. Another difficulty is that the strong organic bases (*n*-butyllithium or lithium diisopropylamide) often used to prepare the ylid are somewhat hazardous to prepare and handle, and require a great deal of care and attention in their use. In the procedure here this problem has been circumvented by using phase-transfer catalysis employing the phosphonium salt itself as the phase-transfer catalyst. This makes feasible the use of aqueous NaOH as the base. Although still dangerous, aqueous NaOH is more convenient, less dangerous, and less expensive than the above-mentioned organic bases.

The reactions that you will be carrying out are the formation of an ylid (benzyltriphenylphosphonium ylid) from a phosphonium salt (benzyltriphenylphosphonium chloride) and the reaction of the ylid with benzaldehyde to form a mixture of *cis*- and *trans*-stilbene.

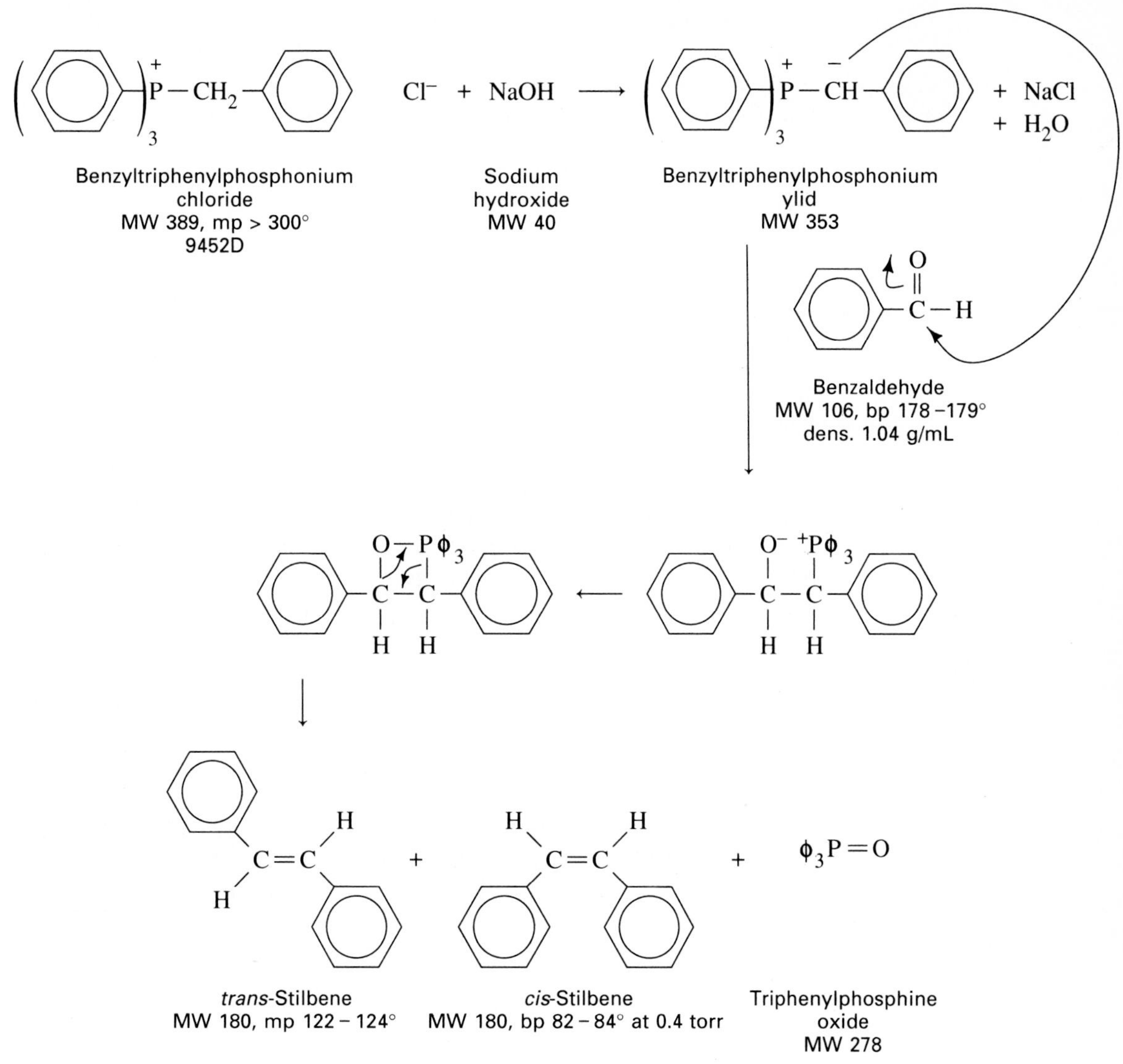

Benzyltriphenylphosphonium
chloride
MW 389, mp > 300°
9452D

Sodium
hydroxide
MW 40

Benzyltriphenylphosphonium
ylid
MW 353

+ NaCl
+ H_2O

Benzaldehyde
MW 106, bp 178–179°
dens. 1.04 g/mL

trans-Stilbene
MW 180, mp 122–124°

cis-Stilbene
MW 180, bp 82–84° at 0.4 torr

Triphenylphosphine
oxide
MW 278

EXPERIMENT 30.1 WITTIG SYNTHESIS
OF *TRANS*-STILBENE

Estimated Time:
2.0 hours

Prelab

1. Calculate the volume of 10 mmol of benzaldehyde and the mass of 10 mmol of benzyltriphenylphosphonium chloride.
2. What is the theoretical yield of the mixed stilbenes?

Special Hazards

10 N NaOH is extremely corrosive to skin and eyes. Dichloromethane is irritating and toxic. Benzyltriphenylphosphonium chloride is also an irritant.

Procedure

Into a 100-mL round-bottomed flask clamped over a magnetic stirring plate place benzaldehyde (10 mmol, _____ g, _____ mL), benzyltriphenylphosphonium chloride (10 mmol, _____ g), methylene chloride (10 mL,) and a small magnetic stirring bar. Establish stirring as vigorously as possible. Add a condenser and establish water flow. Then add through the funnel 7.5 mL of 10 N aqueous sodium hydroxide. The solution should warm spontaneously and turn yellow. Add a variable transformer, and a heating mantle between the flask and stirring plate, then heat gently as required to maintain reflux for 30 minutes. It is essential to keep the magnetic stirring vigorous throughout the reflux. After the reflux, cool the solution in an ice bath, and pour the mixture into a small separatory funnel. Add water to help separate the layers.

Remove the aqueous phase, dry the organic phase over Na_2SO_4, decant or filter, add some boiling chips, and evaporate the CH_2Cl_2 on a steam bath in the hood to obtain a syrupy residue. This residue is a mixture of *trans*- and *cis*-stilbenes plus triphenylphosphine oxide. Add 8 mL of absolute ethanol to the residue, stir thoroughly, and warm slightly if necessary to dissolve, then cool the solution in an ice bath for 10 minutes. Collect the precipitated *trans*-stilbene by suction filtration and recrystallize from a small volume of ethanol (6–10 mL) to which a few drops of water are added at the boiling point to approach the cloud point.

Collect the purified product by suction filtration, wash with a little ice-cold ethanol, allow to air dry, and place in a labeled vial. Record the mass and melting point of the *trans*-stilbene.

REFERENCES

MAERCKER, A., "The Wittig Reaction," *Organic Reactions* (1965) *14*, 270.

SCHLOSSER, M., G. MULLER, and K. F. CHRISTMAN, "Cis-Selective Olefin Syntheses," *Angewandte Chemie, International Edition in English* (1966) *5*, 667.

WITTIG, G., and G. GEISSLER, "Zur Reaktionsweise des Pentaphenyl-phosphors und einige Derivate," (Reaction Pathways of Pentaphenylphosphorus and Some Derivatives) *Justus Liebigs Annalen der Chemie* (1953), *580*, 44.

WITTIG, G. and U. SCHÖLLKOPF, "Uber Triphenyl-phosphin-methylene als Olefinbildende Reagenzien," (Methylenetriphenylphospine as a Reagent for Alkene Formation) *Chemische Berichte* (1954), *87*, 1318.

SECTION **31**

Reactions and Spectroscopy of Aldehydes and Ketones

Overview

Let's imagine that you are doing medical research and have heard about a wonderful and rare medicinal herb found only in remote regions of Nepal. Determined to find the active component of this herb, you fly to Katmandu and trek through out-of-the-way valleys with your trusty Sherpa guides to find some of these herbs. Bringing them back to your lab, you grind them up, extract with various solvents, and separate the compounds by column chromatography. Finally, you isolate 100 mg (a tip of a spatula full) of a colorless powder that has the pharmacological activity you are interested in. Now, how do you determine the structure of that compound?

This lab will introduce you to some of the principles of structure determination, an essential skill for doing any kind of chemical or biochemical research. In fact, after running *any* reaction you need to know whether you have the desired product, some other product, or just unreacted starting material. In our lab, taking a melting point or boiling point usually suffices to identify product or starting material. But what if you are making a compound that has never been made before?

Although approximately 5 million organic compounds are known, there are many times that number possible, and most graduate students in organic synthesis make several new compounds. To determine the structure of a new compound the standard procedure includes taking IR and NMR spectra and sending a sample for elemental analysis.

Before the discovery of spectroscopy (IR in the 1930s, NMR in the 1950s) chemists developed a lot of "spot tests" to determine functional groups on unknown compounds. Although these tests are now largely unnecessary because of spectroscopy, they are still needed occasionally and illustrate some useful principles. One thing they will certainly do is to give you a greater appreciation for spectroscopy.

These tests tell, for example, whether an unknown contains a hydroxyl group and if it is 1°, 2°, or 3° (Lucas test), or whether it contains an amino group, and if it is 1°, 2°, or 3° (Hinsberg test). About 30 of these tests are commonly used, although several hundred are known. They usually involve some visible change such as a color change or precipitation.

It should be remembered that, although fairly dependable, these tests are not 100% reliable—occasionally, a false positive or negative may turn up. For this reason we usually like to run several different tests and consider the ''preponderance of the evidence''. It is also wise to run tests on known compounds at the same time as running them on the unknown. By running both a positive and negative standard on known compounds you will gain the experience of judging positive and negative results and will be sure that your reagent is working properly.

Since we are only taking one lab period for this experiment, we will limit the unknown to the category of aldehydes and ketones. You will carry out three of the most common tests used with these compounds: the Tollens test, the iodoform test, and the 2,4-dinitrophenylhydrazine (2,4-DNP or 2,4-DNPH) test. In the next experiment you will deal with a broader range of functional groups and will decide as you go along which functional group tests to perform.

In this experiment, the Tollens test will tell you whether you are dealing with an aldehyde or a ketone. Aldehydes are easily oxidized to carboxylic acids by mild oxidizing agents such as silver ion (Ag^+, possibly complexed with ammonia). The silver ion, meanwhile, is reduced to metallic silver and forms a silver mirror on the wall of the test tube. This is how people used to make mirrors and Christmas tree ornaments in the ninetheenth century, before modern vacuum-deposition methods became available. Ketones, of course, are not easily oxidized and give a negative result (no mirror).

$$\underset{\text{An aldehyde}}{R-\overset{\overset{\displaystyle O}{\|}}{C}-H} + \underset{\substack{\text{Tollens} \\ \text{reagent}}}{Ag(NH_3)_2^+} \longrightarrow \underset{\text{A carboxylate}}{R-\overset{\overset{\displaystyle O}{\|}}{C}-O^-} + \underset{\substack{\text{Silver} \\ \text{mirror}}}{Ag(s)}$$

The iodoform test tells whether the compound is a methyl ketone (or methyl carbinol which also reacts).

$$\underset{\text{A methyl ketone}}{R\overset{\overset{\displaystyle O}{\|}}{C}CH_3} + I_2 \overset{OH^-}{\longrightarrow} \underset{\text{A carboxylate}}{R\overset{\overset{\displaystyle O}{\|}}{C}-O^-} + \underset{\substack{\text{iodoform} \\ \text{(yellow precipitate)}}}{CHI_3}$$

The mechanism of this reaction is that hydroxide pulls off one of the acidic hydrogens next to the carbonyl, then the enolate attacks I_2. This happens three times, until finally all three hydrogens on the methyl group are replaced by iodines. At this time hydroxide can attack the carbonyl and displace the good leaving group, CI_3^-.

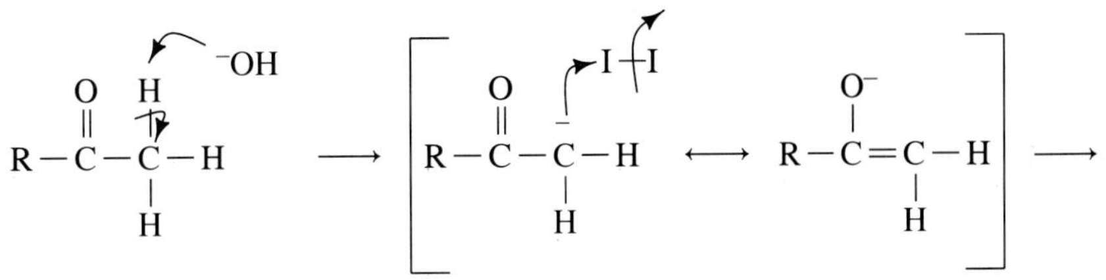

Enolate ion

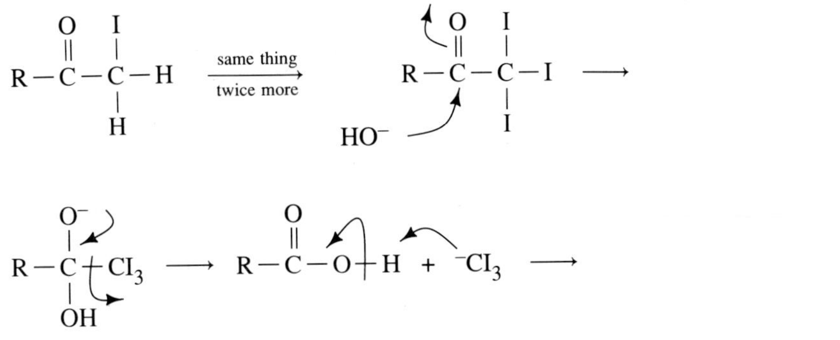

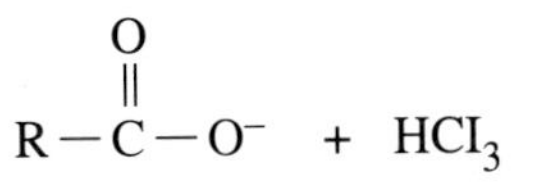

iodoform

A positive test for a methyl ketone (or methyl carbinol), then, consists of a heavy precipitate of yellow iodoform.

The 2,4-dinitrophenylhydrazine (2,4-DNPH) test is very general for both aldehydes and ketones, giving a yellow, orange, or red precipitate. The reason we do this test is to make a derivative of your unknown that is crystalline and easy to purify. Comparing the melting point of this derivative with literature values will confirm (or disprove) the suspected structure. The reagent condenses with the carbonyl group to form an imine with loss of water (the oxygen from the carbonyl and the two hydrogens from the attacking nitrogen).

Overall reaction:

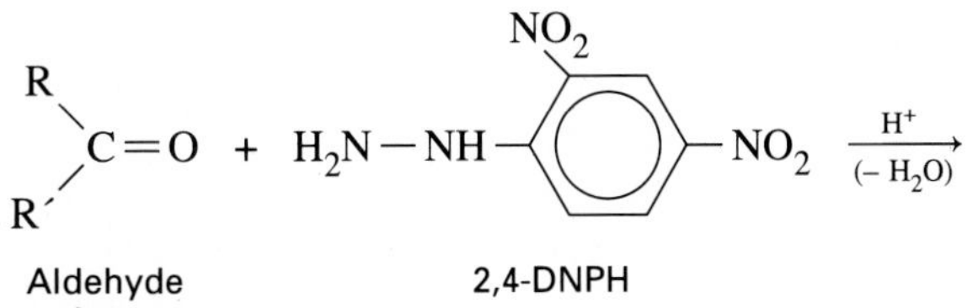

Aldehyde
or ketone

2,4-DNPH

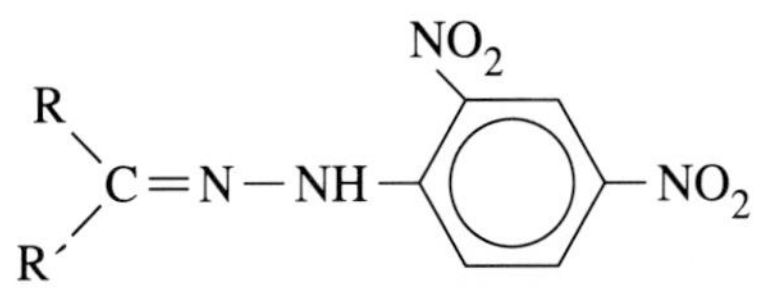

2,4-DNPH derivative

The mechanism involves protonation of the carbonyl oxygen and attack by the highly nucleophilic hydrazine nitrogen, followed by elimination of water.

You will also be given IR and NMR spectra for your unknown. A thorough discussion of these is left to the textbook and lecture, and you may wish to refer back to the infrared lab in Section 13.

An excellent way to confirm the identity of an unknown is to match up the fingerprint region of its IR spectrum with a literature spectrum, usually from the *Aldrich Library of IR Spectra*. Matching the NMR using the *Aldrich Library of NMR Spectra* is also helpful, although sometimes not definitive. Although IR fingerprint region bands are unique to a compound, occasionally several compounds will have similar NMR spectra.

QUESTIONS

1. Where would you expect an aldehyde proton ($-$CHO) to appear in the NMR spectrum?

2. Where would a proton next to a carbonyl ($-\overset{\overset{\text{H}}{|}}{\underset{|}{\text{C}}}-\overset{\overset{\text{O}}{||}}{\text{C}}-$) appear in the NMR?

3. Draw out the predicted NMR spectra for the following two compounds.

(a) $CH_3-\langle\bigcirc\rangle-CHO$ (b) $CH_3CH_2\overset{\overset{\text{O}}{||}}{C}-\overset{\overset{CH_3}{|}}{\underset{\underset{CH_3}{|}}{C}}-CH_3$

4. An unknown liquid gives a precipitate with 2,4-DNPH reagent, a negative Tollens test, and a positive iodoform test. In the infrared spectrum its only notable feature is a strong peak at 1710 cm^{-1}. The NMR spectrum reads as follows: δ 1.0 t, 3H; 2.2 s, 3H; 2.4 q, 2H. What is it?

5. A hydrazine nitrogen (R_2N-NR_2) is much more nucleophilic than an amine nitrogen. What is a possible explanation?

6. Another derivative of aldehydes and ketones is the oxime formed using hydroxylamine (NH_2OH) and HCl. Propose a mechanism and structure for formation of the oxime of cyclohexanone.

Concept Map

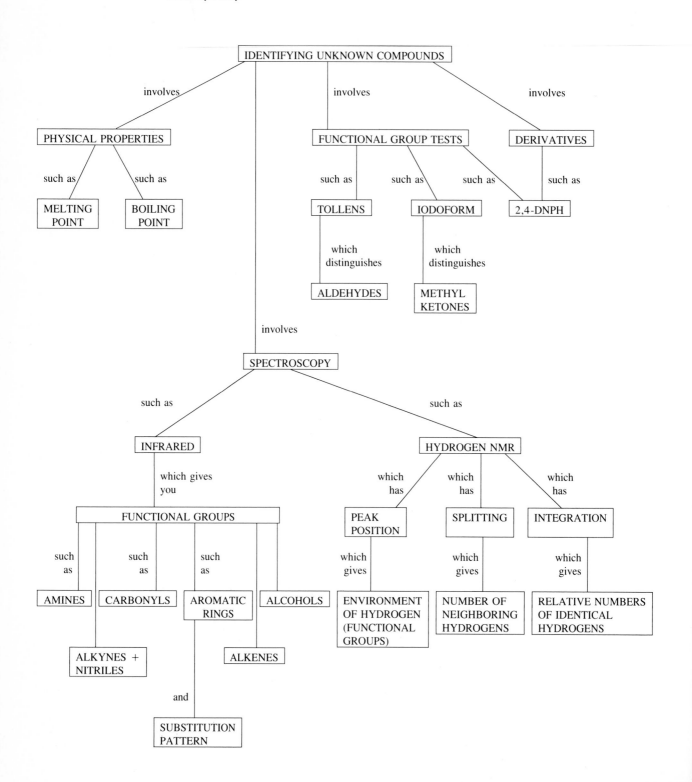

EXPERIMENT 31.1 IDENTIFICATION OF AN UNKNOWN ALDEHYDE OR KETONE

Estimated Time:
2.5 hours

Special Hazards

In this experiment you will be working with several dangerous chemicals, though in small quantities. Exercise all standard precautions including wearing goggles and avoiding skin contact or breathing fumes. Treat all unknowns as potentially toxic and flammable. Do not use old bottles of Tollens reagent; the reagent may develop explosive compounds on prolonged storage.

Procedure

Choose one of the bottles of unknowns in the lab. Record its number and physical appearance and obtain its IR and NMR spectra from the instructor. Examine the spectra to determine whether you have an aldehyde or a ketone, and to note any other obvious features.

If your unknown is a solid, record its melting point. If it is a liquid, record its boiling point using the following ultramicro boiling point procedure, with the apparatus shown in Figure 31-1. To determine the boiling point of the unknown, take a standard capillary melting-point tube, closed at one end. Using a clean 10-μL gas chromatographic syringe, place 3–4 μL of the unknown in the tube. Place the capillary closed-end down inside a test tube or centrifuge tube and centrifuge briefly to force the liquid to the bottom.

Make a small glass bell by heating 3-mm (o.d.) soft-glass tubing with a micro burner, removing it from the flame and quickly drawing it out to a capillary fine enough

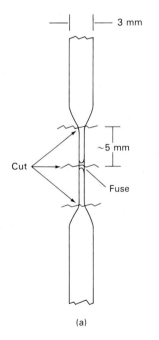

Figure 31–1 Making and Using a Glass Bell for Ultramicro BP Determination

(a)   (b)

to fit inside the melting-point tube. Cautiously break the fine capillary and fuse one end closed. Cut off the open end to a length of approximately 0.5 cm to form a glass bell. Insert this bell open-end down into the melting-point tube containing the liquid sample and tap it to the bottom. The system now resembles that shown in Figure 31-1.

Insert the melting-point tube into a melting-point apparatus that accepts upright tubes, such as a Thomas–Hoover apparatus. Heat relatively quickly and observe a stream of bubbles start to emerge from the open (lower) end of the glass bell as the air inside expands and escapes. As the bubbles become more frequent, slow the rate of heating to approximately 2° per minute.

As the oil temperature approaches and exceeds the boiling point of the unknown liquid, the stream of bubbles becomes constant. When there is almost one continuous, unbroken bubble emerging from the glass bell, shut off the heat, maintain stirring, and watch as the system cools down for the temperature at which bubbling ceases and the unknown liquid is sucked back into the capillary tube. This is your boiling point. Since all the air has been driven from the bell and it contains only molecules of the unknown, when these condense from vapor to liquid the volume decreases and the liquid unknown is sucked into the bell. Repeat the heating and cooling cycle until a reproducible value is obtained.

Whether you have a solid or liquid unknown, perform the following three tests (in any order) on known compounds (and on your unknown) for both positive and negative tests.

Tollens test.   To prepare the reagent, measure into a clean test tube 2 mL of 5% aqueous silver nitrate and 1 mL of 10% aqueous sodium hydroxide solution. A precipitate of silver oxide (Ag_2O) forms immediately. Add 2 M aqueous ammonia dropwise with shaking until the precipitate just dissolves. To have a sensitive reagent, an excess of ammonia must be avoided. When the precipitate has barely dissolved, dilute with water to 10 mL. This is the Tollens reagent containing $Ag(NH_3)_2^+OH^-$.

Rinse out three small clean test tubes with 10% NaOH and label them "POS" (positive), "NEG" (negative), and "UNK" (unknown). In the POS tube place 1 drop of 0.1 M glucose solution, in the NEG place 1 drop of methyl ethyl ketone (2-butanone), and in the third tube place 1 drop of unknown. To each tube add Tollens reagent (1 mL), let stand at room temperature for about 10 minutes, and observe any change. A silver mirror or black precipitate of Ag is a positive test. Dispose of any excess Tollens reagent by washing down the sink with lots of water, because on prolonged standing Tollens reagent can form an explosive precipitate. Sometimes Tollens tests give false positive or negative results. A false positive can occur for any easily oxidized compound. A false negative usually occurs because the unknown is insoluble in the reagent. If your NMR shows an aldehyde peak but you get a negative Tollens test, warm the tube in the steam bath for a few minutes to increase the solubility and rate of reaction.

Iodoform test.   Label three test tubes POS, NEG, and UNK. In the POS tube place 4 drops of 2-butanone, in the NEG tube place 4 drops of cyclopentanone, and in the third tube place 4 drops of the unknown. To each tube add 2 mL of water. (If the unknown is insoluble in water, use 2 mL of 1,2-dimethoxyethane instead of water, follow the procedure below and dilute with 10 mL of water at the end.)

To each tube add 2 mL of 10% aqueous sodium hydroxide solution and 3 mL of 0.5 M iodine–potassium iodide solution. In a positive test the brown color of the reagent disappears and a substantial precipitate of yellow iodoform appears. As a double check on a positive test, take out a little of the precipitate, dry it on a filter paper, and test its melting point. Iodoform melts at 119.°

2,4-Dinitrophenylhydrazine (2,4-DNPH) test. In a test tube, place 5 mL of the 2,4-DNPH reagent solution, which also contains sulfuric acid as the catalyst and ethanol as the solvent. Add 5 drops of the unknown aldehyde or ketone, or the equivalent amount of solid dissolved in a minimum of ethanol. Shake the tube vigorously. If no precipitate forms immediately, let the solution stand for a few minutes. Collect the product by suction filtration, wash with a little ethanol, and record the melting point. The yellow, orange, or red product should be fairly clean. To purify it further, take some of your crude product and recrystallize it in a test tube from 95% ethanol, using a steam bath for a heat source. In some cases it may be necessary to add water to the cloud point at boiling to cause precipitation on cooling. If the 2,4-DNPH derivative is insoluble in hot 95% ethanol, then ethyl acetate may work instead as a recrystallization solvent.

TABLE 31–1 POSSIBLE ALDEHYDE AND KETONE UNKNOWNS

Name(s)	mp (°C)	bp (°C)	Structure	mp of 2,4-DNP derivative
1. Acetophenone	19–20	202		238–240
2. Benzaldehyde	—	178–185		237
3. 4-Bromoacetophenone	51	—		230–237
4. 4-*t*-Butylcyclohexanone	47–50			128
5. *d*-Carvone	—	227–230		191 red
6. Cinnamaldehyde	—	248		255d red

Compound	mp	bp	Structure	Derivative
7. 3,3-Dimethyl-2-butanone (pinacolone)	—	106	CH_3C (O) $C(CH_3)(CH_3)CH_3$	125 orange-yellow
8. 1,3-Diphenylacetone	32–34	330	(phenyl)CH_2C (O) CH_2(phenyl)	100
9. Ethyl levulinate	—	205	CH_3C (O) CH_2CH_2C (O) OCH_2CH_3	102
10. d-Fenchone	5	192–194	(bicyclic ketone structure)	140
11. Furfural (2-furaldehyde, 2-furancarboxaldehyde)	—	162	(furan)CHO	212–214 yellow or 230 red or 185 mixed
12. n-Heptanal (heptaldehyde)	−43	153	$CH_3CH_2CH_2CH_2CH_2CH_2CHO$	108 yellow
13. o-Hydroxybenzaldehyde (salicylaldehyde)	1–2	197	(benzene with OH and CHO)	248–252d red
14. p-Hydroxybenzaldehyde	117–119	—	(benzene with HO and CHO)	260 orange (monohydrate)

TABLE 31–1 POSSIBLE ALDEHYDE AND KETONE UNKNOWNS (*cont.*)

			Structure
15. 4-Hydroxy-4-methyl-2-pentanone (diacetone alcohol)	—	166	202–203
16. *l*-Menthone	—	207	146 orange
17. 4-Methoxybenzaldehyde	—	248	250–254
18. 4-Methoxybenzophenone	60–63	354–356	180 dark orange
19. 4-Methylbenzaldehyde	—	204–205	232–234
20. 4-Methylbenzophenone	55–60	—	200–202

#	Name	mp (°C)	bp (°C)	Structure	Derivative
21.	3-Methyl-2-butanone	—	93	CH_3—CH—C—CH_3 with CH_3 and O	117–120
22.	4-Methyl-2-pentanone (methyl isobutyl ketone)	−80	117–118	CH_3CCH_2CH with O and CH_3, CH_3	95 orange red
23.	2-Octanone	−16	173	$CH_3C(CH_2)_5CH_3$ with O	58 orange
24.	3-Pentanone (diethyl ketone)	−40	102	$CH_3CH_2CCH_2CH_3$ with O	156 orange
25.	Phenylacetaldehyde	−10	195	(phenyl)CH_2CH with O	121
26.	Propiophenone	18	218	(phenyl)CCH_2CH_3 with O	189–191 red
27.	Vanillin	81–83	—	OCH_3, OH, HC with O on ring	268–271

QUESTIONS

1. Photocopy, fill out, and turn in an Unknown Report Sheet on your compound. (The possible unknowns are listed in Table 31-1.)

2. Did you obtain any anomalous, confusing, or incorrect results in the functional group tests? If so, check the description of the tests in the classic books by Pasto and Johnson and by Shriner, Fuson, and Curtin listed in the References. Then describe several compounds that do not behave as expected in each test. Speculate on why your compound might have reacted unusually if it did.

3. Give the complete mechanism for formation of the 2,4-dinitrophenylhydrazone of your unknown.

4. Did the melting point of your derivative match the literature value in the table, which was obtained from the *CRC Handbook of Tables for Organic Compound Identification*? Give your value and the literature value. What could be the reason for any difference?

REFERENCES

General

CHEMICAL RUBBER COMPANY, *CRC Handbook of Tables for Organic Compound Identification*, CRC Press, Boca Raton, Fla., revised annually.

PASTO, D. J. and C. R. JOHNSON, *Organic Structure Determination*, Prentice-Hall, Englewood Cliffs, N.J., 1969.

SHRINER, R. L., R. C. FUSON, and D. Y. CURTIN, *Systematic Identification of Organic Compounds*, Wiley, New York, 1964.

Spectroscopy

DYER, J., *Applications of Absorption Spectroscopy*, Prentice-Hall, Englewood Cliffs, N.J, 1965.

POUCHERT, C. J., *Aldrich Library of FI-IR Spectra*, Aldrich Chemical Co., Milwaukee, Wis., 1983.

POUCHERT, C. J., *Aldrich Library of NMR Spectra*, 2nd ed., Aldrich Chemical Co., Milwaukee, Wis., 1985.

SADTLER LABS, *Nuclear Magnetic Resonance Spectra*, Sadtler Research Laboratories, Inc., Philadelphia, Pa, 1966–present.

SADTLER LABS, *Standard Grating Spectra* (IR), Sadtler Research Laboratories, Philadelphia, Pa., 1966-1976.

SILVERSTEIN, R. M., G. C. BASSLER, and T. C. MORRILL, *Spectrometric Identification of Organic Compounds*, 4th ed., Wiley, New York, 1981.

If all else fails

Chemical Abstracts, American Chemical Society, Columbus, Ohio, 1907–present.

ALDEHYDE OR KETONE
UNKNOWN REPORT SHEET

Name _____

Unknown No. _____   Date _____

_____

Name(s) of Unknown | Structure of Unknown

Appearance: _____   Melting Point: _____

Odor: _____   Boiling Point: _____°C at _____torr

_____

_____ Major Peaks in IR Spectrum (cm^{-1}) _____ Interpretation _____

1. _____

2. _____

3. _____

4. _____

5. _____

_____ Peaks in NMR Spectrum (δ) _____ Interpretation _____

1. _____

2. _____

3. _____

4. _____

5. _____

6. _____

	Test Performed	Results	Interpretation
1.			
2.			
3.			
4.			
5.			

	Possibilities Considered	mp or bp (circle one)	Lit. mp of 2,4-DNPH deriv.
1.			
2.			
3.			
4.			
5.			

mp of 2,4-DNPH of unknown _____

Comments and reasoning:

SECTION 32

Qualitative Organic Analysis

Overview

In Experiment 31.1 you identified an unknown aldehyde or ketone by measuring the melting or boiling point, performing several functional group tests, analyzing the IR and NMR spectra, and making a derivative. In this experiment you will have a broader range of unknowns which can include any functional group. To identify these unknowns you will take the melting point or boiling point, test the solubility to narrow down the functional group class, and perform a flame test to determine if the compound is aliphatic or aromatic. Then you will carry out a sodium fusion test to analyze for the presence of nitrogen, sulfur, and halogens. After these tests and after examining your IR and NMR spectra, you will determine which specific functional group tests are appropriate for your compound and carry them out.

Normally, the final stage of unknown identification is making a derivative. Because of time constraints in this 3-hour experiment, you will not be making a derivative. However, your unknown will be one of the compounds listed in the table at the end of the overview. From the melting point or refractive index plus spectra plus the other tests you perform, the identification should be definite.

The procedure for taking a refractive index has been described in Section 10. Instructions for performing the 2,4-DNP, iodoform, and Tollens tests for ketones and aldehydes have been described in Section 31, as was basic NMR interpretation. The other tests are discussed in turn on the following pages.

Solubility. The solubility of an unknown compound tells a lot about the functional group(s) present. You will be testing solubility in water, 5% aqueous HCl, 5% aqueous NaOH (strong base), 5% aqueous $NaHCO_3$ (weak base), and concentrated H_2SO_4. The principles of solubility are the same as those discussed in Section 3 in connection with extraction, and will be reviewed here.

For a compound to be soluble in water, it must be either ionic or relatively small (five or six carbons or less) or polyfunctional (containing several polar functional groups). If a compound is soluble in water, it is tested with litmus or universal indicator paper to determine whether it is acidic (a carboxylic acid or phenol), basic (an amine), or neutral (possibly containing other functional groups). A compound that is not soluble in water to begin with (because it is too large and nonpolar) will become water soluble

311

if it develops a charge. For example, a basic amine will become water soluble when protonated, and a carboxylic acid will become water soluble when deprotonated. Thus amines are soluble in 5% aqueous HCl, and carboxylic acids are soluble in 5% aqueous NaOH (strong base) and 5% aqueous $NaHCO_3$ (weak base). Phenols, which are weak acids, need strong base to deprotonate them, so they are soluble in 5% aqueous NaOH but not in 5% aqueous $NaHCO_3$. Here are the relevant reactions:

$$R_3N + H^+ \longrightarrow R_3\overset{+}{N}H$$

An amine dissolving in acid

$$RCOOH + HCO_3^- \longrightarrow RCOO^- + H_2CO_3$$

A carboxylic acid dissolving in bicarbonate solution

$$RCOOH + OH^- \longrightarrow RCOO^- + H_2O$$

A carboxylic acid dissolving in hydroxide solution

A phenol dissolving in hydroxide solution

If a compound is not soluble in water, 5% aqueous HCl, 5% aqueous NaOH, or 5% aqueous $NaHCO_3$, it must be large and neutral. Testing with concentrated H_2SO_4 will determine if it has any functional groups other than halides. Only hydrocarbons and halides do not react with (and dissolve in) concentrated sulfuric acid. Most other compounds, such as alcohols, ketones, amides, and alkenes, are protonated and dissolve in H_2SO_4. The scheme for solubility testing is outlined in the flowchart.

Note that if you have reached the end of the scheme at some point it is not necessary to carry out the other solubility tests. For example, if your compound is water soluble, test its pH and you are done. Only if your compound is insoluble in water, 5% HCl, 5% NaOH, and 5% $NaHCO_3$ must you continue on to testing with sulfuric acid.

Elemental analysis.   There is a relatively simple way to determine qualitatively whether nitrogen, sulfur, or halogens are present in your compound. This method involves decomposing the unknown organic substance in the presence of metallic sodium. This destroys the organic molecule and results in any nitrogen, sulfur or halogens present being converted to NaCN, Na_2S, or NaX, respectively. These inorganic compounds can then be tested for in the following way.

To test for nitrogen, the solution is treated with Fe^{2+} and Fe^{3+}. If there is any cyanide present, it will form a dark blue precipitate of Prussian Blue.

$$18CN^- + 3Fe^{2+} + 4Fe^{3+} \longrightarrow Fe_4[Fe(CN)_6]_3$$

Prussian Blue

To test whether there is sulfur present, lead ion is added. If sulfur is present, an immediate heavy precipitate of black or brown lead sulfide is produced.

$$Pb^{2+}(aq) + Na_2S(aq) \longrightarrow PbS(s) + 2Na^+(aq)$$

For halogens we will use the standard silver nitrate test:

$$Ag^+(aq) + NaX(aq) \longrightarrow AgX(s) + Na^+(aq) \qquad X = Cl, Br, I$$

Flow diagram for solubility tests

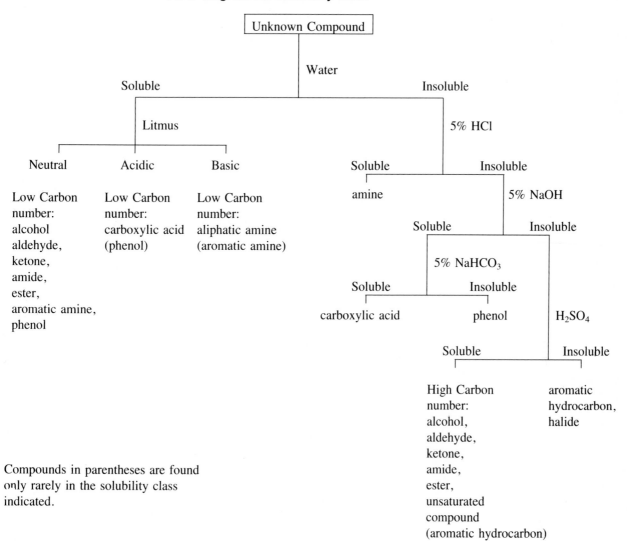

Compounds in parentheses are found only rarely in the solubility class indicated.

By examining the color of the precipitate and its solubility in aqueous ammonia we can determine whether the halogen is chloride, bromide, or iodide.

Lucas test (for alcohols).   The Lucas test (for alcohols) uses a solution of zinc chloride ($ZnCl_2$) in concentrated hydrochloric acid to differentiate between the lower primary, secondary, and tertiary alcohols. The mechanism is a typical S_N1, in which the zinc assists in the heterolytic cleavage of the C—O bond to form a carbocation, as illustrated below.

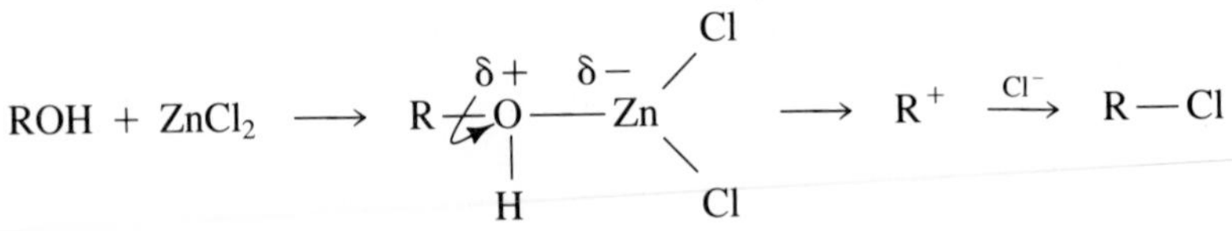

$$ROH + ZnCl_2 \longrightarrow R \overset{\delta+}{-\!\!\!\!\!\diagup}\, \overset{\delta-}{\underset{H}{O}} - Zn \overset{Cl}{\underset{Cl}{\diagdown}} \longrightarrow R^+ \xrightarrow{Cl^-} R-Cl$$

The carbocation finally picks up a chloride ion to form an alkyl chloride.

The test consists of recording the time taken by the reaction before the appearance of the insoluble alkyl chloride. Tertiary, allylic, and benzylic alcohols react within a few minutes at room temperature because they form carbocations easily. Secondary alcohols take longer to react at room temperature, whereas primary alcohols do not react.

Chromic acid test (for alcohols). The chromic acid test relies on the fact that primary and secondary alcohols are easily oxidized by Cr(VI), while tertiary alcohols are not. For primary and secondary alcohols the reaction is

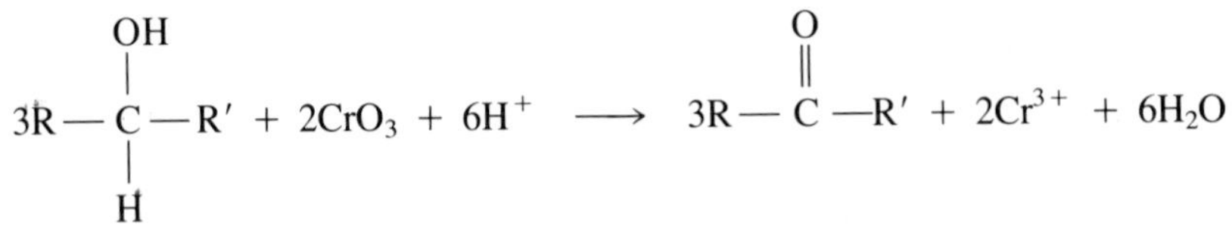

$$3R - \overset{OH}{\underset{H}{\underset{|}{\overset{|}{C}}}} - R' + 2CrO_3 + 6H^+ \longrightarrow 3R - \overset{O}{\overset{\parallel}{C}} - R' + 2Cr^{3+} + 6H_2O$$

A positive test is noted by watching for the appearance of the blue-green Cr^{3+} ion.

This reaction is also useful synthetically as the Jones oxidation, which converts a secondary alcohol to a ketone. Under these conditions primary alcohols are oxidized first to aldehydes and then to carboxylic acids.

Bromine in carbon tetrachloride test (for alkenes or alkynes). This test, which you may have already performed in Experiment 6.1 to test the cyclohexene you made, involves adding Br_2 across a double bond. Decolorization of the red-brown Br_2 means a positive test for an alkene or alkyne.

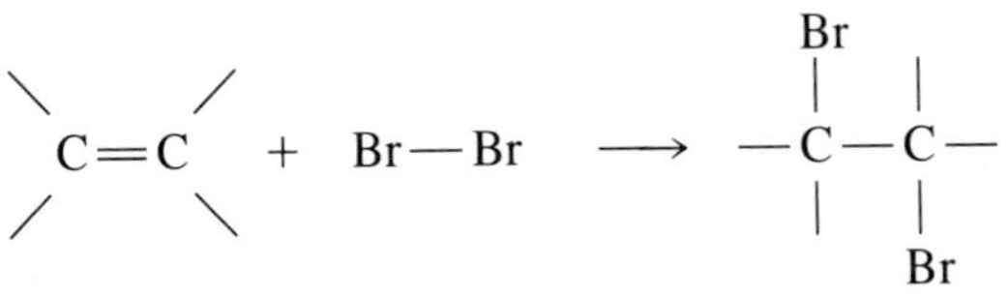

$$\overset{\diagdown}{\underset{\diagup}{C}} = \overset{\diagup}{\underset{\diagdown}{C}} \quad + \quad Br - Br \quad \longrightarrow \quad \overset{\overset{Br}{|}}{-\overset{|}{C}} - \overset{|}{\underset{\overset{|}{Br}}{C}} -$$

Alkene Bromine Vicinal
(colorless) (red-brown) dibromide
         (colorless)

Permanganate test (for alkenes and alkynes). Another way to detect unsaturation is by using the permanganate ion. As the Mn(VII) oxidizes the alkene or alkyne, the purple color of the permanganate disappears and a brown precipitate of MnO_2 appears.

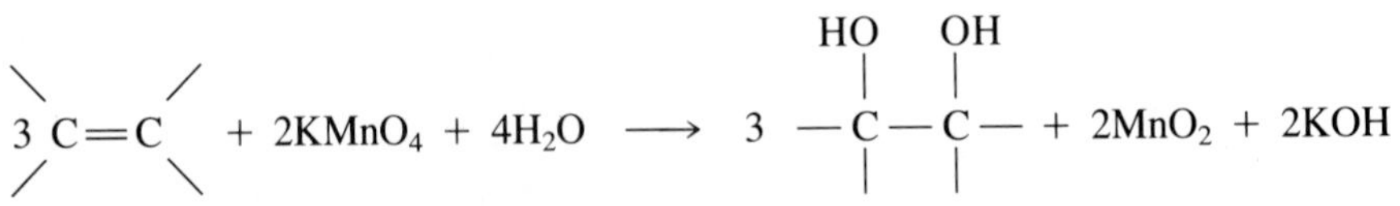

$$3 \overset{\diagdown}{\underset{\diagup}{C}} = \overset{\diagup}{\underset{\diagdown}{C}} + 2KMnO_4 + 4H_2O \longrightarrow 3 \; \overset{\overset{HO}{|}}{-\overset{|}{C}} - \overset{\overset{OH}{|}}{\underset{|}{C}} - + 2MnO_2 + 2KOH$$

Hinsberg test (for amines). Primary and secondary amines react with *p*-toluenesulfonyl chloride to give *p*-toluenesulfonamides, as shown below.

$$CH_3 - \langle\!\bigcirc\!\rangle - \overset{\overset{\displaystyle O}{\|}}{\underset{\underset{\displaystyle O}{\|}}{S}} - Cl \ + \ NH_2 - R \ \longrightarrow \ CH_3 - \langle\!\bigcirc\!\rangle - \overset{\overset{\displaystyle O}{\|}}{\underset{\underset{\displaystyle O}{\|}}{S}} - NHR \ + \ HCl$$

| p-Toluenesulfonyl chloride | A primary amine | A primary p-toluenesulfonamide |

$$CH_3 - \langle\!\bigcirc\!\rangle - \overset{\overset{\displaystyle O}{\|}}{\underset{\underset{\displaystyle O}{\|}}{S}} - Cl + \ HN\!\!\overset{\displaystyle R}{\underset{\displaystyle R'}{}} \ \longrightarrow \ CH_3 - \langle\!\bigcirc\!\rangle - \overset{\overset{\displaystyle O}{\|}}{\underset{\underset{\displaystyle O}{\|}}{S}} - N\!\!\overset{\displaystyle R}{\underset{\displaystyle R'}{}} \ + \ HCl$$

A secondary amine   A secondary p-toluenesulfonamide

Tertiary amines do not react with p-toluenesulfonyl chloride.

The test is conducted in the presence of NaOH to remove the HCl produced. The NaOH also distinguishes between primary and secondary p-toluenesulfonamides because the primary ones have an acidic hydrogen on the nitrogen which is removed by the base, causing the product to dissolve.

$$CH_3 - \langle\!\bigcirc\!\rangle - \overset{\overset{\displaystyle O}{\|}}{\underset{\underset{\displaystyle O}{\|}}{S}} - NHR + NaOH \ \longrightarrow \ CH_3 - \langle\!\bigcirc\!\rangle - \overset{\overset{\displaystyle O}{\|}}{\underset{\underset{\displaystyle O}{\|}}{S}} - \overset{-}{N} - R \quad \overset{Na^+}{+} \quad H_2O$$

Water soluble

Secondary p-toluenesulfonamides do not dissolve. So by observing first whether there is any reaction, and then whether the product dissolves, you can distinguish between primary, secondary, and tertiary amines by this test.

Neutralization equivalent test (for carboxylic acids). When a carboxylic acid is titrated accurately with a known strength base, the neutralization equivalent (or molecular weight per COOH group) can be calculated, according to the formula

$$\text{N.E. (of acid)} = \frac{\text{mass of sample (mg)}}{\text{mL of NaOH} \times \text{molarity of NaOH}}$$

This is the only quantitative test described in this lab. It should yield a value accurate within 5%.

Silver nitrate in ethanol and sodium iodide in acetone tests (for halides). As you know, alkyl halides can react by either S_N1 or S_N2 mechanisms. Silver ions remove halides, forming carbocations that can go on to S_N1 reactions.

$$R - X + Ag^+ \ \longrightarrow \ \overset{\delta+}{R}\cdots X \cdots \overset{\delta+}{Ag} \ \longrightarrow \ R^+ + AgX$$

$$R^+ \ \longrightarrow \ \text{alcohols, ethers, alkenes, etc.}$$

The easier it is to form a carbocation, the faster the reaction will be and the faster the silver halide will precipitate. Therefore, tertiary, allylic, and benzylic halides react faster than secondary, which in turn are faster than primary. Aryl and vinyl halides are of course unreactive. Halides also react in order of leaving-group ability; $I^- > Br^- > Cl^-$. If a precipitate is obtained, its identity can be determined from its color and solubility in nitric acid. Silver chloride is white, while silver bromide is off-white or pale yellow, and silver iodide is yellow. These salts will not dissolve when the mixture is acidified.

The sodium iodide in acetone test complements the silver nitrate test because it proceeds by an S_N2 mechanism.

$$Na^+ I^- + R{-}X \longrightarrow I{-}R + Na^+ X^-$$

Therefore, to react, the halide must be primary or secondary. Halides on very hindered positions such as neopentyl type carbons will not react. Nor will tertiary halides. Since sodium iodide is soluble in acetone, while sodium chloride and bromide are not, a positive test consists of the appearance of a precipitate of NaCl or NaBr. This reaction can also be useful in synthesis for replacing a chloride or bromide with iodide; it is called the Finkelstein reaction.

Bromine water test (for phenols).   Phenols react rapidly with bromine in water to produce brominated phenols, as shown for phenol itself below.

This reaction occurs easily because the phenol is electron-rich and the solvent (water) stabilizes the intermediate charged species in the electrophilic aromatic substitution.

QUESTIONS

1. What does the solubility test tell you about an unknown?
2. What does the sodium fusion test tell you about an unknown?
3. If your sodium fusion test shows the presence of nitrogen in your unknown, which additional test(s) should be performed on it?
4. An unknown solid is soluble in 5% aqueous NaOH but not in 5% aqueous $NaHCO_3$. The sodium fusion test reveals the presence of chloride, which is confirmed by the Beilstein test. However, both the silver nitrate in ethanol and sodium iodide in acetone tests fail to give a precipitate. The compound gives a precipitate in bromine water. The IR spectrum has peaks at 3200–3300 (broad), 1600–1460 (three peaks), 830, and 700 cm^{-1}. The NMR shows a broad singlet at δ 5.5 integrating for one hydrogen and a multiplet in the region δ 7.2–7.7 integrating for four hydrogens. Show your interpretation for each piece of data, and the structure and name of the unknown.
5. If a 256-mg sample of an unknown carboxylic acid requires 14.7 mL of a 0.102 M NaOH solution to reach a phenolphthalein endpoint, what is the equivalent weight of the acid?

EXPERIMENT 32.1 IDENTIFICATION OF AN UNKNOWN ORGANIC COMPOUND

Estimated Time:
3.0 hours

Special Hazards

In this experiment you will be working with quite a few dangerous chemicals, although in small quantities. Be sure to use all standard precautions, including wearing goggles and avoiding contact with chemicals as much as possible. Concentrated H_2SO_4 is extremely corrosive; avoid contact and rinse immediately with water if contact occurs.

Sodium metal reacts violently with water, forming caustic sodium hydroxide and flammable hydrogen gas which may be ignited by the heat of the reaction. Keep sodium under paraffin oil until ready for use and do not allow it to touch water. Bromine is highly corrosive and toxic; use only with adequate precautions in the fume hood. Be aware of the location of the bottle of 5% aqueous sodium thiosulfate for rinsing skin in case of a bromine spill.

Note on Planning and Time Management

For this experiment you will need to complete the following procedures; estimated times for each are shown.

Melting point or refractive index	10 min
Solubility test	30 min
Sodium fusion test	45 min
Flame test	5 min
Functional group tests (1 to 3)	15 min each

You can see that the sodium fusion and solubility tests are the most time consuming and therefore the most important to do correctly the first time. There will be plenty of time for all tests if you plan ahead and do not dawdle.

Procedure

Choose one of the bottles of unknowns in the lab and record its number. If your aldehyde or ketone unknown in the previous experiment was a solid, choose a liquid for this unknown so that you will have the opportunity to take a micro boiling point as described in Experiment 31.1. Perform the following tests in any order: melting point or refractive index, solubility, flame, and sodium fusion.

Solubility test.   First make a micro stirring rod from a capillary tube in the following way. Take a capillary tube held in both hands and rotate it quickly in the flame of a Bunsen burner until it softens. Now *before pulling, lift it out of the flame*, then pull quickly and straight apart, leaving a thin straight filament of glass. Cut this in the middle using your thumbnail or a file and *quickly* melt the end closed in the flame without allowing the filament to bend. This process is illustrated in Figure 32-1. It takes some practice to do this well, and you may want to have your instructor demonstrate this technique. Be sure that the end of your finished micro stirring rod can reach *inside* a capillary tube, because that is where you will be stirring.

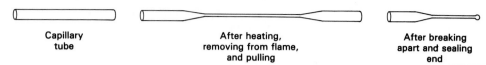

Capillary
tube

After heating,
removing from flame,
and pulling

After breaking
apart and sealing
end

Figure 32–2   Making an Ultra-Micro Stirring Rod

Now to test the solubility of your organic compound take a capillary tube open on both ends (if one end is closed, cut off that end with a file). Put a small amount of your compound in it (about 2–3 mm in the tube). If it is a liquid, capillary action will pull it in. If it is a solid, press the capillary on top of the unknown, being careful not to pack it too tightly. Now using a dropper or Pasteur pipette, touch a drop of water to the end of the capillary tube containing the sample. The water will be drawn in by capillary action. Be sure that there is no air between the sample and the water, and that there is much more water than sample.

Since the definition of solubility is arbitrarily set at 3%, there should be about 33 times as much solvent as solute present, although this may not be possible. Perhaps 5–10 times as much solvent as solute is enough in most cases. Now reach in and stir back and forth with your micro stirrer. Observe whether the organic compound dissolves in the water. If in doubt you can test in another way (that uses more sample) by placing a little in a test tube and adding some water on top. We are trying to get the maximum information from the smallest amount of sample.

If the compound is soluble in water, test its pH. If it is not soluble, proceed to test the next solvent (5% HCl) in a similar way. You may need to make up a new capillary with unknown or you may be able to use the old one by touching it to a paper towel to draw out the previous solvent, leaving the unknown. If your unknown is soluble in 5% aqueous HCl, you are done. If not, proceed to 5% aqueous NaOH. If your unknown is soluble in 5% NaOH, test a new sample with 5% NaHCO$_3$. If your sample was insoluble in 5% aqueous NaHCO$_3$, test it with concentrated H$_2$SO$_4$ (*caution!*).

Record your results and use the flow diagram for the solubility tests to determine your compound class.

Flame test.   Take a little of your compound on a spatula and hold it in the flame of a Bunsen burner (well away from any other flammable materials). Record whether your compound burns or not, and whether it has a sooty (smoky) flame. A sooty flame is often the mark of an aromatic compound, whereas an aliphatic compound generally burns with a clean flame. Ionic compounds such as salts will not burn.

Sodium fusion test (elemental analysis).   Using forceps, remove a small cube of sodium metal (3–4 mm on an edge or about the size of a small pea, *no larger*) from its storage under oil and dry it on a paper towel. Avoid touching the sodium metal with your fingers; it will react with any moisture, forming caustic NaOH. Place the dry cube of sodium in a small test tube held with a test-tube clamp. Being sure to have your goggles on and, while pointing the tube away from yourself and others, heat it in a flame until the sodium vapors rise about 2–3 cm in the tube. Remove the tube from the flame and quickly add a little of the unknown organic compound (2–3 drops of a liquid or or a small tip of a spatula full of solid). Try to drop the sample in vertically so that it hits the sodium; otherwise, it may evaporate from the hot glass surface before reacting. A brief flame may occur as the sample is added. If the unknown is very volatile, it should be mixed with an equal amount of powdered sucrose before adding it to the sodium. Repeat the heating of the tube and addition of another portion of the unknown.

After the second portion of unknown is added, heat the tube to redness in the flame. Let it cool slightly, then add about 1 mL of ethanol to decompose any excess sodium. Evaporate the excess alcohol and heat the tube until red-hot, then plunge it quickly into a small beaker containing 15 mL of cold water. The tube should crack, releasing its contents into the water. Break up any lumps with a stirring rod and heat to boiling to complete the solution process. Gravity filter this solution to remove insoluble materials and store the clear solution in a vial to run the following three tests on. Special notes: If the tube did not crack on plunging into water, place some water inside the tube and boil it carefully (not pointing it towards anyone), then filter as above and dilute to 15 mL to make the test solution.

Test for Sulfur. Acidify a 1- to 2-mL sample of the solution with acetic acid. Add a few drops of 0.15 M lead acetate solution. An immediate black or brown precipitate of lead sulfide indicates the presence of sulfur.

Test for Nitrogen. Test the pH of a 1-mL sample of the fusion solution with indicator paper. The pH should be about 13. If it is definitely above 13, adjust it by adding a small drop of 3 M H_2SO_4; if below, add a small drop of 6 M NaOH and recheck the pH. When the pH is close to 13, add 2 drops each of a saturated solution of ferrous ammonium sulfate and of 5 M potassium fluoride. Boil gently for about 30 seconds, cool, and add 2 drops of 5% aqueous $FeCl_3$. Carefully add 3 M sulfuric acid to the mixture dropwise with shaking until the precipitate of iron hydroxide just dissolves. Avoid excessive acid. The appearance of a brilliant blue color (Prussian Blue) at this stage indicates the presence of nitrogen. A slightly blue or blue-green solution may be a weak positive due to incomplete reaction with sodium. Nitrogen originally in a highly oxidized state such as a nitro group may give weak results because it is difficult to reduce it all the way to a cyano group during the decomposition with sodium.

Test for Halogens. In a small test tube acidify 1 mL of the alkaline fusion solution with dilute nitric acid, and test with indicator paper. If nitrogen or sulfur has been found, take the tube to the fume hood and boil it gently for a couple of minutes to expel HCN or H_2S. (*Caution:* Both are highly toxic gases). If no nitrogen or sulfur was present, this step is not necessary.

Add a few drops of aqueous silver nitrate solution. A white or yellow precipitate is a silver halide, indicating a halogen in the original compound. If a precipitate is obtained, centrifuge the tube for about 30 seconds, being sure to balance the centrifuge with an equally filled tube before turning it on. Decant off the supernatant solution and wash the precipitate with water, then with 1 mL of concentrated ammonia solution. If the precipitate is white and dissolves easily in the ammonia solution, it is AgCl. If it is pale yellow and does not dissolve easily, it is AgBr. If it is yellow and insoluble, it is AgI. Since AgF is water soluble, fluoride is not detected by this test.

At this stage, after recording the melting point or boiling point, solubility, and elemental analysis for your unknown, take to the instructor the data you have obtained so far. If the data are correct, you will be told so and given your IR and NMR spectra. If any item is in error, you will be asked to repeat that test. Melting and boiling points will be considered correct if they are within 10° either way from the literature values, so you will need to consider this range in examining the table of possible unknowns.

After all your results so far are correct and you have obtained your spectra, take a few minutes to interpret them. Considering both spectral data and your test results so far, speculate as to the functional groups present in your unknown. Pick two or three reasonable functional group tests to perform on your compound and carry them out. The individual tests are described below. Remember that negative test results are valuable, too. Also, bear in mind that none of these tests is 100% reliable. There are quite a few compounds that may give anomalous results on any one test. Again be sure to look at the preponderance of the evidence for your overall interpretation. If you are in doubt about the accuracy of results of any tests, run positive and negative standards for comparison.

Lucas test (for alcohols). Place 2 mL of the Lucas reagent (consisting of $ZnCl_2$ in HCl) in a small test tube. Add 3–4 drops or the corresponding quantity of solid unknown. Shake vigorously and let the tube stand at room temperature. Since this test depends on the appearance of the alkyl chloride as distinct liquid phase, the test can only be used with alcohols that are soluble in the reagent. Primary alcohols lower than hexyl dissolve, while those higher than hexyl do not. For primary alcohols the aqueous phase remains clear. Secondary alcohols will react in 2–5 minutes, forming a cloudy solution of insoluble alkyl chloride. In the case of tertiary, allylic, or benzylic alcohols, there is almost immediate separation of two phases, due to formation of the insoluble alkyl chloride. If it is not clear whether the alcohol is secondary or tertiary, the test may be repeated using concentrated HCl instead of the Lucas reagent. With concentrated HCl, tertiary alcohols react right away to form the insoluble alkyl chloride, while secondary alcohols do not react.

Chromic acid test (for alcohols). In a small test tube, dissolve 1 drop of liquid unknown or the corresponding amount of solid in reagent-grade acetone (1 mL). Check the time, add 1 drop of the chromic acid reagent, and shake. Primary and secondary alcohols react within 10 seconds to give an opaque blue-green suspension. Tertiary alcohols do not react. Other compounds that are easily oxidized will also react, such as aldehydes, phenols, and enols.

Bromine in carbon tetrachloride (for alkenes and alkynes). In a small test tube in the fume hood dissolve 3 drops or 0.1 g of the unknown in 1 mL of CCl_4 (*caution*). Add 1 drop at a time with shaking a solution of 0.2 M bromine in carbon tetrachloride (*caution*) until the red-brown bromine color persists. Immediately after the addition, exhale over the mouth of the tube. Note whether the moisture causes a cloud of evolved HBr gas to become visible.

Interpretation: If more than 2 drops of the Br_2/CCl_4 solution are decolorized without evolution of HBr, the test is positive for an alkene or alkyne. Other compounds, such as aldehydes, ketones, amines, and phenols, react by substitution, forming HBr.

Permanganate (for alkenes and alkynes). In a small test tube dissolve 1 drop or the corresponding amount of solid unknown in acetone (2 mL). While shaking, add a 1% aqueous solution of potassium permanganate dropwise. The test is positive if more than 1 drop of the permanganate solution is reduced, as shown by the vanishing of the purple color and formation of a precipitate of brown MnO_2.

Hinsberg test (for amines). In a test tube mix about 0.1 mL (or 0.1 g) of the unknown, 5 mL of 10% aqueous NaOH, and 0.3 g of *p*-toluenesulfonyl chloride (use caution). Stopper the tube and shake occasionally for 3–5 minutes. At the end of this time remove the stopper and warm the solution on the steam bath for 1 minute. Check that the solution is still basic; if not, add more NaOH. If there is a solid or liquid residue in the solution, separate it by filtration or by use of a pipette.

Check the solubility of the residue in 5 mL of water and in dilute HCl. Acidify the original solution using 6 *M* HCl. If no precipitate forms immediately, scratch the inside of the tube and cool it in an ice bath.

Interpretation: Primary amines generally give no significant amount of solid or liquid residue after the initial reaction. When the solution is acidified, a *p*-toluenesulfonamide should precipitate. Secondary amines usually yield solid *p*-toluenesulfonamides that do not dissolve in water or dilute HCl.

Tertiary amines should not react. The residue for these is the original liquid or solid amine, which should dissolve in dilute HCl. If the test gives an oily residue, note its density relative to water (whether it floats or sinks). If it floats, it is probably the original amine; if it sinks, it is probably a *p*-toluenesulfonamide that failed to crystallize.

Neutralization equivalent (for carboxylic acids). Weigh about 200 mg of the unknown accurately (to the nearest mg). Dissolve it in 50–100 mL of water, ethanol, or a mixture of both, depending on its solubility. After adding 2–3 drops of phenophthalein solution as an indicator, titrate to a faint persistent pink color with a standard solution of 0.100 *M* aqueous NaOH. It may improve visibility to place a blank white sheet of paper under the solution. Record the volume of NaOH solution used. Calculate the neutralization equivalent of the acid. This test should be repeated two or three times to ensure reproducibility.

Silver nitrate in ethanol (for halides). Place 2 mL of 0.1 *M* in ethanolic silver nitrate in a test tube. Add 1 drop of a liquid or a corresponding amount of a solid dissolved in a minimum of ethanol. Record the time, shake the mixture, and let it stand. If there is no precipitate after 5 minutes, heat the tube to boiling and boil it for 30 seconds. If a precipitate appears, record its color and see if it dissolves when 2 drops of 1 *M* nitric acid are added with shaking.

Halides that react at room temperature:

Chlorides: 3°, allylic, and benzylic
Bromides: 1°, 2°, and 3° alkyl (except geminal di- and tribromides), allylic, and benzylic
Iodides: all aliphatic except vinyl

Halides that react on heating:

Chlorides: 1° and 2° alkyl
Bromides: alkyl geminal di- and tribromides, some activated aryl halides such as 2,4-dinitrohalobenzenes

Halides unreactive on heating:

Chlorides: alkyl geminal di- and trichlorides, most aryl and vinyl halides

Sodium iodide in acetone test (for halides). Place 1 mL of the sodium iodide–acetone reagent in a small test tube. Add the halogen-containing compound, 2 drops of liquid or 0.1 g of solid, dissolved in a minimum of acetone. Shake the mixture and let it stand for 3 minutes. Observe whether a red-brown color or precipitate forms. If there is no reaction after 3 minutes, place the tube in a beaker of water at 50° (made by mixing hot and cold water). After 6 minutes in the 50° bath, remove the tube and cool it to room temperature. Note whether any reaction occurred.

Bromine water (for phenols). Dissolve 3 drops or 0.1 g of the unknown in 10 mL of water. If it is insoluble in water, add barely enough ethanol to cause it to dissolve. Check the pH with indicator paper. Add saturated bromine water 1 drop at a time until the color of bromine persists. Note whether a precipitate forms.

Interpretation: If the initial pH was below 7 and the bromine is decolorized as a white precipitate forms, the test is positive for a phenol. Aromatic amines also react, but the initial pH would be above 7.

TABLE 32–1 SOLIDS (LISTED IN ORDER OF INCREASING MELTING POINT)

	MP	Name(s)	Structure	Comments
1.	24	Methyl anthranilate		
2.	25	*t*-Butyl alcohol (2-methyl-2-propanol)	$(CH_3)_3COH$	
3.	40–42	Phenol		
4.	58–60	Diphenyl disulfide (phenyl disulfide)		
5.	62–64	Chloracetic Acid	$ClCH_2COOH$	
6.	64–68	4-Bromophenol		
7.	67–69	Stearic Acid	$CH_3(CH_2)_{16}COOH$	
8.	69–72	Biphenyl		bp 255
9.	95–96	1-Naphthol		

TABLE 32–1 (cont)   SOLIDS (LISTED IN ORDER OF INCREASING MELTING POINT)

MP	Name(s)	Structure	Comments
10. 112	4-Nitrophenol		
11. 113–115	Acetanilide		
12. 122–123	Benzoic acid		
13. 122–123	2-Naphthol		
14. 122–124d	*trans*-Stilbene		
15. 125–127	2-Aminopyrimidine		
16. 130d	D-Maltose		
17. 134–136	Phenacetin (*p*-ethoxyacetanilide)		
18. 140–142	Acetylsalicylic acid		

TABLE 32–1 (cont) SOLIDS (LISTED IN ORDER OF INCREASING MELTING POINT)

MP	Name(s)	Structure	Comments
19. 147–149	Cholesterol	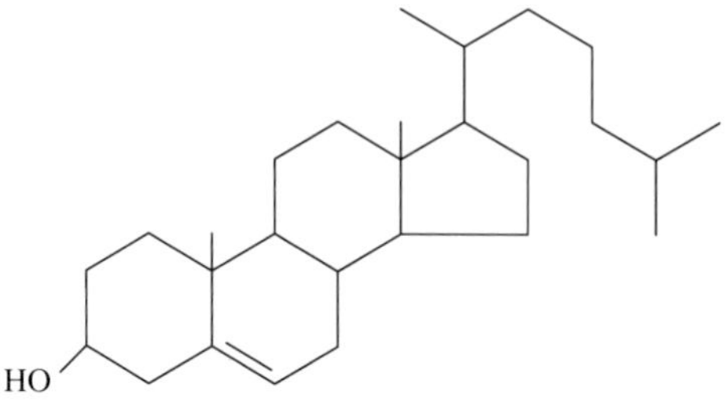	
20. 165–166	4-Bromoacetanilide	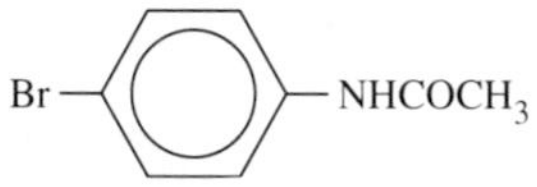	
21. 165–168	Thiosalicylic acid	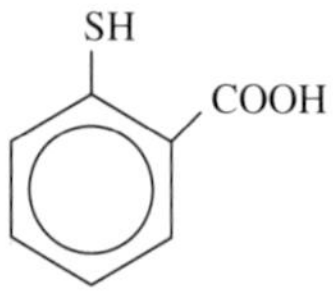	
22. 169–172	Acetaminophen	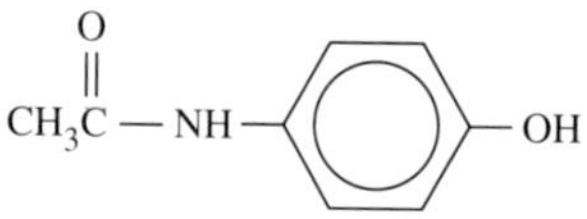	
23. 208	(—)—Borneol	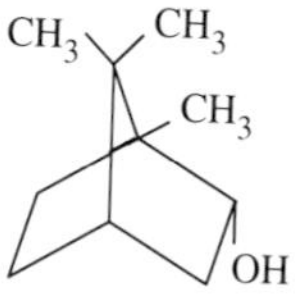	
24. 214	4-Hydroxybenzoic acid	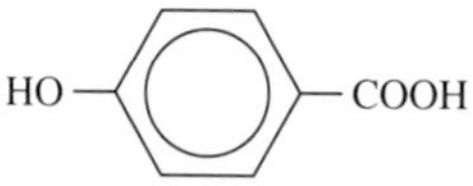	
25. 216	Anthracene	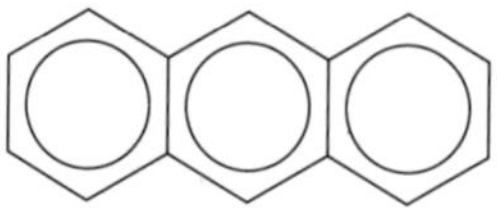	

TABLE 32–1 (cont)   LIQUIDS (LISTED IN ORDER OF INCREASING BOILING POINT at 760° Torr)

BP	Name(s)	Structure	n_D^{20}	Comments
1. 69–73	Iodoethane (ethyl iodide)	CH_3CH_2I	1.5130	
2. 83	*t*-Butyl alcohol (2-methyl-2-propanol)	$(CH_3)_3COH$	1.3860	mp 25
3. 83	Cyclohexene		1.4465	
4. 98	2-Butanol (*sec*-butyl alcohol)	$CH_3 - CH - CH_2CH_3$ with OH above CH	1.3970	
5. 100–101	1-Bromobutane (*n*-butyl bromide)	$CH_3(CH_2)_3Br$	1.4390	
6. 108	Isobutyl alcohol (2-methyl-l-propanol)	$HOCH_2CH(CH_3)_2$	1.3960	
7. 118	1-Butanol (*n*-butyl alcohol)	$CH_3(CH_2)_3OH$	1.3985	
8. 129	2-Chloroethanol	$ClCH_2CH_2OH$	1.4412	
9. 142	Isopentyl acetate (isoamyl acetate)	$(CH_3)_2CHCH_2CH_2OCOCH_3$	1.400	
10. 145	4-Methylpyridine (4-picoline)	CH_3- pyridine ring with N	1.5050	
11. 145–146	*cis*-Cyclooctene		1.4698	
12. 154	Ethyl lactate	$CH_3 - CH - COOEt$ with OH above CH	1.4130	
13. 159–160	Camphene			

TABLE 32–1 (cont)   LIQUIDS (LISTED IN ORDER OF INCREASING BOILING POINT at 760° Torr)

BP	Name(s)	Structure	n_D^{20}	Comments
14. 184–185	Benzylamine	$\bigcirc$—CH$_2$NH$_2$	1.5424	
15. 187	4-Fluoroaniline	F—$\bigcirc$—NH$_2$	1.5395	
16. 188	Benzonitrile	$\bigcirc$—CN	1.5280	
17. 196–197	Di-*n*-butyl sulfide (butyl sulfide)	(n-Bu)$_2$S	1.4952	
18. 199	Diethyl malonate	CH$_2$(CO$_2$Et)$_2$	1.4135	
19. 199–200	*o*-Toluidine (2-methylaniline)	CH$_3$ $\bigcirc$—NH$_2$	1.5709	
20. 205	Benzyl alcohol	$\bigcirc$—CH$_2$OH	1.5403	
21. 210	*n*-Butyl phenyl ether	$\bigcirc$—O(CH$_2$)$_3$CH$_3$	1.4970	
22. 237	Quinoline	(quinoline structure, N)	1.6256	
23. 250	*p*-Phenetidine (*p*-ethoxyaniline)	EtO—$\bigcirc$—NH$_2$	1.5609	
24. 256	Methyl anthranilate	NH$_2$ $\bigcirc$—COOCH$_3$	1.5824	mp 24
25. 268–269	Nonanoic acid	CH$_3$(CH$_2$)$_7$COOH	1.4309	mp 31

QUESTIONS

1. Photocopy and fill out the unknown report sheet, including your interpretations of spectral peaks and test results. Report the name and structure of your unknown. (The possible unknowns are listed in Table 32.1.)

2. Did you obtain any anomalous or misleading results? If so, check the books listed at the end of Section 31 by Pasto and Johnson or by Shriner, Fuson, and Curtin to find explanations for exceptions to various tests. Explain as thoroughly as you can what may have happened.

3. Try to match your spectra with literature spectra in the library, in the *Aldrich Libraries* of IR and NMR spectra. If you find a match, photocopy the literature spectrum and staple it to your own.

4. An unknown organic liquid gave a precipitate of Prussian Blue in the sodium fusion test, and gave a positive Hinsberg test in which the product was soluble in base. The hydrogen NMR reads as follows (δ values): 1.0 s, 9H; 1.6 broad s, 2H (D_2O exchangeable); 1.4 t, 2H; 2.6 t, 2H. Show your interpretation of each piece of data and give the structure of the unknown.

5. An unknown organic solid was found to be soluble in 5% aqueous $NaHCO_3$ and burned with a sooty flame. It gave negative 2,4-DNP, Beilstein, and permanganate tests. Its NMR read as follows (δ values): 1.2 d, 3H; 2.9 q. 1H; 7.3-7.8 m, 5H; 11.2 s, 1H. Interpret each piece of data and show the structure of the unknown.

6. Draw resonance structures for the anion of the *p*-toluenesulfonamide of a primary amine. Why are *p*-toluenesulfonamides of primary amines acidic, whereas those from secondary amines are not?

7. Draw a concept map for this experiment.

QUALITATIVE ORGANIC ANALYSIS
UNKNOWN REPORT SHEET

Name _____

Unknown No. _____   Date _____

Name(s) of Unknown | Structure of Unknown

Appearance: _____   Melting Point: _____

Odor: _____   Boiling Point: _____°C at _____torr

| Major Peaks in IR Spectrum (cm^{-1}) | Interpretation |

1. _____

2. _____

3. _____

4. _____

5. _____

| Peaks in NMR Spectrum (δ) | Interpretation |

1. _____

2. _____

3. _____

4. _____

5. _____

6. _____

Test Performed	Results	Interpretation
1.		
2.		
3.		
4.		
5.		

Possibilities Considered	mp or bp (circle one)
1.	
2.	
3.	
4.	
5.	

Comments and reasoning:

SECTION 33

Grignard Reactions and Multistep Syntheses

Overview

The Grignard reaction is a very reliable and widely used method for forming carbon-to-carbon bonds. Grignard reagents were discovered in the early twentieth century by a French chemist, Victor Grignard, who added magnesium metal to solutions of alkyl or aryl halides in ether. He noticed that the magnesium dissolved and considerable heat was generated. On further investigation he discovered that the magnesium inserts itself into the carbon–halogen bond. The reaction to form a Grignard reagent (also called an alkylmagnesium halide or arylmagnesium halide) is shown below. The net oxidation state of each atom or group is written above it to make clearer the redox reaction occurring.

$$\overset{+1 \quad -1}{R-X} \quad \xrightarrow[\text{Et}_2\text{O}]{\text{Mg}^\circ} \quad \overset{-1 \quad +2 \quad -1}{R-Mg-X}$$

During this process the oxidation state of the magnesium goes from zero to $+2$ (it loses two electrons). The oxidation state of the carbon, meanwhile, goes from $+1$ (assuming that it is attached to one halogen and three carbons) to -1, so the carbon is reduced by (gains) two electrons. Since the carbon-to-magnesium bond is a very polar (almost ionic) bond, we can think of a Grignard reagent as a carbanion, that is, carbon with a negative charge. This negatively charged Grignard reagent will be attracted to any positive center, for example, a carbonyl carbon or a carbon attached to a halide.

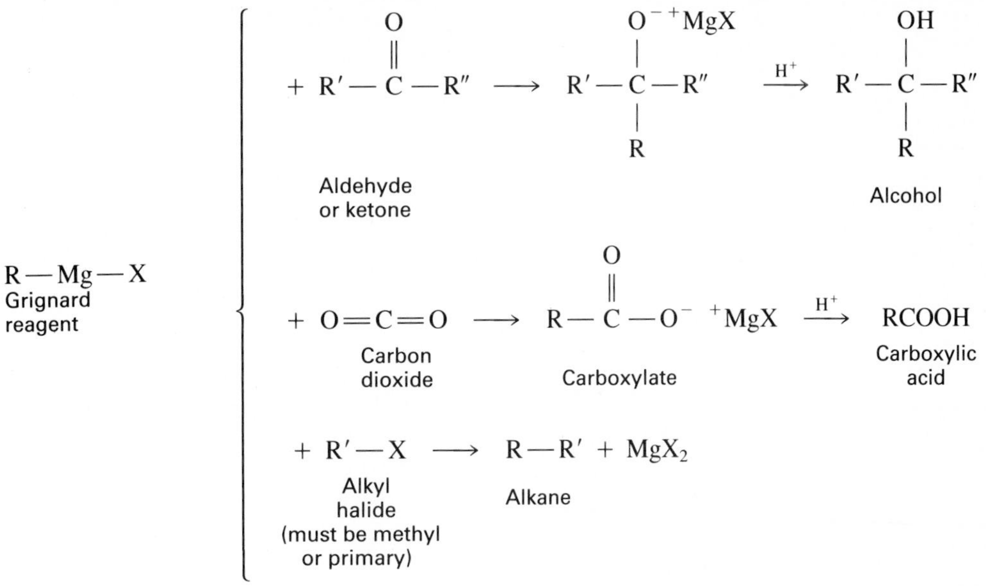

$$R - Mg - X$$
Grignard
reagent

Aldehyde
or ketone

Alcohol

Carbon
dioxide

Carboxylate

Carboxylic
acid

$$+ R' - X \longrightarrow R - R' + MgX_2$$

Alkyl
halide
(must be methyl
or primary)

Alkane

Therefore, the Grignard reaction is a good way to make alcohols, carboxylic acids, and some alkanes. When a Grignard reagent is mixed with an ester, attack occurs twice because the product of the first attack is a ketone. For Experiment 33.1 the mechanism is

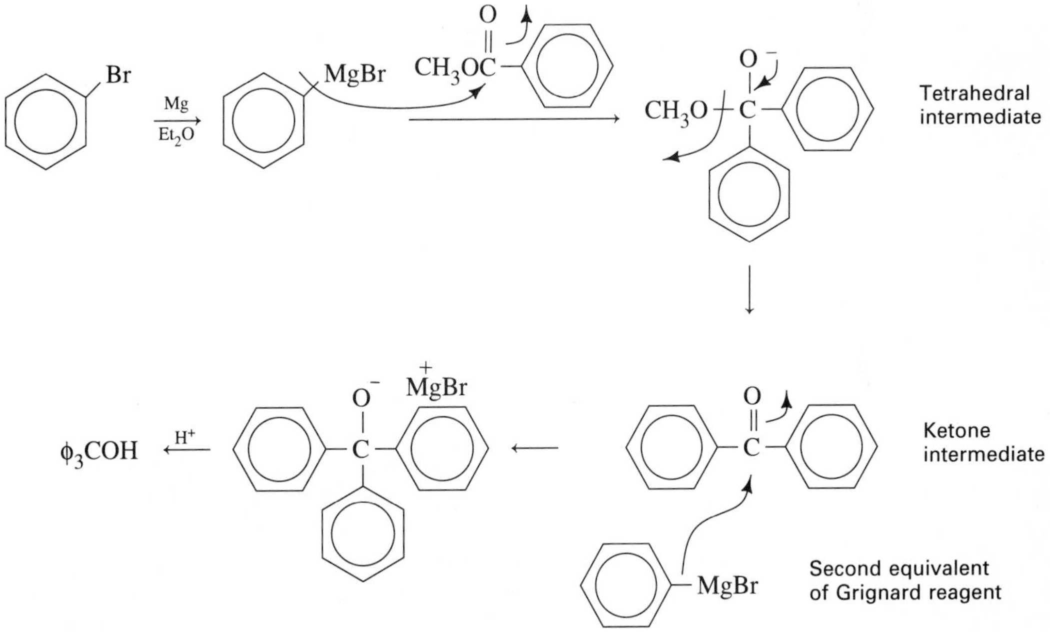

Tetrahedral
intermediate

Ketone
intermediate

Second equivalent
of Grignard reagent

ϕ_3COH

The first part of Experiment 33.1 is making the ester to be attacked. Making an ester from a carboxylic acid and an alcohol with a strong acid catalyst is a very old and reliable reaction known as Fischer esterification, named for Emil Fischer, a very productive German chemist in the late nineteenth and early twentieth century. The mechanism of acid-catalyzed esterification is shown here.

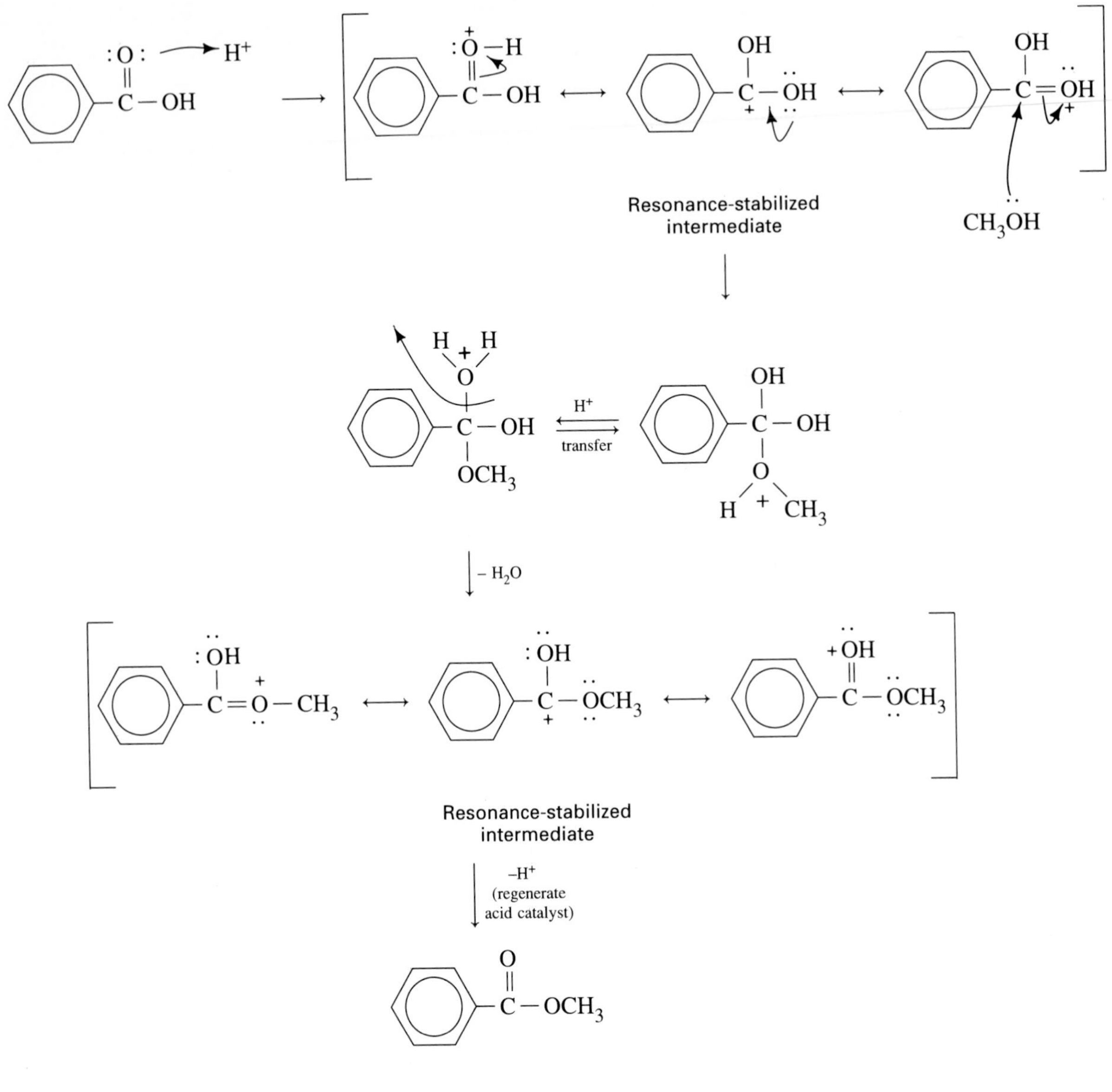

Resonance-stabilized
intermediate
transfer
Resonance-stabilized
intermediate
–H+
(regenerate
acid catalyst)

The second part of Experiment 33.1 is the Grignard reaction of methyl benzoate and phenylmagnesium bromide shown previously. The third part of this experiment consists of purifying the triphenylmethanol produced in the Grignard reaction and performing a quick reaction using it and HBr.

The overall four-step sequence involving preparation of the ester and the Grignard reagent, their reaction together, and treatment of the product with HBr shows the importance of good technique and high yields in a multistep synthesis. For example, in a three-step synthesis, if the yield is 90% at each step, the overall yield will be $(0.90)^3 = 73\%$, while if the yield on each step is 50% the overall yield drops to $(0.50)^3 = 13\%$. This is diagrammed below.

$$A \xrightarrow{90\%} B \xrightarrow{90\%} C \xrightarrow{90\%} D \qquad 73\% \text{ overall}$$

$$A \xrightarrow{50\%} B \xrightarrow{50\%} C \xrightarrow{50\%} D \qquad 13\% \text{ overall}$$

Some published syntheses are 40 steps or longer. This means starting with several kilograms for the first step and winding up with a few milligrams at the end, if you are skillful and lucky. One way to reduce this problem is by using *convergent synthesis*. In other words, rather than building up the molecule one bit at a time, two fairly equal-sized pieces are built and hooked together.

For example, let's say that compound E could be built up by a stepwise (linear) synthesis in four steps and each step has an 80% yield:

$$A \xrightarrow{80\%} B \xrightarrow{80\%} C \xrightarrow{80\%} D \xrightarrow{80\%} E$$

The overall yield for this scheme is $(0.80)^4$ or 41%. If, on the other hand, two pieces could be built up independently and hooked together to form E, the scheme would look as shown below. This scheme gives an overall yield of 64% (0.80 multiplied by 0.80) based on either starting material F or H, a great improvement.

$$F \xrightarrow{80\%} G$$
$$+ \xrightarrow{80\%} E \qquad 64\% \text{ overall}$$
$$H \xrightarrow{80\%} I$$

An additional advantage of a convergent synthesis is that if any intermediate is used up, it is easier and cheaper to replace than in a linear synthesis. For example, intermediate C or D above would take two or three steps to make, while intermediate G or I would take only one.

QUESTION

1. Give the expected product from each of these reactions:

(a) $CH_3Cl \xrightarrow[Et_2O]{Mg}$ ⬚ $\xrightarrow{CH_3CH_2CH_2Br}$ ⬚

(b) $CH_3 \!-\! \overset{\displaystyle O}{\overset{\displaystyle \|}{C}} \!-\! OCH_2CH_3 \xrightarrow[\text{(2) mild } H^+,\, H_2O]{\text{(1) 2 } CH_3MgBr}$ ⬚

(c) $Ph_3COH \xrightarrow{\text{conc. HCl}}$ ⬚

Concept Map

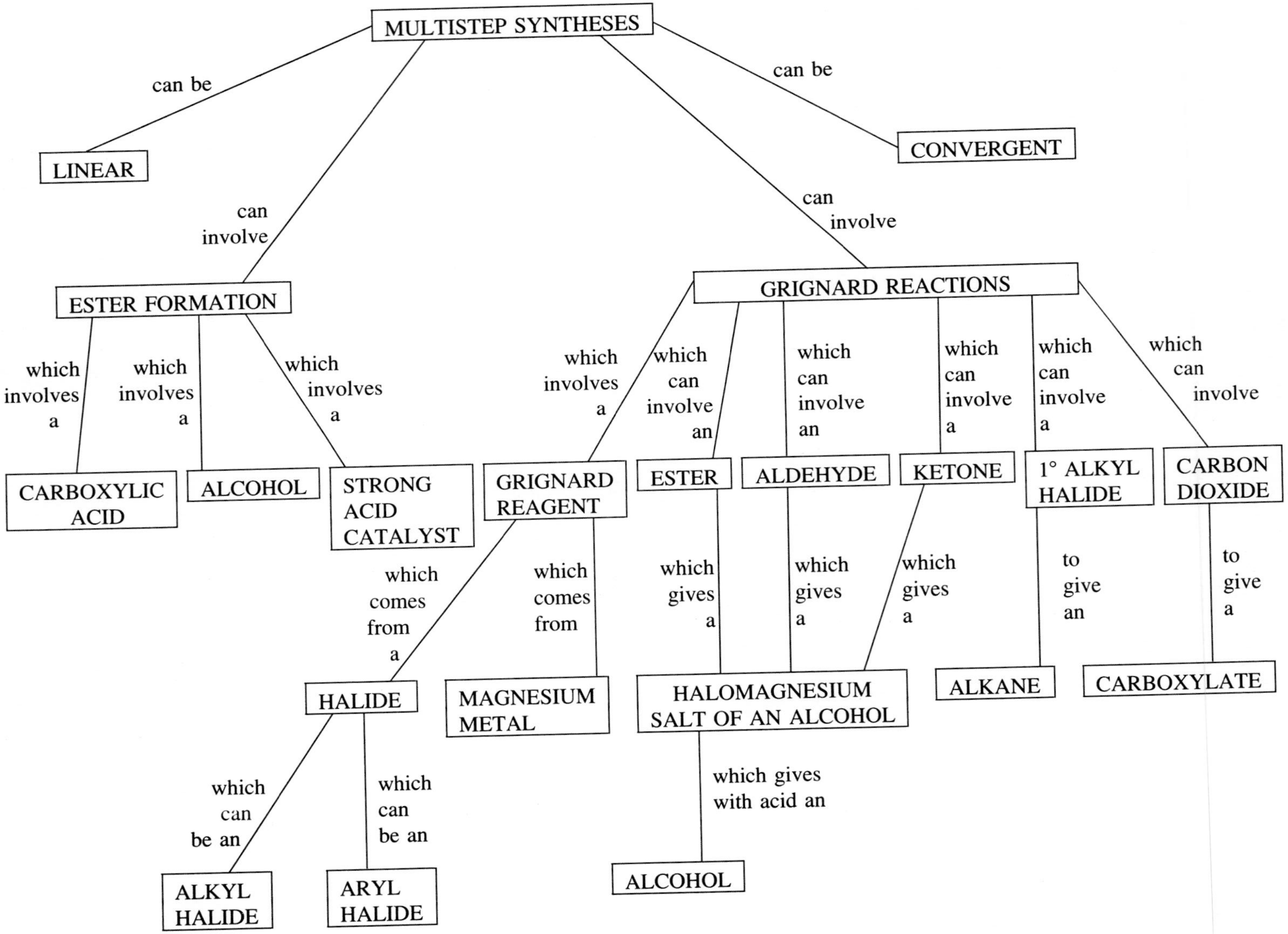

MULTISTEP SYNTHESES
— can be → CONVERGENT
— can be → LINEAR
— can involve → GRIGNARD REACTIONS
— can involve → ESTER FORMATION

GRIGNARD REACTIONS
— which can involve → CARBON DIOXIDE — to give a → CARBOXYLATE
— which can involve a → 1° ALKYL HALIDE — to give an → ALKANE
— which can involve a → KETONE — which gives a → HALOMAGNESIUM SALT OF AN ALCOHOL
— which can involve an → ALDEHYDE — which gives a → HALOMAGNESIUM SALT OF AN ALCOHOL
— which can involve an → ESTER — which gives a → HALOMAGNESIUM SALT OF AN ALCOHOL
— which involves a → GRIGNARD REAGENT

HALOMAGNESIUM SALT OF AN ALCOHOL
— which gives with acid an → ALCOHOL

GRIGNARD REAGENT
— which comes from → MAGNESIUM METAL
— which comes from a → HALIDE

HALIDE
— which can be an → ARYL HALIDE
— which can be an → ALKYL HALIDE

ESTER FORMATION
— which involves a → STRONG ACID CATALYST
— which involves a → ALCOHOL
— which involves a → CARBOXYLIC ACID

334

EXPERIMENT 33.1 SYNTHESIS OF TRIPHENYLMETHANOL BY REACTION OF METHYL BENZOATE AND PHENYLMAGNESIUM BROMIDE

Estimated Time:
7.0 hours for parts a, b, and c

Prelab

1. Write out the products from the reaction of phenylmagnesium bromide and water. Bear in mind the fact that opposite partial charges attract.

$$\text{Ph}\overset{\delta-\ \delta+}{-\text{MgBr}} + \overset{\delta+\ \delta-}{\text{H}-\text{OH}} \longrightarrow \boxed{}$$

2. What volume of water would destroy all the Grignard reagent used in this experiment?
3. Calculate the number of millimoles of benzoic acid in 10 g and of methanol in 25 mL.
4. During the distillation of methyl benzoate, at what temperature do you expect to collect the product?
5. Calculate the theoretical yield for the formation of methyl benzoate and for the Grignard reaction.
6. What product would you expect from reaction of triphenylmethanol and HBr, and why?

Special Hazards

Ether is very flammable and must be kept in closed containers at least 15 feet away from any flames. In the procedure for the second lab period, everyone should flame dry and cool their apparatuses at the same time during the first 15 minutes of lab, before any ether is opened.

If the product of the Grignard reaction needs to be scraped out of the round-bottomed flask at the beginning of Procedure 33.1c, a thin metal spatula should be used, not a glass stirring rod, because of the danger of cuts.

Hydrobromic acid is a strong irritant and contact with skin or clothing should be avoided. Rinse thoroughly with water if this occurs.

Procedure 33.1a: Esterification of Benzoic Acid and Methanol

Estimated Time:
2.5 hours

$$\underset{\substack{\text{Benzoic acid}\\ \text{MW 122, mp }122°}}{\text{Ph}-\overset{\overset{\text{O}}{\|}}{\text{C}}\text{OH}} + \underset{\substack{\text{Methanol}\\ \text{MW 32.0, bp }65°\\ \text{dens. 0.79 g/mL}}}{\text{CH}_3\text{OH}} \xrightarrow[-\text{H}_2\text{O}]{\text{H}^+} \underset{\substack{\text{Methyl benzoate}\\ \text{MW 136, bp }198-200°\\ \text{dens. 1.09 g/mL}}}{\text{Ph}-\overset{\overset{\text{O}}{\|}}{\text{C}}-\text{OCH}_3}$$

Using a powder funnel, in a l00-mL round-bottomed flask place benzoic acid (8.0 g) and anhydrous methanol (20 mL), then carefully pour concentrated sulfuric acid (2 mL) down the walls of the flask. Add a boiling chip, swirl to mix the components, add a condenser and heating mantle with variable tansformer, and reflux for 1 hour.

Cool the solution, decant it into a separatory funnel, and add water (40 mL). Rinse the flask with ether (30 mL) which will be used for extraction, and add this to the separatory funnel. Shake well, venting frequently, and draw off the aqueous layer, which contains the sulfuric acid and most of the leftover methanol. Wash the ether extract with 5% aqueous $NaHCO_3$ (25 mL) to remove unreacted benzoic acid, then with saturated aqueous NaCl solution, and dry over anhydrous Na_2SO_4. After 10 minutes, decant the dry ether solution into a beaker. Rinse the drying agent with a little ether (about 5 mL) and add it to the beaker to improve recovery. Remove the ether on a steam bath in the fume hood. Now pour the crude methyl benzoate into a 50-mL round-bottomed flask and set up an air-cooled semimicro distillation as described in Experiment 15.1. The ester is so high-boiling that it condenses easily. As in any distillation, discard the first 3–4 drops, then collect the product (bp 198–199° C at 1 atm). Weigh the product and calculate the yield. Store the ester until the next period in a tightly stoppered container over a little Na_2SO_4.

Procedure 33.1b: Attack of Phenylmagnesium Bromide on Methyl Benzoate

Estimated Time:
2.5 hours

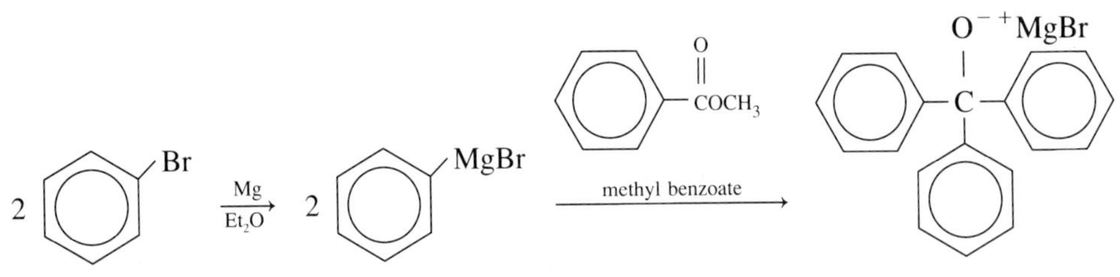

Bromobenzene
MW 157, bp 156°
dens. 1.50 g/mL

Phenylmagnesium bromide (phenyl Grignard)

methyl benzoate

Bromomagnesium salt of triphenylmethanol

Preparation of Grignard Reagent. Clamp a dry 250-mL round-bottomed flask securely, then place in it clean magnesium turnings (2 g, 82 mmol). Add two or three small crystals of iodine; during the heating these will vaporize and coat the magnesium metal, making it more reactive. On top of the flask mount a Claisen head with a pennyhead stopper in one side and a condenser on the other. Be sure to grease all joints lightly when assembling ground-glass apparatus. Place a $CaCl_2$ drying tube in a one-hole rubber stopper or thermometer adapter on top of the condenser. This setup is shown in Figure 33-1.

Be sure that all ether and solvents are at least 15 feet away, then flame dry the flask and magnesium using a cool flame. Be careful not to overheat the magnesium since it could form an unreactive oxide coating or even burn if strongly overheated. The purpose of the flame drying is to drive off any residual water in the apparatus, so heating the glass to just over the boiling point of water is sufficient. On cooling, the flask pulls in dry air through the calcium chloride tube. *Cool the flask to room temperature before proceeding and be sure that all flames are extinguished.*

and add it to the separatory funnel. Be aware that heat is given off when the acid is added and pressure may build up from the ether vapor, so the flask should not be stoppered and the separatory funnel should be vented often.

Wash the ether layer with 10% sulfuric acid (20 mL) to remove any remaining magnesium salt, then wash the ether layer with saturated aqueous sodium chloride and dry it over anhydrous sodium sulfate. Gravity filter to remove the drying agent. To cause the product to precipitate, add ligroin (also called petroleum ether, bp 60–80°, 25 mL) and concentrate the solution on a steam bath in the hood. When you see the first crystals of triphenylmethanol precipitate, remove the flask from the steam bath and let it cool to room temperature, then place it in an ice bath to complete crystallization. Collect the product, wash with cold ligroin, air dry, weigh, and record the melting point (lit. mp 160–163°).

Reaction of triphenylmethanol with HBr. Dissolve some of your triphenylmethanol (0.5 g) in warm glacial acetic acid (10 mL) on the steam bath. Add 47% hydrobromic acid (2 mL; *use caution*), heat for 5 minutes on a steam bath, cool in ice, collect the product by vacuum filtration, wash it with ligroin, air dry, weigh, and record the melting point.

To test if the product contains a halide, perform the Beilstein test by heating a copper wire to redness in a flame, letting it cool to room temperature, touching it to your sample, and placing it back in the flame. If a halide is present in your product, it will give a green flame. Some sparklers and colorful fireplace logs contain copper halides.

QUESTIONS

1. Propose a reasonable mechanism for reaction of triphenylmethanol and HBr, including the structure of the product. See if you can find a literature melting point for your proposed product by checking three common sources: The *CRC Handbook of Chemistry and Physics*, the *CRC Handbook for Organic Compound Identification*, and the *Aldrich Catalog/Handbook of Fine Chemicals*. What evidence do you have that the product formed is what you say?

2. Triphenylmethanol can also be formed by reaction of phenylmagnesium bromide with either benzophenone or diethyl carbonate (shown below). Give mechanisms for both reactions and tell how many equivalents of the Grignard reagent would be needed.

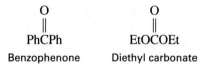

Benzophenone   Diethyl carbonate

3. In the preparation of methyl benzoate, what was the function of each of the following steps?
  (a) Washing the organic layer with $NaHCO_3$ solution.
  (b) Washing the organic layer with saturated aqueous NaCl solution.
  (c) Adding calcium chloride or sodium sulfate to the organic layer.

REFERENCES

ECKERT, T. S., "An Improved Preparation of a Grignard Reagent," *Journal of Chemical Education* (1987) *64*, 179.

KHARASCH, M. S., and O. REINMUTH, *Grignard Reactions of Nonmetallic Substances*, Prentice-Hall, Englewood Cliffs, N.J., 1964.

SECTION 34

Organolithium Compounds

Overview

Lithium reagents are similar to Grignard reagents in that the carbon atom attached to the lithium atom can be considered to have a negative charge. This is, of course, reasonable since lithium is the most electropositive element.

A compound such as *n*-butyllithium, for example, is both an extremely strong base and a reasonable nucleophile because of this negative charge on carbon.

$$CH_3CH_2CH_2CH_2^- Li^+$$
n-Butyllithium

n-Butyllithium is commonly used as a strong base for several reasons. For one, it is conveniently supplied as a 1.6 *M* solution in hexanes. Also, since *n*-BuLi is missing a hydrogen on an sp^3 carbon, it is a stronger base than most other compounds. Recall that hydrogens on carbons are generally not very acidic, and the more the *p* character of the hybridized carbon, the less acidic the hydrogen. If the original hydrocarbon is a weak acid, its conjugate base will be a strong base. Table 34-1 shows the pK_a values of some typical organic compounds.

Note that an alkane (such as butane), has a pK$_a$ value in the range 50–60 (it is a *very* weak acid), so its conjugate base will be correspondingly strong. A hydrogen on a carbon in a benzylic position, for example, has a lower pK$_a$ value (about 41) and is more acidic. A dibenzylic proton would be even more acidic, with a pK$_a$ value of approximately 34. Such a proton is found in the molecule fluorene. In fact, fluorene (pK$_a$ 23) is significantly more acidic than diphenylmethane, probably because its planar geometry and linked rings allow enhanced resonance stabilization of the anion. This name (fluorene) seems peculiar because there is no fluorine in the molecule; the name probably arose because of its fluorescence under ultraviolet light. In any case, a dibenzylic hydrogen from fluorene can be removed easily by *n*-BuLi in an acid–base reaction.

H H   +   *n*-BuLi ⟶ H   Li$^+$ ⟷ other resonance structures

Fluorene   + Butane

TABLE 34–1   ACIDITIES OF ORGANIC ACIDS

Name	Structure	Approximate pK_a
Carboxylic acid	RCOOH	4–6
Phenol	R—⟨benzene ring⟩—OH	7–11
1,3-Dicarbonyl	$\overset{O}{\overset{\|}{R\,C}}\,CH_2\,\overset{O}{\overset{\|}{C\,R}}$	9–11
Alcohol	ROH	16–18
Carbonyl	$R-\overset{O}{\overset{\|}{C}}-CH_3$	20–25
Fluorene	⟨fluorene structure with H H⟩	23
Alkyne	$R-C\equiv C-H$	25
Dibenzylic	⟨phenyl⟩—CH$_2$—⟨phenyl⟩	34
Alkene	$R-CH=CH_2$	40
Benzylic	⟨phenyl⟩—CH$_3$	41
Alkane	$R-H$	50–60

The anion of fluorene, once formed, can act as a nucleophile, much as a Grignard reagent would. In this experiment, you will let it attack CO_2 (in the form of dry ice) to form the anion of a carboxylic acid.

Addition of acid will protonate the carboxylate salt to yield the free carboxylic acid.
Overall reaction:

Fluorene
MW 166, mp 114 – 116°
bp 298°

(1) n-BuLi
(2) CO_2
(3) H^+

Fluorene-9-carboxylic acid
MW 210, mp 228 – 230°

EXPERIMENT 34.1 SYNTHESIS OF FLUORENE-9-CARBOXYLIC ACID

Estimated Time:
2.0 hours

Prelab

1. Calculate the millimoles of fluorene in 0.250 g and of n-BuLi in 1.4 mL of a 1.6 M solution.
2. What is the theoretical yield of fluorene-9-carboxylic acid?

Special Hazards

Ether is highly flammable. n-Butyllithium is an extremely strong base, and dry ice can freeze skin quickly. Avoid contact and wash with water immediately if contact occurs.

Procedure

In a dry 50-mL round-bottomed flask, place anhydrous ether (10 mL) and add cautiously by syringe 1.4 mL of 1.6 M n-BuLi in hexanes. In portions add fluorene (0.250 g, _____ mmol). The solution should develop an orange color. Add a condenser and heating mantle and reflux gently for 45 minutes. Be sure that water flow through the condenser is adequate and heating is very gentle, since ether boils at 35°. For use in the workup later, place a 100-mL beaker in a drying oven or flame dry it in an area of the lab well removed (at least 15 feet) from any flammable solvents.

At the end of the reflux period, remove the heating mantle, shut off the condenser water and cool the flask in an ice bath. Cool the dried beaker and, using tongs, place a few small chunks of dry ice (5–10 g total) in the beaker. Pour the contents of the round-bottomed flask onto the dry ice and allow the mixture to come to room temperature. Add 30 mL of 5% aqueous NaOH and 20 mL of additional ether. Pour the two-phase mixture into a separating funnel, shake and vent, and draw off the aqueous (lower) layer, which contains the fluorene-9-carboxylate anion.

Discard the ether in an appropriate waste container, replace the aqueous phase in the separatory funnel, and wash with another 20-mL portion of ether. Draw off the aqueous phase again, cool it in an ice bath, then carefully acidify by addition of concentrated HCl, checking with pH paper. As the solution is acidified, the desired water-insoluble acid precipitates. Collect the precipitate by suction filtration, recrystallize from methanol to which water is added at the boiling point, collect again, wash with a little water, and air dry. Record the mass and melting point of the fluorene-9-carboxylic acid obtained.

REFERENCE

BURTNER, R. R. and J. W. CUSIC, "Antispasmodics. 1. Basic Esters of Some Arylacetic Acids," *Journal of the American Chemical Society* (1943) *65*, 262.

Hydrolysis of Nitriles to Carboxylic Acids

Overview

Nitriles can be hydrolyzed to carboxylic acids using either acid or base catalysis. In the acid-catalyzed mechanism, the nitrogen is protonated twice during the course of the hydrolysis, allowing two molecules of water to attack the carbon. Nitrile hydrolysis duplicates most of the mechanism of amide hydrolysis, since a protonated amide is an intermediate.

$$R-C\equiv N: \xrightarrow{H^+} \left[R-\overset{+}{C}\equiv N-H \longleftrightarrow R-\overset{+}{C}=NH \right] \xrightarrow{H_2\ddot{O}} R-\underset{\underset{H}{\overset{+}{O}}}{\overset{|}{C}}=NH$$

Nitrile   Protonated nitrile

$$\Big\downarrow \sim H^+$$

$$R-\underset{\underset{H}{\overset{|}{\overset{+}{O}}}}{\overset{\overset{OH}{|}}{C}}-NH_2 \xleftarrow{H_2\ddot{O}:} \left[R-\overset{\overset{+}{O}H}{\overset{||}{C}}-N\overset{H}{\underset{H}{\diagdown}} \longleftrightarrow R-\overset{\overset{:\ddot{O}H}{|}}{\underset{+}{C}}-N\overset{H}{\underset{H}{\diagdown}} \longleftrightarrow R-\overset{\overset{OH}{|}}{C}=\overset{+}{N}\overset{H}{\underset{H}{\diagup}} \right]$$

Protonated amide

$$\Big\downarrow \sim H^+$$

$$R-\underset{\overset{|}{OH}}{\overset{\overset{OH}{|}}{C}}\overset{+}{NH_3} \longrightarrow \left[R-\underset{\overset{|}{OH}}{\overset{\overset{:\ddot{O}H}{|}}{C}}+ \longleftrightarrow R-\overset{\overset{+}{O}H}{\overset{||}{\underset{:\ddot{O}H}{C}}} \longleftrightarrow R-\overset{\overset{OH}{|}}{\underset{+OH}{\overset{||}{C}}} \right]$$

Protonated carboxylic acid

$$\Big\downarrow -H^+$$

$$R-\overset{\overset{O}{||}}{C}-OH$$

Carboxylic
acid

344

The base-catalyzed reaction occurs by attack of hydroxide ions on the slightly positive nitrile carbon.

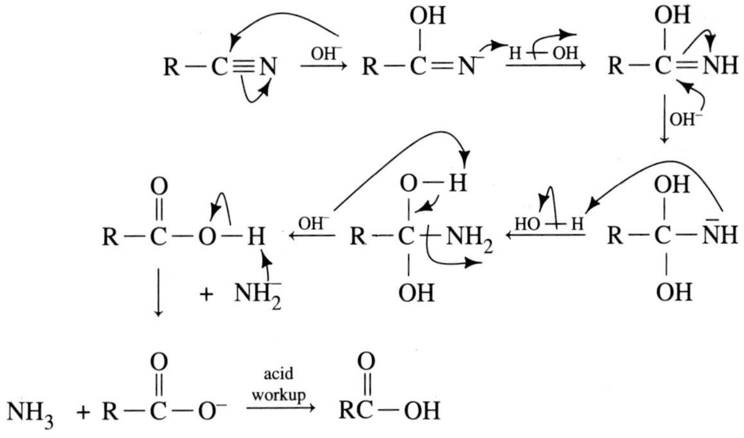

$$R-C\equiv N \xrightarrow{\text{OH}^-} R-\overset{\overset{\displaystyle OH}{|}}{C}=N \xrightarrow{H-OH} R-\overset{\overset{\displaystyle OH}{|}}{C}=NH$$

$$\xrightarrow{\text{OH}^-}$$

$$R-\overset{\overset{\displaystyle O}{\|}}{C}-O-H \xleftarrow{\text{OH}^-} R-\overset{\overset{\displaystyle O-H}{|}}{\underset{\underset{\displaystyle OH}{|}}{C}}-NH_2 \xleftarrow{HO-H} R-\overset{\overset{\displaystyle OH}{|}}{\underset{\underset{\displaystyle OH}{|}}{C}}-\bar{N}H$$

$$+ \ NH_2^-$$

$$NH_3 \ + R-\overset{\overset{\displaystyle O}{\|}}{C}-O^- \xrightarrow[\text{workup}]{\text{acid}} R\overset{\overset{\displaystyle O}{\|}}{C}-OH$$

Carboxylate

In Experiment 35.1 you will hydrolyze benzonitrile to benzoic acid.

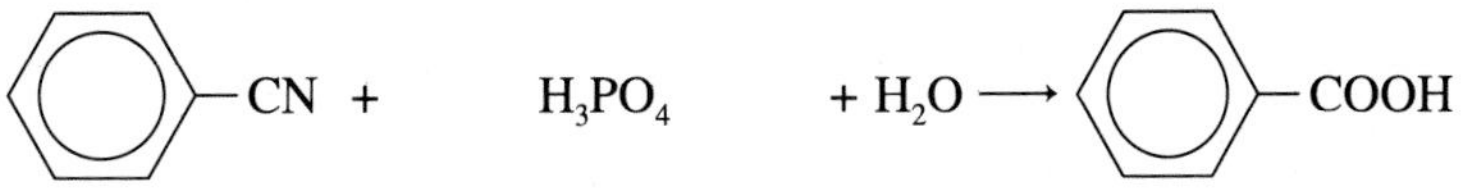

$$\text{C}_6\text{H}_5{-}CN \ + \ H_3PO_4 \ + \ H_2O \longrightarrow \text{C}_6\text{H}_5{-}COOH$$

Benzonitrile	Phosphoric acid 85%	Benzoic acid
MW 103, bp 188°		MW 122, mp 122–123°
dens. 1.01 g/mL		

QUESTIONS

1. Draw the mechanism for the hydrolysis of benzonitrile under the following conditions.
   (a) Acid catalysis
   (b) Base catalysis
2. Show the products of the following reactions.
   (a)

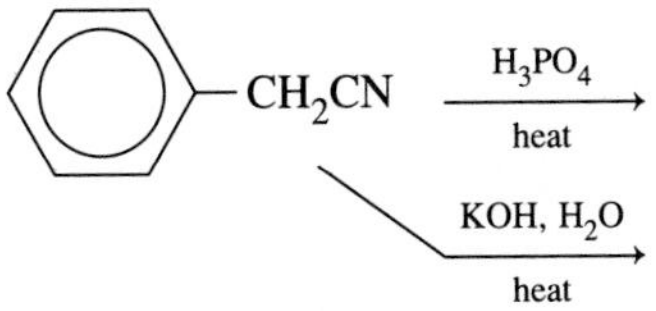

$$\text{C}_6\text{H}_5{-}CH_2CN \xrightarrow[\text{heat}]{H_3PO_4}$$

$$\xrightarrow[\text{heat}]{KOH, \ H_2O}$$

   (b) $ClCH_2CH_2 \ CH_2CH_2CN \xrightarrow[\text{heat}]{H^+, \ H_2O}$

$$\xrightarrow[\text{heat}]{OH^-, \ H_2O}$$

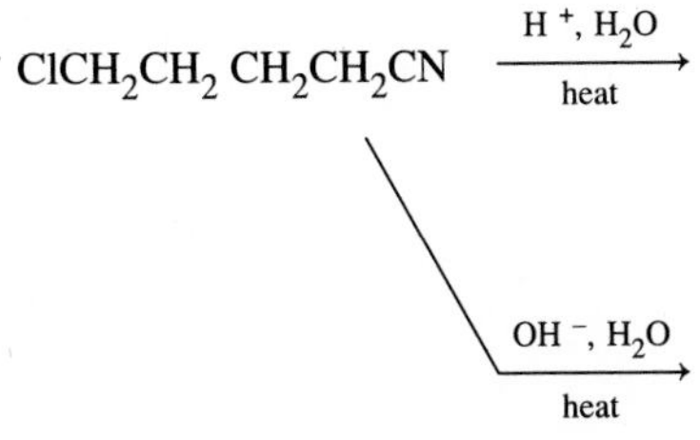

EXPERIMENT 35.1 HYDROLYSIS OF BENZONITRILE TO BENZOIC ACID

Estimated Time:
2.5 hours

Prelab

1. Calculate the millimoles and volume corresponding to 1.0 g of benzonitrile.
2. What is the theoretical yield of benzoic acid?

Special Hazards

Concentrated phosphoric acid is highly corrosive; wash immediately with a lot of water if contact occurs. Benzonitrile is an irritant.

Procedure

In a 50-mL round-bottomed flask, place benzonitrile (1.0 g, _____ mmol, _____ mL) and 85% phosphoric acid (10 mL). Reflux the mixture for 45 minutes, cool the solution in an ice bath, pour into ice water, and collect the product by suction filtration. Recrystallize from water, collect again, wash with water, and air dry. Record the mass and melting point of the product.

REFERENCE

BERGER, G. and S. C. J. OLIVIER, "Une nouvelle méthode de saponification des amides et des nitriles" (A New Method of Saponification of Amides and Nitriles), *Recueil des Travaux Chimiques des Pays-Bas* (1927) *46*, 600.

SECTION **36**

Base-Catalyzed Hydrolysis of Esters: Soap-Making

Overview

Have you ever made soap or watched it being made? This basic necessity was produced almost 5000 years ago from potash (KOH, from wood ashes) and animal fat. Soap production constitutes a major industry today, and we often take the availability of this commodity for granted. In our ancestors' times, however, and even in many third-world countries today, soapmaking was and is a major household event. To make soap at home, one starts with a large kettle and melts in it some lard, grease, coconut oil, or some other fatty substance. To this is added a 50% solution of lye (NaOH) in water. The mixture is stirred and heated for an hour or more, then poured into flat boxes to solidify into soap.

Figure 36–1 Making Soap

The chemical process occurring during soap making is known as saponification, the base-catalyzed hydrolysis of ester groups.

Fats consist largely of triglycerides, triesters of the alcohol glycerol. Upon treatment with strong aqueous base and heat, the ester groups are cleaved to form the free alcohol (glycerol) and the sodium salts of the carboxylic acids.

$$
\begin{array}{c}
\text{R}-\overset{\displaystyle O}{\overset{\|}{\text{C}}}-\text{O}-\text{CH}_2 \\[2mm]
\text{R}-\overset{\displaystyle O}{\overset{\|}{\text{C}}}-\text{O}-\text{CH} \\[2mm]
\text{R}-\overset{\displaystyle O}{\overset{\|}{\text{C}}}-\text{O}-\text{CH}_2
\end{array}
\quad + \quad 3\text{NaOH} \quad \longrightarrow \quad 3\text{R}-\overset{\displaystyle O}{\overset{\|}{\text{C}}}-\text{O}^-\text{Na}^+ \quad + \quad
\begin{array}{c}
\text{HO}-\text{CH}_2 \\
\text{HO}-\text{CH} \\
\text{HO}-\text{CH}_2
\end{array}
$$

| Triglyceride (Fat) | Soap (sodium salt of a fatty acid) | Glycerol |

In this scheme the R groups are residues of long-chain fatty acids, and can differ from one another. The most common fatty acids found in triglycerides contain from 12 to 18 carbons; a table of common fatty acids appears in Table A16-1.

Hydrolysis of an ester can be accomplished with the help of either acid or base catalysis. In the acid-catalyzed mechanism, an oxygen atom is protonated to aid attack by the incoming nucleophile. Transfer of the extra proton to the leaving group then aids its departure.

Mechanism of acid-catalyzed ester hydrolysis:

$$
\text{R}-\overset{\displaystyle O}{\overset{\|}{\text{C}}}-\text{OR}' \xrightarrow{\text{H}^+}
\left[
\text{R}-\overset{\displaystyle +\text{OH}}{\overset{\|}{\text{C}}}-\text{OR}' \longleftrightarrow \text{R}-\overset{\displaystyle \text{OH}}{\overset{|}{\underset{+}{\text{C}}}}-\text{OR}' \longleftrightarrow \text{R}-\overset{\displaystyle \text{OH}}{\overset{|}{\text{C}}}\!=\!\overset{+}{\text{OR}'}
\right]
$$

Ester   Protonated ester

$$\downarrow \text{H}_2\ddot{\text{O}}$$

$$
\text{R}-\overset{\displaystyle \text{OH}}{\underset{\displaystyle \text{OH}}{\text{C}}}\!-\!\overset{\displaystyle \text{H}}{\underset{+}{\text{O}}}\!-\!\text{R}'
\quad \xleftarrow{\sim\text{H}^+} \quad
\text{R}-\overset{\displaystyle \text{OH}}{\underset{\displaystyle \underset{H\,\overset{+}{}\,H}{O}}{\text{C}}}\!-\!\text{OR}'
$$

$$\downarrow \begin{array}{l} -\text{ROH} \\ \text{loss of alcohol}\end{array}$$

$$
\left[
\text{R}-\overset{\displaystyle +\text{OH}}{\underset{\displaystyle \text{OH}}{\overset{\|}{\text{C}}}} \longleftrightarrow \text{R}-\overset{\displaystyle \text{OH}}{\underset{\displaystyle \text{OH}}{\text{C}+}} \longleftrightarrow \text{R}-\overset{\displaystyle \text{OH}}{\underset{\displaystyle +\text{OH}}{\overset{\|}{\text{C}}}}
\right]
\xrightarrow{-\text{H}^+} \text{RCOOH}
$$

Protonated carboxylic acid   Carboxylic acid

The net result is that the ester is split up into its parent carboxylic acid and alcohol, with insertion of water.

For the base-catalyzed mechanism of ester hydrolysis, hydroxide ion attacks the slightly positive carbon of the carbonyl group, forming a negatively charged intermediate that undergoes displacement of the alkoxide group. The alkoxide removes the acidic proton from the carboxylic acid produced to form an alcohol and a carboxylate ion. On acidification the carboxylate ion regains its proton and can be isolated as the carboxylic acid.

Base-catalyzed ester hydrolysis:

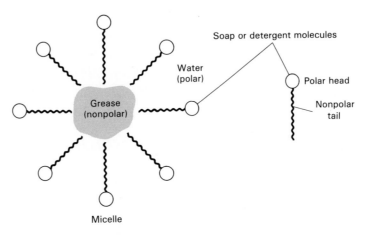

Soaps work by forming micelles when they encounter hydrophobic particles of dirt or grease. The hydrophobic tails of the soap molecules surround the dirt, while the ionic heads protrude into the water, solubilizing the entire micelle, as shown in Figure 36-2.

Figure 36–2   A Soap Micelle

In hard water (water containing large concentrations of Ca^{2+}, Mg^{2+}, and Fe^{3+} ions) soaps are ineffective because these ions form insoluble precipitates (soap scum or bathtub ring) with the fatty acid residues.

One solution is to add washing soda (sodium carbonate, Na_2CO_3) to the water before adding the soap. This causes precipitation of the offending ions as their carbonates, so they do not react with the soap.

$$Na_2CO_3 + Ca^{2+} \longrightarrow CaCO_3(s) + 2Na^+$$

Another solution to washing in hard water is to use detergents. Detergents are sodium salts of long-chain sulfonic acids, which act like soaps but do not form precipitates in the presence of calcium, magnesium, or iron ions.

$$RSO_3^- Na^+ + Ca^{2+} \longrightarrow \text{no reaction}$$

Some typical detergents are shown below.

$$CH_3(CH_2)_{10}CH_2OSO_3^- Na^+$$

Sodium lauryl sulfate

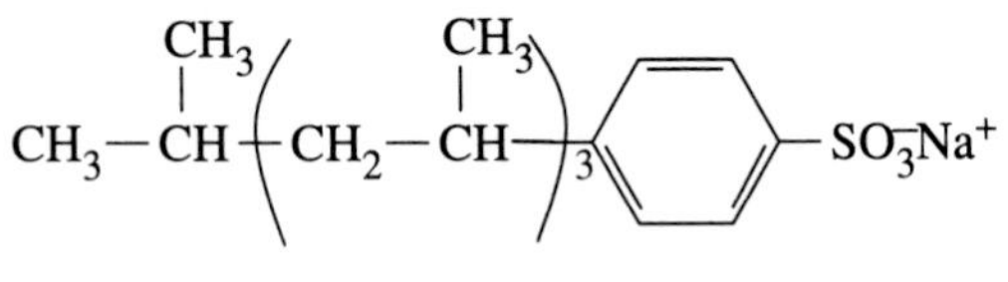

An alkylbenzenesulfonate (ABS)

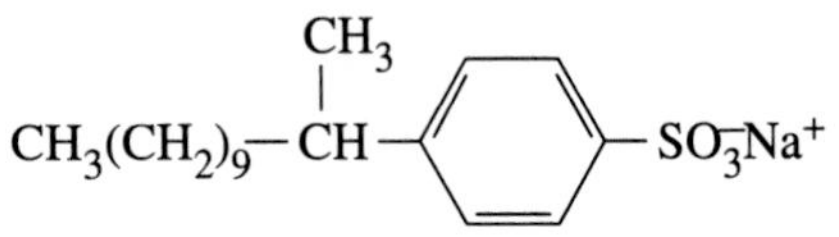

A linear alkylsulfonate detergent (LAS)

When highly branched detergents such as ABS were introduced, they caused significant pollution to rivers and streams because bacterial enzymes could not degrade these compounds. In 1966 the ABS detergents were replaced with biodegradable linear alkylsulfonate (LAS) detergents, which greatly reduced the pollution problem.

EXPERIMENT 36.1 SOAP FROM SHORTENING

Estimated time:
2.0 hours

Procedure

Making the soap. In a 50-mL beaker place 4 g of a commercial solid shortening (such as Crisco). To this add 15 mL of 95% ethanol and stir with slight warming to dissolve.

In a separate 100-mL beaker, place 2 g of sodium hydroxide. Add water (10 mL) and stir to dissolve; the solution will become hot. Pour the ethanolic solution of shortening into the beaker containing the NaOH, then add a magnetic stir bar and place the mixture on a hot plate/stirrer set on low heat. Heat the mixture for 30 minutes with magnetic stirring. It may be desirable to place a large watch glass covering the beaker to prevent splashing.

Meanwhile, in a 150-mL beaker, dissolve 12 g of sodium chloride in 50 mL of water, then cool the solution in an ice bath. At the end of the heating period, pour the saponification solution into the cold salt solution. Stir to help the soap precipitate. Collect the product by suction filtration. To remove residual NaOH, replace the soap in the beaker, stir with a little ice-cold water, and refilter. *Caution:* Do not use too much water or water at room temperature since the soap may dissolve. Dry and press it into a small cake using a paper towel. A small amount of this soap will be used for testing; allow the remainder to dry in your drawer until the next lab period.

Testing the soap.   In a 250-mL Erlenmeyer flask, dissolve 0.30 g of the soap in 20 mL of distilled water. Stopper the flask and shake vigorously for about 10 seconds to create a foam; observe this foam for 1 minute. Now add 6 drops of 5% $MgSO_4$ solution, shake, and record its effect. Finally, add trisodium phosphate (1 g), shake again, and observe and record the results.

Testing a detergent.   In a 250-mL Erlenmeyer flask, dissolve 0.30 g of a solid commercial detergent in 20 mL of water. Stopper, shake for 10 seconds, and observe the foam for 1 minute. Add 6 drops of 5% $MgSO_4$ solution, shake, and record the effect.

Ester Synthesis

Overview

Esters are extremely common compounds; we consume and use them every day. Some esters are important components of foods, drugs, solvents, and fabrics. Many of the compounds found in flavors (especially fruit flavors) are esters; some of these are listed in Table 37-1.

Esters can be prepared by simple and catalyzed dehydration of an alcohol and a carboxylic acid and/or by treating an alcohol with a reactive acid derivative such as an acid chloride or anhydride.

$$\overset{\overset{\textstyle O}{\|}}{R\,C\,OH} + R'OH \xrightarrow{H^+} \overset{\overset{\textstyle O}{\|}}{R\,C\,OR'} + H_2O$$

$$\overset{\overset{\textstyle O}{\|}}{R\,C\,Cl} + R'OH \longrightarrow \overset{\overset{\textstyle O}{\|}}{R\,C\,OR'} + HCl$$

$$\overset{\overset{\textstyle O}{\|}}{R\,C\,O}\overset{\overset{\textstyle O}{\|}}{C\,R} + R'OH \longrightarrow \overset{\overset{\textstyle O}{\|}}{R\,C\,OR'} + RCOOH$$

TABLE 37–1   SOME ESTERS FOUND IN FLAVORS

Name	Structure	Found in:
Benzyl acetate	$CH_3 - \overset{\overset{\textstyle O}{\|}}{C} - O - CH_2 -$ ⬡	Jasmine, peach
Ethyl butyrate	$CH_3CH_2CH_2 - \overset{\overset{\textstyle O}{\|}}{C} - OCH_2CH_3$	Pineapple
Isoamyl acetate	$CH_3 - \overset{\overset{\textstyle O}{\|}}{C} - OCH_2CH_2CH{\overset{\textstyle CH_3}{\underset{\textstyle CH_3}{}}}$	Banana (also sting pheromone of honeybee)

TABLE 37–1   SOME ESTERS FOUND IN FLAVORS (cont.)

Name	Structure	Found in:
Isobutyl propionate	$CH_3CH_2-\overset{\overset{\displaystyle O}{\|\|}}{C}-OCH_2CH\overset{\displaystyle CH_3}{\underset{\displaystyle CH_3}{\diagup}}$	Rum
Isopentenyl acetate	$CH_3-\overset{\overset{\displaystyle O}{\|\|}}{C}-O-CH_2CH=C\overset{\displaystyle CH_3}{\underset{\displaystyle CH_3}{\diagup}}$	Juicy Fruit
Methyl anthranilate	$\overset{NH_2}{\underset{\text{(benzene ring)}}{}}\quad\overset{\overset{\displaystyle O}{\|\|}}{C}-OCH_3$	Grape

One of the simplest and oldest known organic reactions is the dehydration of a carboxylic acid and an alcohol to form an ester. This reaction is called the Fischer esterification in honor of Emil Fischer, a brilliant and very productive German chemist in the late nineteenth and early twentieth centuries. The dehydration is catalyzed by strong acids, with the following mechanism:

Protonated carboxylic acid

Protonated ester   Ester

This reaction proceeds quite well in the absence of any other acid-sensitive functional groups. This reaction, however, often requires a lengthy reflux period in the presence of strong acid, and milder conditions may at times be preferable. These conditions may be achieved by using a more reactive acid derivative, commonly the acid chloride or anhydride. These reactions occur by similar mechanisms.

Acetylation of an alcohol with an acid chloride:

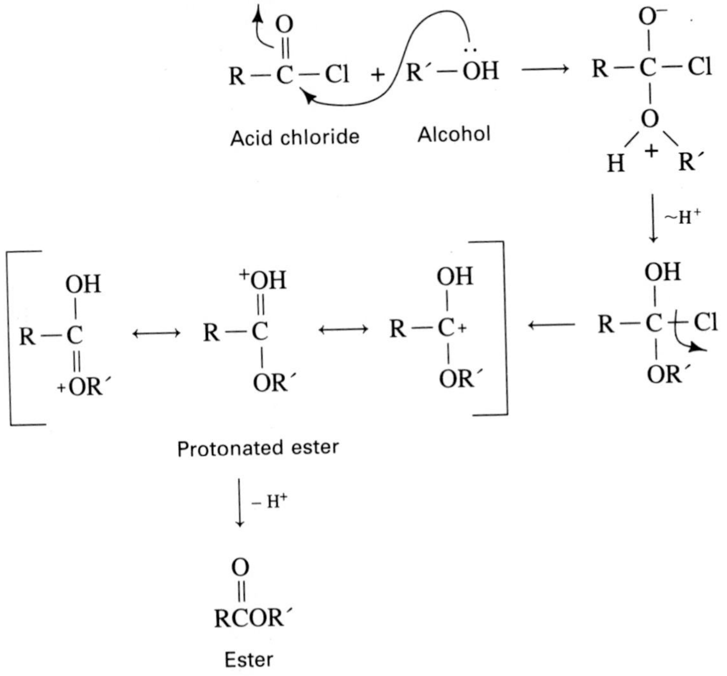

Protonated ester

Ester

Acylation of an alcohol with an acid anhydride:

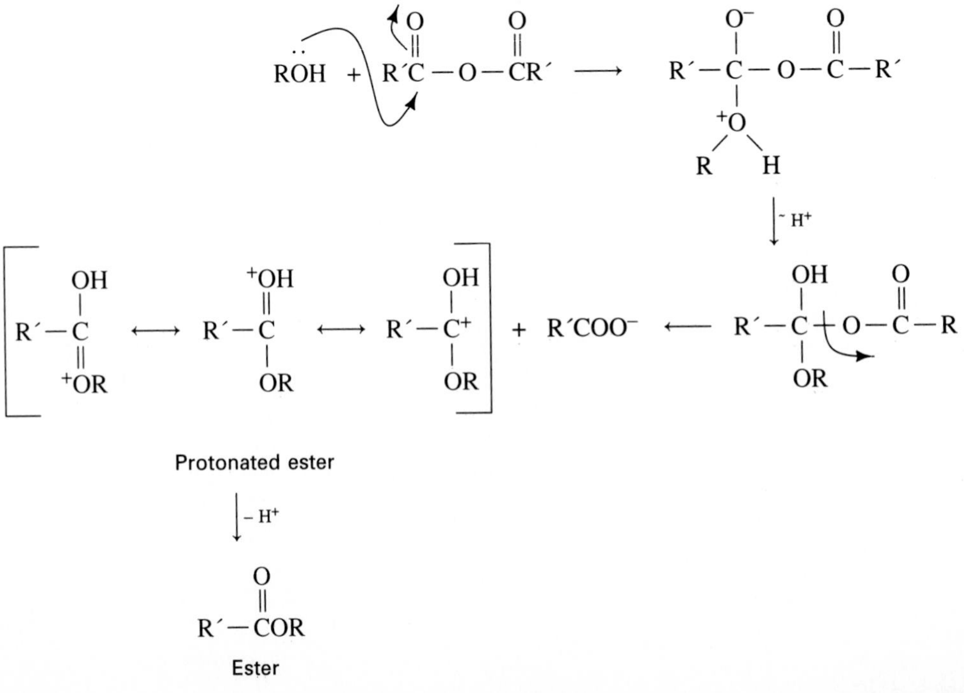

Protonated ester

Ester

Experiment 37.1 is the Fischer esterification of acetic acid with isoamyl alcohol to produce isoamyl acetate, an ester that smells like bananas when concentrated, and like pears when in dilute aqueous solution.

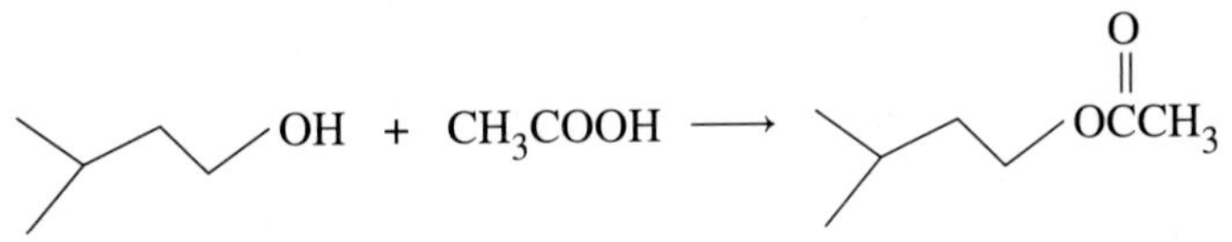

Isoamyl alcohol
(isopentyl alcohol,
3-methyl-1-butanol)
MW 88.2, bp 130°
dens. 0.81 g/mL

Acetic acid
MW 60, bp 118°
dens. 1.06 g/mL

Isoamyl acetate
(isopentyl acetate)
MW 130, bp 142°
dens. 0.88 g/mL

You will know you have succeeded in your synthesis when the entire lab reeks of bananas.

Another simple example of Fischer esterification is found in Experiment 37.2, the synthesis of ethyl acetate from ethanol and acetic acid.

$$CH_3COOH \ + \ CH_3CH_2OH \ \xrightarrow{H_2SO_4} \ CH_3\overset{\overset{\textstyle O}{\|}}{C}-O-CH_2CH_3$$

Acetic acid
MW 60, bp 118°
dens. 1.06 g/mL

Ethanol
MW 46, bp 78°
dens. 0.80 g/mL

Ethyl acetate
MW 88, bp 77°
dens. 0.90 g/mL

This is the type of reaction that is carried out on an industrial scale, because ethanol and acetic acid are cheap and ethyl acetate is a useful solvent for paints, varnishes, adhesives, and so on.

Experiment 37.3 illustrates the preparation of aspirin by acetylation of salicylic acid, which is the industrial process by which aspirin is prepared.

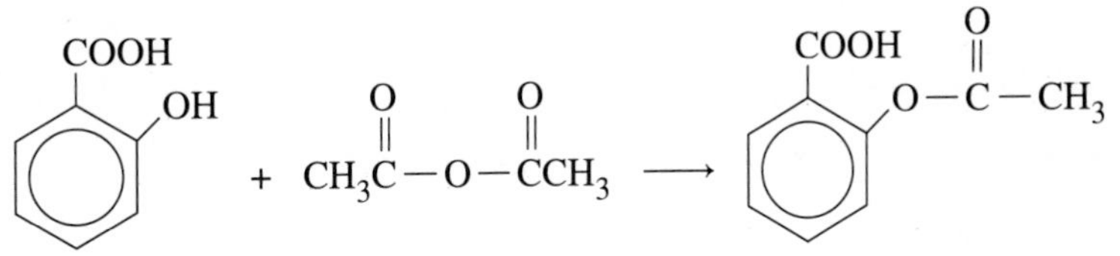

Salicylic acid
MW 138, mp 158–160°

Acetic anhydride
MW 102, bp 138–140°
dens. 1.08 g/mL

Acetylsalicylic acid
(aspirin)
MW 180, mp 140–142°

In recent years studies have shown that aspirin works by inhibiting prostaglandin synthesis. Prostaglandins are a family of compounds that mediate the body's immune response; they cause inflammation, fever, and menstrual cramps. Some examples of prostaglandins are shown below.

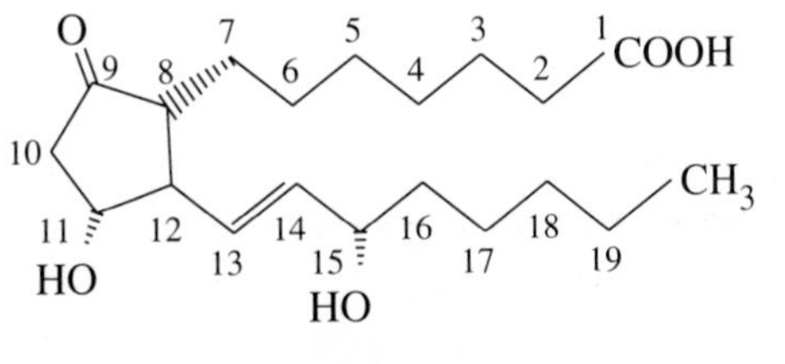

Prostaglandin E_1
(PGE$_1$)

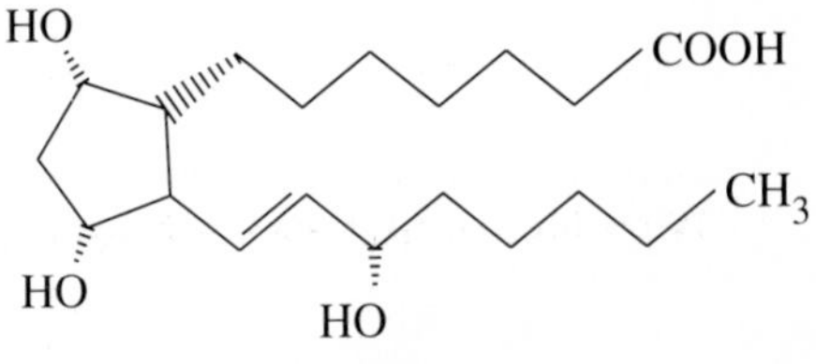

Prostaglandin $F_{1\alpha}$
(PGF$_{1\alpha}$)

Prostaglandins arise biogenetically from oxidation of arachidonic acid, a 20-carbon fatty acid containing four double bonds. Relevant to our discussion of esterification, the mechanism by which aspirin inhibits prostaglandin formation is by acetylation of a serine hydroxyl group at the active site of the enzyme *prostaglandin endoperoxide synthase*. This enzyme normally catalyzes both the oxidation of arachidonic acid to PGG_2 and the subsequent reduction of the hydroperoxide group to an alcohol in PGH_2. These two activities may be classified separately as a fatty acid oxygenase and a peroxidase, although they occur within the same enzyme molecule.

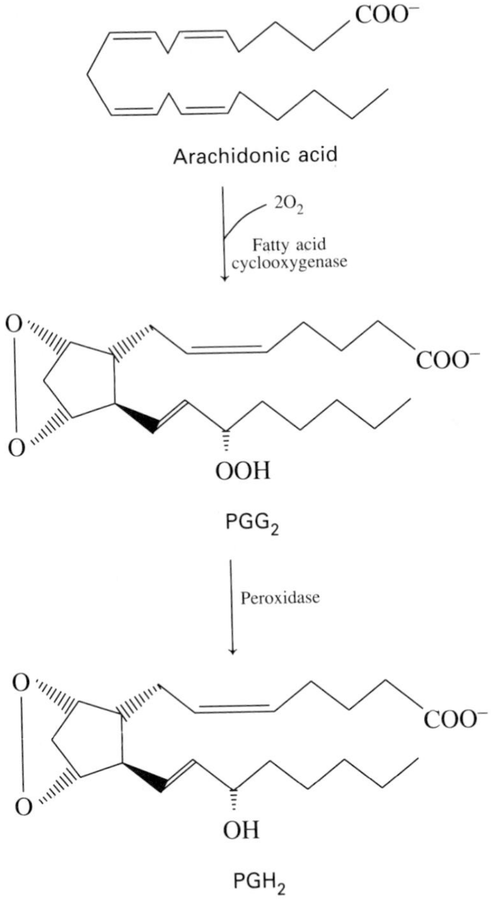

Arachidonic acid

$2O_2$

Fatty acid cyclooxygenase

PGG_2

Peroxidase

PGH_2

By acetylation of the active site serine hydroxyl, aspirin prevents the formation of PGG_2 (it inhibits fatty acid cyclooxygenase).

| Enzyme with active serine residue | Aspirin | | Acylated enzyme (inactive) | Salicylic acid |

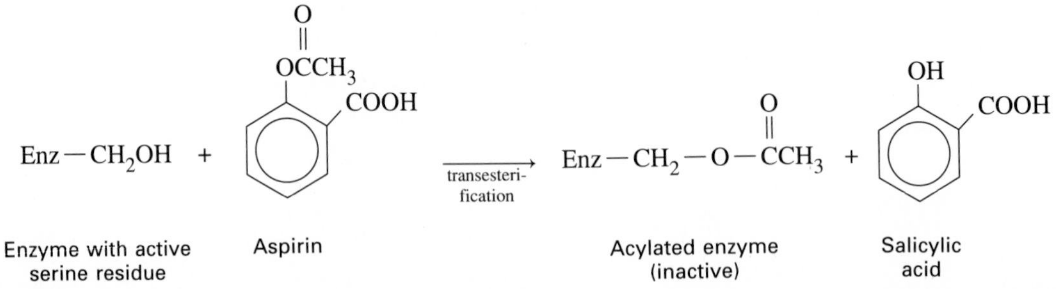

Further examples of ester formation using acetic anhydride and acetyl chloride are found in Experiments 37.4 to 37.6.

For Experiment 37.4:

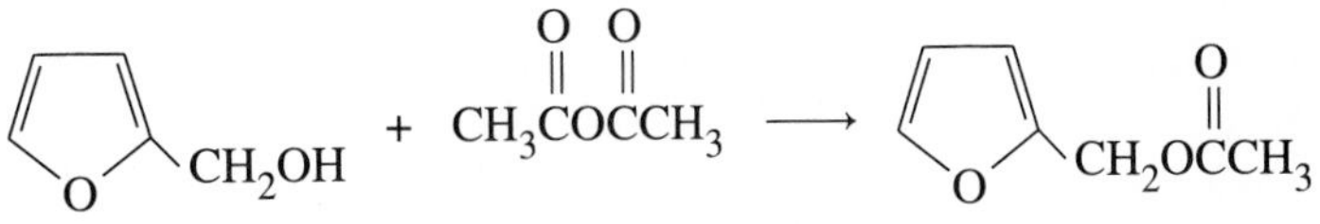

Furfuryl alcohol
MW 98.1, bp 170°
dens. 1.14 g/mL

Acetic anhydride
MW 102, bp 138 – 140°
dens. 1.08 g/mL

2-Furylmethylacetate
(furfuryl alcohol acetate)
MW 140, bp 175 – 177°
dens. 1.12 g/mL

For Experiment 37.5:

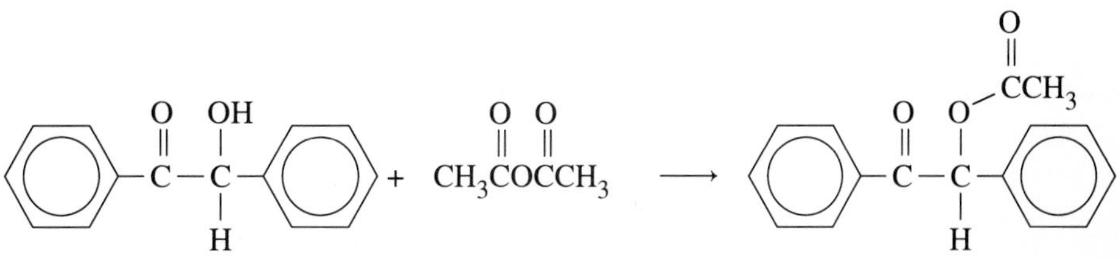

Benzoin
MW 212, mp 134 – 136°

Acetic anhydride

Benzoin acetate
MW 254, mp 81 – 83°

For Experiment 37.6:

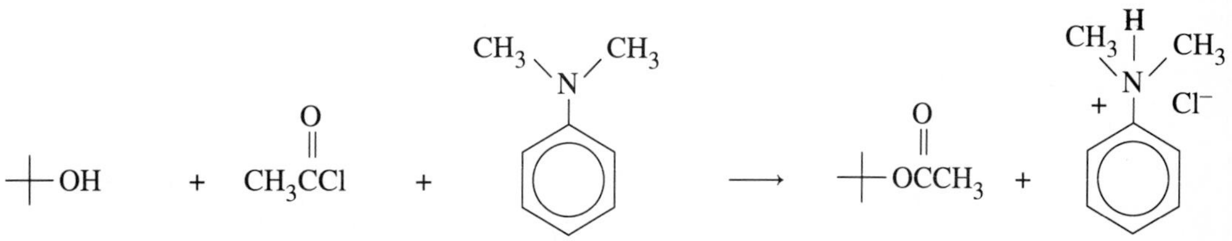

t-Butyl alcohol
(2-methyl-2-propanol)
MW 74, bp 83°
dens. 0.79 g/mL

Acetyl chloride
MW 78.5, bp 52°
dens. 1.10 g/mL

N, N-Dimethylaniline
MW 121, bp 193 – 194°
dens. 0.96 g/mL

tert-Butyl
acetate
MW 116, bp 98°
dens. 0.86 g/mL

In Experiment 37.6, dimethylaniline acts as a base to neutralize the HCl produced by the esterification of the alcohol and acyl chloride. Other bases commonly used for this purpose include triethylamine and pyridine.

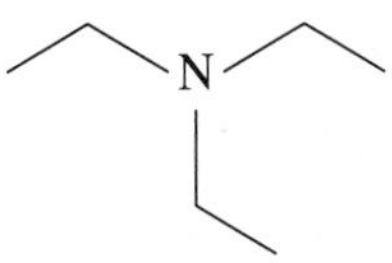

Triethylamine

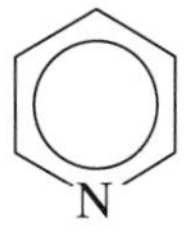

Pyridine

If pyridine is used, care must be taken to avoid exposure, since it has been found to be mutagenic.

APPLICATION: THE PAINFUL HISTORY OF ANALGESICS, ANTIPYRETICS, AND ANTI-INFLAMMATORY AGENTS

Pain is unfortunately one of the more common of life's experiences. On the positive side, pain warns us when injury is occurring (such as touching a hot object or stepping on a tack) or when a disease is present, so that we may seek treatment.

On the other hand, some pains do not seem to serve much purpose and people have always tried to relieve these. Willow bark was used by the ancient Greeks, Native Americans, and Europeans in the Middle Ages as an effective analgesic (pain-relieving), antipyretic (fever-reducing), and anti-inflammatory (swelling-reducing) drug. It came to the attention of the European community in 1763, when the clergyman Edward Stone presented a paper to the Royal Society of London describing its use in the treatment of malaria symptoms.

Organic chemists soon isolated the active principle and named it salicin after *salix*, the Latin name for the willow tree. Salicin is a glycoside of salicyl alcohol.

Salicin

Derivatives including salicylic acid were prepared and tested. The acidity of the compound, however, caused severe irritation of the mucous membranes of the digestive tract. Attempts to use sodium salicylate instead failed because of its unpleasant sweet taste. In 1893, however, Felix Hofmann, while working for the German chemical firm of Bayer, synthesized acetylsalicylic acid, which proved to possess the desirable pharmacological properties of salicylic acid without severe irritation or unpleasant flavor.

Salicylic acid

Sodium salicylate

Acetylsalicylic acid
(aspirin)

The success of this product is attested to by the fact that over 20 million pounds per year of aspirin is consumed in the United States. This story illustrates a common theme in drug development; isolation of an active substance from a natural product, followed by synthesis and testing of derivatives for effectiveness without undesirable side effects.

Several other analgesics have been widely used. Acetanilide, for example, was introduced to medical use in 1886 but was soon found to be too toxic. A derivative, phenacetin, was introduced in 1887. It was discovered in 1949 that the active metabolite of phenacetin is acetaminophen, and since that time the use of acetaminophen itself has become widespread.

Acetanilide   Phenacetin   Acetaminophen

Acetaminophen is comparable to aspirin as an analgesic and antipyretic. It is only slightly anti-inflammatory and does not possess the antirheumatic activity of the salicylates. However, since it is a neutral compound, it does not produce the gastric irritation or bleeding that may occur with salicylates.

In 1964, ibuprofen was patented as an anti-inflammatory agent.

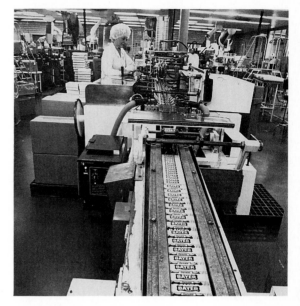

Ibuprofen

It has the advantage of activity similar to that of aspirin with fewer side effects, and it is now widely used in tablets such as Motrin and Advil.

Aspirin Factory (courtesy of The Bayer Company, Glenbrook Laboratories, Division of Sterling Drug, Inc.)

QUESTIONS

1. Show how to prepare each of the esters shown in Table 37-1 using a Fischer esterification.
2. Why is it necessary to add an acid catalyst such as H_2SO_4 in a Fischer esterification when there is already a carboxylic acid present?
3. Could this technique be adapted for the synthesis of amides? Explain your reasoning.
4. Give examples of a carboxylic acid and of an alcohol for which Fischer esterification would not work well.
5. The IR and NMR spectra of isoamyl acetate are shown in Figures 37-1 and 37-2. Assign as many IR peaks as possible and all of the NMR peaks.

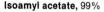
Isoamyl acetate, 99%

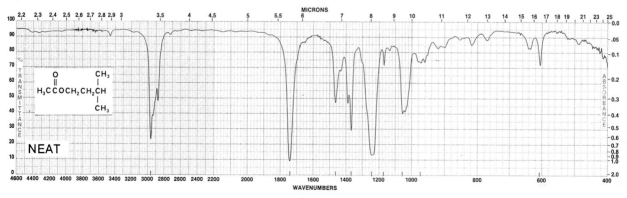

Figure 37–1 IR Spectrum of Isoamyl Acetate

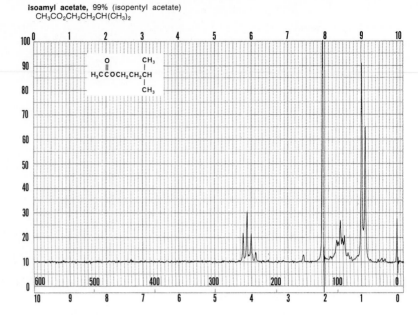

Isoamyl acetate, 99% (isopentyl acetate)
CH₃CO₂CH₂CH₂CH(CH₃)₂

Figure 37–2 NMR Spectrum of Isoamyl Acetate

REFERENCE

BOYER, P. D., ed., *The Enzymes*, 3rd ed., Vol. XVI: *Lipid Enzymology,* Academic Press, New York, 1970, p. 543.

EXPERIMENT 37.1 SYNTHESIS OF BANANA FRAGRANCE (ISOAMYL ACETATE)

Estimated Time:
3.0 hours

Note: There are three steps to this procedure; a 1-hour reflux, two washings, and a simple distillation. To complete this procedure comfortably in 3 hours, it is important to be well organized and plan ahead.

Prelab

1. When washing isoamyl acetate with water or with 5% aqueous $NaHCO_3$, which layer will be the upper one?
2. What reaction occurs between $NaHCO_3$ and an acid?
3. Calculate the volumes of 50 mmol of isoamyl alcohol and of 100 mmol of acetic acid.

Special Hazards

Concentrated sulfuric and acetic acids are highly corrosive; avoid contact and wash thoroughly with water if contact occurs. Isoamyl alcohol and isoamyl acetate are flammable and irritating.

Procedure

In a 50-mL round-bottomed flask in the fume hood, place isoamyl alcohol (50 mmol, _____ g, _____ mL) and acetic acid (100 mmol, _____ g, _____ mL). Cautiously add 20 drops of concentrated sulfuric acid. Add boiling chips, a condenser, heating mantle, and variable transformer, establish a slow constant water flow through the condenser, and bring the mixture to reflux for 1 hour. At the end of the reflux period, cool the flask in an ice bath and pour the contents into a small separatory funnel. Wash the reaction mixture with one 20-mL portion of water, then with one 20-mL portion of 5% aqueous sodium bicarbonate. After adding the bicarbonate, stir the layers together until gas evolution has ceased before stoppering the separatory funnel; the bicarbonate is neutralizing any residual acetic or sulfuric acid present. Vent often while shaking to avoid pressure buildup. As in all extractions, pay attention to which layer is which and save all washings in a large beaker until the product is in hand.

Dry the crude ester over anhydrous $MgSO_4$, Na_2SO_4, or $CaCl_2$ for 5 minutes, then decant it into a clean, dry, 50-mL round-bottomed flask. Set up for simple distillation and collect the fraction boiling from 137° to 142°. Weigh the product and report the percent yield. Describe the odor of the product.

REFERENCE

MINER LABORATORIES, *Organic Syntheses Collective Volume 1* (1941) 285.

EXPERIMENT 37.2 ETHYL ACETATE FROM ETHANOL AND ACETIC ACID

Estimated Time:
2.5 hours

Prelab

1. Calculate the grams and milliliters corresponding to 100 mmol of ethanol and 200 mmol of acetic acid, and the theoretical yield of ethyl acetate.

Special Hazards

Acetic and sulfuric acids are corrosive. Both ethanol and ethyl acetate are flammable. Ethyl acetate is an irritant, while ethanol is toxic.

Procedure

In a 50-mL round-bottomed flask, place anhydrous ethanol (100 mmol, _____ g, _____ mL), acetic acid (200 mmol, _____ g, _____ mL) and concentrated sulfuric acid (20 drops). Reflux the mixture for 1 hour, cool slightly, and rearrange for simple distillation. Collect the fraction boiling at approximately 75–85°, place the crude ester in a separatory funnel, and wash with 5% NaHCO₃ to remove any acetic acid. Bubbles will form; swirl and be sure that the effervescence has subsided before stoppering the separatory funnel, and vent frequently when shaking. Wash with saturated aqueous NaCl, dry over Na₂SO₄ or CaCl₂, decant, and redistill fractionally to obtain pure ethyl acetate.

REFERENCE

Vogel, A. I., "Physical Properties and Chemical Constitution. Part XIII. Aliphatic Carboxylic Esters," *Journal of the Chemical Society* (1948), 624.

EXPERIMENT 37.3 SYNTHESIS OF ASPIRIN BY ACETYLATION OF SALICYLIC ACID

Estimated Time:
1.5 hours

Prelab

1. Calculate the millimoles corresponding to 1.00 g of salicylic acid and the grams and millimoles corresponding to 3.0 mL of acetic anhydride.
2. What is the theoretical yield of acetylsalicylic acid?

Special Hazards

Acetic anhydride is corrosive and lachrymatory. Concentrated sulfuric acid is highly corrosive. Salicylic acid is toxic. Avoid skin contact or ingestion of these reagents.

Procedure

In a 50-mL Erlenmeyer flask in the fume hood, place salicylic acid (1.00 g, _____ mmol). Cautiously add acetic anhydride (3.0 mL, _____ g, _____ mmol) and 3 drops of concentrated sulfuric acid. Swirl gently and warm on a steam bath or hot-water bath for 5–10 minutes with occasional swirling. All of the salicylic acid should dissolve. Allow the flask to cool to room temperature; the acetylsalicylic acid should precipitate. If not, scratch the inner walls of the flask and/or add seed crystals of acetylsalicylic acid. When the precipitation is complete, add ice-cold water (20 mL) and stir vigorously to break up and wash the mass of crystals. Collect the aspirin by suction filtration and wash with a little cold water.

To purify the product, put it in a test tube and recrystallize from 1–2 mL of ethyl acetate. Collect by suction filtration, wash with a few drops of ice-cold ethyl acetate, air dry, and record the mass and melting range.

REFERENCE

SMITH, M. J. H. and P. K. SMITH, *The Salicylates*, Interscience, New York, 1966.

EXPERIMENT 37.4 ACETYLATION OF FURFURYL ALCOHOL

Estimated Time:
2.0 hours

Prelab

1. Calculate the volumes of 100 mmol of furfuryl alcohol and of 110 mmol of acetic anhydride, and the theoretical yield of furfuryl acetate.

Special Hazards

Furfuryl alcohol is an irritant and toxic on ingestion. Acetic anhydride is corrosive and a lachrymator.

Procedure

In a 50-mL round-bottomed flask, place furfuryl alcohol (100 mmol, _____ g, _____ mL) and acetic anhydride (110 mmol, _____ g, _____ mL). Add boiling chips, a reflux condenser, and a heating mantle, and reflux the mixture for 1 hour. After cooling in an ice bath, pour the dark reaction mixture into a separatory funnel. Add ether (30 mL) and wash the organic phase with 30 mL of 5% aqueous Na_2CO_3, then 30 mL of water, followed by 30 mL of brine (saturated aqueous NaCl). Dry the organic phase over $CaCl_2$, decant, and remove the ether on a steam bath in the hood. Simple distillation of the residue affords furfuryl acetate.

REFERENCE

ZANETTI, J. E., "Esters of Furfuryl Alcohol," *Journal of the American Chemical Society* (1925) *47*, 535.

EXPERIMENT 37.5 MICROSCALE ACETYLATION OF BENZOIN WITH ACETIC ANHYDRIDE

Estimated Time:
1.0 hour

Prelab

1. Fill out a reagent table in your lab notebook.

Procedure

Exercise standard precautions with all chemicals. In a 15 x 125 mm test tube in the fume hood, place benzoin (_____ mmol, 100 mg), acetic acid (0.10 mL, _____ g,

_____ mmol, about 10 drops), acetic anhydride (0.10 mL, _____ g, _____ mmol, about 6 drops) and concentrated H_2SO_4 (1 drop). The solution turns yellow on addition of the sulfuric acid. Heat the mixture in a steam bath or boiling water bath for 15 minutes. Add ice-cold water (10 mL) and cool in an ice bath. Stir occasionally and cool for a few minutes until precipitation is complete. If an oil forms, it will solidify on recrystallization. Decant the supernatant liquid, and recrystallize the crude product from approximately 0.5–1 mL of methanol, cooling in an ice bath. Record the mass and melting point.

REFERENCE

CORSON, B. B. and N. A. SALIANI, *Organic Syntheses Collective Volume 2* (1943) 69.

EXPERIMENT 37.6 PREPARATION OF *t*-BUTYL ACETATE FROM *t*-BUTYL ALCOHOL AND ACETYL CHLORIDE

**Estimated Time:
3.0 hours**

Note: This procedure involves a reflux, an extraction, and a distillation. To complete it comfortably in 3 hours requires good organization and efficient work.

Special Hazards

Acetyl chloride is corrosive and produces HCl gas on contact with water. *N,N*-Dimethylaniline is highly toxic and an irritant. All organic substances used in this experiment are flammable.

Procedure

Calculate the masses and volumes of 100 mmol each of *t*-butyl alcohol, N,N-dimethylaniline, and acetyl chloride. In a 100-mL round-bottomed flask, place *t*-butyl alcohol (100 mmol, _____ g, _____ mL), *N,N*-dimethylaniline (100 mmol, _____ g, _____ mL), and 20 mL of dry ether. Add a Claisen head, reflux condenser, and addition funnel. In the addition funnel, place acetyl chloride (100 mmol, _____ g, _____ mL). Cool in an ice bath, then add the acetyl chloride from the funnel into the flask over the course of 1–2 minutes. After the addition the mixture is heated gently to reflux for 1 hour using either a steam bath or a heating mantle with a Variac on a low setting. The flask is then cooled in an ice bath and the contents are suction filtered to remove the solid *N,N*-dimethylaniline hydrochloride. The ether solution is washed with 5% HCl (20 mL), water (20 mL), and brine (20 mL), then dried over Na_2SO_4, decanted, and distilled, to give the sweet-smelling *t*-butyl acetate (lit. bp 98°).

REFERENCE

ABRAMOVITCH, B., J. C. SHIVERS, B. E. HUDSON, and C. R. HAUSER, "A General Method for the Synthesis of *t*-Butyl Esters," *Journal of the American Chemical Society* (1943) *65*, 986.

SECTION 38

Amide Synthesis

Overview

Amide and ester formation are two of the most common synthetic reactions. They are also among the easiest, requiring only an amine or alcohol as the nucleophile, and an acid chloride or anhydride as the electrophile. These four possibilities are shown below in words and in structures.

1. An amine and an acid chloride form an amide.

$$RNH_2 + R'\overset{O}{\overset{\|}{C}}-Cl \longrightarrow RNH\overset{O}{\overset{\|}{C}}R' + HCl$$

2. An amine and an acid anhydride form an amide.

$$RNH_2 + R'-\overset{O}{\overset{\|}{C}}-O-\overset{O}{\overset{\|}{C}}-R' \longrightarrow RNH-\overset{O}{\overset{\|}{C}}-R' + R'COOH$$

3. An alcohol and an acid chloride form an ester.

$$ROH + R'\overset{O}{\overset{\|}{C}}Cl \longrightarrow RO\overset{O}{\overset{\|}{C}}R' + HCl$$

4. An alcohol and an acid anhydride form an ester.

$$ROH + R'-\overset{O}{\overset{\|}{C}}-O-\overset{O}{\overset{\|}{C}}-R' \longrightarrow RO\overset{O}{\overset{\|}{C}}-R' + R'COOH$$

The mechanism is that the unshared pair on the nucleophile attacks the positive carbonyl carbon leading to displacement of a leaving group ($-Cl$ or $-O\overset{O}{\overset{\|}{C}}R$). In Experiment 38.1 you will use aniline and acetic anhydride to form an amide.

Mechanism:

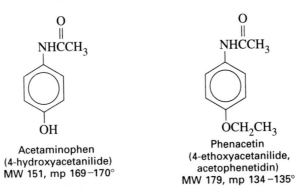

Overall reaction:

Aniline
MW 93, bp 194°
dens. 1.02 g/mL

Acetic
anhydride
MW 102, bp 138–140°
dens. 1.08 g/mL

Acetanilide
MW 135, mp 114°

Acetanilide is an effective analgesic (pain reliever) and antipyretic (fever reducer) for both human and veterinary use, although in recent years it has been largely replaced by acetaminophen (found in Tylenol) and phenacetin.

Acetaminophen
(4-hydroxyacetanilide)
MW 151, mp 169–170°

Phenacetin
(4-ethoxyacetanilide,
acetophenetidin)
MW 179, mp 134–135°

On prolonged or excessive use, acetanilide can cause a serious blood disorder called methemoglobinemia. In this disorder, the central iron atom in hemoglobin is oxidized from Fe(II) to Fe(III) to give methemoglobin. Methemoglobin cannot carry oxygen in the bloodstream, resulting in a type of anemia. Phenacetin and acetaminophen cause the same disorder, but to a much lesser extent. Since they are also somewhat more effective as analgesic and antipyretic drugs than acetanilide, their use is preferable.

Many over-the-counter preparations, as shown in Table 38.1, use a combination of one or more analgesic drugs and caffeine, which serves to "perk up" the user. Many drug companies sell a combination pain reliever called an "APC tablet." Empirin is an example of this type of tablet, which usually contains aspirin, phenacetin, and caffeine (represented by the acronym APC).

TABLE 38–1 GRAMS OF MAJOR ANALGESICS AND CAFFEINE IN SOME COMMON OVER-THE-COUNTER PREPARATIONS

	Aspirin	Phenacetin	Caffeine	Acetaminophen
Anacin	0.4	—	0.03	—
Aspirin[a]	0.3	—	—	—
Bufferin	0.3	—	—	—
Cope	0.4	—	0.03	—
Empirin	0.2	0.1	0.03	—
Excedrin[b]	0.2	—	0.06	0.1
Tylenol	—	—	—	0.3

[a]In a 5-grain (325 mg) tablet.
[b]Also contains 0.1 g of salicylamide, the amide of salicyclic acid, which is also an analgesic.

Experiment 38.2 is an acetylation of the essential amino acid glycine with acetic anhydride.

$$NH_2-CH_2-COOH + CH_3\overset{O}{\overset{||}{C}}O\overset{O}{\overset{||}{C}}CH_3 \longrightarrow CH_3\overset{O}{\overset{||}{C}}-NHCH_2COOH + CH_3COOH$$

Glycine
MW 75, mp 245°d

Acetic anhydride

Acetylglycine
(acetamidoacetic acid,
aceturic acid)
MW 117, mp 207–209°

Acetic
acid

In Experiment 38.3 an acid chloride is prepared from *m*-toluic acid using thionyl chloride, then allowed to react with an amine (diethylamine) to form the amide. This amide is the insect repellent N,N-diethyl-*m*-toluamide found in products such as OFF!

m-Toluic acid
MW 136
mp 108–110°

Thionyl chloride
MW 119
bp 79°
dens. 1.63 g/mL

m-Toluoyl chloride
MW 155
bp 86° at 5 torr
dens. 1.17 g/mL

Diethylamine
hydrochloride
MW 110
mp 227–230°

N,N-Diethyl-*m*-toluamide
MW 191
bp 160° at 19 torr
dens. 1.00 g/mL

A similar reaction to amide formation between an acid chloride and an amine is sulfonamide formation between a sulfonyl chloride and an amine. This reaction, often used for making solid derivatives of amines, is illustrated in Experiment 38.4.

p-Toluenesulfonyl chloride
(tosyl chloride)
MW 191, mp 67–69°

Methylamine
MW 31, bp 48°
dens. of 40%
aq. sol. 0.90 g/mL

N-Methyl-*p*-toluenesulfonamide
MW 185, mp 77–78°

This type of reaction forms the basis of the Hinsberg test for amines, described in more detail in the section on qualitative analysis.

APPLICATION: MOSQUITO REPELLENTS

Using a 1′ × 1′ × 2′ cage filled with thousands of hungry, blood-feeding mosquitoes, repellents are tested for efficiency. The ''Arm-in-the-Cage'' test shown above was pioneered by the Johnson Wax Entomology Research Center in Racine, Wisconsin. It has become an industry standard for repellent efficacy testing accepted by the EPA and used for more than 30 years. A scientist's arm is treated with a repellent formulation and inserted into the cage. The purpose of the ''Arm-in-Cage'' test is to see how long the repellent remains effective. The length of repellency depends on the concentration of DEET (N,N-diethyl-*meta*-toluamide) in the product and thoroughness of coverage. Top photo is an untreated arm vs. a treated arm in the bottom photo. (Photo courtesy of the OFF! Biting Insect Research Team)

As far as is currently known, mosquitoes find their victims primarily by sensing the carbon dioxide breathed out and the convection currents of warm, humid air produced by warm-blooded animals. Repellents seem to act by jamming the insect's moisture sensors.

A desirable mosquito repellent must possess several qualities. It must be effective for several hours, nontoxic and nonirritating to skin, free of unpleasant odor, and harmless to clothing. Many thousands of compounds have been tested and several have been found to meet these criteria.

Repellents are most effective when used in combination, since the effectiveness of each repellent compound varies according to the species of mosquito encountered, weather conditions, and the person wearing it. The following two mixtures are particularly effective: dimethyl phthalate, 2-ethyl-1,3-hexanediol, and either indalone or dimethyl carbate.

Dimethyl phthalate

$$\text{CH}_2\text{CH}_3$$
$$\text{HOCH}_2\text{CHCHCH}_2\text{CH}_2\text{CH}_3$$
$$\text{OH}$$

2-Ethyl-1, 3-hexanediol

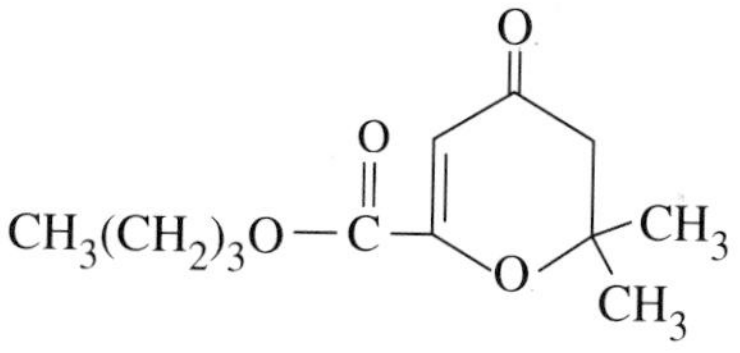

Indalone
(butopyranoxyl)

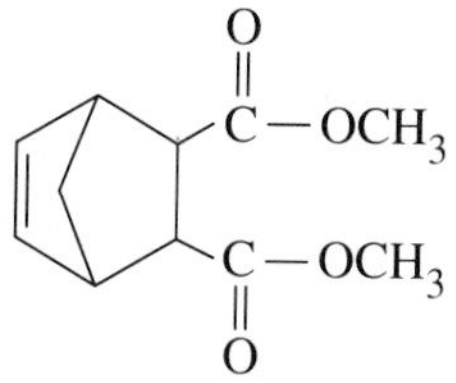

Dimethyl carbate
(dimethyl bicyclo[2.2.1]-
5-heptene-2,3-dicarboxylate)

Dimethyl phthalate may also be mixed into lotions or creams.

Impregnation of clothing is especially effective, and the U.S. Armed Services uses the mixture M-1960 as the standard clothing impregnant against mosquitoes, fleas, ticks, and chiggers.

MIXTURE M-1960

Compound	Parts by weight
Benzyl benzoate	3
N-Butylacetanilide	3
2-Ethyl-2-butyl-1,3-propanediol	3
Tween 80	1

The Tween 80 functions as an emulsifier and solubilizer.

QUESTIONS

1. Show how to make each of the following compounds using the appropriate acid chloride.

(a) 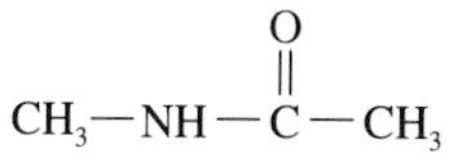 (b) 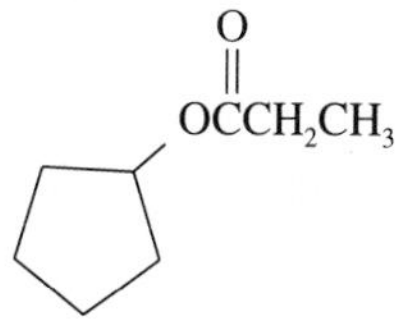 (c)

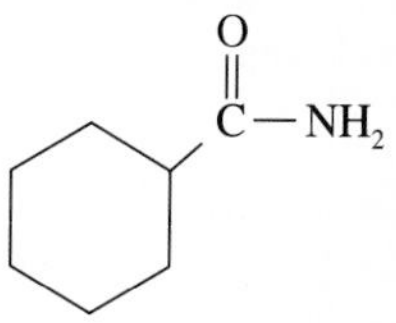

(d) 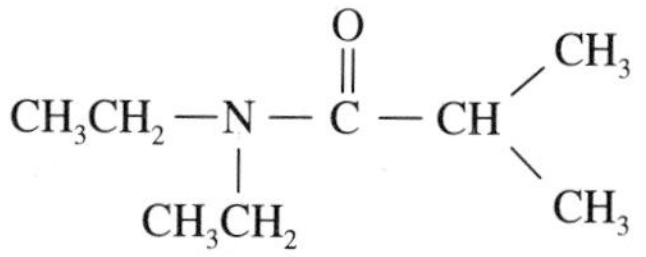 (e)

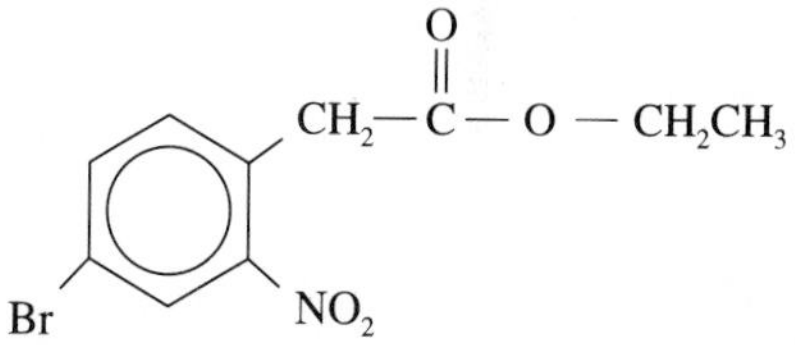

2. Show how to make each of the compounds in Question 1 using an acid anhydride.

3. Draw the predicted IR and NMR spectra for acetanilide, indicating which structural feature gives rise to each peak.

4. Propose a synthesis of the artificial sweetener saccharin (structure below) from benzene.

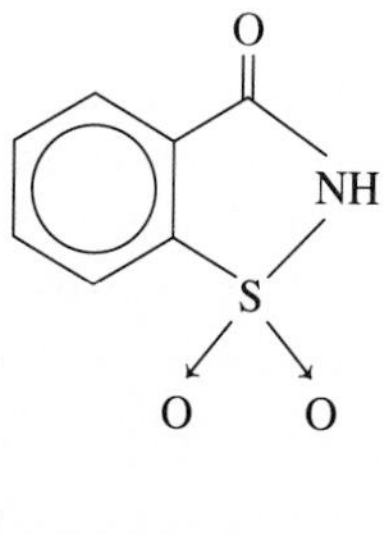

5. Acid chlorides are versatile intermediates since the carbonyl carbon has a large partial positive charge and is very susceptible to nucleophilic attack with displacement of chloride. Many other products could have been made from the *m*-toluoyl chloride produced in the first step of this procedure. What would be the products of the following reactions?

$$NH_3$$

$$CH_3$$

$$\underset{\underset{N}{\overset{N}{||}}}{}\;\;\overset{H}{\underset{|}{N}}-CH_3$$

$$O \\ || \\ CCl$$

$$CH_3CH_2OH$$

6. Draw a concept map for this experiment.

EXPERIMENT 38.1 ACETANILIDE FROM ANILINE AND ACETIC ANHYDRIDE

Estimated Time:
1.0 hour

Prelab

1. Calculate the volume of 10 mmol of aniline.
2. Calculate the theoretical yield of acetanilide.
3. If you obtained an 85% crude yield of acetanilide, about how much water would be needed to recrystallize it? (Hint: see procedure)

Special Hazards

Aniline is highly toxic on ingestion and is a suspected carcinogen on long-term exposure to high concentrations. Avoid breathing its vapors or contact with skin. Use of gloves is recommended; if skin contact occurs, wash thoroughly with soap. When mixing acid and water, always add acid to water, not vice versa.

Procedure

Dissolve aniline (_____ mL, 10 mmol) in a mixture of water (25 mL) and conc. HCl (0.8 mL, 10 mmol). Measure out acetic anhydride (1.2 mL, 12 mmol) and prepare a solution of sodium acetate (1.0 g of anhydrous NaOAc, 12 mmol) in water (5 mL). While stirring, add the acetic anhydride to the solution of aniline hydrochloride and immediately add the sodium acetate solution. The acetate ion serves to deprotonate the anilinium ion so that the aniline can act as a nucleophile and attack the acetic anhydride.

Cool the mixture in an ice bath and collect the product. Recrystallize from water. It is reported in the Merck Index that acetanilide has a solubility of 1 g in 20 mL of boiling water and 1 g in 185 mL of cold water. Collect and air dry the product and record its mass and melting point.

EXPERIMENT 38.2 ACETYLATION OF GLYCINE WITH ACETIC ANHYDRIDE

Estimated Time:
1.0 hour

Prelab

1. Calculate the millimoles corresponding to 1.0 g of glycine. What are the milli-moles and volume corresponding to 2.5 g of acetic anhydride?
2. What is the theoretical yield of acetylglycine?

Special Hazards

Acetic anhydride is corrosive and lachrymatory.

Procedure

In a 25-mL Erlenmeyer flask, place glycine (1.0 g, _____ mmol) and 4 mL of water. Swirl the flask and warm it just enough to dissolve the glycine. Add acetic anhydride (2.5 g, _____ mmol, _____ mL), swirl to mix, and keep swirling occasionally for 10 minutes. Cool the flask in an ice bath for at least 15 minutes to allow the acetylglycine to precipitate. Collect the product by suction filtration, wash with a few milliliters of water, air dry, and record the mass and melting point.

REFERENCE

HERBERT, R. M. and A. SHEMIN, *Organic Syntheses Collective Volume 2* (1943) 11.

EXPERIMENT 38.3 SYNTHESIS OF THE INSECT REPELLENT N,N-DIETHYL-*m*-TOLUAMIDE

Estimated Time:
4.0 hours for parts a and b

Prelab

1. What are the masses of 25 mmol of *m*-toluic acid and of 20 mmol of diethylamine hydrochloride? What is the volume of 30 mmol of thionyl chloride?
2. What reaction will occur if any water (such as wet glassware) comes in contact with $SOCl_2$? With *m*-toluoyl chloride?

3. Remember that an aspirator in top condition will provide a pressure of about 20 torr, but if it is old or partly clogged, it will not be as effective. What is the expected boiling point of DEET during distillation if your aspirator is giving you a pressure of 40 torr? 80 torr?

Special Hazards

Thionyl chloride reacts violently with water, forming strong acids. Use gloves and handle only in the fume hood.

Procedure 38.3a: Preparation of *m*-Toluoyl Chloride and *N,N*-Diethyl-*m*-Toluamide

Estimated Time:
2.0 hours

In a dry 50-mL round-bottomed flask in the fume hood, place *m*-toluic acid (25 mmol, _____ g) and thionyl chloride (30 mmol, _____ mL; *use caution*!). Add a condenser and heating mantle, establish water flow through the condenser, and reflux the mixture gently for 20–30 minutes. Cool the flask in an ice bath, stopper it, and leave it in the fume hood. This flask now contains *m*-toluoyl chloride.

In a 250-mL Erlenmeyer flask, place 30 mL of 3.0 *M* aqueous NaOH. Weigh out diethylamine hydrochloride, (20 mmol, _____ g) and sodium lauryl sulfate (1.0 g). Add the diethylamine hydrochloride in portions to the Erlenmeyer flask with swirling. When this addition is complete, add the sodium lauryl sulfate. This Erlenmeyer now contains free diethylamine, since the NaOH has neutralized the HCl.

Transfer the previously prepared *m*-toluoyl chloride from the round-bottomed flask to a dry 125-mL separatory funnel supported on an iron ring above the flask of diethylamine. Open the stopcock slightly to establish a moderate rate of addition (about 1 drop every 1 or 2 seconds) and continue to swirl the Erlenmeyer flask. After the addition is complete, swirl occasionally for 5 minutes more, then stopper the flask securely, wrap the stopper with Parafilm, and place it in an upright position in your drawer for storage.

Procedure 38.3b: Extraction and Distillation of DEET

Estimated Time:
2.0 hours

Place the mixture in a 250-mL separatory funnel and extract with two 25-mL portions of diethyl ether. Wash the combined ether extracts with saturated aqueous NaCl solution (20 mL) to remove most of the water, then dry the ether over anhydrous $MgSO_4$ or Na_2SO_4. After drying, decant the ether into a 10-mL beaker, add some boiling chips, and remove the ether on a steam bath in the hood. Record the mass of the crude DEET obtained.

Set up an apparatus for vacuum distillation using an aspirator (see Section 2), establish vacuum and water flow through the condenser, and distill the DEET (lit. bp 160° at 19 torr). Record the boiling range and yield of your product and take an IR spectrum and/or refractive index to verify its identity.

REFERENCE

WANG, B. J-S., "An Interesting and Successful Organic Experiment," *Journal of Chemical Education* (1974) *51*, 631 (synthesis of *N,N*-diethyl-*m*-toluamide).

QUESTION

1. If you took an IR spectrum, find a literature spectrum for DEET and compare it with yours. Assign as many peaks as you can and state whether any impurities appear to be present.

EXPERIMENT 38.4 MICROSCALE SULFONAMIDE FORMATION BY REACTION OF *p*-TOLUENESULFONYL CHLORIDE AND METHYLAMINE

Estimated Time:
1.0 hour

Prelab

1. Calculate the millimoles in 100 mg of *p*-toluenesulfonyl chloride and in 0.50 mL of 40% (by weight) aqueous methylamine (density 0.90 g/mL).
2. Calculate the theoretical yield of N-methyl-*p*-toluenesulfonamide.

Special Hazards

Methylamine is a flammable and corrosive compound that also has a stench. Use caution and handle only in the fume hood. *p*-Toluenesulfonyl chloride is corrosive and moisture sensitive.

Procedure

In a 15 x 125 mm test tube in the fume hood, place *p*-toluenesulfonyl chloride (100 mg, _____ mmol), 0.5 mL of ethanol, and 40% aqueous methylamine (0.50 mL, _____ mmol, by pipette or burette). Add a boiling stone and heat the tube to near boiling in a steam bath in the hood for 10 minutes. Cool the tube in an ice bath to precipitate the product, add water (1 mL), then suction filter. Recrystallize the crude N-methyl-*p*-toluenesulfonamide from 3 mL of 10:1 water-ethanol, filter, air dry, and record the mass and melting range.

REFERENCE

DE BOER, T. J., and H. J. BACKER, *Organic Syntheses Collective Volume 4* (1963) 943.

SECTION 39

Aldol Condensations

Overview

The term "aldol condensation" applies to a variety of well-known reactions involving aldehydes, ketones, and esters. The general scheme is that the (negative) enolate of one attacks the slightly positive carbon of the carbonyl group on another, giving a β-hydroxy carbonyl compound. If the enolate came from an aldehyde this product is an *ald*ehyde-alcoh*ol*, or "aldol," product.

Enolate

α, β-Unsaturated carbonyl compound

Enolate

Aldol product

This general type of condensation, with attack of an anion on a carbonyl compound, resulting in formation of a double bond with the overall elimination of water, can occur between several types of compounds. Nitro and cyano compounds, for example, also possess acidic α-hydrogens that can be removed in base, leading to condensation with aldehydes.

$$RCH_2NO_2 + R'CHO \xrightarrow{OH^-} \overset{R}{\underset{O_2N}{>}}C=CHR'$$

$$RCH_2CN + R'CHO \xrightarrow{OH^-} \overset{R}{\underset{NC}{>}}C=CHR'$$

Other types of carbonyl compounds, such as esters, carboxylic acids, and acid anhydrides, may also act as nucleophiles once their α-hydrogens have been removed to form enolates. The names of these reactions are shown along with examples of each type of reaction.

Claisen:

$$CH_3COOCH_2CH_3 + CH_3COOCH_2CH_3 \xrightarrow{OH^-} CH_3COCH_2COOCH_2CH_3$$

Perkin:

$$ArCHO + (RCH_2\overset{O}{\overset{\|}{C}}O-)_2O \xrightarrow{RCOO^-Na^+} ArCH=C\overset{COOH}{\underset{R}{<}}$$

Knoevenagel:

$$RCHO + CH_2(CO_2Et)_2 \longrightarrow RCH=C(CO_2Et)_2$$

Doebner:

$$RCH_2CHO + \overset{COOH}{\underset{COOH}{\overset{|}{\underset{|}{CH_2}}}} \xrightarrow{base} RCH_2-CH=C\overset{COOH}{\underset{COOH}{<}}$$

$$\xrightarrow{-CO_2} RCH_2-CH=CH-COOH$$

Dieckmann:

$$EtO_2C-(CH_2)_3-CH_2CO_2Et \longrightarrow \text{(cyclopentanone ring)}-CO_2Et$$

In the Perkin reaction the anhydride undergoes hydrolysis after the condensation, and in the Doebner reaction one of the carboxylic acid groups is lost in a decarboxylation, which is not surprising since there is a carbonyl group beta to it.

In Experiment 40.1 we will use benzaldehyde and acetone in an aldol condensation, so the mechanism is:

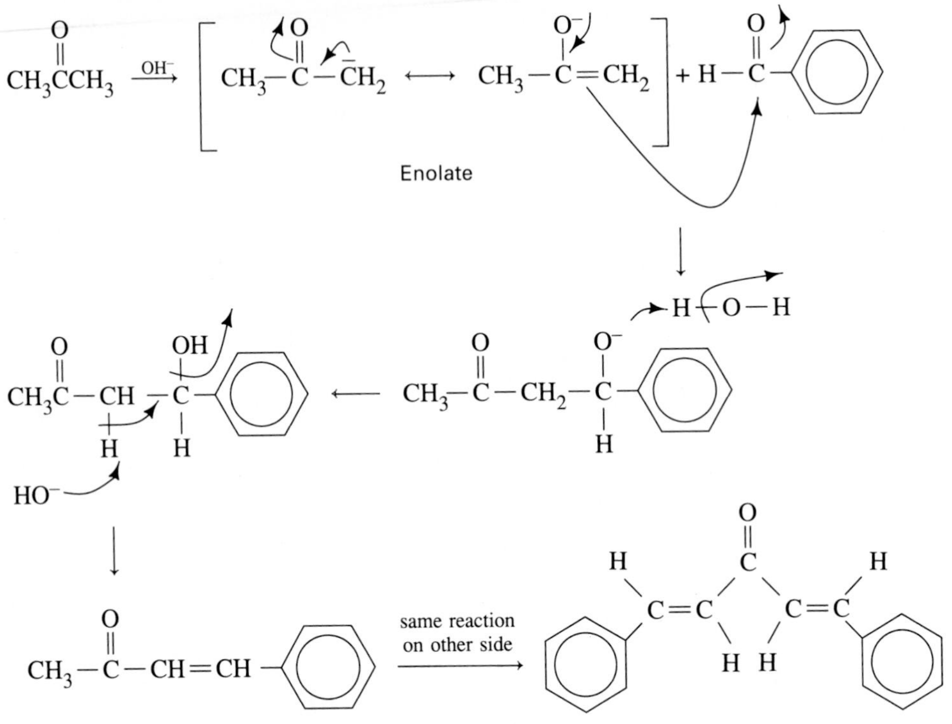

Enolate

Benzalacetone

Dibenzalacetone

Note the trans, trans configuration in the product, which is formed because the *trans*-alkene is more stable thermodynamically than the *cis*-alkene, and all the steps are reversible.

This reaction will be done on a micro scale, using approximately 100 mg total of starting materials.

Overall reaction:

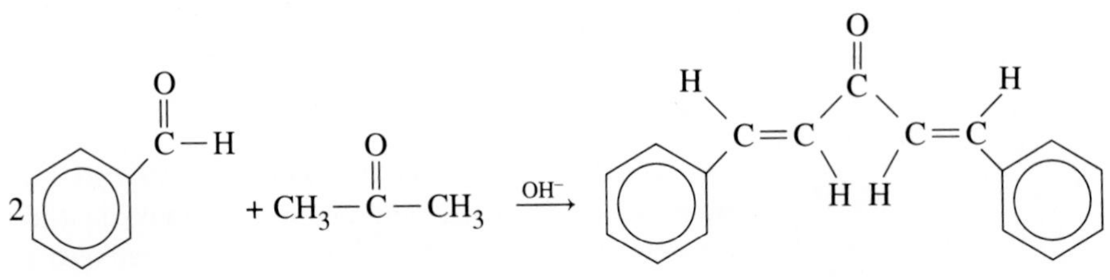

Benzaldehyde
MW 106, bp 178°
dens. 1.04 g/mL

Acetone
MW 58, bp 56°
dens. 0.79 g/mL

Dibenzalacetone
(1, 5-diphenyl-1, 4-pentadien-3-one)
MW 234, mp 110 − 111°

Experiment 40.2 is an example of a condensation of an aromatic aldehyde with a nitro compound.

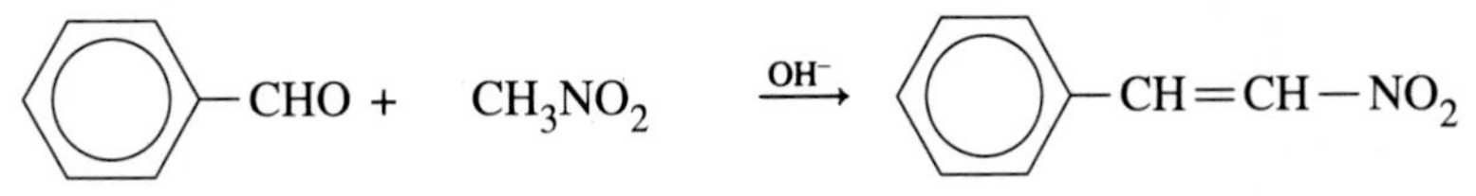

Benzaldehyde + Nitromethane → β-Nitrostyrene

Benzaldehyde
MW 106, bp 178°
dens. 1.04 g/mL

Nitromethane
MW 61, bp 101°
dens. 1.13 g/mL

β-Nitrostyrene
MW 149, mp 56–58°

In Experiment 40.3, benzaldehyde condenses with a nitrile.

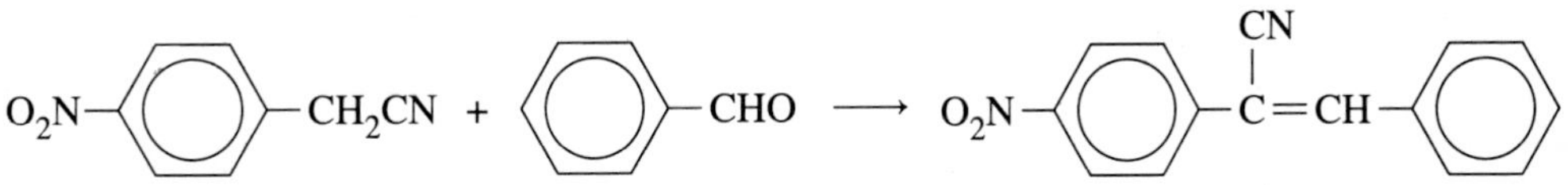

(4-Nitrophenyl)acetonitrile
(*p*-nitrobenzyl cyanide)
MW 162, mp 115 – 116°

Benzaldehyde

α-(4-Nitrophenyl)-
cinnamonitrile
MW 250, mp 179–180°

Experiment 40.4 is an example of a Claisen-type condensation between benzaldehyde and ethyl acetate.

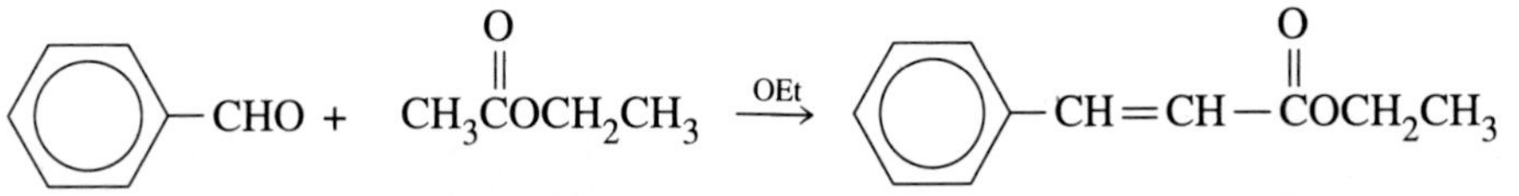

Benzaldehyde
MW 106, bp 178°
dens. 1.04 g/mL

Ethyl acetate
MW 88.1, bp 77°
dens. 0.90 g/mL

Ethyl cinnamate
mw 176, mp 6.5–7.5 °
bp 271°, dens. 1.05 g/mL

Many biological examples of aldol and Claisen reactions are known. Collagen, the most abundant protein in many vertebrates and invertebrates, provides mechanical strength to bone, tendon, cartilage, and skin. A great deal of its strength arises from extensive cross-linking of lysine residues. The first step of this cross-linking is the conversion of the terminal aminomethyl groups on two lysine residues to aldehydes, a reaction catalyzed by the enzyme *lysyl oxidase*. The two aldehydes then undergo aldol condensation to give a strong covalent cross-link. The conjugated aldehyde thus formed may then undergo Michael addition of a histidine side chain and/or condensation of the aldehyde with an amine to form an imine (Schiff's base) and provide linkage of up to four side chains.

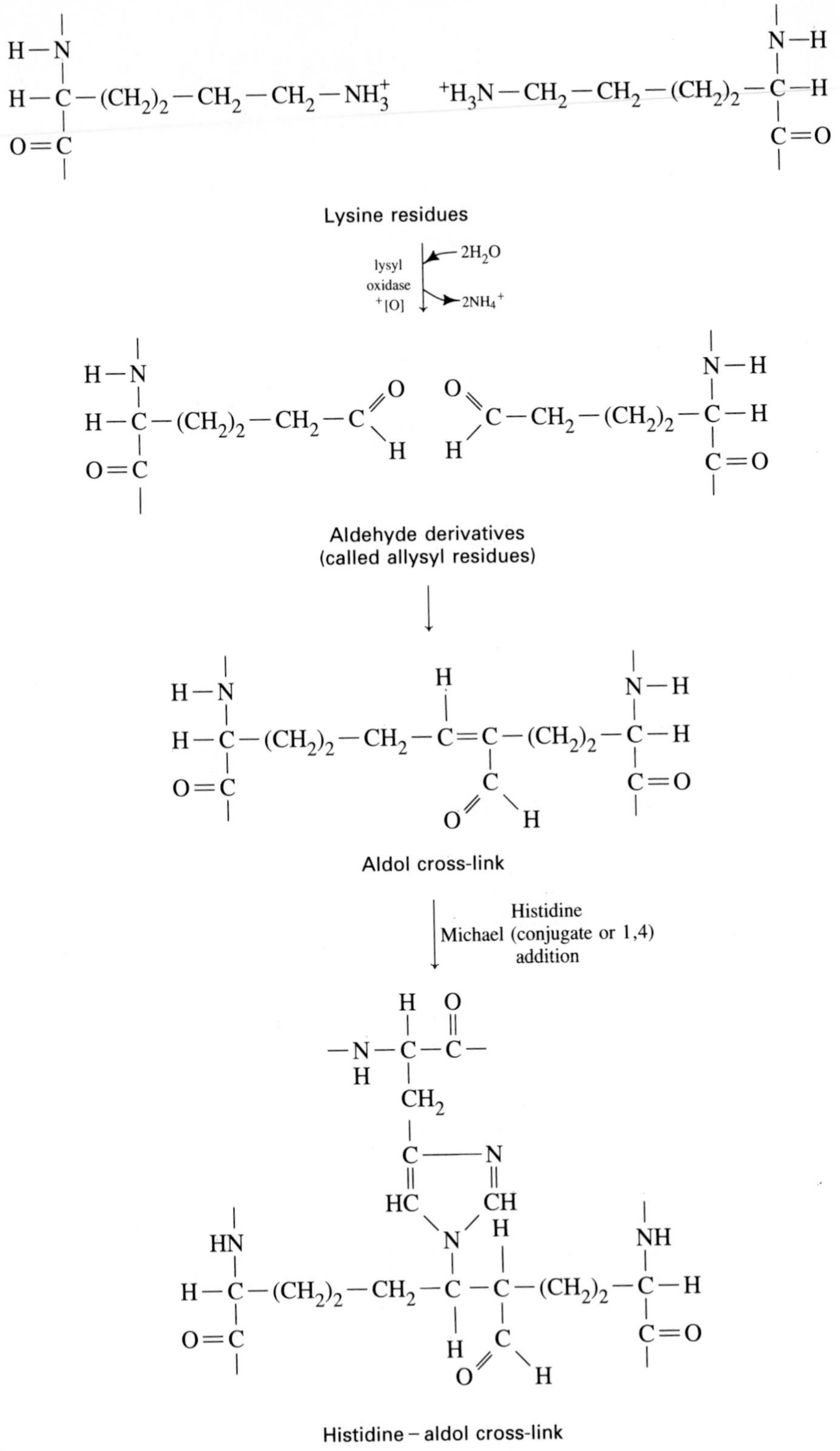

$$H-N$$
$$H-C-(CH_2)_2-CH_2-CH_2-NH_3^+ \quad {}^+H_3N-CH_2-CH_2-(CH_2)_2-C-H$$
$$O=C \qquad\qquad\qquad\qquad\qquad\qquad\qquad\qquad\qquad C=O$$
$$N-H$$

Lysine residues

lysyl oxidase $+[O]$ $\longrightarrow$ $-2H_2O$
$\longrightarrow$ $2NH_4^+$

$$H-N$$
$$H-C-(CH_2)_2-CH_2-C\!\!\overset{\displaystyle O}{\underset{\displaystyle H}{<}} \quad \overset{\displaystyle O}{\underset{\displaystyle H}{>}}\!\!C-CH_2-(CH_2)_2-C-H$$
$$O=C \qquad\qquad\qquad\qquad\qquad\qquad\qquad\qquad\qquad C=O$$
$$N-H$$

**Aldehyde derivatives
(called allysyl residues)**

$$H-N \qquad\qquad\qquad\qquad H$$
$$H-C-(CH_2)_2-CH_2-C=C-(CH_2)_2-C-H$$
$$O=C \qquad\qquad\qquad\qquad C \qquad\qquad\qquad C=O$$
$$O \quad H$$
$$N-H$$

Aldol cross-link

Histidine
Michael (conjugate or 1,4)
addition

$$-N-C-C-$$
$$H \quad CH_2$$
$$C-N$$
$$HC \quad CH$$
$$N \quad H$$

$$HN \qquad\qquad\qquad\qquad\qquad\qquad NH$$
$$H-C-(CH_2)_2-CH_2-C-C-(CH_2)_2-C-H$$
$$O=C \qquad\qquad\qquad H \quad C \qquad\qquad C=O$$
$$O \quad H$$

Histidine – aldol cross-link

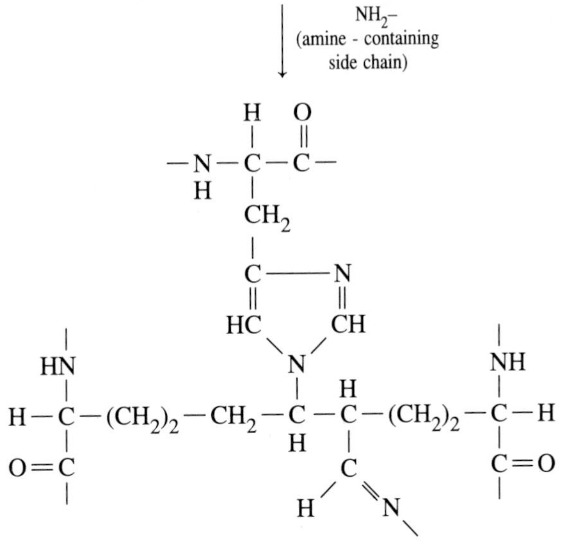

$$NH_2^-$$
(amine - containing side chain)

$$-N-C-C-$$

Four residues cross-linked

In the citric acid cycle, the equilibrium between the enolate anion of D-fructose-1,6-diphosphate and dihydroxyacetone phosphate (DHAP) plus D-glyceraldehyde-3-phosphate is an aldol reaction (or retro-aldol, depending on direction).

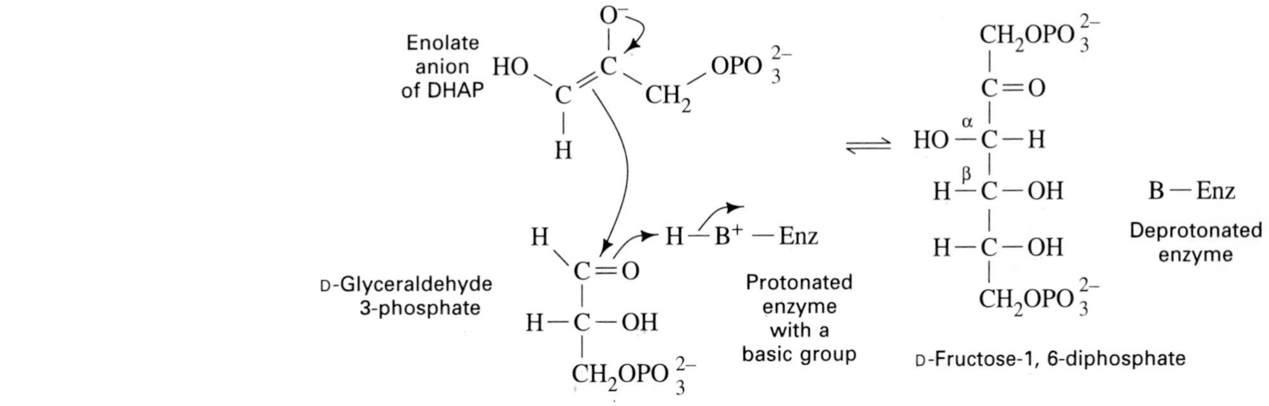

Enolate anion of DHAP

D-Glyceraldehyde 3-phosphate

Protonated enzyme with a basic group

B—Enz Deprotonated enzyme

D-Fructose-1, 6-diphosphate

Citric acid is formed by a Claisen-type condensation between oxaloacetate and acetyl coenzyme A. The enzyme *citrate synthetase* removes an acidic hydrogen (pk_a 8.5) from the methyl group. This enolate anion attacks the carbonyl group of oxaloacetate to give citryl CoA. Hydrolysis of the thiolester bond to coenzyme A provides the driving force for the reaction (ΔG for hydrolysis $= -30$ kJ/mol) and releases citric acid.

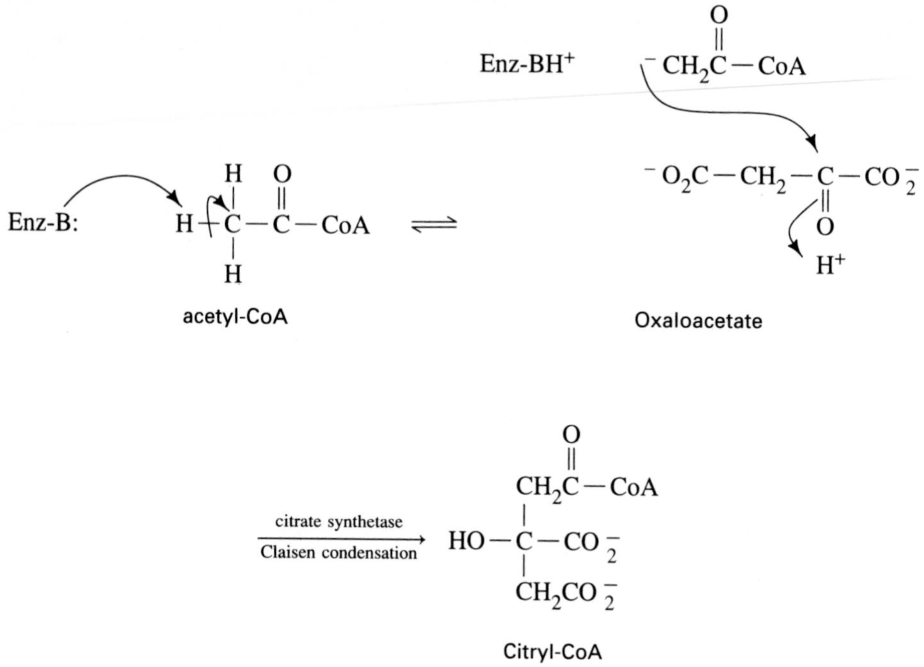

Enz-BH⁺ becomes $Enz\text{-}BH^+$

$$Enz\text{-}B: \quad H\text{---}\overset{\overset{H}{|}}{\underset{\underset{H}{|}}{C}}\text{---}\overset{\overset{O}{\parallel}}{C}\text{---}CoA \quad \rightleftharpoons$$

acetyl-CoA

$$^-CH_2\overset{\overset{O}{\parallel}}{C}\text{---}CoA$$

$$^-O_2C\text{---}CH_2\text{---}\overset{\overset{}{\underset{\underset{O}{\parallel}}{C}}}\text{---}CO_2^-$$

$$H^+$$

Oxaloacetate

$$\xrightarrow[\text{Claisen condensation}]{\text{citrate synthetase}} \quad HO\text{---}\overset{\overset{CH_2\overset{\overset{O}{\parallel}}{C}\text{---}CoA}{|}}{\underset{\underset{CH_2CO_2^-}{|}}{C}}\text{---}CO_2^-$$

Citryl-CoA

QUESTIONS

1. What are the predicted products of the following aldol reactions?

(a) $CH_3\overset{\overset{O}{\parallel}}{C}CH_3 \ + \ 2H\text{---}\overset{\overset{O}{\parallel}}{C}\text{---}H \xrightarrow{\text{NaOH}}$

(b) $2CH_3\overset{\overset{O}{\parallel}}{C}H \xrightarrow{\text{NaOH}}$

(c) 2 $\xrightarrow{\text{KOH}}$

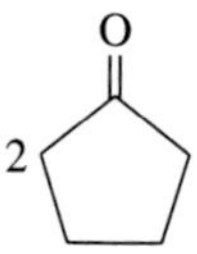

2. How would you make the following compounds using aldol-type condensations?

(a) 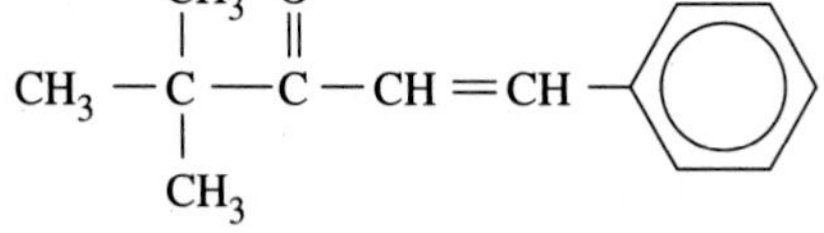 $CH_3\text{---}\overset{\overset{CH_3}{|}}{\underset{\underset{CH_3}{|}}{C}}\text{---}\overset{\overset{O}{\parallel}}{C}\text{---}CH\text{=}CH\text{---}$

(b)

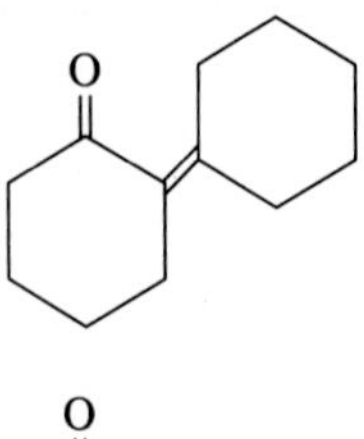

(c)

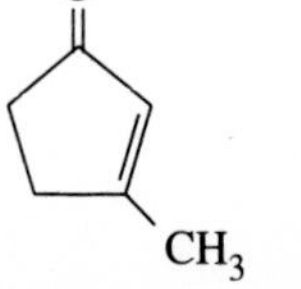

(d) $CH_3\overset{\overset{O}{\parallel}}{C}CH_2\overset{\overset{O}{\parallel}}{C}\text{---}OEt$

(e) $CH_3\text{---}CH\text{=}CH\text{---}COOH$

(f) 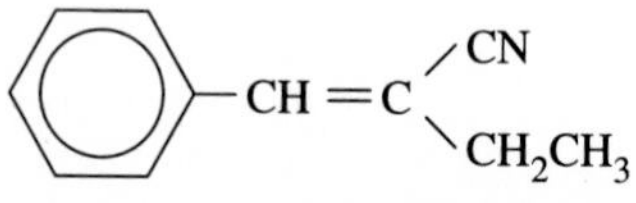 $\text{---}CH\text{=}C\overset{\diagup CN}{\diagdown CH_2CH_3}$

Concept Map

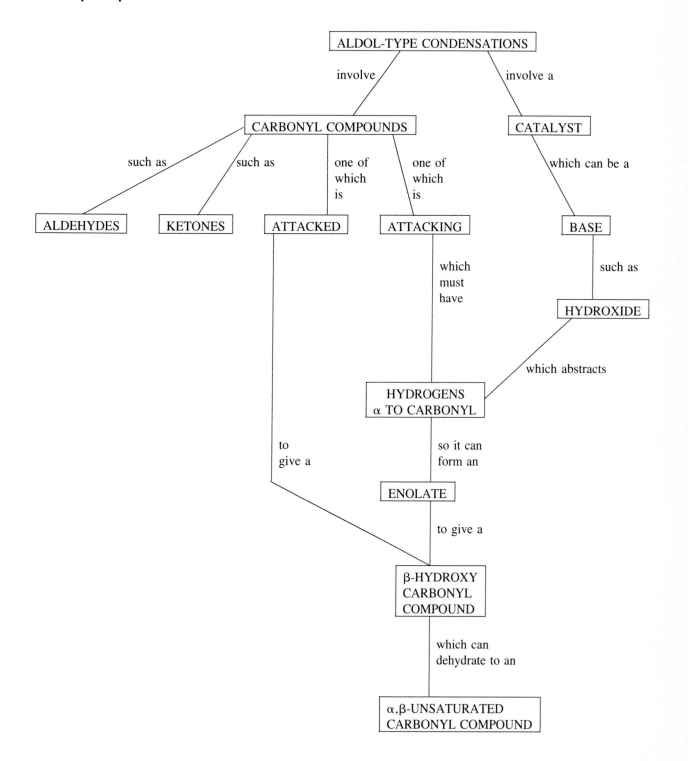

EXPERIMENT 39.1 MICROSCALE ALDOL CONDENSATION OF ACETONE AND BENZALDEHYDE

Estimated Time:
1.0 hour

Prelab

1. Calculate the theoretical yield of dibenzalacetone, based on 0.10 mL of the stock solution of benzaldehyde and acetone described in the procedure. (*Hint:* There is no limiting reagent; benzaldehyde and acetone are in the correct stoichiometric proportions.) Assume volumes are additive in mixing liquid.
2. If the cap were left off the stock solution of benzaldehyde and acetone for a long time, what would you expect to happen, and how would this affect the results?

Special Hazards

Benzaldehyde and acetone are flammable. Sodium hydroxide is highly corrosive; avoid contact and wash with water if contact occurs.

Procedure

Into a micro test tube (1 x 8 cm), pipet as accurately as possible 1.5 mL of a stock solution that contains

NaOH	10g
H$_2$O	100 mL
EtOH	80 mL

Into the same tube pipet 0.10 mL of a stock solution containing

Benzaldehyde: 50.0 mL
(MW 106, dens. 1.04 g/mL)

Acetone: 18.0 mL
(MW 58, dens. 0.79 g/mL)

While holding the test tube firmly in your hand, flick it with your finger to mix the reagents and/or agitate it briefly on an electric agitator, called a vortex mixer. Agitate occasionally for the next 15 minutes and note the precipitation of the yellow dibenzalacetone. Centrifuge the tube briefly (30 seconds to 1 minute) and remove the supernatant liquid using a Pasteur pipette. (*Caution:* Be sure the centrifuge is balanced with another equally filled tube before turning it on.)

To recrystallize the product, warm the tube on a steam-bath and add a few drops (less than 1 mL) of ethyl acetate to dissolve the product. Remove the water remaining as a layer under the ethyl acetate with a Pasteur pipette. Let the tube cool for a minute or two (small tubes cool quickly), then place it in an ice bath. Shake it occasionally to help induce crystallization. When crystallization is complete, scrape the product into a Büchner funnel and rinse it with a little ice-cold methanol. Let the product dry by standing on a filter paper about 10 minutes, stirring with a spatula occasionally. Record the melting point and the mass, using an analytical balance.

REFERENCES

CONRAD, C. R., and M. A. DOLLIVER, *Organic Syntheses Collective Volume 2* (1943) 167.

HATHAWAY, B. A., "An Aldol Condensation Experiment Using a Number of Aldehydes and Ketones," *Journal of Chemical Education* (1987) *64*, 367.

EXPERIMENT 39.2 CONDENSATION OF BENZALDEHYDE AND NITROMETHANE TO FORM NITROSTYRENE

Estimated Time:
1.5 hours

Prelab

1. Calculate the grams and millimoles corresponding to 1.00 mL of nitromethane and 1.60 mL of benzaldehyde.
2. What is the theoretical yield of nitrostyrene?

Special Hazards

Methanol, benzaldehyde, and nitromethane are all flammable. 10 *N* NaOH is extremely corrosive and 5% HCl is somewhat corrosive; wash immediately with a large volume of water if contact occurs.

Procedure

In a 50-mL Erlenmeyer flask, place a magnetic stir bar, methanol (20 mL), nitromethane (1.00 mL, ____ g, ____ mmol), and benzaldehyde (1.60 mL, ____ g, ____ mmol). Cool and stir the solution in an ice bath atop a magnetic stirring plate. Add 10 *N* NaOH (2.0 mL or 40 drops, *use caution*). Continue stirring in the ice bath for 5 minutes to complete the reaction; a white precipitate forms. Add water (20 mL) to obtain a clear solution; pour this with swirling into a beaker containing 20 mL of 5% aqueous HCl. Collect the pale yellow precipitate of nitrostyrene by suction filtration and wash with water. Recrystallize from ethanol/water by dissolving in a few milliliters of ethanol, bringing to a boil, and adding water to incipient cloudiness. If an oil forms on cooling, use the standard techniques of cooling in an ice bath, scratching the glass surface, and seeding to achieve crystallization. Suction filter, then wash the product with water and air dry. Record the mass and melting point.

REFERENCE

WORRALL, D. E., *Organic Syntheses Collective Volume 1* (1941) 413.

EXPERIMENT 39.3 MICROSCALE CONDENSATION OF A NITRILE AND AN AROMATIC ALDEHYDE: PREPARATION OF α-(4-NITROPHENYL) CINNAMONITRILE

Estimated Time:
1.0 hour

Prelab

1. Calculate the millimoles coresponding to 100 mg of 4-nitrophenylacetonitrile and 0.07 mL of benzaldehyde.
2. What is the theoretical yield of α-(4-nitrophenyl) cinnamonitrile?

Special Hazards

Benzyltrimethylammonium hydroxide is corrosive and toxic. Ethanol and benzaldehyde are flammable.

Procedure

In a 25-mL Erlenmeyer flask, place 4-nitrophenylacetonitrile (100 mg, _____ mmol), benzaldehyde (0.07 mL or approximately 3 drops, _____ mmol, _____ g) and ethanol (2 mL). Warm the flask briefly on a hot plate, just enough to dissolve the solid. In the fume hood add 1 drop of a 40% methanolic solution of the base benzyltrimethylammonium hydroxide (Triton B) and swirl. The solution turns dark purple, then slightly green as a precipitate forms. Collect the final precipitate, wash with methanol to remove the green color, and recrystallize from 1-propanol.

REFERENCE

ALLEN, C. F. H., and G. P. HAPP, "The Thermal Reversibility of the Michael Reaction 1. Nitriles," *Canadian Journal of Chemistry* (1964) *42*, 641.

EXPERIMENT 39.4 CLAISEN CONDENSATION OF ETHYL ACETATE AND BENZALDEHYDE

Estimated Time:
2.5 hours

Prelab

1. How many grams and millimoles of ethyl acetate correspond to 9.2 mL? To 2.0 mL of benzaldehyde? To 3.0 mL of 21% (by weight) ethanolic sodium ethoxide (density 0.87 g/mL)?
2. What is the theoretical yield of ethyl cinnamate?
3. At approximately what temperature would you expect to collect ethyl cinnamate (lit. bp 271° at 760 torr) during distillation under aspirator vacuum (approx. 20–40 torr)?

Special Hazards

Ethyl acetate and benzaldehyde are flammable; avoid exposure to heat or flames. Sodium ethoxide and acetic acid are corrosive; rinse copiously with water if contact occurs.

Procedure

In a 100-mL beaker clamped securely in an ice bath atop a magnetic stirring plate, place a Teflon-coated magnetic stirring bar, ethyl acetate (20 mL, _____ mmol, _____ g, previously dried over K_2CO_3) and 6.0 mL of a 21% ethanolic solution of sodium ethoxide. The solution is stirred and the temperature checked to make sure it is below 10°. Benzaldehyde (4.0 mL, _____ g, _____ mmol) is added dropwise over the course of 5 minutes (at a rate of approximately 1 drop every 3 seconds), making sure that the temperature stays in the range 0–10°. The solution is stirred an additional 45 minutes to form a yellow paste, then 40 mL of 10% aqueous acetic acid is added.

The mixture is stirred and poured into a separatory funnel; two phases should appear. If the phases do not separate well, add some solid NaCl to salt out the organic phase. Remove the aqueous (lower) layer, wash the remaining organic layer with water (20 mL) and brine (20 mL), then dry it over Na_2SO_4. Decant the solution away from the drying agent, add boiling chips, and distill under aspirator vacuum to obtain ethyl cinnamate. Record the boiling range and mass of the product.

REFERENCE

MARVEL, C. S., and W. B. KING, *Organic Syntheses Collective Volume 1* (1941) 252.

SECTION 40

Amine Synthesis

Overview

Amines can be prepared in several ways, including direct displacement of a halide by nitrogen.

$$R\ddot{N}H_2 + R'{-}X \longrightarrow R{-}\overset{\overset{\displaystyle H}{|}}{\underset{\underset{\displaystyle H}{|}}{N^+}}{-}R' + X^- \equiv R{-}\overset{\overset{\displaystyle H}{|}}{N}{-}R\cdot HX \xrightarrow{\text{heat}} \xrightarrow{\text{OH}^-} R{-}\overset{\overset{\displaystyle H}{|}}{N}{-}R' + H_2O + X^-$$

Since a hydrohalic acid (HX) is produced in this reaction, either an extra equivalent of amine or another added base must be present to consume the acid.

This reaction has the additional drawback that it may be difficult to stop at the desired number of alkyl groups on nitrogen since reaction can continue until four groups are attached, forming a quaternary ammonium salt.

$$NH_3 \xrightarrow{RX} \underset{+\ HX}{RNH_2} \xrightarrow{RX} \underset{+\ HX}{R_2NH} \xrightarrow{RX} \underset{+\ HX}{R_3N} \xrightarrow{RX} R_4N^+X^-$$

Sometimes by control of the reaction conditions an acceptable yield of the desired compound can be achieved. For example, in Experiment 40.1 the hydroiodide precipitates after monoalkylation.

$$\underset{\substack{m\text{-Toluidine}\\ \text{MW 107, bp 203}^\circ\\ \text{dens. 1.00 g/mL}}}{\text{(structure: NH}_2,\ \text{CH}_3)} + \underset{\substack{\text{Iodoethane}\\ \text{(ethyl iodide)}\\ \text{MW 156, bp 69–73}^\circ\\ \text{dens. 1.95 g/mL}}}{CH_3CH_2I} \longrightarrow \underset{}{\text{(structure: NHCH}_2\text{CH}_3,\ \text{CH}_3)} \cdot HI \xrightarrow{\text{NaOH}} \underset{\substack{N\text{-Ethyl-}m\text{-toluidine}\\ \text{MW 135, bp 221}^\circ\\ \text{dens. 0.96 g/mL}}}{\text{(structure: NHCH}_2\text{CH}_3,\ \text{CH}_3)}$$

Another method of preparing amines which avoids the danger of overalkylation is the Gabriel synthesis. The Gabriel synthesis of primary amines employs potassium phthalimide as the nitrogen source, yielding a phthaloyl-protected amine as the product. In Experiment 40.2 the phthalimide anion displaces the primary bromide ion of 2-bromoacetophenone in an S_N2 reaction.

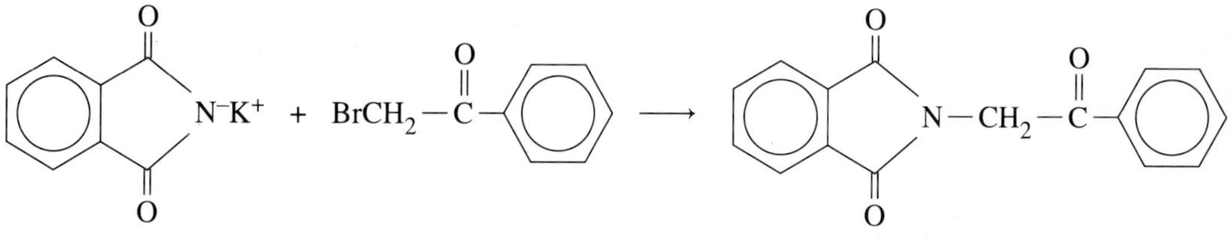

Potassium phthalimide
MW 185

2-Bromoacetophenone
(phenacyl bromide)
MW 199, mp 48 – 51°

Phthalimidoacetophenone
MW 265, mp 165 – 167°

Creation in one step of a protected amine may be an advantage in a multistep synthesis because the relatively inert phthaloyl group can be kept on as long as needed, while other functional groups undergo oxidation or other reactions that would destroy unprotected amines. When desired the phthaloyl group can be removed under normal conditions for amide hydrolysis or under mild conditions by hydrazine. Hydrazine is particularly effective for this reaction because it is highly nucleophilic and forms a six-membered ring while spitting out the amine in a double transamidation.

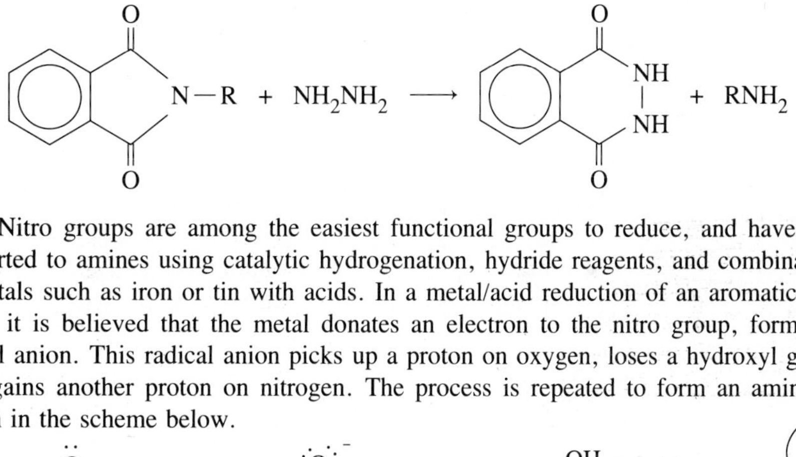

Nitro groups are among the easiest functional groups to reduce, and have been converted to amines using catalytic hydrogenation, hydride reagents, and combinations of metals such as iron or tin with acids. In a metal/acid reduction of an aromatic nitro group it is believed that the metal donates an electron to the nitro group, forming a radical anion. This radical anion picks up a proton on oxygen, loses a hydroxyl group, then gains another proton on nitrogen. The process is repeated to form an amine, as shown in the scheme below.

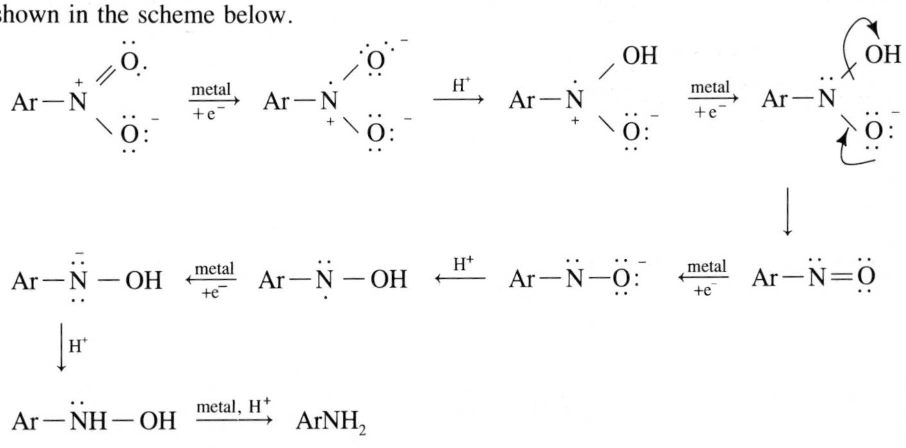

In this reaction the role of Sn is uncertain, since the reduction can also be carried out using Sn^{2+}. Experiment 40.3 is the reduction of nitrobenzene to aniline using iron and HCl. The aniline will then be acetylated with acetic anhydride to yield acetanilide, which is easier to isolate and safer to handle than aniline.

Nitrobenzene	Aniline	Acetanilide
MW 123, mp 5–6°	MW 93.1	MW 135
bp 210–211°	bp 184°	mp 113 –115°
dens. 1.20 g/mL	dens. 1.02 g/mL	

This experiment illustrates the common process of carrying an intermediate through another reaction step without purification and demonstrates the ease of reduction of nitro groups of acetylation of amines.

EXPERIMENT 40.1 PREPARATION OF N-ETHYL-*m*-TOLUIDINE FROM *m*-TOLUIDINE AND ETHYL BROMIDE

Estimated Time:
2.0 hours

Prelab

1. Calculate the millimoles corresponding to 4.00 g of *m*-toluidine and 2.5 mL of iodoethane.
2. What is the theoretical yield of N-ethyl-*m*-toluidine?

Special Hazards

m-Toluidine and ethyl iodide are both highly toxic; use only in the fume hood and avoid contact.

Procedure

In a 150-mL beaker in the fume hood, place *m*-toluidine (4.00 g, _____ mmol) and iodoethane (2.5 mL, _____ g, _____ mmol, dispensed from a pipette or burette in the fume hood). Heat the mixture to the boiling point of ethyl iodide on a hot plate or steam bath in the hood. The mixture first melts, then becomes a solid mass as reaction takes place. To remove unreacted starting materials, cool the flask in an ice bath, add ether (20 mL), stir, and suction filter. The collected material is the hydrodide of N-ethyl-*m*-toluidine. To free the N-ethyl-*m*-toluidine from its solid hydriodide, place the crystalline product in a beaker, add 60 mL of 5% aqueous NaOH, stir well, and pour the two-phase mixture into a separatory funnel. Remove the aqueous (lower) layer. Dry the organic layer over anhydrous Na_2SO_4, decant, and distill under aspirator vacuum to obtain purified N-ethyl-*m*-toluidine. Record the mass and melting range of the product and, if possible, take an IR spectrum or refractive index.

REFERENCE

BUCK, J. S., and C. W. FERRY, *Organic Syntheses Collective Volume 2* (1943) 290.

EXPERIMENT 40.2 MICROSCALE GABRIEL SYNTHESIS OF PHTHALIMIDOACETOPHENONE

Estimated Time:
1.0 hour

Prelab

1. Calculate the millimoles corresponding to 100 mg of 2-bromoacetophenone and 100 mg of potassium phthalimide.
2. What is the theoretical yield of phthalimidoacetophenone?

Special Hazards

2-Bromoacetophenone is corrosive and a lachrymator. Chloroform is a suspected carcinogen; use only in the fume hood and avoid breathing the vapor.

Procedure

In a 15 x 125 mm test tube, place 2-bromoacetophenone (phenacyl bromide, 100 mg, _____ mmol), potassium phthalimide (100 mg, _____ mmol), and dimethylformamide (1.0 mL or 40 drops). The tube is warmed in a steam bath or boiling water bath for 15 minutes with occasional swirling. The tube is cooled in an ice bath in the fume hood, and approximately 3 mL of chloroform is added. (*Caution:* Avoid breathing fumes.) Add 2 mL of water and mix thoroughly by Pasteur pipette to extract the product into the chloroform. The chloroform (lower) layer is removed by Pasteur pipette and placed in a fresh tube. One milliliter of 5% aqueous NaOH is added and the contents are mixed using a Pasteur pipette to extract any unreacted phthalimide. The NaOH (upper) layer is removed by pipette and discarded, and a similar washing is carried out using 1 mL of water. After the water is removed, a small scoop of anhydrous Na_2SO_4 is added to dry the chloroform. The solution is filtered into a small beaker and evaporated to dryness on a steam bath in the fume hood. The residue is cooled in an ice bath to obtain the crude solid product. To remove remaining ether-soluble impurities, the crude phthalimidoacetophenone is triturated (crushed and stirred) with 2 mL of ether, which is decanted and discarded. Then the product is placed on a filter paper and allowed to air dry. The mass and melting range are recorded (lit. mp 165–167°).

REFERENCE

SHEEHAN, J. C. and W. A. BOLHOFER, ''An Improved Procedure for the Condensation of Potassium Phthalimide with Organic Halides,'' *Journal of the American Chemical Society* (1950) 72, 2786.

EXPERIMENT 40.3 SYNTHESIS OF ACETANILIDE FROM NITROBENZENE

Estimated Time:
2.0 hours

Prelab

1. Calculate the millimoles corresponding to 4.0 g of Fe, 3.0 mL of 5% aqueous HCl (density 1.02 g/mL), 2.0 mL of nitrobenzene, 2.8 mL of acetic anhydride (density 1.08 g/mL), and 1.0 g sodium acetate.
2. What is the theoretical yield of acetanilide?

Special Hazards

Nitrobenzene is highly toxic; the aniline formed as an intermediate is highly toxic and a cancer suspect agent. The oxalic acid used in cleanup is corrosive and toxic. Avoid skin contact or ingestion of these materials.

Procedure

In a 50-mL round-bottomed flask, place iron powder (4.0 g, _____ mmol), 10 mL water, 5% aqueous HCl (3.0 mL, _____ mmol), and nitrobenzene (2.0 mL, _____ g, _____ mmol). Add a water-cooled condenser, heating mantle, and variable transformer, and heat the mixture to reflux. Boil for 30 minutes, shaking the flask occasionally to ensure good mixing.

Cool the reaction flask in an ice bath and pour the contents through a Buchner funnel under suction. Oily drops of aniline should appear floating on the top of the filtrate, and the distinctive shoe polish odor of nitrobenzene should be gone. To increase the yield, rinse the flask with 10 mL of water and pour this through the filter cake. Discard the filter cake in the appropriate waste container, and transfer the filtrate to a 100-mL beaker.

You will now acetylate the aniline to form acetanilide. Add sodium acetate (1.0 g, _____ mmol) and stir to dissolve. Place some ice (about 20 mL) into the beaker and stir to cool the mixture. Now comes the exciting part. Add acetic anhydride (2.8 mL, _____ mmol) and stir the solution. In a few seconds a copious precipitate of acetanilide forms. Collect this product by suction filtration, wash with water, and recrystallize from water. Record the mass and melting range of the acetanilide obtained (lit. mp 113–115°).

To clean the round-bottomed reaction flask of the tenacious residue of iron oxides, add a scoop of oxalic acid and some water and let it soak for a few minutes. Oxalic acid chelates and dissolves the ferric ions; it is a handy material for removing rust stains in bathrooms, and so on. Exercise due caution, since it is very toxic on ingestion.

REFERENCE

Reeve, W. and V. C. Lowe, "Preparation of Acetanilide from Nitrobenzene," *Journal of Chemical Education* (1979) *56*, 488.

Diazonium Reactions

Overview

Diazonium salts are very useful for synthesizing substituted aromatic compounds. The diazo group can be considered to be creating a positive charge on carbon as the excellent leaving group N_2 (highly stable nitrogen gas) departs. The diazo group can be replaced with nucleophiles, including halogens, hydroxyl groups, and cyano groups. It can also couple with an electron-rich ring. These reactions are summarized below.

391

The mechanism of coupling is the same as any other electrophilic attack on an electron-rich ring:

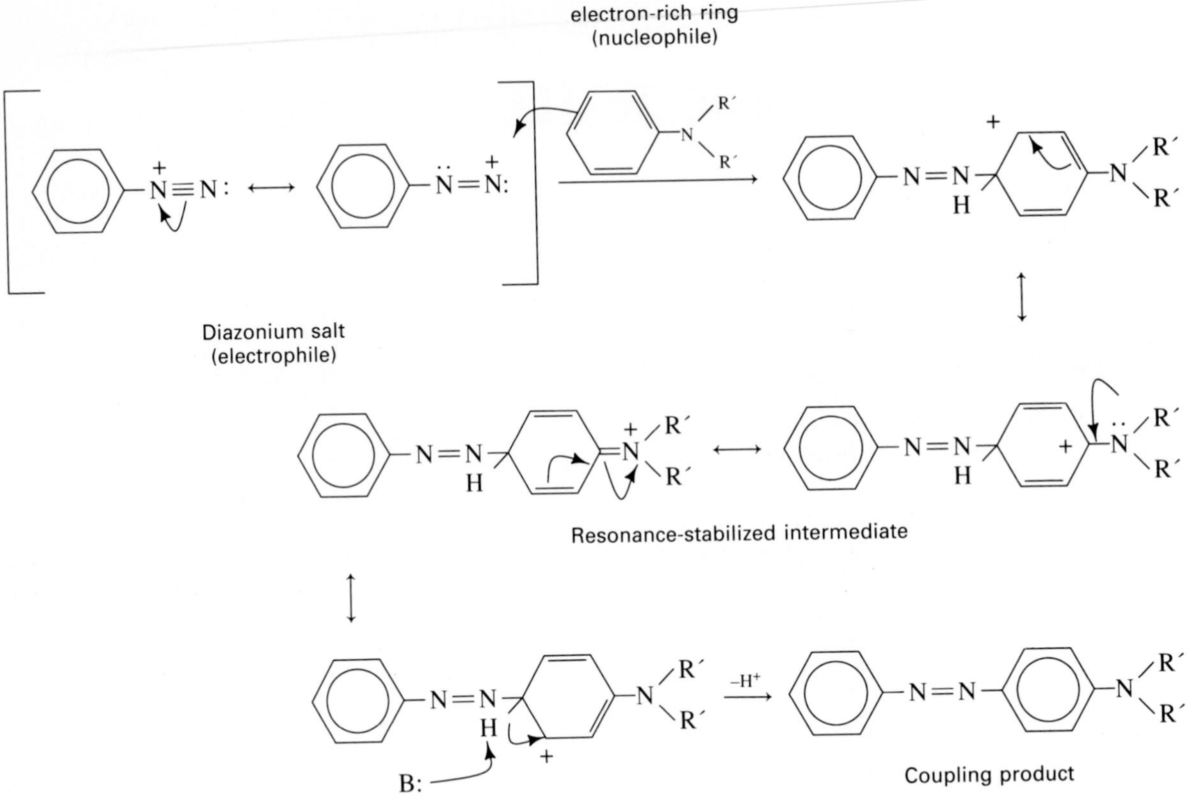

In Experiment 41.1 you will form the diazonium salt of sulfanilic acid and couple it with N,N-dimethylaniline.

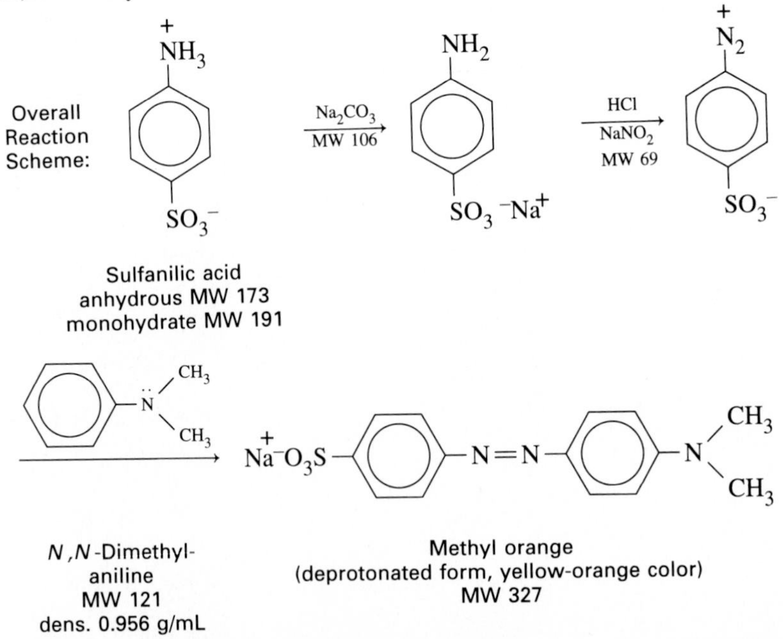

Overall Reaction Scheme:

Sulfanilic acid
anhydrous MW 173
monohydrate MW 191

N,N-Dimethyl-
aniline
MW 121
dens. 0.956 g/mL

Methyl orange
(deprotonated form, yellow-orange color)
MW 327

Diazonium salts are often unstable or even explosive, but sulfanilic acid forms an unusually stable one that will keep for a few hours, and we will form and use it quickly without drying it out or isolating it. The product you make, methyl orange, is both an acid-base indicator and a dyestuff. In weak acid solution one of the azo nitrogens is protonated to give the structure shown here:

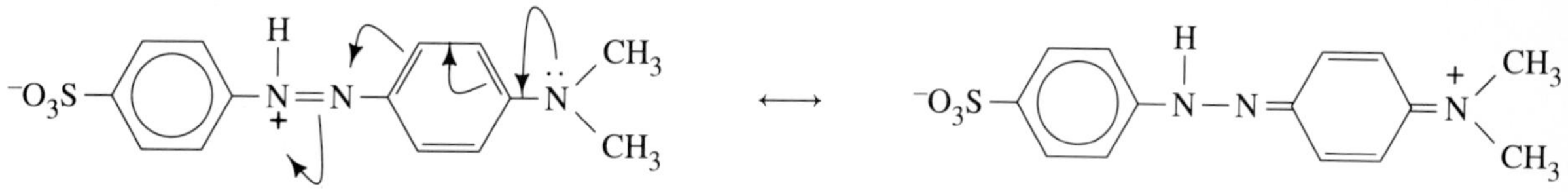

Methyl orange
(protonated form, also called Helianthin, red color)
MW 305

Methyl orange in the protonated form has a pK_a of approximately 3.8, and changes color from red to yellow in the range pH 3.2–4.4, so it is a handy indicator for titrations with endpoints in this range. Also, either form of methyl orange can be used to dye cloth. The highly polar sulfonate groups bind to the many ionic groups (carboxyl and amino) present on the surface of the proteins wool or silk, attaching the dye securely. Cotton, which consists of cellulose (a polymer of glucose), does not have ionic groups on its surface, but does have a lot of —OH groups. It must be dyed with other dyestuffs which form hydrogen bonds, such as indigo, which puts the blue in blue jeans.

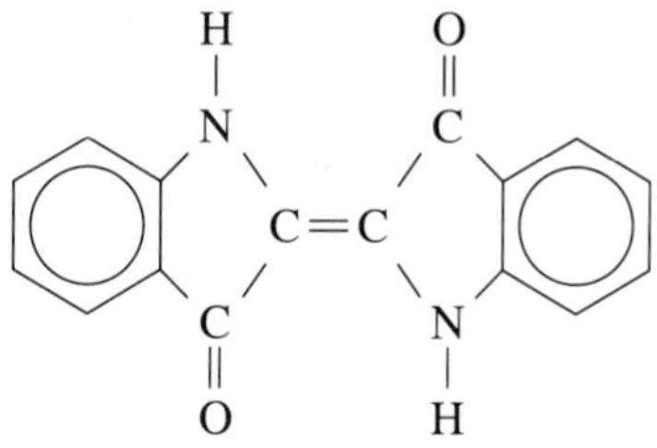

Indigo

From this discussion, it is apparent that if at this stage you are seriously considering putting on an orange robe and going to a cave in Tibet, you should use a material such as wool or silk and methyl orange in its basic form.

In Experiment 41.2 an aromatic amine is converted to an iodide through a diazonium salt.

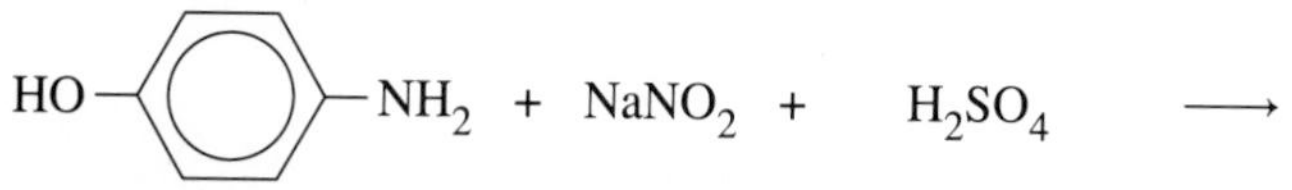

p-Aminophenol	Sodium	Sulfuric acid
MW 109	nitrite	MW 98
mp 188–190°	MW 69	dens. 1.84 g/mL

Diazonium	Potassium	*p*-Iodophenol
salt	iodide	MW 220
	MW 166	mp 93–95°

APPLICATION: NATURAL AND SYNTHETIC DYES

Archeologists have found fragments of cloth over 4,000 years old that were dyed with the naturally occurring plant materials indigo and woad. The red dye madder, now better known by its chemical name, alizarin, was also used in ancient Egypt, India, Arabia, Persia, Greece, Rome, and Germany.

Alizarin

Tyrean purple, derived from *Murex brandaris*, was considered the color of nobility in ancient Phoenecia.

Kermes or kermesic acid, derived from an oak louse, has been used in the Orient for over 6,000 years to give a brilliant scarlet color when used with an alum mordant.

Kermesic acid

A similar anthraquinone compound, carminic acid (derived from sun-dried cochineal bugs), was highly prized by the Aztecs for the beautiful red it produced.

Carminic acid

These so-called vat dyes produce exceptionally fast (long-lasting) coloration on cotton and can also dye almost any other fiber (wool, silk, nylon, acrylic, and polyester) to some degree.

Vat dyes are complex organic molecules that are insoluble in water but may be converted to a water-soluble form by reduction of carbonyl groups (using $NaHSO_3$ and NaOH). Once solubilized, the molecules penetrate the fibers of the fabric. On chemical reoxidation to the insoluble form, the color is trapped in the fabric.

William Henry Perkin prepared mauve, the first synthetic dye, in 1856.

Mauve

Mauve is classified as a *basic* dye because it dissociates into an anion and a colored cation. Soon other basic dyes, such as fuchsin, methyl violet, and aniline blue, were developed.

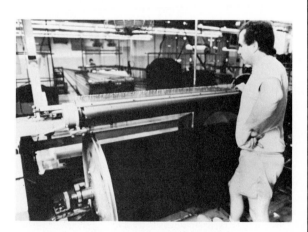

Dyeing Machine (courtesy of Greenwood Mills)

Basic dyes, being ionic, are soluble in water. The cations bind strongly to the anions of acidic residues ($-COO^-$ or $-SO_3^-$) on the surfaces of acrylic fibers to provide long-lasting color.

Acid dyes (those containing acidic functional groups) are used on nylon, wool, silk, and on modified acrylic and polypropylene fibers, plus their blends. Acid dyes may be in any of the following classes: azo, anthraquinone, nitro, pyrazolone, quinoline, and triphenylmethane.

Often in dyeing a *mordant* is used, which is usually a metal hydroxide. A mordant forms a link between the fabric and the dyestuff. The color of the fabric depends both on the dye used and on the mordant. For example, alizarin dyes red in the presence of alum mordant, violet with a iron mordant, and brownish-red with a chromium mordant.

In 1869, von Baeyer prepared synthetic alizarin, the first synthesis of a natural dye. In 1897, he succeeded in synthesizing indigo. Since that time, many hundreds of synthetic dyes have been produced which give excellent color and fastness and are quite inexpensive compared to natural dyes.

QUESTIONS

1. Predict the products of the following reactions:

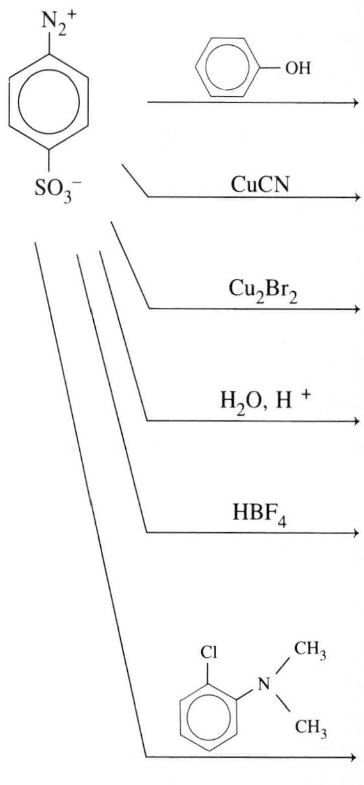

2. Explain how cloth could be dyed two different colors using methyl orange as the only dyestuff.

3. Draw a compound similar to methyl orange which might have a slightly different color. Suggest a synthesis for this compound from readily available starting materials. Check the *Aldrich Catalog/Handbook of Fine Chemicals* and list the prices of your starting materials and reagents. Considering these costs, plus equipment costs, plus your time and effort, at what price per kilogram could you sell your product and still make a profit?

4. Draw a concept map for this experiment.

EXPERIMENT 41.1 SYNTHESIS OF METHYL ORANGE: DIAZONIUM COUPLING REACTION OF SULFANILIC ACID AND *N,N*-DIMETHYLANILINE

Estimated Time:
2.5 hours

Prelab

1. Complete a reagent table and calculate the theoretical yield.

Special Hazards

Methyl orange is a dyestuff. If it gets on your clothes, it will leave permanent stains. Careful handling and use of lab coats are recommended. Diazo compounds are often carcinogenic on ingestion.

Many diazonium salts are unstable or even explosive when dry, but the one you make here is unusually stable. You will also keep it in solution and use it promptly.

Procedure

A. Diazotization of sulfanilic acid.   In a 25-mL Erlenmeyer flask, place anhydrous sodium carbonate (2.5 mmol, _____ g), sulfanilic acid (5 mmol, _____ g anhydrous or _____ g monohydrate) and water (10 mL). Heat to boiling on a hot plate to dissolve the solids. Cool to room temperature and add sodium nitrite (6 mmol, _____ g). Stir until dissolved, then pour the solution into a 100-mL beaker containing 15 mL of a mixture of ice and water and 1 mL of concentrated HCl. Soon the diazonium salt of sulfanilic acid separates as a powdery white precipitate. Keep the suspension cooled in an ice bath until ready for the next step. We will not isolate this diazonium salt but will use it directly as a suspension in the coupling reaction.

B. Methyl Orange (*p*-Sulfobenzeneazo-4-dimethylaniline sodium salt).
In a test tube mix dimethylaniline (0.7 mL, _____ mmol) and glacial acetic acid (0.5 mL, _____ mmol). Add this solution with stirring to the suspension of diazotized sulfanilic acid. Stir vigorously and in a few minutes the red, acid-stable form of the dye should separate. Keep the mixture cooled in the ice bath for another 10 minutes to complete the reaction, then add 10% sodium hydroxide solution (8 mL) to form the orange sodium salt. Collect the pasty precipitate by suction filtration and recrystallize from water.

C. Dyeing cloth.   In a 250-mL beaker, prepare a dye bath containing water (100 mL), sodium sulfate (0.5 g), concentrated sulfuric acid (3 drops), and methyl orange (0.2 g of your product above, which can still be wet). Heat the solution to boiling, then immerse a piece of wool or silk cloth or yarn in it for 5 minutes. For comparison also immerse a square of cotton cloth for the same time. Remove and rinse the samples and compare their colors. Now you know what the expression "dyed in the wool" means.

D. Indicator properties.   While waiting for the dye bath to heat up or the cloth to dye, place a few crystals of methyl orange in a test tube and dissolve them in a little water. Alternately add a few drops of dilute hydrochloric acid and dilute sodium hydroxide solution, observing and recording the color change.

When finished with parts A–D, weigh the still-slightly-wet methyl orange and store it in a beaker in your desk to dry. You will get an accurate dry weight next lab period. Correct your yield for the 0.2 g used to dye the cloth by figuring out what fraction of the wet weight is dry weight, taking that fraction of 0.2 g and adding it to the dry weight of the remaining product.

REFERENCES

ABRAHART, E. N., *Dyes and Their Intermediates*, Pergamon Press, Oxford, 1968.

ALLEN, R. L. M., *Colour Chemistry*, Appleton-Century-Crofts, New York, 1971.

GORDON, P. R., and P. GREGORY, *Organic Chemistry in Colours,* Springer-Verlag, Berlin, 1982.

HAMER, F. M., *The Cyanine Dyes and Related Compounds*, Interscience, New York, 1964.

HARPER, R. S., JR., and R. M. REINHARDT, ''Chemical Treatment of Textiles,'' *Journal of Chemical Education* (1984) *61*, 368.

EXPERIMENT 41.2 *p*-IODOPHENOL FROM *p*-AMINOPHENOL USING A DIAZONIUM SALT

Estimated Time:
1.5 hours

Prelab

1. Calculate the millimoles corresponding to 1.00 g of *p*-aminophenol, 0.72 g of NaNO$_2$, and 2.00 g of KI.
2. What is the theoretical yield of *p*-iodophenol?

Special Hazards

p-Aminophenol is toxic and irritating. Sodium nitrite is also toxic in large doses (MLD orally in dogs, 330 mg/kg), although it is used as a preservative in small quantities in meats. *p*-Iodophenol is an irritant. See the warning on diazonium salts in ''Special Hazards'' for Experiment 41.1.

Procedure

In a 50-mL beaker in an ice bath atop a magnetic stirring plate, place 5 mL of water and 5 mL of 3 *M* H$_2$SO$_4$. Add *p*-aminophenol (1.00 g, _____ mmol) and a Teflon-coated magnetic stir bar. Stir to dissolve the *p*-aminophenol and check to make sure that the temperature is below 10°. In a separate 25-mL Erlenmeyer flask, dissolve sodium nitrite (0.72g, _____ mmol) in water (2 mL) and add this solution dropwise to the beaker over the course of 2–3 minutes. Stir for 10 minutes. Rinse out the 25-mL Erlenmeyer flask and in it dissolve potassium iodide (2.00 g, _____ mmol) in water (3 mL) and cool this in the ice bath. Add this solution to the beaker and maintain stirring. After 5 minutes add a tiny amount (0.01 g) of copper bronze and warm the beaker on a steam bath; nitrogen gas will be evolved. When the evolution of N$_2$ has ceased and the *p*-iodophenol has separated as a heavy dark oil, cool the beaker in an ice bath, add ether (30 mL), and pour the contents into a separatory funnel. Take care not to drop the stir bar into the separatory funnel. Rinse the beaker with a little ether and add this to the separatory funnel to improve the recovery. Remove the aqueous phase and wash the organic phase with water containing a few crystals of sodium thiosulfate (to reduce I_2 to colorless, water-soluble I^-). Wash the organic phase with brine, dry it over Na_2SO_4, decant, add boiling chips, and remove the ether in the steam bath. Recrystallize the crude product from ligroin (bp 90–110°, about 30 mL). During the recrystallization a small amount of tarry residue does not dissolve, and the hot solution is decanted away from this material. Record the mass and melting point of the recrystallized *p*-iodophenol.

REFERENCE

DAINS, F. B., and F. EBERLY, *Organic Syntheses Collective Volume 2* (1943) 355.

Isocyanate Formation and Reactions

Overview

Isocyanates are highly reactive species because of the large amount of positive character on the carbon attached to both a nitrogen and an oxygen.

$$\overset{\delta- \quad \delta+ \; \delta-}{R-N=C=O}$$

site for nucleophilic attack

Because of the presence of this highly electrophilic carbon, isocyanates react easily with alcohols to form compounds called urethanes.

$$R-N=C=O \; + \; R'-\overset{..}{O}H \longrightarrow \left[\begin{array}{c} R-N=C \\ \\ H \end{array} \overset{O^-}{\underset{O-R'}{\overset{}{\big\backslash}}} \longleftrightarrow R-\overset{-}{N}-C \overset{O}{\underset{O-R'}{\overset{}{\big\backslash}}} \right]$$

$$\Big\downarrow {\sim}H^+$$

$$\underset{RNH-C-OR'}{\overset{O}{\overset{\|}{}}}$$

A urethane

Reaction of an isocyanate with water is a special case of this reaction, in which case a transitory carbamic acid is formed. Carbamic acids are unstable (none have been isolated) and spontaneously lose CO_2, forming amines.

$$R-N=C=O + H_2O \longrightarrow \left[\underset{\text{A carbamic acid}}{\overset{\overset{\textstyle H \quad O}{\overset{|}{}\;\;\overset{\|}{}}}{R-N-C-OH}} \right] \overset{-CO_2}{\longrightarrow} RNH_2$$

399

Primary alcohols react at room temperature, and secondary and tertiary alcohols react much more slowly.

In a manner similar to their reaction with alcohols, isocyanates react with amines to give substituted ureas.

$$R-N=C=O + R'NH_2 \longrightarrow R-NH-\overset{\overset{\displaystyle O}{\|}}{C}-NH-R'$$

A substituted urea

Because of this high reactivity with alcohols and amines, and because living tissues such as lungs and eyes contain large numbers of alcohol and amine groups which are necessary for proper functioning, low-molecular-weight gaseous isocyanates are quite toxic. The disaster in Bhopal, India, in 1984, for example, which claimed over 3000 lives, occurred when water was released into a large tank containing methyl isocyanate. The ensuing reaction built up heat and pressure as N-methylcarbamic acid formed and released CO_2. When the contents of the tank escaped, a cloud of gaseous methyl isocyanate swept over the sleeping city. Some of the quickest thinkers wrapped wet towels around their faces and escaped serious injury; many others were not so lucky.

$$CH_3-N=C=O + H_2O \longrightarrow \left[CH_3NH-\overset{\overset{\displaystyle O}{\|}}{C}-OH \right] \longrightarrow CH_3NH_2 + CO_2$$

| Methyl isocyanate bp 37–39° | N-Methylcarbamic acid | Methylamine | Carbon dioxide |

There are three related methods for forming isocyanates, all of which involve rearrangement of a nitrene. These are known as the Curtius, Hofmann, and Schmidt rearrangements. A nitrene, of course, is similar to a carbene in that it has no charge and four unshared electrons around the nitrogen. Nitrenes can be formed by loss of a good leaving group from nitrogen.

In the Curtius rearrangement, an acid chloride reacts with azide ion to form an acyl azide. This acyl azide loses nitrogen, forming a nitrene that undergoes rearrangement.

$$R-\overset{\overset{\displaystyle O}{\|}}{C}-Cl + NaN_3 \longrightarrow \left[R-\overset{\overset{\displaystyle O}{\|}}{C}-N=\overset{+}{N}=\overset{-}{N} \longleftrightarrow R-\overset{\overset{\displaystyle O}{\|}}{C}-\overset{-}{N}-\overset{+}{N}\equiv N \right]$$

Acid chloride   Sodium azide   Acyl azide

$$\downarrow {-N_2}$$

$$R-N=C=O \longleftarrow R-\overset{\overset{\displaystyle O}{\|}}{C}-\ddot{N}:$$

Acyl nitrene

The Hofmann and Schmidt reactions form acyl nitrenes in other ways but proceed through the same rearrangement. In the Hofmann reaction an amide is treated with Br_2 and NaOH to form a bromoamide, which eliminates HBr to form a nitrene.

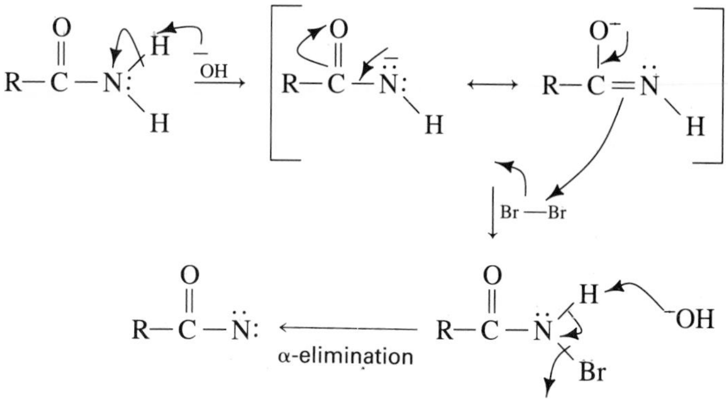

Nitrene

In the Schmidt reaction a carboxylic acid condenses with hydrazoic acid to form an acyl azide, which reacts with water to give the carbamic acid and the amine.

$$\mathrm{RCOOH + HN_3 \xrightarrow[-H_2O]{H_2SO_4} RC(=O)-N_3 \xrightarrow{-N_2} RC(=O)-\ddot{N}: \longrightarrow}$$

$$\mathrm{R-\ddot{N}=C=O \xrightarrow{H_2O} [R-NHCOOH] \xrightarrow{-CO_2} RNH_2}$$

In this experiment you will carry out the Curtius rearrangement by treating undecyl chloride with sodium azide.

$$\mathrm{CH_3-(CH_2)_{10}-C(=O)-Cl + NaN_3 \longrightarrow \xrightarrow{\Delta} CH_3-(CH_2)_{10}-N=C=O}$$

Undecyl chloride (lauroyl chloride), MW 219, bp 134–137° at 11 torr, dens. 0.946 g/mL

Sodium azide, MW 65

Undecyl isocyanate, MW 197, bp 103° at 3 torr

Once the isocyanate has been formed, it will be allowed to react with methanol to form the urethane.

$$\mathrm{CH_3-(CH_2)_{10}-N=C=O + CH_3OH \longrightarrow CH_3(CH_2)_{10}NHC(=O)OCH_3}$$

Undecyl isocyanate

Methanol, MW 32, bp 65°, dens. 0.79 g/mL

Methyl *N*-undecylcarbamate, MW 229, mp 42–43°

The first synthetic organic isocyanate was prepared by Wurtz in 1849, when he treated alkyl halides with salts of cyanic acid (H—N=C=O).

$$\mathrm{RX + K^{+}\,{}^{-}N=C=O \longrightarrow R-N=C=O + KX}$$

After World War II commercial interest increased with the discovery of processes for manufacturing polymeric foams, fibers, coatings, and elastomers using isocyanates. Polyfunctional isocyanates, usually with two or three isocyanates groups in the molecule, have been particularly useful in the tailoring of molecules with specific properties, and form the cornerstone of a large branch of the plastics industry today.

Polyurethanes, formed by reaction of di- or triisocyanates with polyols, possess remarkable chemical and abrasion resistance.

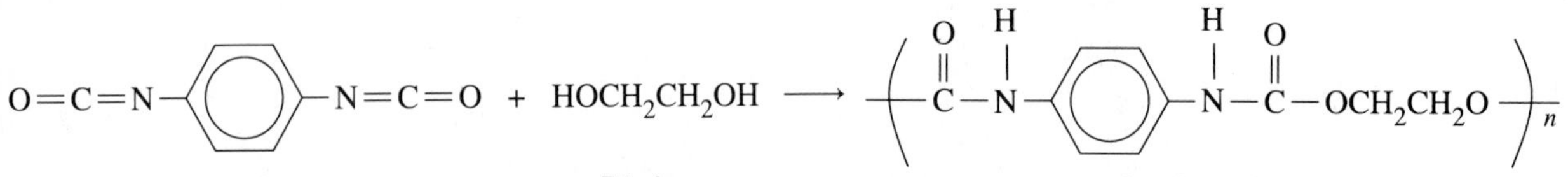

1,4-Phenyldisocyanate — Ethylene glycol — A polyurethane

Formation of a simple polyurethane

Two of the highest-volume isocyanates currently in use are toluene diisocyanate (TDI, usually a mixture of the 2,4 and 2,6 isomers) and diphenylmethane-4,4′-diisocyanate (MDI).

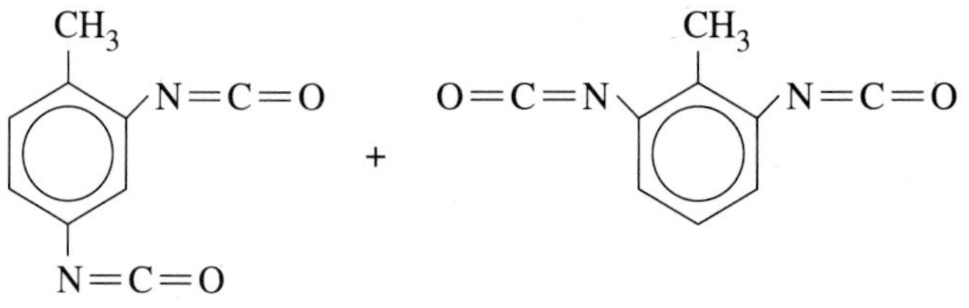

65 : 35 mixture
TDI

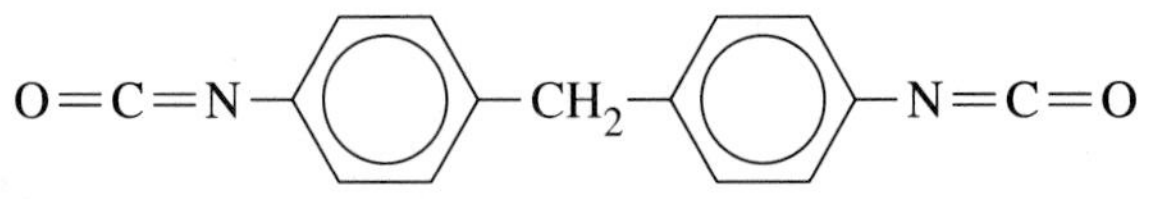

MDI

Recently, world production of TDI and MDI has topped 2 million metric tons annually. Commercially, isocyanates are produced by reaction of amines with phosgene accompanied by loss of 2 moles of HCl.

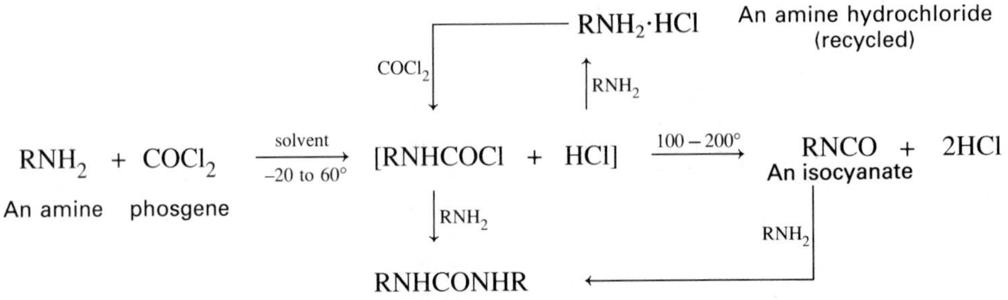

A urea
(unwanted
by-product)

APPLICATION: USES OF POLYURETHANES

Polyurethanes have found wide use in such diverse areas as home insulation, automotive bumpers and moldings, fabric coatings, recreational surfaces (such as running tracks), contact lenses, and craniofacial reconstruction. The favorable properties of polyurethanes result from their long-chain structures. The strength of the polymer depends on the isocyanate, polyol, catalyst, surfactant, temperature, and method of mixing used.

A representative curve of elastic modulus versus temperature is shown in the figure. At low temperatures the material is hard and stiff (glassy). As the temperatures rises past the *glass transition temperature*, T_g, the polymer becomes rubbery and eventually flows as a liquid. Research is currently under way to find methods of reinforcing polyurethane elastomers to increase stiffness and reduce

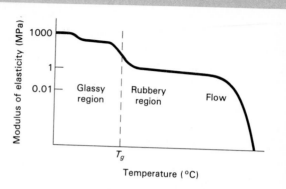

Elastic Modulus Versus Temperature

thermal expansion. Fibers of aramid, boron, glass, and graphite, and flakes of mica are possible additives.

Because of their high reactivity, isocyanates have found a wide variety of other uses. Nonpolymer uses for isocyanates include insecticides, herbicides, explosives, and many biologically active products.

QUESTIONS

1. What would be the products of the following reactions?

(a)

$$\text{(phenyl)}-\overset{\overset{\displaystyle O}{\|}}{C}Cl \xrightarrow{\text{NaN}_3} \boxed{} \xrightarrow{\Delta} \boxed{}$$

(b) $CH_3CH_2NH_2 \xrightarrow{\text{COCl}_2} \boxed{}$

(c) $CH_3—\overset{\overset{\displaystyle H}{|}}{\underset{\underset{\displaystyle CH_3}{|}}{C}}—N=C=O$

$\xrightarrow{H_2O}$

$\xrightarrow{ \bigcirc—NH_2 }$

$\xrightarrow{CH_3OH}$

2. Design an efficient synthesis of TDI from toluene.

3. What would be the structure of the polymer formed between MDI and 1,4-butanediol?

4. How would you design a scrubber to remove methyl isocyanate from escaping gases if pressure built up in a storage tank?

REFERENCE

RANNEY, M. W., *Isocyanates Manufacture*, Noyes Data Corporation, Park Ridge, N.J., 1972.

EXPERIMENT 42.1 FORMATION OF UNDECYL ISOCYANATE BY CURTIUS REARRANGEMENT AND ITS REACTION WITH METHANOL

Estimated Time:
2.5 hours

Special Hazards

Lauroyl chloride is corrosive and a lachrymator. Sodium azide is highly toxic on ingestion and can explode if heated. Toluene is flammable. Use due caution.

Procedure

In a 50-mL Erlenmeyer flask, dissolve sodium azide (NaN_3, 2.30 g, _____ mmol) in 10 mL water with swirling (endothermic dissolution). Add a Teflon-coated magnetic stirring bar and stir the flask in an ice bath atop a stirring plate. In a separate 50-mL beaker, prepare a solution of lauroyl chloride (5.0 mL, _____ g, _____ mmol) in 10 mL of acetone. Over the course of 2–3 minutes, add the solution of acid chloride dropwise to the sodium azide solution. This corresponds to an addition rate of 4–5 drops per second. After the addition, stir the cloudy solution for another 20 minutes to complete the reaction. Efficient stirring must be maintained during this time. Meanwhile, prepare a 100-mL beaker containing 20 mL of toluene on a hot plate in the hood, maintained at a gentle boil. At the end of the stirring period, decant the reaction mixture into a separatory funnel. Be careful not to drop the stir bar into the separatory funnel. If this occurs, it may be removed using a stir-bar retriever (a magnet on a long handle). Draw off the aqueous (lower) layer from the separatory funnel. Dry the organic layer (the acyl azide) over Na_2SO_4 in a 25-mL Erlenmeyer flask, then add it dropwise to the boiling toluene.

As the undecyl azide dissolves in the hot toluene, rearrangement occurs and nitrogen gas is evolved. Keep heating until evolution of N_2 virtually ceases (about 15–20 minutes). Cool the solution to near room temperature and filter if necessary to remove insoluble material. Set up for distillation under aspirator vacuum and collect the toluene, insulate the top of the flask with foil, and heat strongly to collect the isocyanate. Foaming may occur; to minimize this, use a 100-mL round-bottomed flask and monitor the distillation closely. Be ready to remove the heat source quickly if necessary. Take an IR spectrum of the isocyanate, which should show a strong characteristic band in the region 2250–2275 cm^{-1}.

Urethane formation.   In a large, dry test tube, place 1.0 mL of anhydrous methanol and 1.0 mL of undecyl isocyanate. Warm the test tube for 5 minutes in a steam bath or hot-water bath. Cool the tube in an ice bath to induce crystallization. Collect the crude product and dry it by blotting between filter papers. Recrystallize from a small volume of petroleum ether (bp 90–110°). Record the mass and melting point (lit. mp 42–43°).

REFERENCES

ALLEN, C. F. H., and A. BELL, *Organic Syntheses Collective Volume 3* (1955) 846.

CHADWICK, D. H., and T. H. CLEVELAND, "Isocyanates, Organic," in H. F. Mark et al., eds., *Kirk-Othmer Encyclopedia of Chemical Technology*, 3rd ed., Vol. 13, 1981, pp. 789–818.

MAGNIEN, E., and R. BALTZLY, "A Re-examination of Limitations of the Hofmann Reaction," *Journal of Organic Chemistry* (1958) *23*, 2029.

SAUNDERS, J. H., and K. C. FRISCH, *Polyurethanes: Chemistry and Technology, Part I: Chemistry*, Interscience, New York, 1962.

SAYIGH, A. A. R., H. ULRICH, and W. J. FARRISSEY, JR., "Diisocyanates," in J. K. Stille and T. W. Campbell, eds., *Condensation Monomers*, Wiley-Interscience, New York, 1972.

Heterocycles

Overview

Heterocyclic compounds are those containing nitrogen, oxygen, or sulfur atoms in carbon-containing rings. Some common heterocylic parent compounds are shown below.

Pyridine

Quinoline

Piperidine

Pyrrole

Indole

Pyrrolidine

Imidazole

Benzimidazole

Pyrimidine

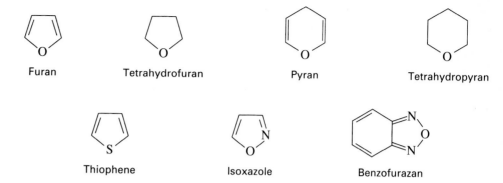

Furan　　　Tetrahydrofuran　　　Pyran　　　Tetrahydropyran

Thiophene　　　Isoxazole　　　Benzofurazan

Many heterocycles are aromatic since N, O, or S atoms may use their unshared pairs of electrons to participate with pi bonding electrons to satisfy Hückel's $4n + 2$ rule. Heterocyclic compounds make up a large proportion of known organic compounds and of biologically active molecules. For example, seven out of the ten largest-selling prescription drugs in the U.S. in 1988 were heterocyclic compounds (Table 43–1).

TABLE 43–1　THE TEN LARGEST-SELLING PRESCRIPTION DRUGS IN THE U.S. IN 1988

(table courtesy of Dr. Norman Schmuff)

Rank	Drug	Hetero-cycle	Manufacturer	Therapeutic class	Revenue $million
1	Zantac	√	Glaxo / Sankyo	H2-antagonist	1,480
2	Tagamet	√	SmithKline	H2-antagonist	1,130
3	Tenormin		ICI	beta-blocker	870
4	Capoten	√	Squibb	acetylcholinesterase inhibitor	780
5	Vasotec	√	Merck	acetylcholinesterase inhibitor	640
6	Adalat	√	Bayer / Takeda	Ca antagonist	590
7	Naprosyn		Syntex	Non-steroidal anti-inflammatory	560
8	Voltaren		Ciba-Geigy	Non-steroidal anti-inflammatory	540
9	Feldene	√	Pfizer	Non-steroidal anti-inflammatory	520
10	Ceclor	√	Eli Lilly	Oral cephalosporin	520

One common method for preparing heterocycles is by condensation of carbonyl compounds with amines accompanied by loss of water. For example, in Experiment 43.1 a pyrrole ring is formed from an amine and two ketone groups.

$$\langle\phi\rangle\!-\!NH_2 + CH_3-\overset{O}{\overset{\|}{C}}-CH_2CH_2-\overset{O}{\overset{\|}{C}}-CH_3 \xrightarrow{H^+}$$

Aniline
MW 93, bp 184°
dens. 1.02 g/mL

2,5-Hexanedione
(acetonylacetone)
MW 114, bp 191°
dens. 0.97 g/mL

2,5-Dimethyl-1-phenylpyrrole
MW 171, mp 50 – 51°

The mechanism for this reaction is the normal acid-catalyzed amine formation between a carbonyl compound and an amine. In this case, however, it occurs twice, accompanied by isomerization of double bonds to form an aromatic ring.

It is also possible for a carbonyl group to condense onto existing aromatic rings in positions allylic to nitrogen. In Experiment 43.2, for example, condensation of acetoacetanilide leads to formation of a quinoline derivative.

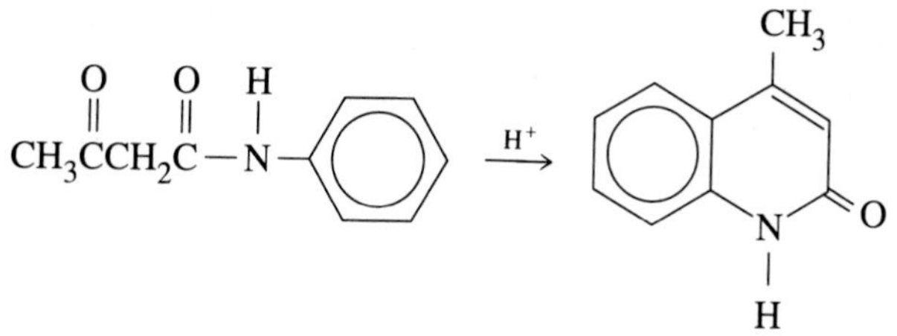

Acetoacetanilide
MW 177, mp 85–86°

4-Methyl-2-quinolinone
(4-methylcarbostyril)
MW 159, mp 224–245°

This reaction proceeds by a mechanism similar to the previous one.

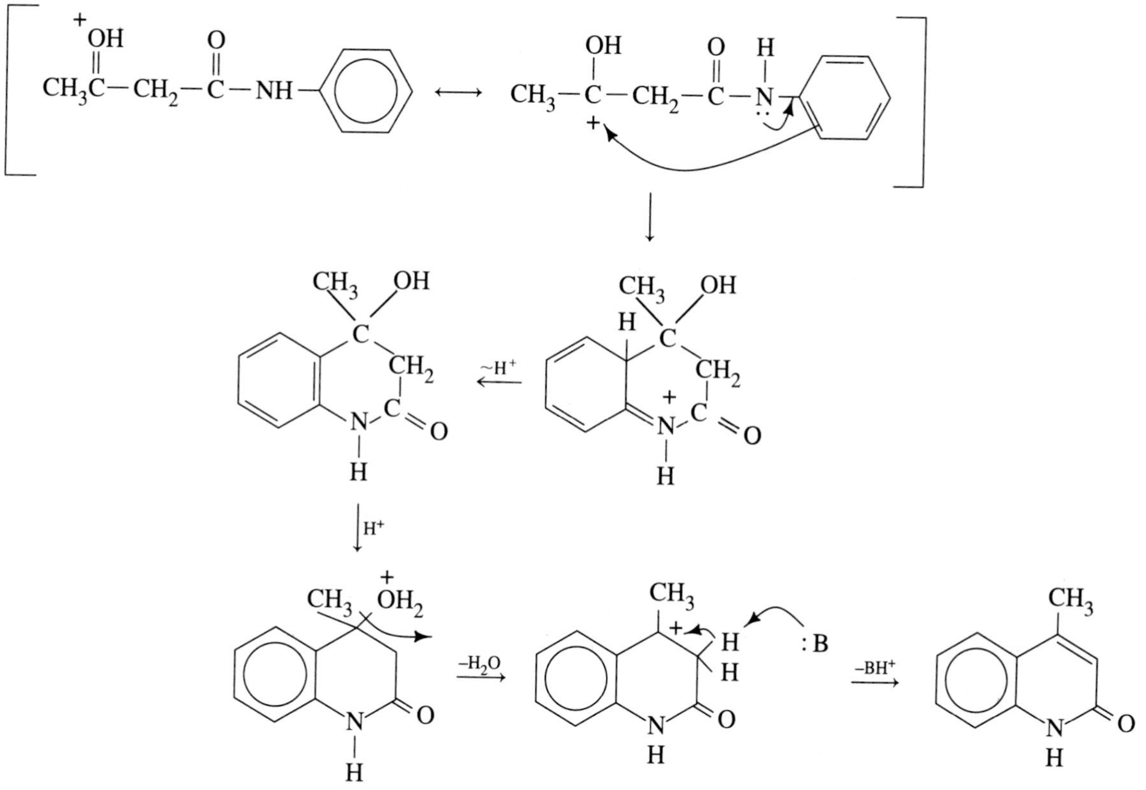

A similar reaction occurs in the formation of benzimidazole from formic acid and *o*-phenylenediamine in Experiment 43.3.

o-Phenylenediamine
MW 108, mp 103–105°

Formic acid
MW 46, bp 100–101°
dens. 1.22 g/mL
mp 8°

Benzimidazole
MW 118, mp 172–173°

This reaction involves an esterification as well as an imine formation.

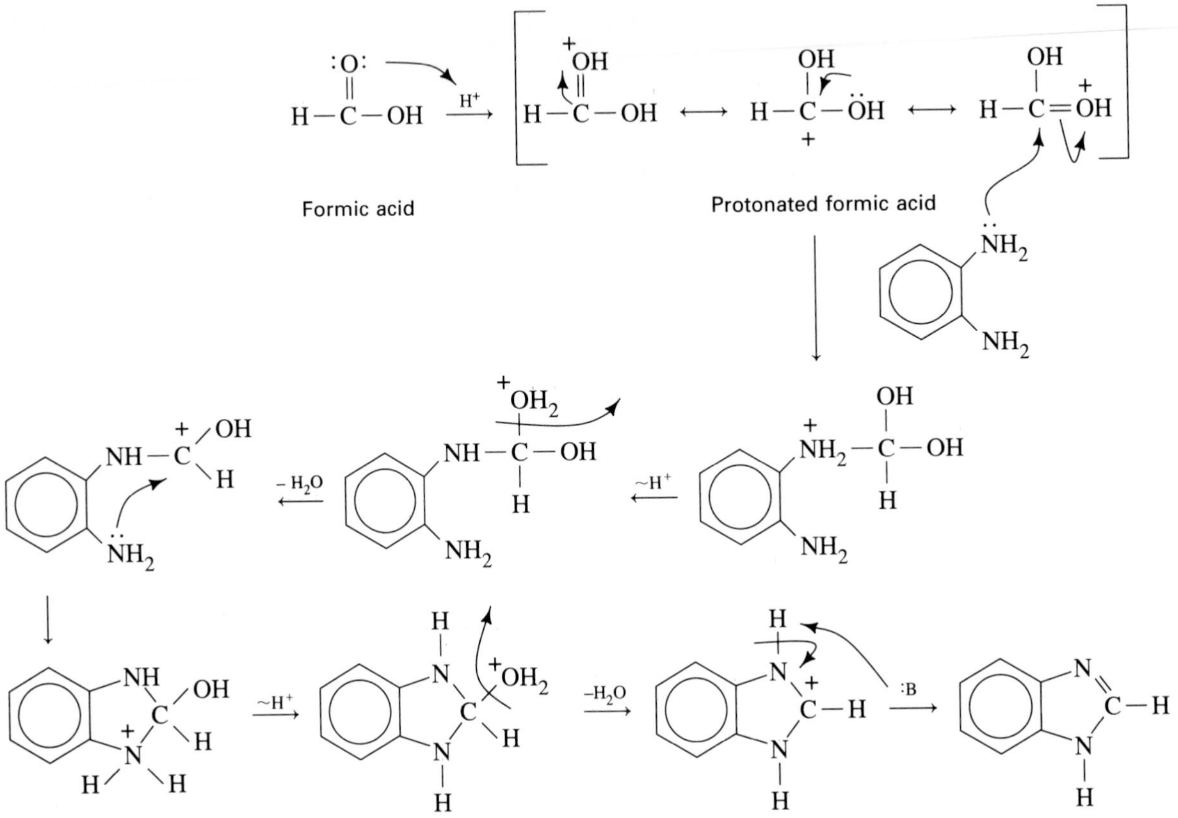

Formic acid

Protonated formic acid

Aromatic amines are particularly easy to oxidize. Aniline, for example, on exposure to air during storage, forms dark-colored products, which can include several stages of oxidation. Several possible oxidation products of aniline are shown below with the oxidation numbers of the nitrogens involved.

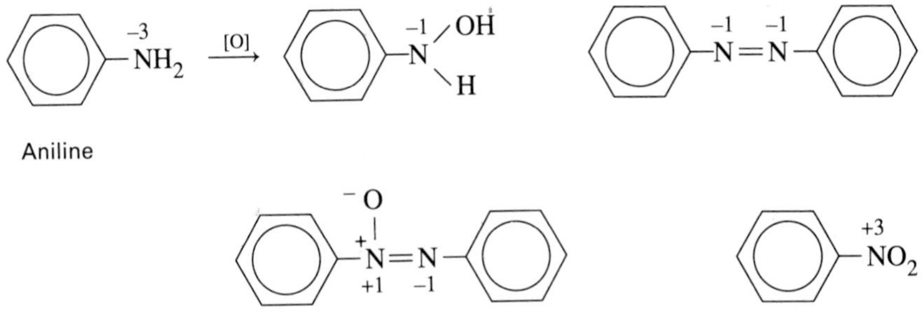

Aniline

In the case of a *o*-nitroaniline it is possible to oxidize the amine using household bleach and to obtain a heterocycle from attack of an oxygen atom from the nitro group. This reaction is carried out in Experiment 43.4.

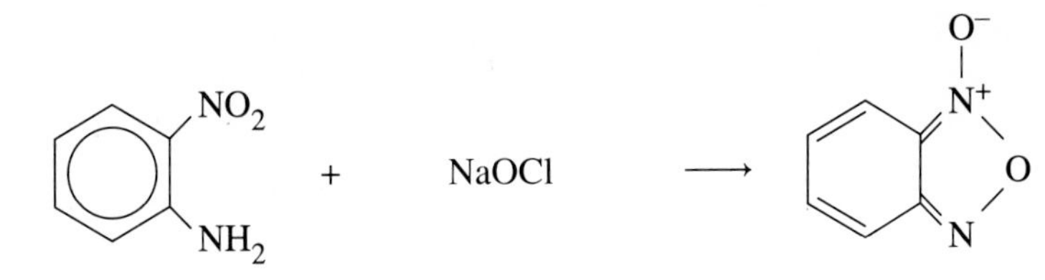

o-nitroaniline
MW 138, mp 71−73°

Sodium hypochlorite
(bleach)
5.25% aq. solution
MW 74.4

Benzofurazan-1-oxide
(benzofuroxan)
MW 136, mp 69−71°

Note that the amine nitrogen has been oxidized from a net oxidation state of −3 (two bonds to H, one to C), to a state of −1 (two bonds to carbon, one to oxygen).

In a base-catalyzed reaction amines can displace alkoxides from esters to form amides, as in the synthesis of barbituric acid from urea and diethyl malonate in Experiment 43.5.

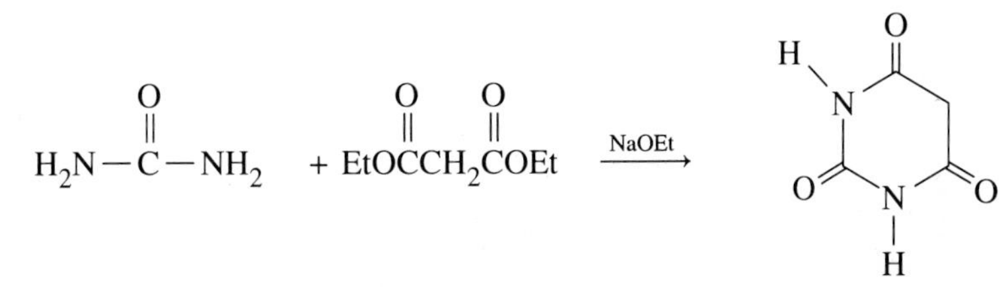

Urea
MW 60, mp 133−135°

Diethyl malonate
MW 160, bp 199°
dens. 1.06 g/mL

Barbituric acid
MW 128, mp 248−252°d

The active methylene group of diethyl malonate can condense with an aldehyde, as in Experiment 43.6. Condensation of salicylaldehyde with diethyl malonate, followed by transesterification of the phenol onto the ethyl ester, affords 3-carbethoxycoumarin.

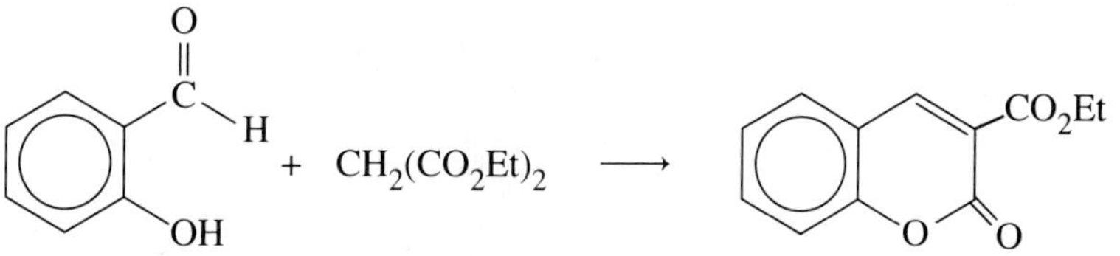

Salicylaldehyde
MW 122, bp 197°
dens. 1.15 g/mL

Diethyl
malonate
MW 160, bp 199°
dens. 1.06 g/mL

3-Carbethoxycoumarin
(ethyl 2-oxo-2H-1-benzopyran-
3-carboxylate)
MW 218, mp 92−93°

QUESTIONS

1. Which of the parent compounds shown at the beginning of this section are aromatic?

2. How would you prepare each of the following?

1-(3-Ethylphenyl)-
2,5-dimethylpyrrole

2-Methylbenzimidazole

5-Phenylbarbituric acid

7-Methoxy-4-methylcarbostyril

REFERENCE FOR ALL EXPERIMENTS IN THIS SECTION

Wolthuis, E., "The Synthesis of Heterocyclic Compounds," *Journal of Chemical Education* (1979) *56*, 343.

EXPERIMENT 43.1 PREPARATION OF 2,5-DIMETHYL-1-PHENYLPYRROLE

Estimated Time:
2.5 hours

Prelab

1. Calculate the millimoles corresponding to 2.0 mL of aniline and 2.5 mL of 2,5-hexanedione.

2. What is the theoretical yield of 2,5-dimethyl-1-phenylpyrrole?

Special Hazards

Aniline is highly toxic and a cancer suspect agent.

Procedure

In a 50-mL round-bottomed flask, place aniline (2.0 mL, _____ g, _____ mmol), 2,5-hexanedione (2.5 mL, _____ g, _____ mmol), 1 drop concentrated HCl, and a boiling chip. Add a reflux condenser and heating mantle and reflux the mixture for 45 minutes. Pour the hot solution into a beaker containing 1.5 mL of water and 1 mL of concentrated HCl and collect the precipitate by suction filtration. Wash with water and recrystallize the crude product from methanol–water. To do this, dissolve the product in methanol (about 15 mL), bring to a boil, and add water dropwise until the cloudiness barely persists. Add a little more methanol to obtain a clear solution and allow it to cool. Collect the product, wash with a little water, and allow it to air dry. Record the mass and melting point.

EXPERIMENT 43.2 PREPARATION OF 4-METHYL-2-QUINOLINONE

Estimated Time:
2.5 hours

Prelab

1. Calculate the millimoles corresponding to 1.77 g of acetoacetanilide, and the theoretical yield of 4-methyl-2-quinolinone.

Special Hazards

Concentrated H_2SO_4 is highly corrosive. Exercise due care.

Procedure

On top of a magnetic stirrer/hot plate, prepare a hot water bath by placing 50 mL of water in a 150-mL beaker and heating to 70–90°. Meanwhile, in a 25-mL Erlenmeyer flask place a magnetic stir bar, 2.0 mL of concentrated H_2SO_4, and acetoacetanilide (1.77 g, _____ mmol). Place the flask in the hot-water bath and stir magnetically for 20 minutes. Pour the hot mixture into 25 mL of water, collect the product by suction filtration, wash with water and then with a little methanol (to hasten drying), allow the product to air dry, then record its mass and melting point.

EXPERIMENT 43.3 PREPARATION OF BENZIMIDAZOLE

Estimated Time:
2.5 hours

Prelab

1. Calculate the millimoles corresponding to 2.7 g of o-phenylenediamine and 5.0 mL of 90% formic acid, density 1.20 g/mL.
2. What is the theoretical yield of benzimidazole?

Special Hazards

o-Phenylenediamine is an irritant and cancer suspect agent. Formic acid is corrosive; if any gets on the skin, *wash with water immediately and thoroughly.*

Procedure

In a 50-mL round-bottomed flask, place a boiling chip, *o*-phenylenediamine (2.7g, _____ mmol) and 90% formic acid (5.0 mL; *use caution*). Reflux the mixture for 1 hour. Using about 5 mL of water, rinse the dark-colored solution into a beaker, cool in an ice bath, and add 8 mL of concentrated ammonia to precipitate the product (which is an organic base). Collect the product by suction filtration and wash with water. Dissolve the crude, moist product in 40 mL of near-boiling water, add 0.5 of activated charcoal, and boil briefly. Gravity filter the hot solution using a heated funnel and a little filter aid on the paper. Allow the solution to cool to near room temperature, then cool it in an ice bath. Collect the product by suction filtration, wash with a little water, and air dry. Record the mass and melting point.

EXPERIMENT 43.4 PREPARATION OF BENZOFURAZAN-1-OXIDE

Estimated Time:
2.0 hours

Prelab

1. Calculate the millimoles corresponding to 1.25 g of NaOH, 2.0 g of *o*-nitroaniline, and 20 mL of 5.25% aqueous NaOCl (0.75 *M*).
2. What is the theoretical yield of benzofurazan-1-oxide?

Special Hazards

o-Nitroaniline is an irritant and highly toxic. Both sodium hydroxide and sodium hypochlorite are corrosive. Avoid contact and wash with large volumes of water if contact occurs.

Procedure

In a 50-mL Erlenmeyer flask, place a small magnetic stirring bar, 1.5 mL of water, NaOH (1.25 g, _____ mmol), 15 mL of methanol, and *o*-nitroaniline (2.0 g, _____ mmol). Warm and stir the mixture on a hot plate/stirrer to dissolve the solids. Once the solution is homogeneous, add an ice bath and cool the mixture with stirring. Mount an addition (separatory) funnel securely above the flask and in it place 20 mL of a commercial bleach solution (5.25% aqueous sodium hypochlorite). While maintaining stirring and cooling, add the bleach solution dropwise during 10–15 minutes at the rate of about 1 drop every 2 seconds, keeping the temperature in the range of 5–10°. After the addition, continue stirring for 5 minutes, then collect the yellow product by suction filtration, wash with water, and recrystallize from methanol–water (add water to a boiling methanol solution of the product to approach the cloud point). Collect the product, allow to air dry, and record the mass and melting range.

Since the melting ranges of the starting material and product are so close, additional confirmation of the identity of the product is necessary. One simple test is to take a *mixed melting point*. To do this, mix an equal quantity of your product and starting material and grind together finely on a small watch glass. The melting range of this mixture should be significantly lower than that of the pure compounds, since now they contaminate each other. Record this mixed melting range and state what it indicates about the identity of your product.

EXPERIMENT 43.5 PREPARATION OF BARBITURIC ACID

Estimated Time:
2.5 hours

Prelab

1. Calculate the millimoles corresponding to 2.0 mL of 21% (by weight) ethanolic NaOEt (density 0.868 g/ml), 2.5 mL of diethyl malonate, and 1.0 g of urea.
2. What is the theoretical yield of barbituric acid?

Special Hazards

While some substituted barbiturates have hypnotic and sedative properties, barbituric acid itself does not.

Procedure

In a dry 50-mL round-bottomed flask, place a Teflon-coated magnetic stirring bar, 20 mL of anhydrous ethanol, and 2.0 mL of a commercial solution of 21% sodium ethoxide in ethanol. Add diethyl malonate (2.5 mL, _____ g, _____ mmol) and urea (1.0g, _____ mmol). Underneath the flask, place a magnetic stirring plate and a heating mantle. Establish stirring (the flask must be over the center of the plate), add a reflux condenser, and boil for 1.5 hours. Pour the hot solution into a beaker and add 15 mL of hot water. Stir to suspend the precipitate and add 1 mL of concentrated HCl to dissolve it. Cool the flask in an ice bath and scratch to induce crystallization, which may be slow. When the product has precipitated, collect by suction filtration, wash with water, allow to air dry, and record the mass and melting point.

EXPERIMENT 43.6 PREPARATION OF 3-CARBETHOXYCOUMARIN

Estimated Time:
2.5 hours

Prelab

1. Calculate the millimoles and volumes corresponding to 1.5 g of salicylaldehyde and 2.2 g of diethyl malonate.
2. Calculate the theoretical yield of 3-carbethoxycoumarin.

Special Hazards

Salicylaldehyde is an irritant and toxic. Piperidine is highly toxic and flammable.

Procedure

In a 50-mL round-bottomed flask, place salicylaldehyde (1.5 g, _____ mmol), diethyl malonate (2.2 g, _____ mL, _____ mmol), 5 mL of ethanol, 3 drops of piperidine, 1 drop glacial acetic acid, and a boiling chip. Add a heating mantle and reflux condenser and boil the mixture for 1.5 hours. Rinse the hot solution into a beaker with 1 mL of ethanol, then with 9 mL of water. If cloudy, reheat to 50–60°. Allow the solution to cool to near room temperature, then cool in an ice bath. Collect the product by suction filtration and wash it with a mixture of 10 mL of water and 5 mL of ethanol. Allow the product to air dry, then record the mass and melting point.

Carbohydrates

Overview

Carbohydrates may be classified as monosaccharides, disaccharides, or polysaccharides, depending on the number of units covalently bound. Monosaccharides can be described by the number of carbons in the chain, so that five carbons is a pentose, six is a hexose, and so on. A chart of the D-pentoses and hexoses appears below. The D, of course, specifies that the OH group on the next-to-the-bottom carbon points to the right in a standard Fischer projection. A handy mnemonic device for remembering the names of the D-aldohexoses is: All Altruists Gladly Make Gum In Gallon Tanks, representing *all*ose, *al*trose, *gl*ucose, *ma*nnose, *gu*lose, *i*dose, *gal*actose, and *ta*lose (Figure 44-1). Of these eight aldohexoses, only glucose, mannose, and galactose occur naturally.

By convention, the more highly oxidized end of the molecule (containing an aldehyde or ketone group) is written at the top of the Fischer projection. A monosaccharide containing an aldehyde group is an aldose, while one with a keto group is a ketose. So, for example, a five-carbon sugar containing an aldehyde group is an aldopentose.

Monosaccharides exist in equilibrium between the open-chain and cyclic forms. Formation of a hemiacetal or hemiketal group can occur easily between the aldehyde (or ketone) and one of the OH groups to form a five- or six-membered ring. Glucose (an aldohexose), for example, is in equilibrium with its cyclic hemiacetal form, called the pyranose form after the pyran ring it contains.

Glucose
(open form)

Glucose
(Hemiacetal or
pyranose form)

On carbon 1 (the aldehyde carbon) a new asymmetric center is created when the ring forms. This chiral center can either have its OH pointing down in the standard

All altruists gladly make gum in gallon tanks

CHO	CHO	CHO	CHO	CHO	CHO	CHO	CHO
HCOH	HOCH	HCOH	HOCH	HCOH	HOCH	HCOH	HOCH
HCOH	HCOH	HOCH	HOCH	HCOH	HCOH	HOCH	HOCH
HCOH	HCOH	HCOH	HCOH	HOCH	HOCH	HOCH	HOCH
HCOH	HCOH	HCOH	HCOH	HCOH	HCOH	HCOH	HCOH
CH₂OH	CH₂OH	CH₂OH	CH₂OH	CH₂OH	CH₂OH	CH₂OH	CH₂OH
D–Allose	D–Altrose	D–Glucose	D–Mannose	D–Gulose	D–Idose	D–Galactose	D–Talose

Figure 44–1 Mnemonic for Aldohexoses   (artist: Eric Kvatek)

Haworth projection of the ring, in which case it is called the α form, or the OH can be up, which is the β form.

Similarly, fructose (which is the 2-ketohexose corresponding to glucose) forms a 5-membered hemiketal, called fructofuranose.

D-Fructose

D-Fructose, furanose form
(Haworth projection)

To visualize the reaction, it would be wise to build a model of the open-chain form and convert it to the ring.

When drawing Haworth projections from Fischer projections (or vice versa) it may be helpful to remember the mnemonic: If it were *left up* to me, I'd get *right down* to work. In other words, for a corresponding carbon, *left* on a Fischer projection means *up* on the Haworth and *right* means *down*. In glucose, for example, the hydroxyl groups on the open-chain form are (starting at carbon 2, the first chiral center) right–left–right–right, or down–up–down–down (DUDD). This makes the configuration of glucose easy to remember, since there is lots of glucose in Milk Dud(d)s. Similarly, galactose can be remembered as DUU.

A sugar containing a hemiacetal group is called a *reducing* sugar, because in the open-chain form it contains an aldehyde group. The aldehyde can be easily oxidized, causing the reduction of another species. This is the basis of Benedict's test. In this test blue Cu^{2+} ions are reduced to Cu^+ by a reducing sugar. The Cu^+ appears in the form of Cu_2O, a brick-red precipitate. Overall, the balanced reaction can be written

$$RCHO + 2Cu^{2+} + 2H_2O \longrightarrow RCOOH + Cu_2O + 4H^+$$

Barfoed's test relies on this same reaction, and on the fact that reducing monosaccharides produce Cu_2O faster than reducing disaccharides.

Carbohydrates react with phenylhydrazine to form osazones, which serve as useful derivatives because of their crystallinity and sharp melting points. This reaction is a dehydration occurring between an aldehyde or ketone and the hydrazine, forming an imine bond.

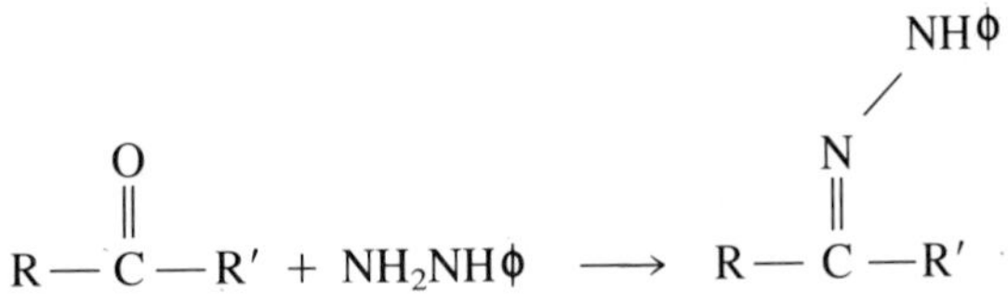

$$\underset{\displaystyle R-\overset{\displaystyle O}{\overset{\|}{C}}-R'}{} + NH_2NH\phi \longrightarrow R-\overset{\overset{\displaystyle NH\phi}{\diagup}}{\underset{\|}{N}}{C}-R'$$

Under the reaction conditions, hydroxyl groups adjacent to the carbonyl groups may be oxidized and undergo further condensation.

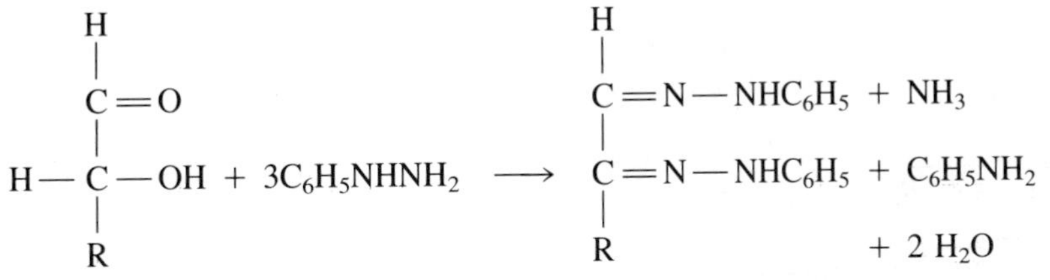

$$
\begin{array}{c}
H \\
| \\
C = O \\
| \\
H - C - OH \\
| \\
R
\end{array}
+ 3C_6H_5NHNH_2 \longrightarrow
\begin{array}{l}
H \\
| \\
C = N - NHC_6H_5 + NH_3 \\
| \\
C = N - NHC_6H_5 + C_6H_5NH_2 \\
| \\
R \qquad\qquad\quad + 2\,H_2O
\end{array}
$$

Osazone

One equivalent of phenylhydrazine has oxidized a secondary alcohol. Because this type of reaction is occurring, several monosaccharides give identical osazones. Glucose, fructose, and mannose, for example, all yield the osazone shown above. The rate of formation and crystalline structure of the osazone yield additional information. The unknown you will be given is one of those listed in Table 44-1.

TABLE 44-1 POSSIBLE CARBOHYDRATE UNKNOWNS

Carbohydrate	Decomposition point	mp of osazone	mp of acetate	Specific rotation
L-Arabinose	160	166	—	+103
Cellobiose	225	198	—	+35
Fructose				−89.5
D-Galactose	170	196	132(β) or 95(α)	+81.7
Glucose (anh.)	146	205	132(β) or 112(α)	+528
Glycogen	240	198	—	—
Lactose	203	200d	100	+52.5
Starch	Very broad	—	—	—
Sucrose	185	205	70	+66.5
D-Xylose	145	163	141	+18.7

EXPERIMENT 44.1 STRUCTURE DETERMINATION OF A CARBOHYDRATE

**Estimated Time:
2.0 hours**

Procedure

Obtain a sample of an unknown carbohydrate from your instructor. Record the unknown number and take its decomposition (melting) point. In a 50-mL Erlenmeyer flask, dissolve 0.25 g of the carbohydrate in 25 mL of deionized water to form a 1% solution, which will be used in several of the tests. Carry out as many of the following tests as are needed for identification. For some of the tests you will prepare three clean test tubes, labeled POS (positive), NEG (negative), and UNK (unknown). Any tests yielding questionable results should also be carried out on an authentic sample of the carbohydrate suspected.

Benedict's test for reducing sugars. In a 250-mL beaker on a hot plate, prepare a boiling-water bath. Prepare three test tubes using 1 mL (20 drops) each of a 1% glucose solution (positive), deionized water (negative), and the 1% solution of your unknown. To each tube, add 5 mL of Benedict's solution. Place the tubes in the boiling water bath for 2–3 minutes, then remove them and record the appearance. A precipitate that is red, brown, or yellow is a positive result for a reducing sugar. No change or a color change in the solution without formation of a precipitate is negative.

Barfoed's test. This test should only be carried out if the Benedict's test was positive, since the Barfoed's test only applies to reducing sugars. Prepare three tubes containing 1 mL each of 1% solutions of glucose (POS for monosaccharide), sucrose (NEG for monosaccharide), and your unknown. To each tube add 5 mL of Barfoed's reagent and place them in the boiling-water bath for 10 minutes. Observe and record the results.

Osazone formation. Place 0.30 g of the unknown carbohydrate in a test tube, 0.6 g of phenylhydrazine hydrochloride (*caution!*), 0.9 g of sodium acetate trihydrate, 0.15 g of sodium bisulfite, and 6 mL of water. Mix well and place the tube in the boiling-water bath for 30 minutes with occasional swirling. At the end of this time, place the tube in an ice bath. When it has cooled, collect the precipitated osazone by suction filtration, wash with 2–3 mL of methanol, and allow it to air dry. Record the decomposition point of the osazone.

Seliwanoff's test for ketoses.   Prepare three test tubes containing (respectively) 10 drops of 1% solutions of fructose (positive), glucose (negative), and the unknown. To each tube add 3 mL of water and 9 mL of Seliwanoff's reagent. Check the time and place the tubes in the boiling water bath for exactly 2.5 minutes. Record the results.

Iodine test for starch.   In three test tubes place (respectively) 1 mL of 1% starch (positive), deionized water (negative), and your unknown. To each tube add 1 drop of iodine solution, observe, and record the results.

Hydrolysis of sucrose.   Carry out this procedure only if you suspect you have sucrose. In three test tubes place 5 mL (respectively) of a 1% solution of sucrose (positive), deionized water (negative), and your unknown. To each tube add 2 drops of concentrated HCl and heat them in the boiling-water bath for 10 minutes. Cool them in ice and neutralize each tube with a few drops of 10% NaOH solution until they are just basic to litmus. This requires about 20 drops per tube. Test each tube with Benedict's reagent as described above and record the results.

Acetate formation.   In a small beaker place 0.4 g of the unknown carbohydrate and 5 mL of 2-methylimidazole. Heat briefly on a steam bath to dissolve. Remove from the heat, add 1.5 mL of acetic anhydride, and swirl intermittently for 10 minutes. Add water (10 mL) and cool in an ice bath for 10 minutes. Collect the precipitated crystals of the acetate by suction filtration, record a crude melting point, and recrystallize from a minimum of ethanol. Collect the purified product by suction filtration, wash with a little ethanol, air dry, and record the melting point.

REFERENCES

Pasto, D. J., and C. R. Johnson, *Organic Structure Determination*, Prentice-Hall, Englewood Cliffs, N.J. 1969.

Shriner, R. L., R. C. Fuson, and D. Y. Curtin, *The Systematic Indentification of Organic Compounds*, Wiley, New York, 1964.

SECTION **45**

Polymers

Overview

The discovery and use of polymers represent some of the greatest advances of organic chemistry in the twentieth century. Synthetic fibers make up a large fraction of our clothing, carpeting, and other fabrics, and plastics have revolutionized the range of materials available for many purposes.

A polymer consists of repeating subunits which may be identical or alternating, in other words of the form -A-A-A-A-A- or -A-B-A-B-A-B-, where A and B are different subunits. Polyethylene, polypropylene, and polystyrene provide examples of the former type, while nylon 6,10 illustrates the latter.

Examples of some polymers and their precursors appear in Table 45-1.

TABLE 45–1 POLYMERS AND THEIR PRECURSORS

Monomer	Polymer
$CH_2 = CH_2$ Ethylene	$+ CH_2 - CH_2 +_n$ Polyethylene
$CH = CH_2$ $\|$ CH_3 Propylene	$+ CH - CH_2 +_n$ $\|$ CH_3 Polypropylene
$CH = CH_2$ Styrene	$+ CH - CH_2 +_n$ Polystyrene

$$NH_2(CH_2)_6NH_2 \ + \ \overset{\overset{\displaystyle O}{\parallel}}{ClC}(CH_2)_8\overset{\overset{\displaystyle O}{\parallel}}{CCl} \ \xrightarrow{\ OH^-\ } \ [-NH-(CH_2)_6NH-\overset{\overset{\displaystyle O}{\parallel}}{C}(CH_2)_8\overset{\overset{\displaystyle O}{\parallel}}{C}-]_n$$

1,6-Hexanediamine   Sebacoyl chloride   Nylon 6,10

Note the amide bonds formed in Nylon 6,10 by loss of HCl from an amine and and acid chloride. Proteins, of course, are also polymers linked by amide bonds with amino acid subunits, where the side chains R vary depending on which of the 20 common amino acids are present.

$$+NH-CH-\overset{\overset{\displaystyle O}{\parallel}}{C}\underset{\displaystyle |}{}+_n$$
$$\underset{\displaystyle R}{|}$$

A protein

In writing polymers this way we are leaving out the first and last fragments of the chain, which must necessarily be different from the internal units because the end pieces will only make one bond each to the chain.

Experiments 45.1 and 45.2 are two polymerizations to form polystyrene and Nylon 6, 10. Both experiments can be carried out conveniently in one three-hour laboratory period. The reaction of styrene occurs by a radical mechanism while the formation of nylon occurs by nucleophilic substitution. For polystyrene the overall reaction is:

The mechanism is as follows.
Initiation

t-Butyl peroxybenzoate   t-Butoxy radical   Benzoyloxy radical

Propagation. Let R• represent either of the radicals formed in the initiation step.

Intermediate radical
stabilized by resonance

Then

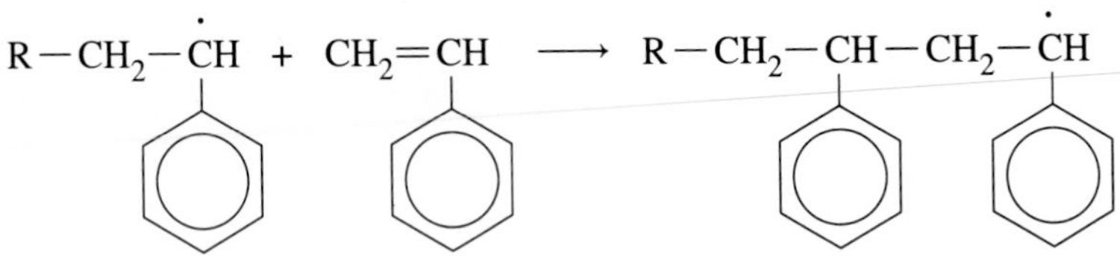

$$R-CH_2-\overset{\cdot}{C}H \;+\; CH_2=CH \;\longrightarrow\; R-CH_2-CH-CH_2-\overset{\cdot}{C}H$$

(with phenyl groups)

This chain-growth step repeats numerous times to give

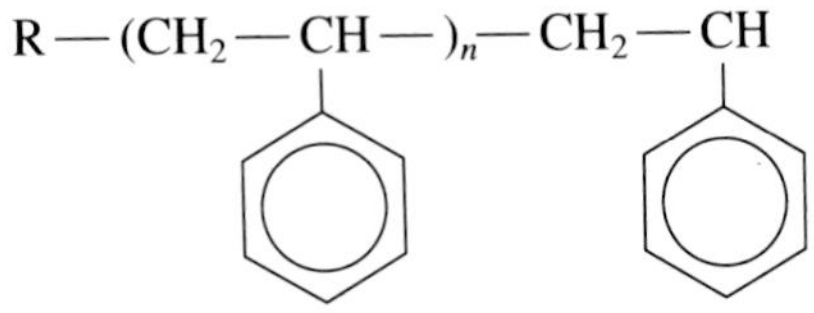

$$R-(CH_2-CH-)_n-CH_2-\overset{\cdot}{C}H$$

(with phenyl groups)

where n is some large integral number of styrene units.

Termination. This reaction terminates with a disproportionation reaction, in which one of two identical radicals removes a hydrogen from the other to form an alkane and an alkene. Disproportionation, of course, means the reaction of two identical molecules to yield two distinct products.

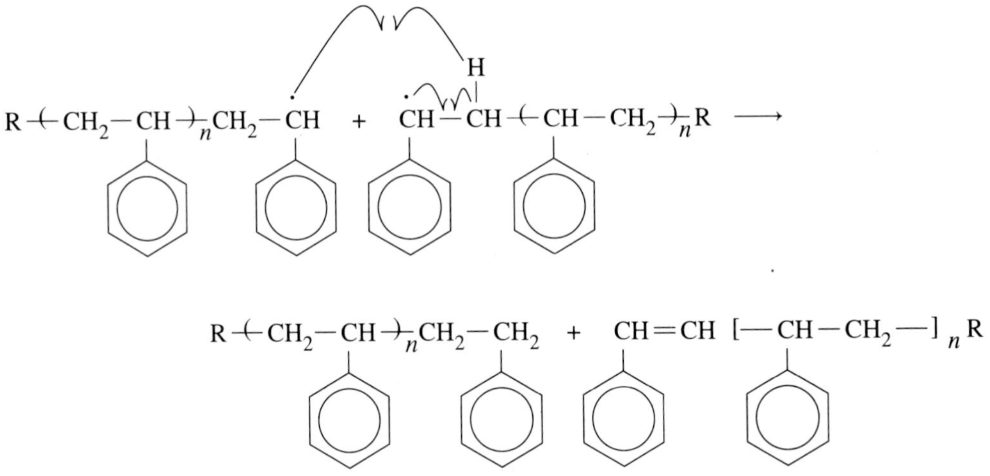

$$R\text{-}\!\!\left(CH_2-CH\right)_{\!n}\!CH_2-\overset{\cdot}{C}H \;+\; \overset{\cdot}{C}H\overset{|}{-}CH\text{-}\!\!\left(CH-CH_2\right)_{\!n}\!R \;\longrightarrow$$

(with phenyl groups)

$$R\text{-}\!\!\left(CH_2-CH\right)_{\!n}\!CH_2-CH_2 \;+\; CH=CH\; [-CH-CH_2-]_n\,R$$

(with phenyl groups)

For the polymerization of nylon 6,10, the overall reaction is

$$n\;ClC\overset{O}{\overset{\|}{\;}}(CH_2)_8\,C\overset{O}{\overset{\|}{\;}}Cl \;+\; n\;NH_2(CH_2)_6NH_2 \;\xrightarrow{\;OH^-\;}\; [-C\overset{O}{\overset{\|}{\;}}(CH_2)_8C\overset{O}{\overset{\|}{\;}}NH(CH_2)_6NH-]_n$$

Sebacoyl chloride   1,6-Hexanediamine
MW 239, bp 168°   MW 116, mp 42–45°   Nylon 6,10
at 12 torr
dens. 1.12 g/mL

The unusual feature of this reaction is that it occurs at the interface between two phases. The acid chloride, dissolved in hexane, stays in a layer atop the diamine and NaOH in water. Where the two layers meet, a reaction occurs to form a film. As this filament is pulled out of the beaker, more forms to give a long, continuous strand. This process is shown in Figure 45-1.

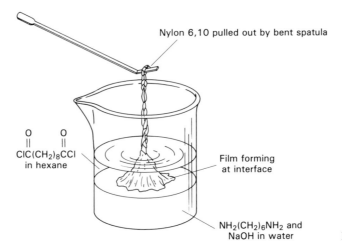

Nylon 6,10 pulled out by bent spatula

$$\overset{O}{\underset{\|}{ClC}}(CH_2)_8\overset{O}{\underset{\|}{CCl}}$$
in hexane

Film forming
at interface

$NH_2(CH_2)_6NH_2$ and
NaOH in water

Figure 45–1 Making Nylon 6,10

Experiment 45.3 illustrates the preparation of a cellulose triacetate film. This type of material is used in photographic film, clear blister packaging, toys, beads, and eyeglass frames.

Cellulose is the polymer of glucose containing β 1,4 linkages.

Glucose ⇒ Cellulose

By soaking cotton balls (cellulose) in acetic acid to expand the fiber structure, then adding excess acetic anhydride to acetylate all free hydroxyl groups, the triacetate can be prepared. On dissolution in an appropriate solvent followed by evaporation of the solvent, this material forms a strong, clear film.

APPLICATION: PLASTICS

Plastics can be classified into two categories, according to how they are affected by heat. *Thermoplastic* resins soften and liquefy on heating and thus may be molded and cooled to yield an immense variety of shapes. They may be remelted and reformed easily.

In contrast, *thermosetting* resins liquefy initially on heating, but on continued heating they undergo extensive covalent cross-linking of the polymer chains, forming solids. Since the reaction has changed the chemical structure permanently, a thermosetting resin cannot be remelted and molded again.

Approximately 90% of the resins produced today are thermoplastics; these fall into the two classes *crystalline* and *amorphous*. Crystalline resins possess relatively sharp melting points, while amorphous resins soften over broad ranges. The following table lists some common crystalline and amorphous thermoplastics.

Thermoplastics are usually purchased in the form of cubic, cylindrical, or spherical pellets 3 mm in diameter, which are shipped in containers ranging from 25-kg bags to railroad hopper cars. The resins are stored in silos until processing.

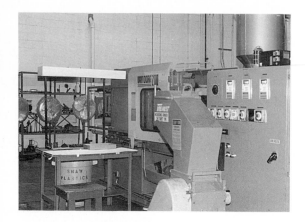

Plastics Manufacturing (courtesy of Shaw Plastics Corporation)

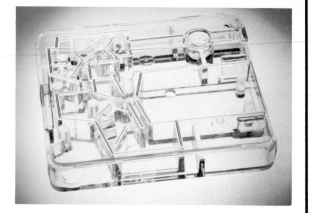

This component for medical diagnostic equipment to analyze blood was injection-molded using acrylic compounds. The piece is about the size of an audio cassette and contains many intricate interior walls and passageways (courtesy of CYRO Industries).

SOME COMMON THERMOPLASTICS

Crystalline	Amorphous
Low-density polyethylene	Acrylonitrile–butadiene–styrene (ABS) terpolymer
High-density polyethylene	Cellulose acetate
Polypropylene	Phenylene-oxide-based resins
Nylon (many types)	Polycarbonates
Polyester (many types)	Poly(methyl methacrylate) (PMM)
	Polystyrene
	Poly(vinyl chloride) (PVC)
	Styrene–acrylonitrile copolymers (SAC)

Approximately 50% of all products manufactured from resins are made by *extrusion* methods. Extrusion is a continuous process for producing an infinite length of material with a particular cross section. Products formed in this way include films and sheets, pipes, tubing, moldings, and coated wire.

In the extrusion process, the resin is melted in an extruder and pumped through a die, which gives it the desired shape. The melt is cooled by passing it through a water trough, then cut to the desired length or wound into a roll or coil.

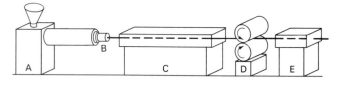

Extruder and Die

QUESTIONS

1. Draw the mechanism and predicted product for the radical polymerization of acrylonitrile, $CH_2 = CH — CN$.

2. If more catalyst were used, how would the average chain length of polystyrene be affected?

3. Kevlar (structure below) is an extremely strong, lightweight plastic used in bulletproof vests and airplane parts. How would you make it?

$$\left[\overset{\overset{\textstyle O}{\|}}{C} - \bigcirc - \overset{\overset{\textstyle O}{\|}}{C} - NH - \bigcirc - NH \right]_n$$

Kevlar

EXPERIMENT 45.1 POLYMERIZATION OF STYRENE

Estimated Time:
1.5 hours

Prelab

1. How does washing styrene with NaOH solution remove the radical inhibitor *t*-butyl catechol?

2. Why is the solubility of polystyrene much greater in xylene than in methanol?

Special Hazards

Styrene is toxic and flammable. *t*-Butyl peroxybenzoate is a strong oxidizing agent and irritant.

Procedure

First the commercial styrene must be washed to remove the inhibitor. Place 10 mL of styrene in a 125-mL separatory funnel and add 20 mL of 1 *M* NaOH solution. Shake and vent several times to ensure thorough mixing. After the layers have separated, discard the lower (aqueous) layer. Wash the styrene layer remaining in the separatory funnel with one 20-mL portion of water to remove any residual NaOH. Dry the styrene over $CaCl_2$ in a small stoppered Erlenmeyer flask for about 5 minutes.

Now to carry out the polymerization, place the styrene in a 100-mL round-bottomed flask and add 25 mL of xylene as a solvent. Add 7–8 drops of the radical initiator *tert*-butyl peroxybenzoate, then place a reflux condenser on top and add a heating mantle with Variac. Making sure that water is flowing through the condenser, heat the mixture to reflux, and allow it to reflux for 30 minutes. A relatively high setting can be used on the variable transformer since the boiling points of xylene and styrene are above 140.°

After the reflux period, cool the solution to room temperature using an ice bath. Pour half of the mixture into a beaker containing 100 mL of methanol. Collect the white polystyrene precipitate using suction filtration, rinse it with a little fresh methanol, and allow it to air dry. Pour the remaining half of the polystyrene solution onto a watch glass and allow the methanol to evaporate; a clear film of polystyrene will appear.

EXPERIMENT 45.2 FORMATION OF NYLON 6,10

Estimated Time:
0.5 hours

Prelab

1. Calculate the millimoles corresponding to 10 mL of 5% (by weight) aqueous 1,6-hexanediamine, 10 mL of 5% (by weight) sebacoyl chloride in hexane (density 0.66 g/mL), and 0.2 g of NaOH. Assume that the densities of 5% solutions are about the same as the densities of the solvents.
2. What is the theoretical yield of nylon 6,10?

Special Hazards

1,6-Hexanediamine, sebacoyl chloride, and sodium hydroxide are all corrosives. Sebacoyl chloride is also a lachrymator. Hexane is flammable.

Procedure

In a 50-mL beaker, place 10 mL of a 5% aqueous solution of 1,6-hexanediamine. Add 0.2 g (one pellet) of solid NaOH and swirl to dissolve. Tilt the beaker and carefully pour down the side 10 mL of 5% sebacoyl chloride in hexane. The organic layer remains above the aqueous layer and the polymer (nylon 6,10) forms at the interface. Reach in with a bent spatula and pull out the film; it continues forming a long strand as it is pulled. Rinse the material with water and dry it on a paper towel. After pulling a few strands, stir the layers vigorously to obtain a ball of nylon. Rinse and dry this material. After examining it, dispose of it in the trash. Do not put the polymer or monomer solutions down the sink since they may clog it.

REFERENCE

SHAKHASHIRI, B. Z., *Chemical Demonstrations*, Vol. 1, University of Wisconsin Press, Madison, Wis., 1983, p. 213.

EXPERIMENT 45.3 PREPARATION OF CELLULOSE TRIACETATE

Estimated Time:
1.5 hours

Prelab

1. In what excess is the 5 mL of acetic anhydride of the amount necessary to triacetylate 1 g of cellulose?

Special Hazards

Acetic acid, sulfuric acid, and acetic anhydride are all corrosive and irritating. Methanol is flammable and toxic; dichloromethane is toxic and irritating.

Procedure

Preparation of cellulose triacetate.   Using a 150-mL beaker on a hot plate, prepare a water bath maintained at 70–80°C. In a 50-mL Erlenmeyer flask, place 5 mL of glacial acetic acid and 1 drop of concentrated H_2SO_4. Add 1.0 g of cotton balls and stir to saturate them with the acid solution. Stopper this flask and immerse it in the water bath for 20 minutes to swell the cotton. Add 5 mL of acetic anhydride to the flask, restopper, and continue heating for another 20 minutes with occasional stirring. At the end of this time the cotton should be dissolved and the acetylation of the hydroxyl groups completed.

Add to the syrupy solution 15 drops water to hydrolyze any remaining acetic anhydride. Use caution and do not stopper the flask—foaming may occur. Place the unstoppered flask in the hot-water bath for 5 minutes to complete the hydrolysis.

Pour the solution into a 250-mL beaker and add about 50 mL of water; cellulose triacetate precipitates. Collect this precipitate by suction filtration. To wash it, resuspend and stir it in another 50 mL of water, then refilter. Place another filter paper on top of the product in the Büchner funnel and press to remove as much water as possible.

Preparation of a film.   In a 50-mL beaker in the fume hood, place 9 mL of dichloromethane and 1 mL methanol. Bring the solvent mixture to a boil on a hot plate in the hood, then add cellulose triacetate gradually to obtain a nearly saturated solution. Remove the beaker from the heat and add some anhydrous Na_2SO_4 to absorb any remaining water. Decant a small amount of the solution onto a watch glass or into a small beaker, and allow the solvent to evaporate. With care, remove the film produced from the surface and tape it in your lab book.

REFERENCE

LAMPMAN, G. M., D. W. FORD, W. R. HALE, A. PINKERS, and C. G. SEWELL, "Polymer Preparations in the Laboratory," *Journal of Chemical Education* (1979) *56*, 626.

SECTION 46

Enzymes in Organic Synthesis

Overview

Enzymes are the envy of the synthetic organic chemist. These remarkable chemical "factories" can bring reagents together in the correct orientation, stretch bonds that are to be broken, and change conformation to release the products formed, freeing the enzyme for another catalytic cycle. These protein catalysts can speed up reactions by factors of up to 10^{20}.

The formation of an enzyme–substrate complex, reaction to form the product, and release of that product may be illustrated as

$$E + S \quad \rightleftharpoons \quad E \cdot S \quad \rightleftharpoons \quad E \cdot P \quad \rightleftharpoons \quad E + P$$

| Enzyme + substrate | Enzyme–substrate complex | Enzyme–product complex | Enzyme + product |

The protein chain of an enzyme adopts a conformation with a pocket suited to fit the substrate. This lock-and-key mechanism allows it to operate on only one type of substrate in the presence of many other compounds.

The catalytic site of an enzyme often contains acidic or basic groups and metal ions to aid in the reaction. As with any catalyst, an enzyme lowers the activation energy of a reaction by providing an alternative pathway (Figure 46-1). It also catalyzes both the forward and reverse reactions. Which direction a reaction takes depends on the ΔG (Gibbs free energy) of the reaction, which in turn depends on concentrations, temperature, enthalpy (H), and entropy (S) changes during the reaction.

Since the first use of rennet (a dried extract of calf stomach containing the enzyme rennen) to curdle milk for cheese making thousands of years ago, human beings have used enzymes for chemical transformations *in vitro*. Enzyme preparations (as well as whole microorganisms) have also been used for centuries in beer and wine making and tanning of leather.

Despite this long history of industrial use, only in recent years have organic chemists begun to use enzyme in laboratory-scale synthesis. This late start was probably due to the expense and difficulty of purifying enzymes, and the impression that their use was outside the scope of the synthetic chemist. Recently, though, many enzymes have become commercially available and chemists have begun to find that their remarkable properties can be extremely useful in the laboratory.

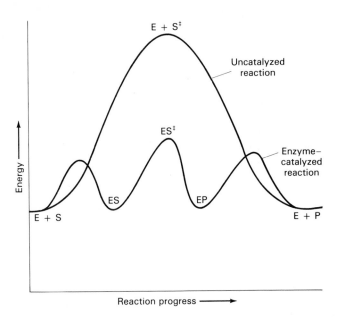

Figure 46–1 Reaction Energy Differential for Enzyme-catalyzed vs. Uncatalyzed Reactions

TABLE 46–1 INTERNATIONAL CLASSIFICATION OF ENZYMES (CLASS NAMES, CODE NUMBERS, AND TYPES OF REACTIONS CATALYZED)

1. Oxido-reductases
   (oxidation-reduction reactions)

   Acting on $-$ CH $-$ OH

   Acting on $-$ C $=$ O
   Acting on $-$ CH $=$ CH $-$

   Acting on $-$ CH $-$ NH$_3^+$

   Acting on $-$ CH $-$ NH $-$
   Acting on NADH; NADPH

2. Transferases
   (transfer of functional groups)
      One-carbon groups
      Aldehydic or ketonic groups
      Acyl groups
      Glycosyl groups
      Phosphate groups
      S-containing groups

3. Hydrolases
   (hydrolysis reactions)
      Esters
      Glycosidic bonds
      Peptide bonds
      Other C $-$ N bonds
      Acid anhydrides

4. Lyases
   (addition to double bonds)

   $-$ C $=$ C $-$

   $-$ C $=$ O

   $-$ C $=$ N $-$

5. Isomerases
   (isomerization reactions)
      Racemases

6. Ligases
   (formation of bond with ATP cleavage)
      C $-$ O
      C $-$ S
      C $-$ N
      C $-$ C

Each enzyme can use a certain range of compounds as substrates; the range may be broad or narrow. The synthetic chemist wants as broad a range as possible while maintaining good regio-, stereo-, and possibly enantioselectivity. The classes of reactions catalyzed by enzymes are given in Table 46-1. Remember that enzymes are named by adding the suffix -ase to the name of the substrate (the molecule on which the enzyme acts).

To choose an enzyme to catalyze a particular reaction, one would check a table of enzymes by reaction class, then look up the known range of substrates. Many enzymes require the presence of coenzymes, as shown in Tables 46-2 and 46-3.

TABLE 46–2   COENZYMES USED IN GROUP-TRANSFERRING REACTIONS

Species transferred	Coenzyme
Hydrogen atoms (electrons)	Nicotinamide adenine dinucleotide (NAD)
Hydrogen atoms (electrons)	Nicotinamide adenine dinucleotide phosphate (NADP)
Hydrogen atoms (electrons)	Flavin mononucleotide (FMN)
Hydrogen atoms (electrons)	Flavin adenine dinucleotide (FAD)
Hydrogen atoms (electrons)	Coenzyme Q (CoQ)
Aldehydes	Thiamine pyrophosphate (TPP)
Acyl groups	Coenzyme A (CoA)
Acyl groups	Lipoamide
Alkyl groups	Cobamide coenzymes
Carbon dioxide	Biocytin
Amino groups	Pyridoxal phosphate
Methyl, methylene, formyl or formimino groups	Tetrahydrofolate coenzymes

Although extracellular, soluble enzymes such as α-chymotrypsin have long been available and relatively inexpensive, only recently has a wide variety of enzymes become available, many costing up to $1,000 per gram. Even though enzymes are used in very small quantities, these prices are still prohibitive if the enzyme is lost each time. The development of *immobilization* methods made the use of these expensive enzymes for chemical processes feasible. Immobilization methods have revolutionized enzyme technology by allowing the easy recovery and reuse of enzymes by binding them onto a solid support.

Common methods of immobilization include:

1. Adsorption on glass, silica, activated clay, and so on
2. Trapping in a polyacrylamide gel
3. Covalent binding to derivatized glass, cellulose, nylon membranes, or synthetic polymers
4. Microencapsulation by trapping in a thread of inert polymer

The activity is rarely decreased significantly and immobilization usually greatly increases the stability (i.e., active lifetime) of the enzyme.

Not only do enzymes operate at very mild conditions of temperature and pH but they also catalyze many reactions that are difficult to achieve by conventional means. Synthetically useful enzymic reactions may be broadly classified into four groups.

**TABLE 46–3 SOME ENZYMES
CONTAINING OR REQUIRING METAL IONS
AS COFACTORS**

Zn^{2+}
  Alcohol dehydrogenase
  Carbonic anhydrase
  Carboxypeptidase
Mg^{2+}
  Phosphohydrolases
  Phosphotransferases
Mn^{2+}
  Arginase
  Phosphotransferases
Fe^{2+} or Fe^{2+}
  Cytochromes
  Peroxidase
  Catalase
  Ferredoxin
Cu^{2+} (Cu^{+})
  Tyrosinase
  Cytochrome oxidase
K^{+}
  Pyruvate phosphokinase
   (also requires Mg^{2+})
Na^{+}
  Plasma membrane ATPase
   (also requires K^{+} and Mg^{2+})

1. *Separation of enantiomers by selective reaction.* In Japan a full-scale industrial plant is in operation, producing L-amino acids by the action of L-amino acid acylase on a chemically synthesized mixture of D- and L-acylamino acids. The unchanged D-acyl-amino acid is racemized and recycled.

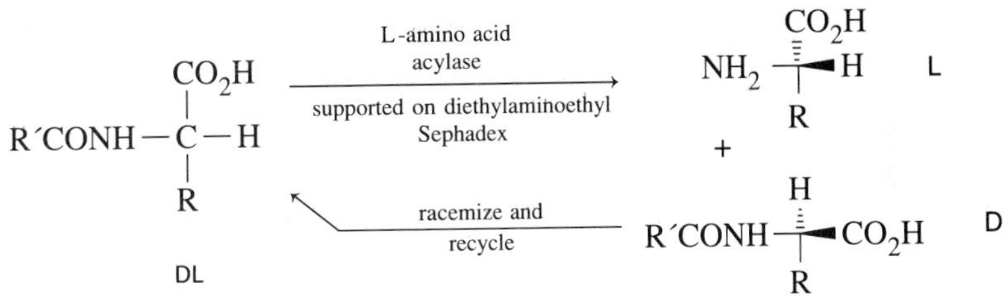

2. *Asymmetric induction (creating a chiral product from an achiral starting material.* Treatment of an aldehyde or unsymmetrical ketone with alcohol dehydrogenase and NADH often results in exclusive formation of the S-alcohol, as shown below with a tritium-labeled aldehyde.

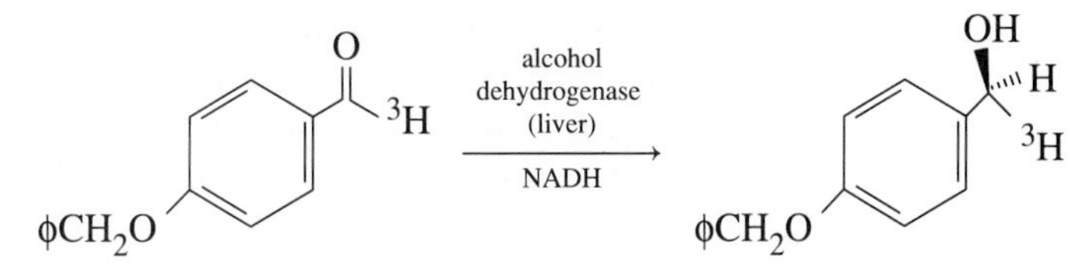

This generalization is known as the Prelog rule, shown below, where L is a large group and S is a small group.

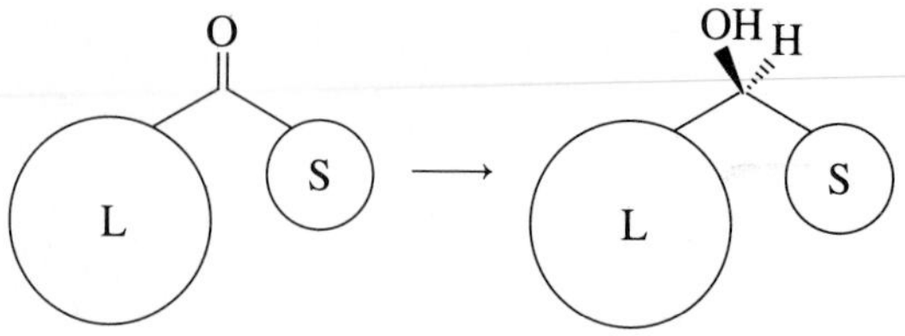

3. *Chemoselective reaction (reaction of only one functional group in a molecule possessing another similar group).* For example, penicillin amidase cleaves the amide group of penicillins to give large quantities of 6-aminopenicillanic acid for synthesis of synthetic penicillins.

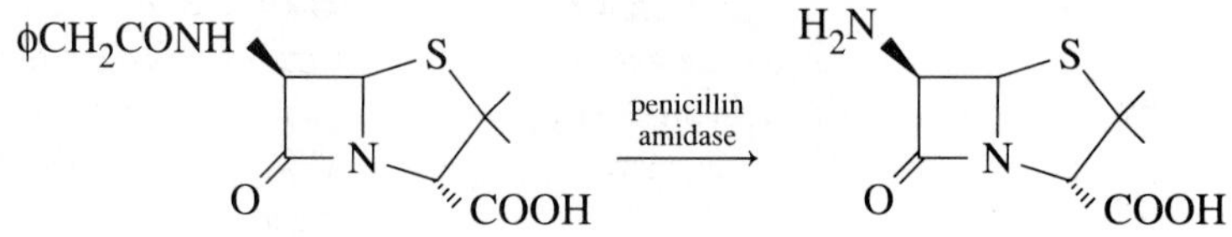

Note that one amide bond is cleaved in the presence of another, strained four-membered cyclic amide (a β-lactam). Using normal acid- or base-catalyzed hydrolysis would cleave the strained β-lactam before the other amide bond.

4. *Functionalization of unactivated carbon.* As you know by now, it is very difficult to introduce a functional group into a saturated hydrocarbon. Radical chlorination and bromination are two methods you have studied to do this. There are organisms, however, which possess enzymes that cannot only replace a C—H bond with a C—OH bond, but can do it with complete stereoselectivity at only one site in a molecule containing many C—H bonds. Most examples known of this type involve hydroxylation of a steroid nucleus by whole microorganisms. A common reaction is 11α-hydroxylation.

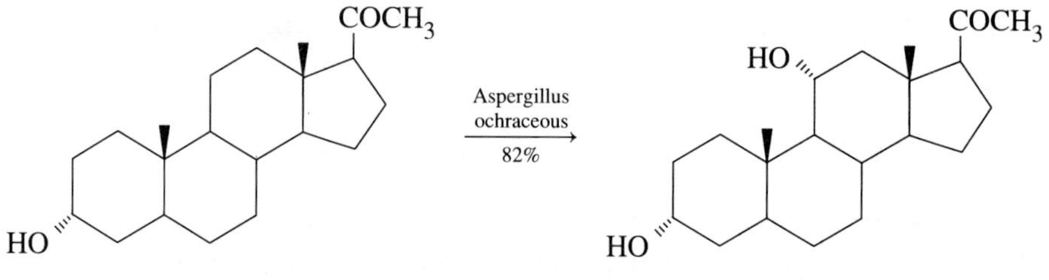

These applications appear to be only the beginning. Advances in immobilizing multienzyme systems and recycling expensive cofactors promise to make enzymes important synthetic tools of the future. In Experiment 46.1 you will carry out the reduction of vanillin to vanillyl alcohol using whole yeast. An enzyme in the yeast (an oxidoreductase, an alcohol dehydrogenase working in reverse), using the NADH and H^+ present, carries out the same reduction as that performed in Experiment 24.1 using the chemical reducing agent $NaBH_4$.

$$\underset{\substack{\text{Vanillin}\\ \text{MW 152, MP 81-83°}}}{\text{(vanillin)}} \xrightarrow[\text{yeast}]{\text{growing}} \underset{\substack{\text{Vanillyl alcohol}\\ \text{MW 154, mp 115°}}}{\text{(vanillyl alcohol)}}$$

Vanillin
MW 152, MP 81–83°

Vanillyl alcohol
MW 154, mp 115°

In this experiment you will not isolate the vanillyl alcohol formed (although that could easily be done) but will monitor the progress of the reaction with TLC. By removing and working up (extracting) aliquots at 10-minute intervals, then developing TLC slides and observing the relative amounts of vanillin and vanillyl alcohol, you can observe how fast the reaction is occurring. This technique of removing small aliquots from a reaction mixture to monitor progress by TLC is very useful for many reactions in that

APPLICATION: WINE MAKING

Wine Chemistry Lab (courtesy of Mirassou Vineyards, San Jose, CA)

In brief (and perhaps inelegant) outline, the process of wine making consists of mixing the juice of freshly crushed grapes (or other fruit) with a desired culture of yeast (which may be the "must" on the grapeskin) and "yeast nutrient" (diammonium phosphate) to speed fermentation. The net process of fermentation converts one glucose molecule into two molecules of ethanol and two molecules of CO_2,

yielding two molecules of ATP for the microorganism's energy metabolism

$$C_6H_{12}O_6 + 2Pi + 2ADP \longrightarrow$$
$$2CH_3CH_2OH + 2CO_2 + 2\ ATP$$

Glucose

Of course, many other chemical changes are occurring during fermentation as well, to give each wine its individual bouquet. Extraneous microorganisms must be carefully excluded to prevent further oxidation of ethanol to acetic acid or formation of other undesirable compounds.

The relative amounts of sugar and alcohol during the fermentation are monitored by pycnometers, which indicate the density of the solution. A higher density indicates more sugar; a lower one, more alcohol. Fermentation ceases naturally at approximately 12–14% alcohol by volume, since this concentration kills the yeast.

The wine can be made more acidic (tart) by adding citric, malic, or tartaric acid. If desired, wine can be made less acidic by addition of sodium bicarbonate. Wineries employ many chemists to monitor the progress of fermentations and to study the reactions and products involved.

it enables you to decide if the desired product is being formed (by comparison with a TLC spot of an authentic sample of the product), if by-products are being formed (extra spots), and when the reaction is complete (the starting material is gone).

EXPERIMENT 46.1 REDUCTION OF VANILLIN BY YEAST

Estimated Time:
2.5 hours

Prelab

1. What are some advantages of enzymic reactions? Some disadvantages?
2. Describe the process of spotting, developing and visualizing a TLC slide.
3. Which should have a higher R_f value, vanillin or vanillyl alcohol, and why?
4. What would the theoretical yield of vanillyl alcohol be from 0.20 g of vanillin?

Special Hazards

Ether is extremely flammable; keep away from heat or flames. Chloroform is a suspected carcinogen on long-term exposure; use only in the fume hood and avoid breathing fumes. UV light can cause cataracts and skin cancer: Do not look at the light or shine it on your skin. Set the TLC slide down and hold the light above it to visualize.

Procedure

In a 250-mL Erlenmeyer flask atop a magnetic stirring plate, place vanillin (0.20 g), sucrose (table sugar, 10 g), water (100 mL), and a Teflon-coated magnetic stirring bar. Stir for a few minutes to dissolve the vanillin (solubility 1 g per 100 mL water at room temperature).

While the mixture is stirring, label six 15 × 125 mm test tubes 0, 10, 20, 30, 40 and 50, respectively, representing the minutes of reaction time that will elapse before the samples are taken.

When the vanillin has dissolved, add to the flask Fleischmann's "active dry" yeast (7 g, one packet or weighed from bulk). Record the time. As soon as possible (within the first minute) after addition, remove by Pasteur pipette a sample of approximately 1 mL and place it in a test tube (not one of the labeled ones). The volume is not critical, since you are only interested in the *relative* amounts of vanillin and vanillyl alcohol present. Add about 1 mL of ether to the tube, and use a Pasteur pipette to mix the phases thoroughly in order to extract the compounds of interest into the ether. Draw both phases into the Pasteur pipette and squirt them out forcefully several times to ensure adequate mixing. When this is finished, carefully draw off the ether (upper) layer and place it in the tube labeled 0. This is the zero time sample. Repeat the taking of a sample and extraction at 10-minute intervals, up to 50 minutes. Add a boiling stone to each tube, warm continuously to remove the ether, and add 4 or 5 drops of chloroform to each tube. The reason for this change of solvent is that ether is too volatile for easy spotting on TLC slides. Some water may remain on the top of the $CHCl_3$ but it does not interfere with the TLC analysis.

TLC analysis.   (If necessary, reread Section 4 for general procedures involving TLC.) Obtain three TLC slides (*Caution:* handle only by the edges!) and on each mark three dots lightly in pencil about 1 cm from the end, as shown in Figure 46-2. Be sure not to disturb the surfaces of the slides. In pencil label the central spot on each slide M (for mixture) and the outer spots with the sample times, as shown. On the middle dot (M) of each slide spot the authentic vanillin–vanillyl alcohol mixture your instructor has prepared. Be sure to keep the spots small and sharp, using either a filed-flat metal syringe needle or a fine glass capillary drawn out in a flame for spotting. It is best to touch the tip of the spotter two or three times to the spot to ensure adequate sample without having a spot that is too large in diameter.

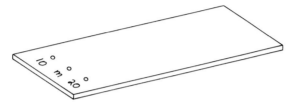

Figure 46–2   Marking a Slide for TLC Analysis

Develop the slides in 1:1 ethyl acetate–toluene in a covered jar. Allow the solvent to rise about two-thirds to three-fourths of the way up the slides before removing them and marking the solvent fronts quickly (before the solvent evaporates) in pencil. After the slides have dried for a couple of minutes, observe them under a UV light and mark any spots visible. (*Caution:* Never look directly at a UV light or shine it on your skin.) Next, place the slides in an iodine chamber for a few minutes, and mark the spots observed. Tape the slides directly into your lab book or make an accurate sketch of them.

QUESTIONS

1. What do the TLC slides tell you about the course of the reaction?

2. What is the approximate half-life of the vanillin under these conditions?

3. What are the approximate R_f values for vanillin and vanillyl alcohol under these conditions?

4. Did you observe any significant quantities of by-products (other spots) formed?

5. What are some advantages of monitoring reactions by TLC?

6. What kinds of reactions can be conveniently monitored by TLC?

REFERENCES

PORTER, R., and S. CLARK, eds., *Enzymes in Organic Synthesis*, Pitman, London, 1985.

BOYER, P. D., *The Enzymes*, 3rd ed., Academic Press, New York, 1971.

JONES, J. B., C. J. SIH, and D. PERLMAN, eds., *Techniques of Chemistry*, Vol. X, Wiley-Interscience, New York, 1976.

SCOTT, D., "Enzymes, Industrial," in H. F. Mark et al., eds., *Kirk–Othmer Encyclopedia of Chemical Technology*, 3rd ed., Vol. 9, Wiley, New York, 1978, p. 173.

SECTION 47

Bio- and Chemiluminescence

Overview

Have you ever seen fireflies twinkling on a warm summer's evening, or observed the blue-green glow as a ship's bow gently cleaves tropical waters at night? These are two examples of *bioluminescence*, the emission of light by living organisms. Thousands of organisms possess the ability to emit light, including some fungi, dinoflagellates, protozoans, bacteria, hydras, sponges, corals, jellyfish, worms, shrimp, clams, snails, insects, squids and fishes. It is believed that in some cases such as fireflies, a distinctive pattern of light emission aids in finding a mate, while in other cases, such as deep-sea fishes, it may aid in vision, defense, or attracting prey.

Fireflies provide the best-studied examples of bioluminescence. In many North American species, it has been shown that courtship begins as the low-flying male emits his flash. A stationary female near the ground responds after a short delay with her own species-specific flash. Depending on the species, fireflies may recognize the light organ pattern, flash pattern, or the interval between male and female flashes.

Some species of lampyrid fireflies appear to use bioluminescence to illuminate their surroundings for improved visibility as they alight. A closely spaced series of flashes emitted just before landing may help them avoid such natural hazards as water and spider webs.

A spectacular sight has been reported of hundreds of fireflies of certain species in Southeast Asia congregating on a single tree and flashing synchronously and rhythmically for hours at a stretch.

Perhaps the most complex known mimicry in the animal kingdom occurs in the genus Photuris. In addition to the flash pattern for their own species, carnivorous Photuris females are able to mimic the codes of other species to attract males which they then devour. Photuris males have responded by developing their own flash pattern mimicry to locate and seduce their own hunting females. I guess this goes to show that dating is not always easy, even for fireflies.

Deep-sea fish also make extensive use of bioluminescence. Examination of the contents of trawl nets has revealed that over 70% of all sea creatures found below 400 meters are bioluminescent. Some deep-sea angler fish, for example, contain cultures of bioluminescent bacteria in their artificial baits to create especially attractive lures.

A fascinating use of bioluminescence for defense (as camouflage) occurs in the use of *counterillumination*, as practiced by the pony fish *Leiognathus equulus*. During daylight, the fish conceals itself from predators looking up from below by emitting light over its entire lower surface. The light emitted is proportional to the ambient light coming from above, so the bottom of the fish blends in with natural light and is extremely difficult to see.

Photoblepharon, the flashlight fish, defends itself by covering the bioluminescent patches below its eyes with a black flap of tissue which it can raise quickly to startle a would-be predator.

Light is, of course, a form of energy, and some chemical reactions emit light instead of, or in addition to, heat. The key to production of light in a reaction is the production of a molecule in a high-energy, electronically excited state. This means that an electron is in a higher-energy orbital than its normal "ground state." As the electron loses energy and falls from its excited state to its ground state, light is emitted.

This process appears below in schematic form. You may be familiar from physics with the equation relating energy (E) to the frequency (ν) of light, $E = h\nu$, where h is Planck's constant, 6.63×10^{-34} Joule-sec. Chemists abbreviate light with the expression $h\nu$.

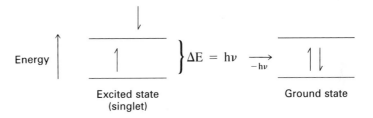

The excited state shown here is a *singlet*, so-called because the spins on the pair of electrons are opposite and cancel one another. This cancellation means only one line is observed in the spectrum. Singlet excited states have very short lifetimes (less than 10^{-9} sec) because it is easy for the higher-energy electron to fall to the lower orbital, the opposite spins being compatible in the same orbital as stated in the Pauli Exclusion Principle. The emission of light by an electron falling from the first excited singlet state to the ground state is called *fluorescence*. If the excited state is a *triplet*, having two electrons with identical spins, its lifetime will be much longer, between approximately 10^{-4} seconds and several minutes. This increased lifespan occurs because it is necessary for reversal of the spin of one electron to occur before both electrons can pair up in the same orbital. This inversion of spin is called "intersystem crossing" which is shown below.

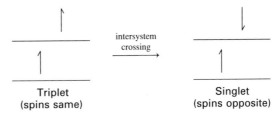

The emission of light by a triplet state undergoing intersystem crossing and falling to the ground state is called *phosphorescence*.

The overall scheme, then, for production of light consists of the creation of a molecule or ion in an excited triplet state, followed by intersystem crossing to a singlet state and emission of light as it drops to the ground state.

$$A \longrightarrow B^* \longrightarrow B^* \xrightarrow{-h\nu} B$$

(triplet)   (singlet)   (ground state)

Bioluminescence generally requires an oxidizable substrate and an enzyme, called a *luciferin*, and a *luciferase*, respectively. The structures of a few of the known luciferins (and their source) appear below.

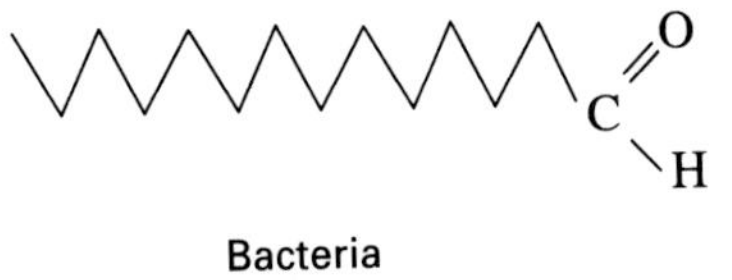

Bacteria

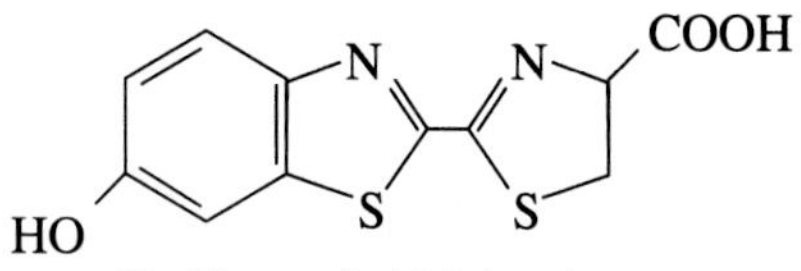

Fireflies and chick beetles

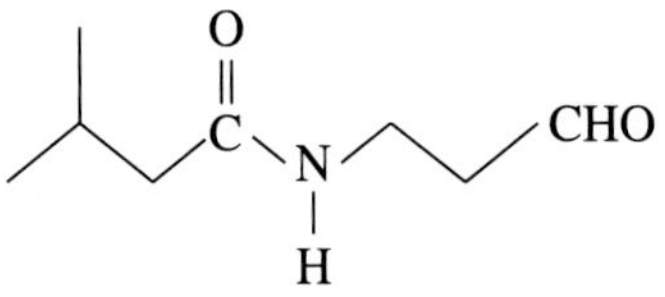

Earthworms

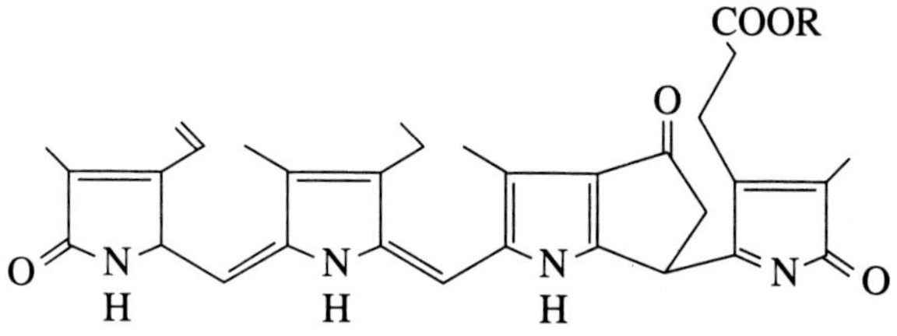

Dinoflagellates

Note the tremendous variability in structure, suggesting independent evolution of bioluminescence in these organisms.

In many cases a luciferin is a low molecular weight heterocyclic compound that undergoes oxidative decarboxylation in the presence of a specific luciferase, yielding an electronically excited oxyluciferin, which then emits light. Schematically, this process may be illustrated as

$$\text{Luciferin} + \text{Luciferase} + O_2 \longrightarrow \text{Oxyluciferin}^* + CO_2 \longrightarrow \text{Oxyluciferin} + h\nu$$

Most bioluminescent systems involve these three organic components: luciferin, luciferase, and oxyluciferin. They also require oxygen as either O_2, peroxide (HOO^-), or superoxide (O_2^{2-}). Some systems also require organic cofactors or metal ions. While all of the reactions are not thoroughly understood, it appears that for fireflies and click beetles, the reaction occurs according to the scheme below.

A chemiluminescent reaction, while not requiring an enzyme, also generally requires oxygen, since it usually produces an electronically excited species through the decomposition of a peroxide. In Experiment 47.1 you will synthesize the molecule luminol which, in the presence of hydrogen peroxide, loses nitrogen and forms an excited triplet state. As this molecule undergoes intersystem crossing and decay to the ground state, blue-green light is emitted.

Synthesis of luminol

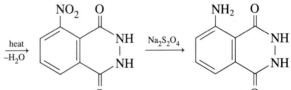

| 3-Nitrophthalic acid mp 222° MW 211 | Hydrazine MW 32.1 | 5-Nitrophthalhydrazide | Luminol (5-aminophthalhydrazide) mp 332° MW 177 |

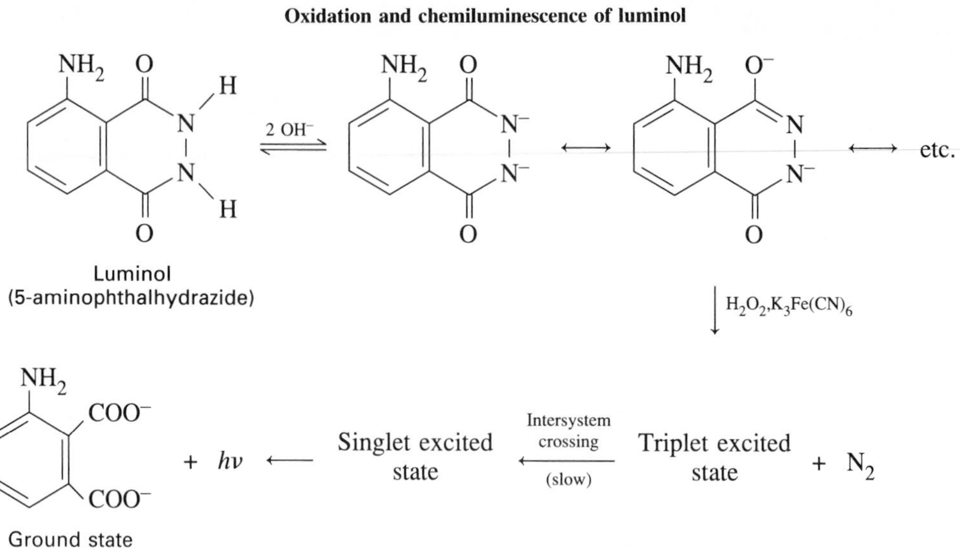

Oxidation and chemiluminescence of luminol

Luminol
(5-aminophthalhydrazide)

$H_2O_2, K_3Fe(CN)_6$

Ground state

$+ \; h\nu \; \longleftarrow$ Singlet excited state $\xleftarrow[\text{(slow)}]{\text{Intersystem crossing}}$ Triplet excited state $+ \; N_2$

To prepare luminol in a two-step synthesis, 3-nitrophthalic acid is first heated with hydrazine to eliminate two moles of water, forming two amide linkages in a geometrically favorable six-membered ring. This product is treated with the mild reducing agent, sodium hydrosulfite, to convert the nitro group to an amino group. This product, luminol, exists largely in zwitterionic form, in which the amino group has removed a proton from the enol form of one amide group. The major driving forces for this proton transfer are the basicity of the amine, mild acidity of the amide, and the strong ionic attraction of the protonated amine for the nearby enolate oxygen.

To oxidize the luminol and produce light, base is first added to form the dianion, which has a large number of resonance structures. Recall that if a product has many resonance structures, it is more stable and, therefore, easier to form. In this case this means that the starting material (luminol) must have been fairly acidic, giving up a proton relatively easily to form the resonance-stabilized anion. Now the amide bonds are cleaved oxidatively to give the anion of 3-aminophthalic acid in its triplet excited state. As the slow process of intersystem crossing to the singlet state occurs and the excited electrons fall to the ground state, beautiful blue-green light is emitted.

Since light intensity can be measured very accurately by modern photometers, light emission forms the basis for several highly precise analytical methods. Chemiluminescence (CL) analyses include gas, solution and solid phase reactions which have been used to analyze a variety of samples of environmental and clinical interest. Very low concentrations of metal ions, inorganic ions, gases, carcinogens, drugs, and biological molecules have all been determined with high precision. Chemiluminescence techniques are among the few methods having detection limits comparable to those in radioimmunoassay.

In fact, chemiluminescent reagents such as luminol derivatives can be used in immunoassays in place of radioisotope labels. CL reagents are cheaper and more stable, and radiation hazards are eliminated. As shown below, a diazonium salt derived from luminol can be covalently bound to a biological analyte (substance being analyzed) such as an antigen. After binding with an antibody, an oxidizing agent is added and the resulting chemiluminescence is used to locate and quantify the antigen-antibody complex.

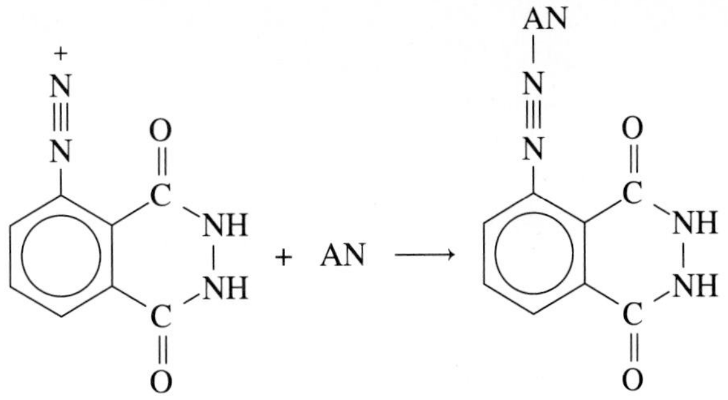

AN = ANALYTE

REFERENCES

Burr, J. G., ed., *Chemi- and Bioluminesence*, Marcel Dekker, New York, 1985.

Gundermann, K-D., *Chemiluminescence in Organic Chemistry*, Springer-Verlag, New York, 1987.

QUESTIONS

1. Why are hydrogen peroxide and potassium ferricyanide strong oxidizing agents?

2. Define the terms ground state, excited state, singlet, triplet, and intersystem crossing.

3. Show the expected products of the following reactions:

(a)

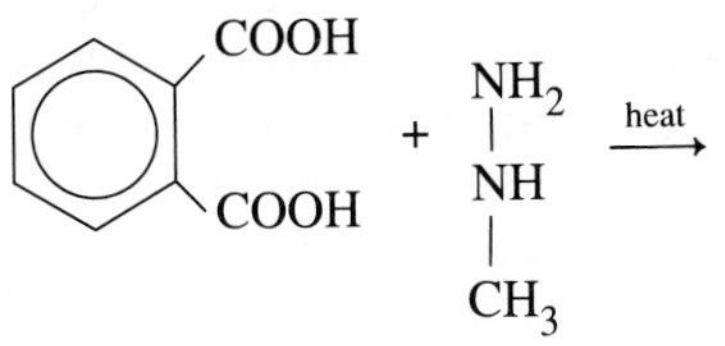

(b)

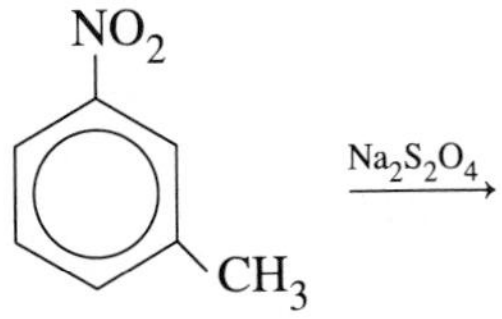

EXPERIMENT 47.1 SYNTHESIS AND CHEMILUMINESCENCE OF LUMINOL

Estimated time:
2.5 hours

Prelab

1. Calculate the mmol corresponding to 0.5 g of 3-nitrophthalic acid and 1 mL of 8% (by weight) aqueous hydrazine (density 1.00 g/mL).
2. What is the theoretical yield of 5-aminophthalhydrazide (luminol)?

Special Hazards

Hydrazine is corrosive and a suspect cancer agent.

Procedure

In a 20 x 150 mm test tube, place 0.5 g of 3-nitrophthalic acid, 1 mL of 8% aqueous hydrazine, and 2 mL of triethylene glycol. Add a boiling chip and thermometer and, in the fume hood, bring the tube to a boil over a burner. Continue boiling to drive off the water present in the tube and then heat strongly to a thermometer reading of 220°. By intermittent gentle heating, maintain a temperature of 210-220° for two minutes. Let the tube air-cool to below 100° and add 10 mL hot water. Cool the tube in ice, then collect the light yellow product (5-nitrophthalhydrazide) by suction filtration, washing with a little water.

Place the wet product in another test tube and add 1 g of sodium hydrosulfite dihydrate and 3 mL of 5% aqueous NaOH. Over a burner in the hood, heat the tube to a boil, stir with a glass rod, and keep the solution near the boiling point for 3 minutes. Add 1 mL of acetic acid and cool the tube in an ice bath. Using suction filtration, collect the pale yellow crystals of luminol and wash with a little water.

To observe the reaction producing light, first dissolve the moist luminol in 10 mL of 5% aqueous NaOH and add 200 mL of water; call this solution A. In another flask, mix 20 mL of 3% aqueous potassium ferricyanide ($K_3Fe(CN)_6$), 20 mL of 3% aqueous hydrogen peroxide (H_2O_2), and 160 mL of water; this is solution B.

Now, in a dark place, put a large Erlenmeyer flask with a funnel on top. Simultaneously pour roughly equal quantities of solutions A and B into the funnel. Swirling the flask will cause the mixture to glow brightly.

Porphyrins

Overview

Without porphyrins there would be no life on Earth. Various substituted porphyrins constitute the "business end" of such indispensable molecules as chlorophyll, hemoglobin, and cytochrome P-450. The essential nutrient vitamin B_{12} contains a closely related corrin ring, which differs from a porphyrin in that two of the pyrrole rings are directly linked. All of these molecules are involved in electron transport: the oxidation and reduction of organic molecules. The porphyrin molecule possesses a cavity magnificently suited for chelating a metal ion.

Porphyrin

A metalloporphyrin

The structure of chlorophyll is shown in Section 5. The structures of the heme portion of hemoglobin and the corrin portion of cobalamin (vitamin B_{12}) are shown below.

Heme

445

Corrin portion of cobalamin (vitamin B$_{12}$)

With the appropriate metal ion and substituents, a metalloporphyrin can be "fine-tuned" as to the ease of gaining or losing an electron, or strength for holding oxygen. Iron in the free state, for example, combines essentially irreversibly with oxygen to form rust, Fe$_2$O$_3$. However, when the iron is held in the heme porphyrin in our red blood cells, the porphyrin nitrogens supply some electron density to the iron so that it can bind O$_2$ reversibly and act as an oxygen carrier.

Porphyrins can be prepared by the condensation of pyrrole with aldehydes in the presence of an acid catalyst. In this experiment you will prepare tetraphenylporphyrin by condensing pyrrole and benzaldehyde in the presence of acetic acid. After collecting the product you will insert Co^{2+} and Cu^{2+} ions to form the corresponding metalloporphyrins.

Pyrrole	Benzaldehyde	*meso*-Tetraphenylporphyrin
MW 67.1	MW 106	MW 615
bp 131°	bp 178–179°	
dens. 0.967g/mL	dens. 1.04 g/mL	

One important group of compounds related to porphyrins is the phthalocyanines. Phthalocyanine is formed by the condensation of four molecules of phthalonitrile. When carried out in the presence of a metal ion, a template effect aids in ring formation as the nitrogens wrap around the metal ion.

When chelated to metals, phthalocyanines often display brilliant colors and great stability. These complexes were discovered by R. P. Linstead in the 1930s. In recent years phthalocyanine pigments have accounted for over 20% of the total value of pigment consumption, particularly for blue and green dyes, because of their moderate price, low solubility, high tinting strength, and low toxicity. They are used in printing inks, textiles and plastics, as well as in paints. Metal phthalocyanines often function as catalysts for oxidation–reduction reactions. Cobalt phthalocyanine, for example, catalyzes the conversion of toluene to benzyl alcohol by oxygen. Either cobalt or iron phthalocyanine can be used in cigarette filters to catalyze the oxidation and removal of nitrogen oxides. Several metal phthalocyanines can catalyze the hydrogenation of carbon monoxide to small (C_1–C_5) hydrocarbons.

REFERENCES

ADLER, A. D., LONGO, R. F., FINARELLI, J. D., GOLDMACHER, J., ASSOUR, J., and KORSAKOFF, L., "A Simplified Synthesis for *meso*-Tetraphenylporphyrin," *Journal of Organic Chemistry* (1967) *32*, 476.

ADLER, A. D., LONGO, R. F., KAMPAS, F., and KIM, J., "On the Preparation of Metalloporphyrins," *Journal of Inorganic and Nuclear Chemistry* (1970) *32*, 2443.

BEREZIN, B. D., *Coordination Compounds of Porphyrins and Phthalocyanine*, Wiley, New York, 1981.

DOROUGH, G. D., MILLER, J. R., and HUENNEKENS, F. M., "Spectra of the Metallo-Derivatives of α,β,γ,δ-Tetraphenylporphine," *Journal of the American Chemical Society* (1951) *73*, 4315.

FALK, J. E., *Porphyrins and Metalloporphyrins*, Elsevier, Amsterdam, 1975.

EXPERIMENT 48.1 SYNTHESIS AND METALLATION OF A PORPHYRIN

Estimated time:
2.5 hours for parts a and b

Prelab

1. Calculate the mmol corresponding to 2.8 mL of pyrrole and 4.0 mL of benzaldehyde.
2. What is the theoretical yield of *meso*-tetraphenylporphyrin?

Special Hazards

Pyrrole and benzaldehyde are flammable. Acetic acid is corrosive and irritating.

Procedure

Preparation of tetraphenylporphyrin. In a 100-mL round-bottomed flask, using a graduated pipette or burette, place pyrrole (2.8 mL, _____ freshly distilled) and benzaldehyde (4.0 mL, _____ g, _____ mmol). Add 40 mL of glacial acetic acid and a few boiling chips. Add a condenser, heating mantle, and variable transformer. Establish water flow through the condenser and heat the mixture to reflux. Allow the solution to reflux for 1 hour, then remove the heat and cool the flask in an ice bath. As it cools, purple crystals of tetraphenylporphyrin should precipitate from the dark solution. Collect the product by vacuum filtration, wash with a little ice-cold methanol, and allow to air dry for 15 minutes as you collect the materials for part B.

Preparation and visible spectra of metalloporphyrins. In the fume hood, prepare an approximately 1.0×10^{-3} M solution of tetraphenylporphyrin by dissolving 0.030 g in 50 mL of glacial acetic acid. *Caution:* Use glacial acetic acid only in the fume hood and avoid exposure to skin or eyes. Label two small beakers *cobalt* and *copper*. Into the two beakers weigh respectively 0.10 g of cobalt(II) acetate and copper(II) acetate. To each beaker add 10 mL of the 1×10^{-3} M tetraphenylporphyrin solution. Add a boiling chip to each and heat both beakers to boiling on a hot plate in the fume hood. When they have been boiling for about 5 minutes, carefully remove them from the heat and cool in an ice bath. For each beaker carry out the following procedure to extract the metalloporphyrin and obtain its visible spectrum.

In the fume hood, pour the contents of the beaker into a separatory funnel, leaving the boiling stone behind. Add toluene (20 mL), shake to extract the metalloporphyrin into the toluene, and vent. Discard the acetic acid layer into a waste beaker and wash the toluene with three 10-mL portions of water to remove any remaining acetic acid. Dry the toluene layer over anhydrous Na_2SO_4, then decant it to fill a solution cell for the visible spectrometer. Run the visible spectrum of the metalloporphyrin over the range 400–700 nm.

Repeat this procedure for the other metalloporphyrin and for 10 mL of the 1×10^{-3} M tetraphenylporphyrin solution to which no metal ion has been added. Label and save the spectra stapled into your notebook.

QUESTIONS

1. Compare the positions and relative intensities of the peaks in your spectra. How do your spectra compare to the literature spectrum of tetraphenylporphyrin in Figure 48-1?

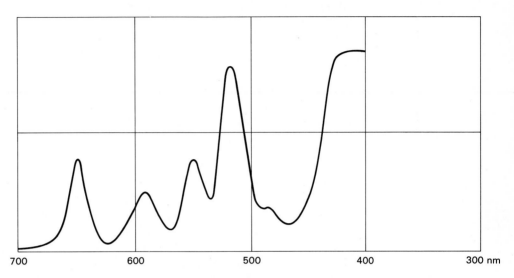

Figure 48–1   Visible Spectrum of Tetraphenylporphyrin

2. How do these absorbencies correlate with the observed colors of the solutions?

3. How would you explain (in terms of electron energy levels) the differences in absorption?

SECTION **49**

Chiral Resolution

Overview

Often during a synthesis of a compound containing a chiral center, a racemic mixture is produced. A racemic mixture, of course, means an equal mixture of two enantiomers. This situation can occur because achiral starting materials and reagents cannot induce net chirality into a molecule. To obtain one enantiomer in purified form we must either use chiral reagents or perform a resolution of the racemic mixture.

There is often a lot of discussion of the differences between natural and synthetic vitamins, drugs, and so on. The major difference between a synthetic and natural version of the same compound is often that the synthetic version exists as a racemic mixture. This means that half of it is the wrong enantiomer (which cannot be used by the body) and that therefore twice as much of the racemic compound must be used to obtain the same effect as the natural material. The FDA requires extensive testing to make sure that the other optical isomer does no harm.

Life requires complex and selective chemical reactions. One way to generate this complexity and selectivity is through use of chiral molecules. Enzymes in the body are, of course, chiral, consisting of chains of L-amino acids. On Earth, life has evolved to use only one set of amino acids (the L-amino acids) and one set of sugars (the D-sugars).

$$
\begin{array}{cc}
\text{COOH} & \text{R} \\
H_2N\!-\!\overset{|}{\underset{|}{C}}\!-\!H & H\!-\!\overset{|}{\underset{|}{C}}\!-\!OH \\
\text{R} & \text{CH}_2\text{OH}
\end{array}
$$

L-Amino acids   D-Sugars

It would seem there was a 50:50 chance that the first organisms would begin using these compounds instead of their enantiomers. Somewhere else in the universe life has probably evolved using the other enantiomers. Perhaps someday if we contact them, we can trade our leftover D-amino acids for their L ones.

One theory about how life evolved on earth to use only these stereoisomers is based on the fact that quartz crystals can form in either right- or left-handed versions and adsorb chiral organic molecules preferentially. This theory postulates that when the crust of the earth was still hot, and molten silica was crystallizing into quartz, by chance one crystal formed in one configuration and this seeded an entire large bed of quartz to crystallize this way. Meanwhile, an organic "soup" was forming in the oceans under the actions of heat, sunlight, and lightning on the primitive atmosphere containing methane, ammonia, water vapor, and molecular hydrogen.

In experiments mimicking these early conditions it has been demonstrated that amino acids, nucleic acids, and numerous other organic molecules are formed. Any chiral molecules formed are, of course, racemic, since the starting reagents are achiral. If one of the enantiomers of a racemic mixture was selectively adsorbed onto the quartz, it would leave an excess of the other enantiomer in solution. Over time, the concentrations in solution would increase and reactions would occur to build more complex molecules. Eventually, a globule of organic material divided (as drops of oil do when they become too large). One globule engulfed another (which means that it had to use the same amino acids), and so on. This is one possible explanation of how life on Earth came to use only one set of enantiomers.

Enantiomers have identical physical properties, such as melting point, boiling point, index of refraction, and solubility. They can be distinguished by the direction they rotate plane-polarized light and by their reactions with other chiral molecules. In Experiment 49.1 you will resolve (separate) a racemic mixture of R- and S-α-phenylethylamine (also called methybenzylamine) by forming a salt with the chiral molecule (+)-tartaric acid. Tartaric acid is very inexpensive, being obtained as a by-product of wine making. It is used in soft drinks, baking powder, and other foodstuffs.

Racemic phenylethylamine
(methylbenzylamine)　　(+)-Tartaric acid

(+)-Amine-(+)-tartrate
Higher solubility in
methanol

(–)-Amine-(+)-tartrate
Lower solubility in
methanol

The salts are diastereomeric since they differ at one asymmetric center (on the amine) and match at two asymmetric centers (on the tartrate).

Diastereomers have different physical properties, and these two can be separated on the basis of their differing solubility in methanol. After preparation, treatment with base regenerates the free amine. The amine is distilled, and its optical rotation is measured in a polarimeter to determine the optical purity.

A polarimeter is an instrument used to measure the optical rotation of organic compounds. It uses a single wavelength of light (usually the sodium D line at 589 nm) from a sodium lamp or an ordinary tungsten bulb with filters. This light passes through a polarizing filter that allows only light vibrating in one plane to pass. The light passes through the sample tube containing the compound of interest dissolved in an achiral solvent, then through a second polarizing filter. The observer rotates the second polarizing filter to obtain minimum transmission (a dark field). If the test compound is achiral, the two filters will be at 90° for a minimum of light to pass through. If, however, the test compound is chiral, the position at minimal light transmission for the second filter will differ from 90° by the amount the solution rotated the light. This setup appears in Figure 49-1.

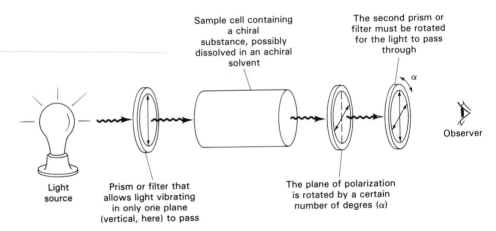

Figure 49–1   Schematic of a Polarimeter

To calculate the specific optical rotation of the compound, one takes the observed rotation (in degrees) and divides by the length of the cell in decimeters (usually 1.0) and the concentration of the solute in g/mL.

$$[\alpha]_D^{20} = \frac{\alpha}{lC}$$

observed rotation

Specific rotation using the sodium D line at 20°C

length of cell in decimeters

concentration of solute in g/mL

Once specific rotation is known, it can be compared to the tabulated specific rotation for that compound to determine optical purity:

$$\text{optical purity} = \frac{\text{calculated } [\alpha]_D^{20}}{\text{literature value of } [\alpha]_D^{20}} \times 100\%$$

The optical purity represents the excess of one enantiomer in the sample. A pure enantiomer has an optical purity of 100% and a racemic mixture 0%.

EXPERIMENT 49.1 RESOLUTION
OF METHYLBENZYLAMINE

Estimated time:
This procedure should be started during one laboratory period, allowing
about an hour, and finished during the next period (2 hours).

Procedure

In a 500-mL Erlenmeyer flask place (+)-tartaric acid (126 mmol, 19.0 g) and methanol (175 mL). Heat the mixture to near boiling to dissolve the tartaric acid. Remove the flask from the heat, then slowly and with caution add racemic α-methylbenzylamine (126 mmol, 16.2 mL); if addition is too fast, the heat from the exothermic acid–base reaction occurring may cause the solution to boil over. Swirl the hot mixture to obtain a clear solution, stopper and insulate the flask, and store it in your drawer for at least 20 hours. If you have available seed crystals of the (−)-amine-(+)-tartrate, these can be added. After standing, collect the crystals by suction filtration. If crystals are prismatic, they are somewhat more optically pure than if they are needles, but proceed in either case. The filtrate contains mostly the more soluble (+)-amine-(+)-tartrate, which may, if desired, be isolated by evaporation, then treated as described below to obtain the (+)-amine. Wash the collected crystals of (-)-amine-(+)-tartrate with methanol, air dry, and weigh. Record the melting range of this salt.

In a 250-mL separatory funnel, place these crystals, 10% NaOH (50 mL), and ether (50 mL). Shake vigorously, venting often, until the crystals dissolve, then draw off the aqueous (lower) layer. Wash the remaining organic layer with brine (20 mL), then dry it over anhydrous Na_2SO_4. Decant into a beaker, add a boiling chip, and remove the ether on a steam bath in the hood. Simple distillation of the residue from a 50-mL round-bottomed flask gives (−)-α-methylbenzylamine, bp 184–186°.

If a polarimeter is available, determine the optical rotation of your product. If you have sufficient material to fill the tube, the determination may be run neat (i.e., without solvent), or a methanol solution of known concentration may be used. The reported specific optical rotation for (−)-α-methylbenzylamine is $\alpha_D^{22} = -40.3°$.

SECTION 50

Computers in Chemistry

Overview

In recent years the use of computers has expanded dramatically in chemistry as in many other technical fields. Both personal computers and mainframes can fulfill a number of useful roles for chemists. Physical and theoretical chemists make extensive use of rapid computational ability to solve complex equations and perform lengthy calculations involving matrices, successive approximations, and differential equations.

Chemists studying kinetics frequently attach a sensing device to a computer so that it can both record the readings and analyze the data. Similarly, a gas chromatography–mass spectrometry (GC-MS) system is often interfaced with a computerized library of known spectra for easy identification of compounds.

By using a large library of known reactions and reagents, complex organic syntheses can be designed by computer. Working backward from the product (retrosynthetically), these programs retrace one step at a time until available starting materials are reached.

Educational software has become available so that students can practice problems at their own speed and observe simulated experiments. It has been shown that students simulating experiments often learn as much or more as those performing them in actuality. Perhaps the ideal is to do both: a simulation on the computer followed by the real experiment. Software is available dealing with the following subjects in organic chemistry:

Nomenclature
Titration
Distillation
Kinetics
Chromatography
Synthesis
Spectroscopy
Unknown identification

In a typical program you will be presented with a brief explanation of the relevant theory, perhaps accompanied by a simulation, then you are asked questions about specific examples. A correct response elicits an encouraging message and a new question, while an incorrect response may allow another try or offer an explanation. People often enjoy learning from computers since they are much more patient than people and easier to admit mistakes to. Of course, they can also be exasperatingly demanding in terms of format.

Programs are available from many sources. One of the largest libraries belongs to Project Seraphim of the National Science Foundation. For a nominal fee they will send copies of diskettes containing the desired programs, and they allow unlimited copying for use within the department. A catalog may be obtained by writing to

Project Seraphim
NSF: Science and Engineering Education
Department of Chemistry
Eastern Michigan University
Ypsilanti, MI 48197

Numerous other software publishers distribute educational chemistry programs, and many chemistry teachers write their own.

All chemists benefit from the information retrieval capabilities of Chemical Abstracts Online. Offered directly to customers by CAS, this service can search a file of over 7 million chemical substances cited since 1965. Chemical Abstracts Service assigns a unique registry number to each substance. A search can be conducted by subject, compound, name, registry number, molecular formula structure, or author. It is possible to specify several parameters in a given search by combining them using the ''and'' and ''or'' commands. For example, if you wanted to find out about reactions of any of five compounds with HCl, you would combine the five registry numbers of the compounds with ''or'' statements, and then ''and'' that group with the registry number for hydrochloric acid.

You would also have the choice of specifying articles only in English, review articles, patents, and published material since a particular date. The program will tell you how many references are found for each individual parameter, and how many are found in the overlap.

If the number of references is reasonable, you can review the listings that contain the chemical abstract number plus the original title, author, and reference. If there are too many references obtained, more parameters may be added. If too few, the scope or time span may be enlarged.

It may be necessary to copy references from the screen or it may be possible to print them locally. CAS also offers the option of printing (including the full abstract if desired) and mailing from a home office for a fee.

EXPERIMENT 50.1 ONLINE CHEMICAL INFORMATION RETRIEVAL

Estimated time:
1–3 hours

Procedure

Pick an appropriate subject you are interested in learning more about. Some examples might be:

Finding a review article within the last two years on a topic such as superconductors, purification of polluted water by ozonation, or anticancer drugs

Finding references to a particular compound and its preparation or uses

Finding a recent review article on a topic such as carbenes or the Wittig reaction

Finding recent statistics on chemical employment or manufacturing

Be sure the subject matter is very narrowly defined, to avoid using excessive computer time and generating hundreds or thousands of references. If the number of references obtained proves too large during the search, you can add extra parameters, such as specifying a review article, an article only in English, or one within the last year. If none or too few references are obtained, the search can be broadened.

If your search involves a particular compound (or several), be sure to locate the Chemical Abstracts Service (CAS) registry number(s) in the Chemical Abstracts Formula Index before starting the search. Often a compound has several names, and using its registry number is the most efficient way to search.

Check your topic with the CAS online consultant who will talk with you for a few minutes about the system and key in the commands during the search. When relevant reference(s) have been obtained, copy down the Chemical Abstract reference numbers or obtain a printout. Find and read the original articles, and write a brief (three– to five–page) description of what you have learned, citing references.

EXPERIMENT 50.2 USE OF EDUCATIONAL SOFTWARE

Estimated time:
1–3 hours

Procedure

Obtain from your instructor or computer consultant a description of the hardware and software available to you. Obtain diskettes containing the desired educational programs and the disk operating system (DOS), a set of internal instructions that enable the computer to run the programs. The following procedure applies to programs, supplied by Project Seraphim, written in Basic A, which will run on any IBM-compatible personal computer. If you are using another type, the procedure may vary slightly; your instructor or consultant will supply the changes if needed.

At an IBM-compatible personal computer, place an MS-DOS diskette into the upper drive (drive A) and close the lever or the catch to hold the diskette in. Turn on the computer and screen using the switches on the back.

A red light goes on as the computer reads the diskette, then it displays:

CURRENT DATE IS ____
ENTER NEW DATE (MM-DD-YY):
You hit the RETURN key. (There is no need to enter the date or time in order to run the program.) The computer displays
ENTER NEW TIME:
You hit RETURN. The computer gives you the prompt
A>
Now remove the MS-DOS disk and put the chemistry disk in drive A.
You type in
BASICA
Then hit RETURN. The computer gives you the prompt
OK

Now you type
LOAD "HELLO"
The computer prompts you
OK
You type
RUN

Now the computer should display a menu of programs on that disk and instructions on how to proceed to run the programs. You hit the up and down arrows to highlight the name of the program desired, then press RETURN.

When you are done, simply remove the disk, shut off the switches, and return the disks.

REFERENCES

HILEMAN, B., "Computers in Undergraduate Chemistry Education," *Chemical and Engineering News* (Oct. 24, 1988), p. 29.

THAKKAR, A. J., et al., *Topics in Current Chemistry*, Vol. 39: *Computers in Chemistry*, Springer-Verlag, 1973.

WEIGERS, K. E., and S. G. SMITH, "The Use of Computer-Based Chemistry Lessons in the Organic Laboratory Course," *Journal of Chemical Education* (1980) *57*, 454.

Choose Your Own Experiment

Special Note. Students should begin searching for a procedure at least three weeks before the scheduled start of this experiment and bring the procedure in for checking by the instructor two weeks ahead. This allows time for the instructor and stockroom manager to locate the required materials and equipment, or for you to choose a different experiment if the original choice is judged impractical or unsafe.

Overview

So far, all the procedures for lab have been specified for you in great detail. Now is the chance for you to pick a subject that particularly interests you and carry out an experiment on it.

First, you should go to the library and find an experimental procedure that looks interesting to you. There are several promising places to look; the list below includes some of the current widely used laboratory texts, plus such general sources for experiments as the *Journal of Chemical Education* and *Collective Organic Syntheses*. You may also use any other sources available. One word of caution: a procedure found only in a patent should not be followed, since it probably leaves out important information. The purpose of a patent is, of course, to prevent other people from using a process, not to enable them to do it.

Laboratory Manuals

AULT, A., *Techniques and Experiments for Organic Chemistry*, 5th ed., Allyn and Bacon, Boston, 1986.

FIESER, L. F., and K. L. WILLIAMSON, *Organic Experiments*, 6th ed., D.C. Heath, Lexington, Mass., 1987.

LEHMAN, J., *Operational Organic Chemistry*, 2nd ed., Allyn and Bacon, Boston, 1988.

MAYO, D. W., R. M. PIKE, and S. S. BUTCHER, *Microscale Organic Laboratory*, Wiley, New York, 1986.

PAVIA, D. L., G. M. LAMPMAN, and G. S. KRIZ, *Introduction to Organic Laboratory Techniques*, 3rd ed., W. B. Saunders, New York, 1988.

ROBERTS, R. M., J. C. GILBERT, L. B. RODEWALD, and A. S. WINGROVE, *Modern Experimental Organic Chemistry*, W. B. Saunders, New York, 1985.

Other Sources

Journal of Chemical Education.

Each issue contains a section of new experiments, some of which are organic.

Organic Syntheses, collective volumes.

These are reliable, tested procedures. By checking the index, you may find a compound or procedure of interest.

SANDLER, S. R. and W. KARO, *Organic Functional Group Preparation*, 2nd ed., Academic Press, New York, 1983.

Contains a large variety of reactions and references, arranged by functional group (three volumes).

Be sure to locate a procedure that can be carried out in the number of laboratory periods specified by your instructor. To do this, draw out a step-by-step procedure and estimate the time required for each step. And remember that you will want to be as prepared and efficient as possible, because lab work can easily take two or three times as long as expected if one is not organized. After photocopying the procedure, take it to your laboratory instructor so that he or she can go over it carefully to decide if you have the requisite equipment and chemicals on hand, and to make sure that the procedure is not too hazardous. Discuss in detail with your instructor the procedure and any special dangers that may be encountered. Be alert for steps that must be carried out in the fume hood, involve strong acids or bases, lachrymators, carcinogens, and so on.

Estimated time:

Plan for one to three laboratory periods, as indicated by the instructor.

Prelab

Before coming to lab, prepare your lab notebook by filling in the date, reaction(s) to be carried out, and a table of reagents and products. Perform any calculations of quantities needed to complete the table, outline the procedure in your notebook, and mentally run through it. If it will take more than one laboratory period, mark reasonable stopping points in the procedure.

List the hazards of the chemicals and equipment involved, and measures that will be taken to minimize the risks. Before beginning the experiment, show your notebook to the lab instructor for initialling.

Special Hazards

As described in the procedure found and elaborated upon by your instructor. Information in hazards of particular compounds should be looked up in the *Aldrich Catalog*, the *Merck Index*, or on safety data sheets. Be extremely cautious, since safety hazards may not be adequately discussed in the procedure you have found. Exercise all standard precautions and safety procedures.

Procedure

Carry out the procedure you have found and discussed with your instructor. If questions arise, do not hesitate to check with the instructor. When you have obtained your final products, turn each product in to your instructor in a vial clearly labeled with your name, the date, the name and structure of the compound, yield and criterion of purity (mp, bp, refractive index, and so on).

QUESTIONS

1. What new reactions and techniques did you learn? How else could these be applied?

2. Were the yield and purity of your products comparable to those described in the procedure? (Bear in mind that in a published procedure researchers often report their best yield out of several runs.)

3. Were there any unexpected snags in the procedure? If so, describe them and how you got around them.

4. Did the author(s) of the procedure write it up well? How could both the experiment and the writeup be improved?

APPENDIX A

NAMES OF COMMON INORGANIC IONS

Cations	Anions
Aluminum (Al^{3+})	Bromide (Br^-)
Ammonium (NH_4^+)	Carbonate (CO_3^{2-})
Barium (Ba^{2+})	Chlorate (ClO_3^-)
Calcium (Ca^{2+})	Chloride (Cl^-)
Chromium(III) or chromic (Cr^{3+})	Chromate (CrO_4^{2-})
Cobalt(II) or cobaltous (Co^{2+})	Cyanide (CN^-)
Copper(I) or cuprous (Cu^+)	Dichromate ($Cr_2O_7^{2-}$)
Copper(II) or cupric (Cu^{2+})	Dihydrogen phosphate ($H_2PO_4^-$)
Hydrogen (H^+)	Fluoride (F^-)
Iron(II) or ferrous (Fe^{2+})	Hydride (H^-)
Iron(III) or ferric (Fe^{3+})	Hydrogen carbonate or bicarbonate (HCO_3^-)
Lead(II) or plumbous (Pb^{2+})	Hydrogen phosphate (HPO_4^{2-})
Lithium (Li^+)	Hydrogen sulfate or bisulfate (HSO_4^-)
Magnesium (Mg^{2+})	Hydroxide (OH^-)
Manganese(II) or manganous (Mn^{2+})	Iodide (I^-)
Mercury(I) or mercurous (Hg_2^{2+})	Nitrate (NO_3^-)
Mercury(II) or mercuric (Hg^{2+})	Nitride (N^{3-})
Potassium (K^+)	Nitrite (NO_2^-)
Silver (Ag^+)	Oxide (O^{2-})
Sodium (Na^+)	Permanganate (MnO_4^-)
Tin(II) or stannous (Sn^{2+})	Peroxide (O_2^{2-})
Zinc (Zn^{2+})	Phosphate (PO_4^{3-})
	Sulfate (SO_4^{2-})
	Sulfide (S^{2-})
	Sulfite (SO_3^{2-})
	Thiocyanate (SCN^-)

APPENDIX B

COMMON ORGANIC FUNCTIONAL GROUPS

Note: R represent an alkyl or aryl group and can also be hydrogen unless R ↑ H is specified. If a molecule contains more than one R group, the substituents can, in general, be different.

Name(s)	Expanded structure	Condensed structure
Acid anhydride (acyl anhydride)	$$\underset{\text{R}-\overset{\displaystyle O}{\overset{\|}{C}}-O-\overset{\displaystyle O}{\overset{\|}{C}}-\text{R}}{}$$	$$\overset{\displaystyle O}{\overset{\|}{(RC)_2O}}$$
Acid chloride (acyl chloride)	$$\text{R}-\overset{\displaystyle O}{\overset{\|}{C}}-\text{Cl}$$	RCOCl
Alcohol (hydroxyl)	R—O—H	ROH, HOR
Aldehyde	$$\text{R}-\overset{\displaystyle O}{\overset{\|}{C}}-\text{H}$$	RCHO
Alkene (double bond, olefin)	$$\underset{\text{R}}{\overset{\text{R}}{}}\!\!\!\diagdown C = C \diagup\!\!\!\underset{\text{R}}{\overset{\text{R}}{}}$$	$R_2C = CR_2$
Alkyne (triple bond)	R—C≡C—R	
Amide	$$\text{R}-\overset{\displaystyle O}{\overset{\|}{C}}-\text{N}\!\!\diagup\diagdown\!\!\underset{\text{R}}{\overset{\text{R}}{}}$$	$RCONR_2$
Amine	$$\underset{\text{R}}{\overset{}{\text{R}-\text{N}-\text{R}}}$$	R_3N

Name(s)	Expanded structure	Condensed structure
Arene (aromatic, benzenoid)		$-R$, C_6H_5-R, $Ph-R$, $\Phi-R$
Carboxylic acid	$R-\overset{\overset{\textstyle O}{\|}}{C}-OH$	$RCOOH$, RCO_2H, $HOOCR$, HO_2CR
Ester	$R-\overset{\overset{\textstyle O}{\|}}{C}-OR$	$RCOOR$, RCO_2R, RO_2CR
Ether	$R-O-R$	ROR, R_2O
Halide	$R-X$ $X = F, Cl, Br, I$ $(R \neq H)$	RX
Ketone	$R-\overset{\overset{\textstyle O}{\|}}{C}-R$ $(R \neq H)$	$RCOR$, R_2CO
Nitrile (cyano)	$R-C\equiv N$	RCN, NCR
Nitro	$R-\overset{+}{N}\overset{O^-}{\underset{O}{}}$	RNO_2, O_2NR
Sulfide	$R-S-R$ $(R \neq H)$	RSR, R_2S
Sulfonic acid	$R-\underset{\downarrow}{\overset{\uparrow}{\underset{O}{\overset{O}{S}}}}-O-H$	RSO_3H
Thiol	$R-S-H$	RSH

APPENDIX C

How to Use the Chemical Literature

The chemical literature is immense, and it would be impossible for one person to read all the new articles appearing monthly in even one branch, such as organic synthesis. For this reason it is essential to be well organized and efficient in a literature search and to make extensive use of secondary sources. Secondary sources are collections of information taken from the primary sources (journals). Secondary sources include abstracts and cross-referencing works as well as textbooks, handbooks, encyclopedias, collections of reactions, and descriptions of reagents. These categories will be discussed briefly individually. This discussion is not intended to be comprehensive; it will deal with only the most common reference works likely to be needed by an undergraduate.

Handbooks

A common question that arises for a chemist is how to find the physical properties of a compound, its melting point, boiling point, density, refractive index, solubility, and so on. If it is a reasonably common compound, the handbooks listed below should suffice.

Aldrich Catalog/Handbook of Fine Chemicals, published annually by the Aldrich Chemical Co., Milwaukee, Wis.

A convenient paperback listing the molecular weight, melting point, boiling point, refractive index, and special hazards for most common organic and inorganic compounds. Also includes references to the IR and NMR spectra as well as the *Merck Index*, Beilstein, and Fieser and Fieser (all described below).

The Merck Index, 10th ed., M. Windholz, ed., Merck & Co., Rahway, N.J., 1983.

This work provides a paragraph of information on most common organic and inorganic compounds. It includes physical properties and descriptions of medical and veterinary uses for compounds as well as their toxicity. Key references on the preparation of compounds are included. Another useful feature found in the *Merck Index* is an extensive appendix on organic "name" reactions, giving a brief description and references for each.

CRC Handbook of Chemistry and Physics, R. C. Weast, ed., CRC Press, Boca Raton, Fla., revised annually.

This huge book contains useful sections on the physical properties of common organic and inorganic compounds. There are also handy tables of the densities of aqueous solutions of acids and bases.

CRC Handbook of Tables for Organic Compound Identification, 3rd ed., Z. Rappaport, ed., CRC Press, Boca Raton, Fla., 1987.

This handbook is extremely helpful in identifying unknown compounds. Over 8000 organic compounds are listed by functional group. Within each group, tables are provided of solids and liquids in order of increasing melting and boiling points, respectively. For each compound, the melting points of common derivatives are given.

Lange's Handbook of Chemistry, 13th ed., J. Dean, ed., McGraw-Hill, New York, 1985.

A general-purpose reference providing access to chemical and physical data used both in laboratories and in manufacturing.

Chemical Technicians' Ready Reference Handbook, 2nd ed., McGraw-Hill, New York, 1981.

Contains detailed instructions on how to carry out each step of normal laboratory procedures.

Handbuch der Organischen Chemie.

As you must have deduced from the title, this work (known more familiarly as Beilstein) was until recent years in German. Most of the literature of organic chemistry before 1940 is in German, since the Germans were the world leaders in the field at that time. With a slight knowledge of German and/or a German/English dictionary, translation is not too difficult. Fortunately, many technical terms are similar in several languages. The most useful dictionary for chemists is the *German-English Dictionary for Chemists*, 3rd ed., A. M. Patterson, ed., Wiley, New York, 1959. Beilstein consists of a main work (Hauptwerk, abbreviated H, covering compounds known in 1909) followed by four supplements (Ergänzungswerke). The first supplement (Erstes Ergänzungswerk, abbreviated EI), covers the literature from 1910 through 1919; EII covers 1920 through 1929; EIII 1930 through 1949; and EIV 1950 through 1959.

There are two types of cumulative indices: a name index (Sachregister) and a formula index (Formelregister). The formula index is particularly useful for an English-speaking user.

Here are a few common chemical terms in German and their English equivalents:

Bildung = formation or structure
Darstellung (Darst.) = preparation
Säure = acid
Schmelzpunkt (F) = melting point
Siedepunkt (Kp) = boiling point

There are several guidebooks describing the use of Beilstein; some of them are:

How to Use Beilstein, Beilstein Institute, Frankfurt-am-Main, Springer-Verlag, Berlin.

A Brief Introduction to the Use of Beilstein's Handbuch der Organischen Chemie, 2nd ed., E. H. Huntress, ed., Wiley, New York, 1938.

The Beilstein Guide: A Manual for the Use of Beilstein's Handbuch der Organischen Chemie, O. Weissbach, ed., Springer-Verlag, New York, 1976.

Organic Chemistry Textbooks

The following introductory organic chemistry texts can provide basic information on reaction mechanisms and expected side reactions. This is only a partial list of the best known texts; there are many others in use as well.

CAREY, F. A., *Organic Chemistry*, McGraw-Hill, New York, 1987.

EGE, S. N., *Organic Chemistry*, 2nd ed., D.C. Heath, Lexington, Mass., 1989.

FESSENDEN, R. J. and J. S. FESSENDEN, *Introduction to Organic Chemistry*, 3rd ed., Brooks/ Cole, Monterey, Calif., 1986.

FINAR, I. L., *Organic Chemistry: The Fundamental Principles*, Vol. 1, 6th ed., Halsted Press, New York, 1986.

Loudon, M., *Organic Chemistry*, 2nd ed., Benjamin-Cummings, Menlo Park, Calif., 1988.

March, J., *Advanced Organic Chemistry: Reactions, Mechanisms, and Structure*, 3rd ed., Wiley, New York, 1985.

McMurry, J., *Organic Chemistry*, 2nd ed., Brooks/Cole, Monterey, Calif., 1987.

Morrison, R. T. and R. N. Boyd, *Organic Chemistry*, 5th ed., Allyn and Bacon, Boston, 1987.

Pine, S. H., J. B. Hendrickson, D. J. Cram, and G. S. Hammond, *Organic Chemistry*, 4th ed., McGraw-Hill, New York, 1980.

Raber, D., and N. Raber, *Organic Chemistry*, West, St. Paul, Minn., 1988.

Roberts, J. D., and M. C. Caserio, *Basic Principles of Organic Chemistry*, 2nd ed., W.A. Benjamin, Menlo Park, Calif., 1977.

Solomons, T. W. G., *Organic Chemistry*, 4th ed., Wiley, New York, 1988.

Streitwieser, A., Jr., and C. H. Heathcock, *Introduction to Organic Chemistry*, 3rd ed., Macmillan, New York, 1985.

Vollhardt, K. P. C., *Organic Chemistry*, W.H. Freeman, New York, 1987.

Wade, L., *Organic Chemistry*, Prentice-Hall, Englewood Cliffs, N.J., 1987.

Encyclopedias and Dictionaries

There are several chemical encyclopedias and dictionaries which may be useful to consult for general information on chemical processes and types of compounds. These often include tables giving specific properties of individual compounds.

McGraw-Hill Encyclopedia of Chemistry, S. Parker, ed., McGraw-Hill, New York, 1983.

Kirk–Othmer Encyclopedia of Chemical Technology, H. F. Mark et al., eds., 3rd ed., Wiley, New York, 1978–1984.

Dictionary of Organic Compounds, 5th ed., J. Buckingham, ed., Chapman & Hall/Methuen, New York, 1982.

Encyclopedia of Chemistry, 4th ed., D. Considine, ed., Van Nostrand Reinhold, New York, 1984.

Glossary of Chemical Terms, 2nd ed., C. Hampel and G. Hawley, Van Nostrand Reinhold, New York, 1982.

Hawley's Condensed Chemical Dictionary, 11th ed., N. I. Sax and R. Lewis, Van Nostrand Reinhold, New York, 1987.

McGraw-Hill Dictionary of Chemical Terms, S. Parker, McGraw-Hill, New York, 1985.

Vocabulary of Organic Chemistry, M. Orchin et al., Wiley, New York, 1980.

Dictionary of Chemistry and Chemical Technology, H. Gross, Elsevier, Amsterdam, 1984.

Facts on File Dictionary of Chemistry, 2nd ed., J. Daintith, ed., Facts on File, New York, 1981.

Collections of Spectra

To locate the IR or NMR spectrum of a particular compound, the most promising approach is to check the Aldrich collection first, then Sadtler. The Aldrich collections are the most convenient because they are more compact and well organized.

Aldrich Library of Infrared Spectra, 3rd ed., C. J. Pouchert, Aldrich Chemical Co., Milwaukee, Wis., 1981.

Aldritch Library of FTIR Spectra, C. J. Pouchert, Aldritch Chemical Co., Milwaukee, Wis, 1985.

Aldrich Library of NMR Spectra, 2nd ed., C. J. Pouchert, Aldrich Chemical Co., Milwaukee, Wis., 1983.

The Sadtler collection is more piecemeal and may require some searching to locate the desired spectrum. It contains more spectra than Aldrich, including UV spectra. The best approach is to consult the Cumulative Formula or Name Index.

Nuclear Magnetic Resonance Spectra, Sadtler Research Laboratories, Philadelphia, Pa., 1966–present.

Standard Grating Spectra (IR), Sadtler Research Laboratories, Philadelphia, Pa., 1966–1976.

Ultraviolet Spectra, Sadtler Research Laboratories, Philadelphia, Pa., 1975–present.

Specific Reactions and Reagents

There are several works designed to help you locate a specific reaction or which describe a particular reagent. Some of these are listed here.

FIESER, L. F. and M. FIESER, *Reagents for Organic Synthesis*, Wiley-Interscience, New York, 1967–1986.

> A continuing series, in several volumes, describing the properties and uses of reagents.

HARRISON, I. T. and S. HARRISON, *Compendium of Organic Synthetic Methods*, Wiley-Interscience, New York, 1971–1977.

> Three-volume set showing how to transform a given functional group into another.

HOUSE, H. O., *Modern Synthetic Reactions*, 2nd ed., W.H. Benjamin, Menlo Park, Calif., 1972.

> Good general description of types of reactions with examples and mechanisms.

Organic Syntheses Collective Volumes 1 to 5.

> These describe reliable, tested procedures that can often be modified for use with other compounds. A collective index covers all five volumes.

PATAI, S., ed., *The Chemistry of the Functional Groups*, Interscience, London, 1964–present.

> A many-volume series, each one specializing in the properties and reactions of a particular functional group.

SANDLER, S. R., and W. KARO, *Organic Functional Group Preparations*, 2nd ed., Academic Press, New York, 1983.

THEILHEIMER, W., *Synthetic Methods of Organic Chemistry*, Interscience, New York, 1948–present.

> An annual series of compilations of synthetic methods, translated from the German.

Primary Journals

Here is a partial listing of the main English-language journals publishing articles in organic chemistry:

> *Accounts of Chemical Research*
> *Angewandte Chemie, International Edition in English*
> *Chemical Reviews*
> *Chemical Society Reviews* (formerly *Quarterly Reviews)*
> *Journal of the American Chemical Society*
> *Journal of the Chemical Society, Perkin Transactions 1 and 2*
> *Journal of Organic Chemistry*
> *Tetrahedron*
> *Tetrahedron Letters*

Abstracts and Cross-Referencing Works

If all else fails, or if you are dealing with an obscure compound, *Chemical Abstracts* (CA) is your best bet. The staff reads articles in over 10,000 journals worldwide and prepares a brief summary of each article. As you might guess, the collection, which began in 1907, is huge by now. The best way to find information in *Chemical Abstracts* is to check the indices systematically and chronologically. There are name, formula, and general subject indices, as well as author and patent indices, which are less useful. Each volume has its own set of indices, and there are 10-year collective indices before 1956 and 5-year collective indices afterward. These 5- and 10-year indices are usually the most promising places to start a search. Even so, it can be a time-consuming process. A much quicker way to search chemical abstracts is to use a computerized, online database search, versions of which are offered by several companies. By specifying particular compounds, partial structures, or authors' names, the abstracts since 1967 can be searched quickly, and the abstracts (with references) printed out. A more detailed description of Chemical Abstracts Service on-line literature searching may be found in Section 49.

Index Medicus

This work, similar to *Chemical Abstracts*, but smaller, covers topics of medical interest. It does contain a great deal of chemical information as well.

Science Citation Index

This publication lists articles that have referenced a particular article. Thus they are a great way to trace the development of a subject forward in time from a given reference; to find out what has been done in the area since the original article was published.

Additional Information on Literature Usage

For a more thorough discussion of the chemical literature, the following sources may be consulted:

ANTHONY, A., *Guide to Basic Information Sources in Chemistry*, Wiley, New York, 1979.

ASH, J., et al., *Communication, Storage and Retrieval of Chemical Information*, Ellis Horwood, Chichester, West Sussex, England, 1985.

BOTTLE, R. T., ed., *The Use of Chemical Literature*, 3rd ed., Butterworths, London, 1979.

BURMAN, C. R., *How to Find Out in Chemistry*, 2nd ed., Oxford University Press, New York, 1966.

HURT, C. D., *Information Sources in Science and Technology*, Libraries Unlimited, Inc., Englewood, Colo., 1989, pp. 106-123.

MAIZELL, R. E., *How to Find Chemical Information: A Guide for Practicing Chemists, Teachers, and Students*, 2nd ed., Wiley-Interscience, New York, 1987.

MELLON, M. G., *Chemical Publications*, 4th ed., McGraw-Hill, New York, 1965.

WOODBURN, H. M., *Using the Chemical Literature: A Practical Guide*, Marcel Dekker, New York, 1974.

Review of Important Terms in Organic Chemistry

Adsorption   Binding, usually noncovalent, of molecules of a gas or liquid to the surface of a solid, such as finely divided charcoal.

Agitation   Stirring, swirling, or mixing to bring reagents into contact.

Air Dry   To allow the solvent to evaporate from the surface of a solid by letting it stand in an open container. The process may be accelerated by spreading the material on a filter paper or paper towel to absorb excess liquid and stirring with a spatula.

Alicyclic   Aliphatic and cyclic.

Aliphatic   Containing linear or branched chains of carbons which may contain multiple bonds and other functional groups, but no rings.

Aliquot   A portion of a sample which is a known fraction of that sample. For example, a 25-mL aliquot may be removed from a 500-mL sample.

Alkali   A base, such as NaOH or KOH.

Allotrope   One of several forms in which an element may exist. For example, two different crystalline forms of a metal are allotropes, and two allotropic forms of oxygen are O_2 and O_3.

Alloy   A mixture or solution of metals that may be either solid or liquid and may contain nonmetals.

Amalgam   An alloy containing mercury.

Aqueous   Dissolved in water.

Aromatic   Normally containing a benzene ring. Other ring systems may be aromatic if they contain $4n + 2$ pi electrons (where n is an integer). Aromatic systems exhibit unusual stability.

Asymmetric Center   A central atom (usually carbon) bonded to four different elements or groups, giving rise to the possibility of enantiomers.

Benzene Ring   A ring of six carbon atoms in which each one is sp^2 hybridized and bonded to one hydrogen atom. These rings exhibit high stability and are found in many compounds. Below are several representations of a benzene ring.

While sometimes drawn with alternating single and double bonds, the bonding between all carbon atoms in the ring is identical due to resonance.

$$Ph-H \quad \phi-H \quad C_6H_6$$

Boiling Point   The temperature at which the vapor pressure of a liquid equals the atmospheric pressure.

Buffer   A solution containing substantial concentrations of both members of a conjugate acid–base pair where both the acid and base are weak. An example is an acetic acid–sodium acetate buffer. A buffered solution is protected from large changes in pH upon addition of small quantities of acid or base.

Carbohydrate   A compound of empirical formula CH_2O, such as glucose, $C_6H_{12}O_6$. This class includes simple sugars and polymers (starches and cellulose).

Carcinogen   A substance that has been found to cause cancer in laboratory animals, usually when administered in large quantities.

Catalyst   An element or compound that speeds up the rate of a reaction (by providing a new pathway and lowering the activation energy), but is not itself changed or consumed.

Caustic   Corrosive, as a strong alkali.

Charcoal   Also known as carbon black or activated carbon, a porous form of carbon with a large surface area in its powdered form. Used to adsorb impurities, for example during recrystallizations.

Chelation   The process whereby an organic molecule containing two or more electron-rich atoms (such as O or N) binds to a metal ion to form a heterocyclic ring possessing coordinate covalent bonds.

Chiral   Not superimposable on its mirror image. Usually possessing an asymmetric center.

Chromatography   A method for separating components of a mixture based on their differing affinities for a mobile and a stationary phase. The most common types are gas chromatography (GC, also known as gas–liquid chromatography or GLC), column, thin-layer, paper, and high-performance liquid chromatography (HPLC).

Contaminant   Any undesired material present in an otherwise relatively pure substance.

D   A prefix indicating that a compound has a stereochemical configuration similar to that of D-glyceraldehyde.

Dehydration   Removal of water or the elements of water (two hydrogen atoms and one oxygen atom).

Deliquescent   Extremely hygroscopic; so much so that on exposure to air a puddle forms of the compound dissolved in water.

Desiccant   A drying agent, such as anhydrous $CaCl_2$, Na_2SO_4, $MgSO_4$, or K_2CO_3.

Desiccator   An airtight container containing a desiccant, used to remove water from samples or to keep them dry.

Dextrorotatory   Abbreviated $(+)$ or d-, rotating plane-polarized light in a clockwise direction as seen by the observer.

Diastereomers   Stereoisomers that are not mirror images of each other. Diastereomers have the same configuration at at least one asymmetric center and the opposite configuration at at least one other.

Distillation   A process for purifying a liquid in which it is heated to the boiling point and vaporized, then cooled and recondensed to liquid form.

Efflorescence   Loss of water molecules from a solid hydrate, resulting in decomposition of a crystal to a powder.

Enantiomer   One of a pair of molecules which are nonsuperimposable mirror images of each other. They usually possess asymmetric centers of opposite configuration.

Endothermic   Absorbing heat from the environment.

Equilibrium   A state of dynamic balance in which the concentrations of reagents do not change with time, since the forward and reverse reactions are occurring at the same rate.

Evaporation   The process whereby molecules leave the surface of a liquid and enter the gas phase.

Exothermic   Releasing heat to the environment.

Extraction   Drawing a desired compound into a solution. For example, ether can extract nonpolar substances from water.

Filtration   Passing a liquid or solution through a porous material such as paper to remove solid materials.

Free Radical   A reactive intermediate containing one or more free or unpaired electrons: for example, chlorine radical, Cl or Cl·.

Fuming   An adjective describing nitric acid containing NO_2 or sulfuric acid containing SO_3, which emit visible clouds of gas on exposure to air.

Halogen   Any of the elements fluorine, chlorine, bromine, iodine, or astatine found in group VIIa.

Heterocyclic   Having a ring containing one or more heteroatoms (noncarbon atoms), such as nitrogen, oxygen, or sulfur.

Hydration   Gain of water or the elements of water.

Hydrocarbon   A compound consisting only of carbon and hydrogen.

Hydrogenation   Adding hydrogen atoms (in pairs) to a molecule. Usually catalyzed by a metal such as Pt, Pd, or Ni.

Hydrolysis   A reaction with water, accompanied by a gain of the elements of water.

Hygroscopic   Attracting water from the air.

Immiscible   Cannot be mixed to form a homogeneous state.

Impurity   *See* Contaminant.

Isomers   Compounds having the same formula, but different structures. Isomers may be constitutional, geometric, or stereochemical.

L   A prefix indicating the compound has a stereochemical configuration similar to that of L-glyceraldehyde.

Lachrymator   A compound of which the vapor causes severe irritation to eyes. Literally, tear-producing.

Levorotatory   The opposite of *d*, abbreviated (-) or *l*-, rotating plane-polarized light in a counterclockwise direction as seen by the observer.

Lipid   A class of relatively nonpolar molecules formed metabolically, including fats, waxes, lecithins, and sterols.

Macromolecule   A large molecule, such as DNA, a protein, or a polymer which has a molecular weight between several thousand and several million.

Molecular Sieve   A zeolite that adsorbs in its cavities molecules up to a certain diameter, such as 3 Å, 4 Å, 13 Å, and so on.

Neutralization   The reaction between equivalent quantities of an acid and a base to produce a salt.

Nucleation   The process by which crystal growth begins, often on a dust particle or a scratch in a glass surface.

Oligomerization   The joining of a small number of units, the formation of a short polymer.

Optical Rotation   The property possessed by chiral substances of turning the plane of polarized light.

Organometallic   Containing a metal atom bonded to an organic group.

Oxidation   Loss of electrons, often involving gain of oxygen or loss of pairs of hydrogen atoms.

Oxidation Number   The formal charge on an element or group imagining that, for each bond, both electrons belong to the more electronegative atom.

Oxidizing Agent   A substance that removes electrons from (oxidizes) another, itself being reduced in the process.

Polymer   A large molecule formed by the bonding of five or more smaller molecules (monomers).

Polysaccharide   A polymer consisting of carbohydrate units; this group includes starches and cellulose.

Precipitation   The formation of solid particles which sink to the bottom of the container.

Pyrolysis   Chemical reaction caused by the application of heat alone. Often a decomposition of a molecule into smaller fragments.

Pyrophoric   Capable of igniting spontaneously in air.

Redox   Involving reduction and oxidation.

Reduction   The gain of electrons, often accompanied by loss of oxygen or gain of a pair of hydrogen atoms.

Refractive Index   A physical constant describing the amount of bending experienced by a ray of light entering a substance.

Regiospecific   Occurring in a particular direction (orientation) on a molecule. For example, Markovnikov's rule specifies the regiospecificity of additions to alkenes.

Rinse   To pour a liquid over an insoluble solid to remove impurities on the surface.

Salting Out   The process of adding a salt (usually NaCl) to an aqueous phase to decrease the solubility of organic compounds.

Separation   Isolating the individual components of a mixture.

Solute   An element or compound which dissolves in a solvent to form a solution.

Solution   A homogeneous mixture formed when a solute dissolves in a solvent. For example, 10 N NaOH, 5% aqueous HCl, and 1 M bromine in acetic acid are solutions.

Solvation   The process of surrounding a molecule or ion of solute with molecules of solvent.

Solvent   A liquid that can dissolve a solute by breaking its intermolecular forces.

Spectroscopy   The study of the interaction of light or electromagnetic waves with matter. Common types include infrared (IR), nuclear magnetic resonance (NMR), and ultraviolet-visible (UV-vis).

Stereochemistry   Study of the three-dimensional structure of molecules in space, including the arrangements of their chiral (asymmetric) centers.

Stereoisomers   Isomeric compounds differing only in the three-dimensional arrangement of their atoms in space. For example, a pair of geometric (cis-trans) or optical (R and S or D and L) isomers.

Stereospecific   Having a specified stereochemical outcome. For example, an S_N2 reaction is stereospecific in that it always results in inversion at carbon.

Sublimation   The change of phase of a substance from solid to gas without passing through a liquid phase. Dry ice (solid CO_2), for example, sublimes at room temperature and pressure.

Substituent   Any group besides hydrogen on an organic molecule.

Suspension   A dispersion of finely divided particles of a solid distributed fairly uniformly throughout a liquid.

Titration   A widely used analytical method involving the addition of a measured volume of a reagent (such as acid or bases) to react with another substance to reach a visible or measurable end point.

Trituration   Grinding a solid material in the presence of a solvent it is insoluble in for the purpose of removing impurities.

Unsaturation   A measurement of the number of pairs of hydrogen atoms that could theoretically be added to an organic compound if multiple bonds and rings were broken. Also known as hydrogen deficiency.

Vapor Pressure   The upward pressure exerted by molecules of a liquid leaving the surface at a particular temperature. When the vapor pressure equals atmospheric pressure, boiling occurs.

Viscosity   The resistance of a liquid to flow when an external force (such as gravity) is applied. Related to the strength of intermolecular forces. Syrups have high viscosity, while low-molecular-weight hydrocarbons, such as hexane, have low viscosity.

Volatility   A description of the vapor pressure of a substance. The higher the vapor pressure at a given temperature, the more volatile a compound is (and the lower its boiling point).

Wash   To remove impurities either by pouring a liquid over a solid product during filtration, or to shake in a separatory funnel with another phase intended to dissolve impurities, but not the desired compound.

Zeolite   A claylike aluminosilicate material with very regular cavities. Often used as a molecular sieve for drying.

INDEX

475